This book, based on the 1989 Beg-Rohu summer school, contains six sets of pedagogical lectures by internationally respected researchers on the statistical physics of crystal growth. Providing a complete course in which the phenomena of shape and growth are viewed from a novel vantage point, the lectures cover not only very recent developments, but also reflect on old problems that have not always received the attention they deserve.

Statistical physicists, condensed matter physicists, metallurgists, and applied mathematicians will find this a stimulating and valuable introduction to this important area of research.

T0297806

COLLECTION ALÉA-SACLAY Monographs and Texts in Statistical Physics 1

GENERAL EDITOR: C. Godrèche

SOLIDS FAR FROM EQUILIBRIUM

SOLIDS FAR FROM EQUILIBRIUM

EDITED BY
C. GODRÈCHE
Service de Physique de l'Etat Condensé, Saclay

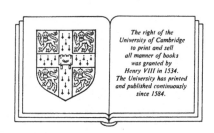

Collection *Aléa Saclay*

The right of the
University of Cambridge
to print and sell
all manner of books
was granted by
Henry VIII in 1534.
The University has printed
and published continuously
since 1584.

Cambridge University Press
Cambridge
New York   Port Chester   Melbourne   Sydney

CAMBRIDGE UNIVERSITY PRESS
Cambridge, New York, Melbourne, Madrid, Cape Town,
Singapore, São Paulo, Delhi, Tokyo, Mexico City

Cambridge University Press
The Edinburgh Building, Cambridge CB2 8RU, UK

Published in the United States of America by Cambridge University Press, New York

www.cambridge.org
Information on this title: www.cambridge.org/9780521177252

First published 1991
First paperback edition 2011

A catalogue record for this publication is available from the British Library

ISBN 978-0-521-41170-7 Hardback
ISBN 978-0-521-17725-2 Paperback

CONTENTS

Preface   xv

Contributors   xvi

Chapter 1   Shape and growth of crystals
P. Nozières

Chapter 2   Instabilities of planar solidification fronts

B. Caroli, C. Caroli, B. Roulet

Chapter 3   An introduction to the kinetics of first-order phase transition

J.S. Langer

Chapter 4   Dendritic growth and related topics

Y. Pomeau, M. Ben Amar

Chapter 5   Growth and aggregation far from equilibrium

L.M. Sander

Chapter 6   Kinetic roughening of growing surfaces

J. Krug, H. Spohn

PREFACE

This book is based on courses given at the 1989 Beg-Rohu summer school entitled: 'Solids far from equilibrium: growth, morphology and defects'. This subject is of great practical importance since it involves the effect of the growth dynamics of solids on their shape, their crystalline quality and consequently on their electronic and mechanical properties. It has therefore been the object of extensive fundamental research by metallurgists and specialists of crystal growth. Lately it has experienced rapid development due to the efforts of condensed matter and statistical physicists motivated by the progress of experimental methods (e.g. growth of solid helium) and by the progress in the understanding of the dynamics of nonlinear systems far from equilibrium.

This volume deals with the various facets of a general problem: shape and growth of crystals, instabilities and pattern formation. The six sets of lectures reflect the spirit of the school: they provide an introduction to the fundamental physics of nonlinear systems far from equilibrium, yet include the most recent developments in the field, with a definite effort towards a pedagogical exposition and a minimal amount of prerequisite. Hence this book is a real course where lectures are designed to be self contained and the calculations are given in detail.

I wish to express my sincere thanks to the many persons and institutions who have contributed to the organization of the school and to the editing of this volume. I owe special gratitude to Cambridge University Press and to my institution, Commissariat à l'Énergie Atomique.

Claude Godrèche
May 1991

CONTRIBUTORS

Professor Philippe Nozières
Collège de France, Paris and Institut Laue-Langevin, Avenue des Martyrs, B.P.156 X, 38042 Grenoble Cedex, France

Professor Christiane Caroli, Professor Bernard Caroli, and
Professor Bernard Roulet
Groupe de Physique de l'École Normale Supérieure, Université Paris VII,
2 place Jussieu, 75251 Paris Cedex 05, France

Professor James Langer
Director, Institute for Theoretical Physics, University of California,
Santa Barbara, California 93106, USA

Professor Yves Pomeau and Professor Martine Ben Amar
École Normale Supérieure, 24 rue Lhomond, 75231 Paris, France

Professor Len Sander
H. M. Randall Laboratory of Physics, University of Michigan, Ann
Arbor, Michigan 48109, USA

Dr Joachim Krug and Professor Herbert Spohn
Theoretische Physik, Theresienstrasse 37, 8000 München 2, Germany

CHAPTER 1

SHAPE AND GROWTH
OF CRYSTALS

P. Nozières

1

Thermodynamics of interfaces

This is an old subject, which goes back to the work of Gibbs. The basic ideas were formulated long ago — nevertheless, some ambiguities remained until very recently. In this brief review, we try to set the problem in a language as simple and precise as possible. This is relatively easy for fluid interfaces, in which every quantity has a well defined smooth average profile: the definition of surface excesses is then straightforward. The situation is somewhat more complicated for crystalline interfaces, as the bulk solid is then modulated with the lattice periodicity. One may view the interface in two ways:

(i) Either on an *atomic* scale, in which the position of every atom is specified. We are thus led to the usual 'terrace-step-kink' model.

(ii) Or on a *coarse grained* scale, in which every property is averaged over a finite volume — for instance a unit cell in such a way as to have constant averages in the bulk. These coarse grained quantities are then treated in a *continuum* language, in much the same way as for fluid interfaces.

In this section, we choose the second approach.

Consider a planar interface between phases 1 and 2, located near the plane $z = 0$. An arbitrary extensive quantity has a profile $\phi(z)$, where ϕ

is the density per unit volume. In the Gibbs description, one deliberately ignores details of that profile. One defines instead an integrated *surface excess* ϕ_s through the following trick:

(i) Choose an arbitrary dividing surface $z = \xi$ somewhere in the interfacial region. Then extrapolate the bulk values ϕ_1 and ϕ_2 up to ξ.

(ii) The surface excess is:

$$\phi_s = \int_{z_1}^{z_2} \phi(z)\,dz - \phi_2(z_2 - \xi) - \phi_1(\xi - z_1), \qquad (1.1)$$

where z_1 and z_2 are well inside each phase. This definition is depicted in Fig. 1a. (ϕ_s is the shaded area.)

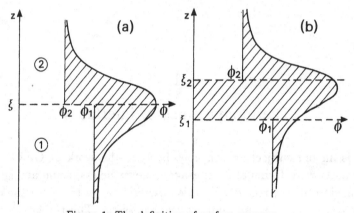

Figure 1: The definition of surface excesses.

It is clear that the value of ϕ_s depends on the choice of ξ. If ξ changes by $\Delta\xi$, then ϕ_s changes by

$$\Delta\phi_s = \Delta\xi\,[\phi_2 - \phi_1]. \qquad (1.2)$$

Only quantities such that $\phi_1 = \phi_2$ have unambiguous surface excesses. Quite generally, any result dealing with surfaces must be invariant upon a change of ξ.

The previous standard definition uses a single dividing surface: the interface has 'zero volume.' It is often more convenient to use two distinct dividing surfaces ξ_1 and ξ_2 (Fig. 1b). The surface excess is then defined as:

$$\phi_s = \int_{z_1}^{z_2} \phi(z)\,dz - \phi_2(z_2 - \xi_2) - \phi_1(\xi_1 - z_1). \qquad (1.3)$$

Upon a change of ξ_1 and ξ_2, the surface excess ϕ_s changes by

$$\Delta\phi_s = \phi_2\Delta\xi_2 - \phi_1\Delta\xi_1. \tag{1.4}$$

The interface then has a finite 'thickness' $\ell_s = \xi_2 - \xi_1$, which of course is a matter of definition.

Our aim now is to define the appropriate independent surface excesses, and to establish thermodynamic relationships between them, dealing both with *mechanical* equilibrium (balance of forces) and with '*chemical*' equilibrium (transfer of matter from phase 1 to phase 2).

1.1 Interface between two fluids

Each phase is isotropic, characterized by a scalar pressure p, a temperature T, chemical potentials $\mu^{(i)}$ for the various constituents. In equilibrium, these parameters are the same on both sides. If phases 1 and 2 are chemically distinct, the pressure p is arbitrary (interface between two immiscible liquids). If on the other hand we allow for mass transfer across the interface (e.g. a liquid-vapour interface), the equality of $\mu^{(i)}$ on both sides implies a constraint on the pressure: we lose one independent variable on the phase coexistence curve. For fluids, this distinction does not matter much — this is not so for solid interfaces.

We define in each bulk phase a set of extensive quantities: energy ε, entropy σ, free energy $f = \varepsilon - T\sigma$, density $\rho^{(i)}$ of species i, grand potential

$$\omega = f - \mu^{(i)}\rho^{(i)}.$$

Thermodynamics tells us that $\omega = -p$ in the bulk. Near the interface, all these scalar quantities display surface excesses

$$\begin{aligned} f_s &= \varepsilon_s - T\sigma_s, \\ \omega_s &= f_s - \mu^{(i)}\rho_s^{(i)}. \end{aligned} \tag{1.5}$$

At this stage, the Gibbs dividing surfaces are arbitrary: surface quantities such as f_s, $\rho_s^{(i)}$ depend on the choice of ξ_1 and ξ_2. Only one combination is intrinsic:

$$\gamma = \omega_s + p\ell_s = f_s - \mu^{(i)}\rho_s^{(i)} + p\ell_s. \tag{1.6}$$

γ is the *surface tension*, which reduces to ω_s in the usual choice of a single Gibbs surface.

We may choose to work with arbitrary ξ_1, ξ_2, checking at each stage
the invariance of our results. In view of this invariance, it is often more
convenient to make a specific choice which simplifies the physics:

▷ For phase equilibrium of a *pure* species, we have a *single* μ and ρ_s.
The natural choice is that of Gibbs: we take a single ξ such that $\rho_s = 0$,
the 'Gibbs interface' has neither volume, nor mass.[†] The surface tension
is just the free energy f_s.

▷ For phase equilibrum of a binary mixture, it is convenient to keep
a single Gibbs surface $\xi_1 = \xi_2 = \xi$. But here only one surface density
can be made zero — for instance that of the solvent in a dilute mixture.
The *solute surface density* remains finite: the corresponding ρ_s may be
viewed as a *surface adsorption*, which is a genuine physical effect.

▷ Another common situation is that of two immiscible pure materials:
phases 1 and 2 are made up of distinct atoms 1 and 2. Here again, a
single ξ cannot cancel both $\rho_s^{(1)}$ and $\rho_s^{(2)}$. Since there is no reason to
select one side more than the other, the natural choice is to choose ξ_1
such that $\rho_s^{(1)} = 0$ and ξ_2 such that $\rho_s^{(2)} = 0$: the interface then has a
finite width ℓ_s.

In the end, all of that is a matter of taste: physics is independent of ξ_1
and ξ_2.

We now turn to the *stress tensor* π_{ij}. While it is scalar in the bulk
($\pi_{ij} = -p\delta_{ij}$), it becomes uniaxial in the interfacial region. Isotropy in
the xy-plane implies

$$\pi_{xx} = \pi_{yy} = \pi_\perp \neq \pi_{zz} \qquad (1.7)$$

(all off diagonal elements are zero by symmetry). We can dispose at once
of π_{zz} by using the condition of local mechanical equilibrium

$$\frac{\partial \pi_{ij}}{\partial x_j} = 0.$$

Since the profile depends only on z, we have $\pi_{zz} =$ const. We recover the
asymptotic condition $p_1 = p_2 = p$. Moreover the *constant* profile $\pi_{zz}(z)$
has no surface excess (anyhow π_{zz} is not an extensive quantity).

In contrast, the transverse stress π_{xx} may be viewed as the flow of
momentum *parallel* to the interface: it behaves as an extensive quantity,

---

[†] Note that the Gibbs choice is not possible if $\rho_1 = \rho_2$, as occurs for
instance at a grain boundary: then either ρ_s or ℓ_s is non-zero.

with a non-zero surface excess $\pi_{\perp s}$. Since π_{xx} is the same on both sides, this *tangential surface stress* does not depend on ξ as long as we choose a single Gibbs surface. More generally, the intrinsic combination is

$$\tilde{\pi}_s = \pi_s + p\ell_s. \tag{1.8}$$

A priori, it is not obvious how $\tilde{\pi}_{\perp s}$ is related to γ.

Thermodynamic relationships between surface quantities

We know that $\omega = -p$ in the bulk: we suspect that ω_s should be equal to $\pi_{\perp s}$, i.e. γ equal to $\tilde{\pi}_{\perp s}$. This is indeed true, but we do not want to use *local* thermodynamic identities as a proof, since presumably they do not hold near the interface. Instead, we shall rely on a *global* application of the first law of thermodynamics.

Assume that the walls of the system are displaced at fixed p, T and $\mu^{(i)}$, and at *fixed interface*. The Gibbs surfaces ξ_1 and ξ_2 do not move. The surface area A just changes by δA. The nominal bulk volumes (measured from ξ_1 and ξ_2) vary by δV_1 and δV_2. The structure of the interface is unchanged, matter being brought in from the bulk in order to fill in the new area δA: γ remains the same. (It is precisely that feature that will break down for solid interfaces.) The net change in free energy is

$$\delta F = f_1 \delta V_1 + f_2 \delta V_2 + f_s \delta A. \tag{1.9}$$

δF must be equal to the *work* δW fed in from outside in this transformation:

▷ mechanical work of the bulk pressure and of the surface stress $\pi_{\perp s}$ as A enlarges;

▷ 'chemical' work of the reservoirs that supply new matter

$$\delta W = \pi_{\perp s} \delta A - p[\delta V_1 + \delta V_2] \tag{1.10}$$
$$+ \mu^{(i)} \big[\rho_1^{(i)} \delta V_1 + \rho_2^{(i)} \delta V_2 + \rho_s^{(i)} \delta A\big].$$

Equating (1.9) and (1.10), we see that the terms involving δV_1 and δV_2 drop out, as they should (remember that $\omega_1 = \omega_2 = -p$). From the surface terms, we infer

$$\pi_{\perp s} = f_s - \mu^{(i)} \rho_s^{(i)} = \omega_s, \tag{1.11}$$

as expected. Note that the argument makes no reference to what happens *in* the interface: it relies on *global* energy conservation. Equation (1.11) is clearly independent of the choice of ξ_1 and ξ_2, valid for any surface thickness ℓ_s.

Assume now that the thermodynamic parameters are changed by $\delta\mu^{(i)}$, δT and δp. In the bulk, we have

$$\begin{cases} \delta f = -\sigma\delta T + \mu^{(i)}\delta\rho^{(i)}, \\ f + p = \mu^{(i)}\rho^{(i)}, \end{cases} \qquad (1.12)$$

from which we infer the Gibbs-Duhem equation

$$\delta p = \rho^{(i)}\delta\mu^{(i)} + \sigma\delta T. \qquad (1.13)$$

Similarly, the change in f_s is due to the work of external forces that change the *interface structure*. Consider a transformation in which the geometry is unchanged: A, V_1, V_2 are untouched, but we allow for a change $\delta\ell_s$ of the interface thickness. Then

$$\delta f_s = -\sigma_s\delta T + \mu^{(i)}\delta\rho_s^{(i)} - p\delta\ell_s \qquad (1.14)$$

(the last term is the mechanical work spent in expanding the *extrapolated* surface layer against the pressure p). It follows that

$$\delta\omega_s = -\sigma_s\delta T - \rho_s^{(i)}\delta\mu^{(i)} - p\delta\ell_s, \qquad (1.15)$$

$$\boxed{\delta\gamma = -\sigma_s\delta T - \rho_s^{(i)}\delta\mu^{(i)} + \ell_s\delta p.} \qquad (1.16)$$

Equation (1.16) is the *surface Gibbs-Duhem equation*, which is usually written for a zero volume interface ($\ell_s = 0$). (Note that σ_s and $\rho_s^{(i)}$, being excess quantities, may well be negative.) It is easily verified that (1.16) is invariant upon a change of ξ_1 and ξ_2.

Curved interfaces: Laplace's law

Assume that the interface is a cylinder with radius R (with phase 1 inside). A simple balance of forces on a surface element $ds = R\,d\theta$ (Fig. 2) shows that the pressure inside is larger:

$$p_1 - p_2 = \frac{\gamma}{R}. \qquad (1.17)$$

This is the well known Laplace capillary pressure. Equation (1.17) may be obtained directly from the local mechanical equilibrium in polar coordinates

$$\frac{\partial\pi_{ij}}{\partial x_j} = 0 \implies \frac{\partial}{\partial r}\left[r\pi_\parallel\right] + \pi_\perp = 0. \qquad (1.18)$$

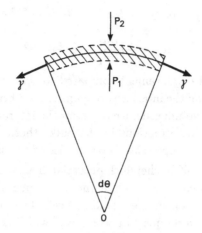

Figure 2: The origin of capillary pressure.

For a zero width interface, $\pi_\perp$ may be approximated as $\gamma\delta(r - R)$. Integrating (1.18) through the singularity yields (1.17). Equation (1.18) has the virtue of being *exact*, which in principle allows an estimate of corrections due to the finite interface width $\sim a$: these corrections are of order a/R.

In fact, such an accuracy raises all sorts of difficulties. Since p_1 and p_2 are different, γ is no longer an intrinsic quantity: changing the Gibbs surface ξ changes the surface tension by an amount $\Delta\gamma/\gamma \sim a/R$. We need *precise* definitions in order to avoid ambiguities — put another way, writing a $1/R$ correction to the surface energy $\sim f_s R$ requires an accurate definition of R! There exists a large literature on the subject [1]: we will not pursue the issue further as such corrections are negligible but for very small samples of atomic size, $R \sim a$.

1.2 Solid interfaces

We consider the interface between two solid phases: the case where one phase is a liquid is simpler and it will be discussed along the way. The situation is here more complicated for a number of reasons:

(i) Usually the solid phase is anisotropic: surface properties depend on the *orientation* of the interface. Moreover, the tangential surface stress may well be anisotropic.

(ii) Second — and more important! — solid phases are *rigid*. The concept of a pressure is no longer valid: equilibrium is described in terms of a *stress* tensor π_{ij} — or equivalently in terms of the *strain* tensor

$$u_{ij} = \frac{1}{2}\Big[\frac{\partial u_i}{\partial x_j} + \frac{\partial u_j}{\partial x_i}\Big], \qquad (1.19)$$

where **u** is the local displacement compared to some reference nominal state. When enlarging the interface area, one *cannot* bring new matter in order to fill the new volume: deformations of the interface are intimately connected to elastic deformations in the bulk: thermodynamics of the capillary layer cannot ignore *elastic* properties of the substrate.

(iii) The concept of a chemical potential makes no sense. When adding one atom at constant volume one must specify *where* it is put: interstitial? compression of the whole lattice? The only unambiguous quantity is the *free energy* per unit volume f, which is a function of the state of deformation.

(iv) Finally, we must worry about atomic rearrangements in the solid (defects, vacancies, impurity clustering in a mixture). If we were *really* at equilibrium, these degrees of freedom would not matter much: they would reach their equilibrium configuration, and they would act to correct surface parameters. Unfortunately, they relax very slowly, and in practical situations one must worry about their diffusion: the formulation becomes very complicated [2].

Here, we ignore atomic rearrangements, and we focus on elastic features: *how does one blend capillarity with elasticity?* We restore a dual Gibbs surface ξ_1, ξ_2, which will prove to be the natural language in dealing with deformations. The central issue is to ascertain the *independent* thermodynamic variables, both in the bulk and at the surface. The situation is different depending on whether we allow for *mass transfer* between phases 1 and 2 or not — as usual, phase equilibrium suppresses one independent variable. We will consider the two situations separately:

(i) If phases 1 and 2 are pure, *distinct* chemical species, we can define separate surface masses $\rho_s^{(1)}$ and $\rho_s^{(2)}$. It will be convenient to attach the Gibbs surfaces to a given atom — or rather to the *extrapolated position of a given atom*, which it would have if the uniform bulk strain were extended to ξ_1 or ξ_2.

(ii) Assume now that species 1 and 2 are the same. If we forbid mass transfer across the interface, a given atom may be assigned to a given side: the situation is essentially the same as in (i), and we can define separately $\rho_s^{(1)}$ and $\rho_s^{(2)}$. If instead we allow for mass transfer, only the *total* surface mass $\rho_s = \rho_s^{(1)} + \rho_s^{(2)}$ is well defined. It will then be convenient to use a single Gibbs surface $\xi_1 = \xi_2 = \xi$, such that $\rho_s = 0$.

In practice, we shall try to keep the Gibbs surfaces ξ_1, ξ_2 arbitrary as long as possible, in order to emphasize the invariance of the theory — only in the final stages shall we specify intrinsic quantities. Our approach follows the work of Andreev and Kosevich [3], as developed in [4].

Bulk phase equilibrium constraint for pure systems

This is a prerequisite to the treatment of surface properties. For fluids, that constraint was $\mu_1 = \mu_2 = \mu$. Here we must be more careful since the chemical potential is an ambiguous concept. Suppose that a mass[†] Δm is transferred from phase 1 to phase 2 at *fixed interface*: the area A is constant, the bottom wall is raised by $\Delta z_1 = \Delta m / \rho_1$, the upper wall by $\Delta z_2 = \Delta m / \rho_2$. The state of both phases is unchanged: same strain, same temperature. The only change in free energy is thus

$$\Delta F = A\big[f_2 \Delta z_2 - f_1 \Delta z_1\big]$$

(the interface is *unaffected*). Phases 1 and 2 will be in equilibrium if ΔE is equal to the work of bulk stress in displacing the walls:

$$W = A\big[\pi_{zz,2}\Delta z_2 - \pi_{zz,1}\Delta z_1\big].$$

We may thus define an 'effective chemical potential'

$$\mu = \frac{f - \pi_{zz}}{\rho}. \qquad (1.20)$$

Phase equilibrium will be achieved if

$$\mu_1 = \mu_2. \qquad (1.21)$$

Note that μ depends both on the local orientation of the interface and on the local state of strain: it may vary from point to point along the interface, in contrast to the fluid case.

Surface stress and surface strain

There is no difficulty in defining a surface *free energy* f_s. Similarly, let $\pi_{ij}(z)$ be the stress tensor profile through the interface (appropriately coarse grained in our continuum description). We know that π_{ij}

[†] For convenience, we consider ρ as a *mass* density: μ is the chemical potential per unit mass.

is symmetric; it goes respectively to $\pi_{ij,1}$ and $\pi_{ij,2}$ deep inside phase 1 or 2 (the bulk stress is assumed uniform). From the mechanical equilibrium condition $\partial\pi_{ij}/\partial x_j = 0$, we infer that π_{iz} is *constant*. Put another way, π_{iz} is the flow of 'i' momentum in the z-direction, which is necessarily conserved, whatever the surface structure. This statement has two crucial consequences:

(i) $\pi_{iz,1} = \pi_{iz,2} = \pi_{iz}$, a statement of *global* force equilibrium;

(ii) π_{iz} has no surface excess.

We conclude that the *genuine surface stress is tangential*:

$$\pi_{\mu\nu,s} \neq 0 \quad \text{if} \quad \mu, \nu = x, y.$$

The 2×2 tensor π_s describes the surface excess of momentum flow parallel to the interface. Note that $\pi_{\mu\nu,s}$ is also symmetric. The tangential nature of surface stress has not been fully appreciated in the past (see for instance [5]). In general, the tangential bulk stress $\pi_{\mu\nu}$ need not be the same in the two phases: as a result, $\pi_{\mu\nu,s}$ is *not* an intrinsic quantity, even if we choose a single Gibbs surface ξ.

While the existence of a surface *stress* is well known, the existence of a surface *strain* is much less familiar, despite the fact it is an essential feature of any consistent thermodynamical description. In order to define it, we start from a nominal reference situation[†] — for instance one in which the bulk tensors π_1 and π_2 both reduce to the same hydrostatic pressure p_0. We then strain the two phases, each point being displaced by an amount $\mathbf{u}(\mathbf{r})$. We use a *Lagrangian* picture in which the displacement is attached to a given *atom* rather than to a given point: $\mathbf{u}$ is a well defined state variable. Well inside the bulk, $\mathbf{u}$ reduces to a displacement and to a *uniform* strain. It is convenient to use an unsymmetrized strain tensor

$$v_{ij} = \frac{\partial u_i}{\partial x_j}. \tag{1.22}$$

(The usual u_{ij} is simply $\frac{1}{2}[v_{ij} + v_{ji}]$.) We want to identify the conserved part and the surface excess of v_{ij}.

For simplicity, we consider only a two dimensional situation, in which $\mathbf{u}$ has two components, a normal one u_z and a tangential one u_x,

---

[†] All quantities, free energy, density, etc. are referred to this reference situation, not to the $p = 0$ state.

which depend only on x and z (the extension to a genuine three dimensional case is *not* trivial). We assume that the interface does not move (no rotation): then the strain may be viewed as a superposition of two parts:

(i) *A tangential strain* $v_{xx} = u_{xx}$. If we exclude an ever *increasing* slip along the interface, u_{xx} is necessarily constant, the same on both sides and with no surface excess: u_{xx} is a conserved state variable, in much the same way as π_{zz} and π_{zx}.

(ii) A displacement which depends only on z, and which is linear at large distances (uniform strain). Taking the Gibbs surface as a reference, we may write

$$u_i = \begin{cases} v_{iz,1}(z - \xi_1) + w_{i,1} & (z \ll \xi_1), \\ v_{iz,2}(z - \xi_2) + w_{i,2} & (z \gg \xi_2). \end{cases} \tag{1.23}$$

w_1 and w_2 are the *extrapolated* displacements at the Gibbs surfaces ξ_1 and ξ_2, not to be confused with the actual displacements, which depend on details of the interfacial profile — see Fig. 3.

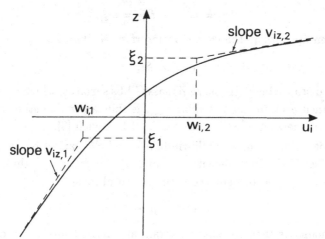

Figure 3: Definition of the surface strain.

The surface strain is defined as any other surface excess:

$$v_{iz,s} = \int_{z_1}^{z_2} \frac{\partial u_i}{\partial z} \, dz - v_{iz,1}(\xi_1 - z_1) - v_{iz,2}(z_2 - \xi_2).$$

We obtain at once

$$v_{iz,s} = w_{i2} - w_{i1}. \tag{1.24}$$

$v_{iz,s}$ is the *relative extrapolated displacement* of the two phases at the dividing surfaces ξ_1, ξ_2 (which as of now are still arbitrary). The component $v_{xz,s}$ is a genuine shear effect: it describes a *slip* of phase 2 relative to phase 1. In contrast, $v_{zz,s}$ is somewhat ambivalent. It may well describe accumulation of matter in the interfacial region — but in a phase equilibrium situation it may also describe an ordinary mass transfer across the interface, such that

$$\rho_1 w_{z1} = \rho_2 w_{z2}.$$

We thus need to consider the two situations separately.

For chemically distinct phases 1 and 2, the displacements w_z imply a change of surface masses at constant ξ_1 and ξ_2:

$$\delta\rho_s^{(1)} = +\rho_1 w_{z,1}, \quad \delta\rho_s^{(2)} = -\rho_2 w_{z,2}. \tag{1.25}$$

If we want to leave the surface masses unchanged — for instance zero — we must correct ξ_1 and ξ_2 as strain proceeds: the Gibbs surface must be attached to an *extrapolated atom*, moving along as it is displaced:

$$\delta\xi_1 = w_{z,1}, \quad \delta\xi_2 = w_{z,2}.$$

$v_{zz,s}$ then appears as *the change in interface thickness*

$$v_{zz,s} = \delta\ell_s. \tag{1.26}$$

ℓ_s is the distance between the zero mass Gibbs surfaces, which is perfectly well defined even for finite strain: $v_{zz,s}$ is thus a genuine state variable. This is the view adopted by Andreev and Kosevich [3].

In a situation of phase equilibrium, $\rho_s^{(1)}$ and $\rho_s^{(2)}$ cannot be ascertained separately. If the bulk densities ρ_1 and ρ_2 are different,[†] the simplest choice is a single Gibbs surface with zero total mass

$$\ell_s = 0, \quad \rho_s = 0.$$

The difficulty is then to separate strain and mass transfer. In order to circumvent this difficulty, we proceed as follows:

(i) Define the Gibbs surface ξ_0 for the initial unstrained state: we thus know the nominal extrapolated bulk masses M_{10} and M_{20} of each phase

$$M_{10} = A\rho_{10}(\xi_0 - z_1), \quad M_{20} = A\rho_{20}(z_2 - \xi_0).$$

[†] This situation $\rho_1 = \rho_2$ occurs for instance at a grain boundary: thermodynamics is then more delicate: see [4].

(ii) A similar definition in the final state yields a Gibbs surface ξ, nominal masses M_1 and M_2. The amount of phase transfer is defined as

$$\Delta M = M_2 - M_{20} = -\left[M_1 - M_{10}\right]$$

(remember that the surface mass is zero).

(iii) $\Delta M = 0$ corresponds to a pure strain situation — but in that case no extrapolated atom ever crosses ξ. Hence

$$w_{z,1} = w_{z,2} = \xi - \xi_0.$$

It follows that $v_{zz,s}$ is *identically zero*: the normal surface strain is *not* an independent thermodynamical variable. It can be absorbed in an appropriate definition of the mass transfer across the interface.

At this stage, we may stop to summarize a few important facts:

a) For an interface between two distinct solid materials, the independent thermodynamic bulk variables that characterize the interface are:

    ▷ a tangential strain $u_{xx} = v_{xx}$;

    ▷ normal stresses π_{zz} and π_{zx}.

These quantities are the same on both sides. They display no surface excess if $\ell_s = 0$.

If one phase is a liquid, π_{zx} is zero: only u_{xx} and π_{zz} remain as independent variables. If furthermore the two phases are in 'chemical' equilibrium, we lose one more variable — for instance π_{zz} which is determined by the constraint (1.21). The thermodynamic state of the interface then depends only on the *tangential strain* u_{xx}.

b) Surface excesses involve:

    ▷ a tangential surface stress $\pi_{xx,s}$;

    ▷ normal strains $v_{zx,s}$ and $v_{zz,s}$.

(Note the complementarity with a).) If one phase is a liquid, $v_{zx,s}$ is a dummy variable: a slip of the liquid with respect to the solid is meaningless since the liquid can flow freely (in contrast, u_{xx} affects the solid substrate). If the two phases are in chemical equilibrium, $v_{zz,s}$ also disappears, since it can be made zero by an appropriate definition of mass transfer: *at the contact between a pure solid and its melt, there is no surface strain at all.*

c) Finally, if the two phases are fluid u_{xx} is also a dummy variable. Then γ depends only on the pressure p — which is itself fixed for two phases in equilibrium.

The assessment of independent thermodynamical variables is not a trivial problem. In the most general case, there are three of them: when deriving, it is crucial to specify what is kept constant.

Energy conservation: relation between surface free energy and surface stress

We consider a system bounded by walls which are initially at z_1 and z_2. The interface area is A. The total free energy is by definition

$$F = f_1 V_1 + f_2 V_2 + f_s A, \qquad (1.27)$$

where the nominal volumes V_1 and V_2 are measured from the Gibbs surfaces ξ_1 and ξ_2 — *a priori* arbitrary.

The system is *closed*: no matter flows through the walls. We then apply an *infinitesimal strain* δu_{ij} by moving these walls, both lateral ($\delta A/A = \delta u_{xx}$) and horizontal (displacements $\delta \mathbf{u}_1(z_1)$ and $\delta \mathbf{u}_2(z_2)$ respectively). If we keep ξ_1 and ξ_2 constant, the nominal phases 1 and 2 each acquire an additional mass

$$\delta M_1 = -A \rho_1 \delta w_{z,1}, \quad \delta M_2 = +A \rho_2 \delta w_{z,2}. \qquad (1.28)$$

The total change in F is accordingly

$$\delta F = \delta F_1 + \delta F_2 + A \delta f_s + f_s \delta A, \qquad (1.29)$$

in which the bulk δF_1 follows from ordinary elasticity theory[†]

$$\delta F_1 = \frac{f_1}{\rho_1} \delta M_1 + V_1 \pi_{ij,1} \delta u_{ij,1}. \qquad (1.30)$$

Since π_{ij} is symmetric, δu_{ij} may be replaced by δv_{ij}. A similar equation holds for δF_2.

The first law of thermodynamics implies that δF is equal to the work δW of *external* forces — here the stress since we bring no matter in:

$$\delta W = A \pi_{iz} \left[\delta u_i(z_2) - \delta u_i(z_1) \right] \qquad (1.31)$$
$$+ \delta A \left[\pi_{xx,2}(z_2 - \xi_2) + \pi_{xx,1}(\xi_1 - z_1) + \pi_{xx,s} \right].$$

Using the definition (1.33), we cast δW in the form

$$\delta W = V_1 \pi_{ij,1} \delta u_{ij,1} + V_2 \pi_{ij,2} \delta u_{ij,2} \qquad (1.32)$$
$$+ A \left[\pi_{iz} \delta v_{iz,s} + \pi_{xx,s} \delta u_{xx} \right].$$

[†] Remember that $\pi \delta u$ is the work per unit *unstrained* volume.

Equating (1.32) and (1.29), we see that the bulk terms drop out — as they should. We are left with a 'surface identity':

$$\delta f_s + f_s \delta u_{xx} = \pi_{xx,s} \delta u_{xx} + \pi_{iz} \delta v_{iz,s} \qquad (1.33)$$
$$- \frac{f_1}{\rho_1} \frac{\delta M_1}{A} - \frac{f_2}{\rho_2} \frac{\delta M_2}{A}.$$

The physical interpretation of (1.33) is obvious. The surface free energy changes for three reasons:

(i) incorporation into the 'surface' of some bulk matter (last two terms);

(ii) work of the *surface stress* $\pi_{xx,s}$ upon *bulk* dilatation u_{xx};

(iii) work of the *bulk stress* π_{iz} upon a change of *surface* strain $\delta v_{iz,s}$. That term is often forgotten: its importance was stressed by Andreev and Kosevich [3].

Equation (1.33) can be made more explicit by using (1.28):

$$\delta f_s + f_s \delta u_{xx} = \pi_{xx,s} \delta u_{xx} + \pi_{iz} \delta v_{iz,s} + f_1 \delta w_{z,1} - f_2 \delta w_{z,2}. \qquad (1.34)$$

In this form, it is easily verified that the identity is invariant[†] upon a change $\Delta \xi_1$, $\Delta \xi_2$.

Equation (1.33) remains valid if ξ_1 and ξ_2 are changed as the shear proceeds: in (1.28), δw_z is just replaced by $(\delta w_z - \delta \xi)$. However it has the unpleasant feature of involving the bulk nominal mass M_1, which is not well defined in a phase equilibrium situation. This flaw can be cured by introducing the effective 'chemical potential' (1.20) instead of f. Equation (1.33) thus becomes

$$\frac{1}{A} \delta[f_s A] = \pi_{xx,s} \delta u_{xx} + \pi_{xz} \delta v_{xz,s} + \pi_{zz} \delta \ell_s \qquad (1.35)$$
$$- \frac{1}{A} [\mu_1 \delta M_1 + \mu_2 \delta M_2].$$

If the two phases are in equilibrium, $\mu_1 = \mu_2$ and only the sum $(\delta M_1 + \delta M_2)$ enters, which is related to the change in surface mass

$$\delta M_1 + \delta M_2 = -\delta \left[A(\rho_s^{(1)} + \rho_s^{(2)}) \right].$$

[†] The additional terms in (1.34) are $\Lambda_2 \Delta \xi_2 - \Lambda_1 \Delta \xi_1$, in which we have $\Lambda = \delta f + f \delta u_{ii} - \pi_{ij} \delta u_{ij} = 0$.

Thus we can write the completely unambiguous *surface* identity:

$$\delta f_s + f_s \delta u_{xx} = \mu_1 \delta \rho_s^{(1)} + \mu_2 \delta \rho_s^{(2)} + \delta u_{xx} [\mu_1 \rho_s^{(1)} + \mu_2 \rho_s^{(2)}] \qquad (1.36)$$
$$+ \pi_{xx,s} \delta u_{xx} + \pi_{xz} \delta v_{xz,s} + \pi_{zz} \delta \ell_s.$$

In this form, the ambiguous surface strain has disappeared.

Rather than f_s we can introduce an intrinsic *surface tension* which does not depend on the choice of ξ_1 and ξ_2:

$$\gamma = f_s - \mu_1 \rho_s^{(1)} - \mu_2 \rho_s^{(2)} - \pi_{zz} \ell_s. \qquad (1.37)$$

(Note the analogy with (1.6).) Equation (1.36) thus becomes

$$\boxed{\begin{aligned} &\delta \gamma + \delta u_{xx}[\gamma - \pi_{xx,s} + \pi_{zz} \ell_s] \\ &= -\rho_s^{(1)} \delta \mu_1 - \rho_s^{(2)} \delta \mu_2 - \ell_s \delta \pi_{zz} - \pi_{xz} \delta v_{xz,s}. \end{aligned}} \qquad (1.38)$$

Equation (1.38) is our basic physical result. It allows direct contact with the much simpler fluid-fluid case. Then:

(i) $\pi_{zx} = 0$;

(ii) δu_{xx} is a dummy variable: its coefficient must vanish, yielding (1.11);

(iii) the remaining terms yield (1.16).

The main complications brought about in the solid case are (i) a shear π_{zx} and (ii) a tangential dilatation δu_{xx}.

Thermodynamic identities

Since γ does not depend on ξ_1 and ξ_2, we are free to choose the most convenient definition. Consider first an interface between two distinct materials. We may set $\rho_s^{(1)} = \rho_s^{(2)} = 0$. The interface width ℓ_s is such that

$$\ell_s - \ell_{s0} = v_{zz,s}.$$

From (1.36) and (1.38), we infer

$$\pi_{xx,s} = f_s + \left.\frac{\partial f_s}{\partial u_{xx}}\right|_{v_{iz,s}},$$
$$\pi_{xx,s} - \pi_{zz} \ell_s = \gamma + \left.\frac{\partial \gamma}{\partial u_{xx}}\right|_{\pi_{zz}, v_{xz,s}}. \qquad (1.39)$$

In contrast to the fluid case, surface free energy and surface stress are *not* equal. They correspond to different processes: area increase at constant *structure* (γ, f_s), or at constant *number of atoms* $(\pi_{xx,s})$. The relation (1.39) was established long ago by Shuttleworth [6], but without specifying the nature of partial derivatives. It is clear that starting from (1.36) one can play the usual game of Legendre transformations, Maxwell relationships, etc.

For two pure phases in equilibrium, $\mu_1 = \mu_2$, we return to the Gibbs choice $\rho_s = 0$ and $\ell_s = 0$. Then π_{zz} disappears from (1.38) — a fortunate simplification since it is no longer an independent thermodynamic variable. Equation (1.38) takes the simpler form

$$\delta\gamma = \left[\pi_{xx,s} - \gamma\right]\delta u_{xx} + \pi_{xz}\,\delta v_{xz,s}.$$

If moreover one phase is a fluid, $\pi_{xz} = 0$. Only one variable remains, and the Shuttleworth relationship holds

$$\pi_{xx,s} = \gamma + \frac{\partial\gamma}{\partial u_{xx}}. \qquad (1.40)$$

In the end, the result is simple — but it is important to assess its validity.

1.3 Equilibrium conditions for a curved solid-fluid interface

There are two such conditions. One, 'mechanical,' expresses the balance of *forces* at the interface. The other, 'chemical,' deals with *mass transfer* between the two phases. For two fluids in contact, these two conditions can be formulated simultaneously: we need only combine the capillary pressure (1.17) with the requirement that $\mu_1 = \mu_2$. Compared to the nominal equilibrium pressure p^* for a planar interface, the equilibrium pressures for a cylindrical interface with radius R are shifted by

$$\delta p_1 = \frac{\rho_1}{\rho_1 - \rho_2}\frac{\gamma}{R}, \quad \delta p_2 = \frac{\rho_2}{\rho_1 - \rho_2}\frac{\gamma}{R}.$$

This is the familiar Gibbs-Thomson displacement.

The situation is completely different at a solid-liquid interface: here, 'mechanical' and 'chemical' equilibrium must be treated separately, as they involve different surface quantities — the surface stress π_s for force equilibrium, the surface energy γ for phase equilibrium. Moreover, the bulk solid stress is not necessarily hydrostatic: equilibrium depends on the full tensor π_{ij}, eventually modified by capillary effects. We shall see that the latter corrections are usually small, negligible in a first approximation — still, the possibility of an externally applied uniaxial stress remains. We examine these various aspects successively, beginning with the simplest case: a solid which is in hydrostatic equilibrium in the absence of capillarity.

Phase equilibrium for hydrostatically stressed solids: surface stiffness

Assume that the system has a fixed total mass. If the solid grows normally by an amount h, the volume increases (per unit area) by

$$\Delta V = h\left[1 - \frac{\rho_C}{\rho_L}\right].$$

If we assume that *the solid grows at constant strain*, the resulting change in free bulk enthalpy $G = F + p_L v$ is

$$\begin{aligned} \Delta G_V &= h\left[f_C - f_L\frac{\rho_C}{\rho_L} + p_L\left(1 - \frac{\rho_C}{\rho_L}\right)\right] \\ &= h[f_C + p_L - \rho_C\mu_L]. \end{aligned} \tag{1.41}$$

To this we must add the change of surface enthalpy ΔG_s. Integrating the displacement h over the (curved) interface, we obtain the net change $\Delta G_V + \Delta G_S$: 'chemical' equilibrium is achieved if ΔG *is zero to first order in* h. We thus have a well defined recipe — but with one weak point: we ignore the change in bulk strain due to capillary effects as growth proceeds. We shall see later that surface stresses do indeed yield a small shift of equilibrium.

Let us first straighten out the case of a planar interface normal to the z-axis. Then $\pi_{xz} = 0$, $\pi_{zz} = -p_L$. There is no change ΔG_S, and the equilibrium condition reduces to

$$\Delta = \frac{f_C + p_L}{\rho_C} - \mu_L = 0, \tag{1.42}$$

which is nothing but (1.20). Unfortunately, this is not enough to treat a curved interface: we then want the *change in* Δ as p_L departs from the nominal equilibrium pressure p_L^*. For an arbitrary crystal strain, this is difficult: the answer is simple only if we start from a pure *hydrostatic* state in the absence of capillary effects:

$$\pi_{ij}^{(0)} = -p_L^*\delta_{ij}.$$

The shift in equilibrium results in an additional strain δu_{ij} in the crystal. To *first order* in this shift, the change of bulk crystal energy is

$$\delta\left[\frac{f_C}{\rho_C}\right] = \frac{1}{\rho_C}\pi_{ij}^{(0)}\delta u_{ij} = p_L^*\frac{\delta\rho_C}{\rho_C^2} \tag{1.43}$$

(remember that $\delta\rho_C/\rho_C = -u_{\ell\ell}$). Inserting (1.43) into (1.42), we find

$$\Delta = \delta p_L \left[\frac{1}{\rho_C} - \frac{1}{\rho_L} \right]. \tag{1.44}$$

In this restricted case, only the liquid pressure enters: details of δu_{ij} have disappeared (because π_{ij}^0 is a unit tensor).

We first consider the cylindrical geometry of Fig. 4. The interface is characterized by its cross section, with a local orientation of the normal $\theta(s)$ (where s is the abscissa along the interface). The surface tension γ is anisotropic, $\gamma(\theta)$ — at this stage we ignore the dependence on tangential strain u discussed at length in the preceding section.

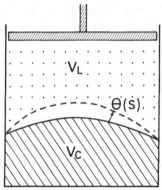

Figure 4: Infinitesimal growth of the crystal at constant total mass.

We then displace the interface by an amount $h(s)$. The change in bulk free enthalpy follows from (1.41) and (1.44):

$$\Delta G_v = \int \mathrm{d}s\, \delta p_L \left[1 - \frac{\rho_C}{\rho_L} \right] h(s). \tag{1.45}$$

The change in surface enthalpy arises from two effects:

▷ *stretching* of a surface element $\mathrm{d}s$, by a factor $(1 + h/R)$ where $1/R = -\mathrm{d}\theta/\mathrm{d}s$ is the local curvature, taken positive if the convexity is in the liquid;

▷ *rotation* of the element $\mathrm{d}s$, by an angle $\mathrm{d}\theta = \mathrm{d}h/\mathrm{d}s$.

As a result

$$\Delta G_s = \int \mathrm{d}s \left[-\gamma h \frac{\mathrm{d}\theta}{\mathrm{d}s} + \frac{\partial\gamma}{\partial\theta} \frac{\mathrm{d}h}{\mathrm{d}s} \right]. \tag{1.46}$$

(Remember that the interface is deformed by adding new matter to the solid at constant strain: 'stretching' has nothing to do with elasticity.)

20   *P. Nozières*

Minimizing the total ΔG with respect to h, we find the local equilibrium condition

$$\delta p_L \left[\frac{\rho_C}{\rho_L} - 1\right] = \frac{\gamma + \gamma''}{R}.$$

(1.47)

We recover a Gibbs-Thomson condition, γ being replaced by the so-called *surface stiffness*

$$\tilde{\gamma} = \gamma + \gamma''.$$

(1.48)

(Remember that (1.47) holds only *in first order* in departures from an hydrostatic stress.) The generalization to a genuine 3D geometry is straightforward: we define a local *curvature tensor* $C_{\mu\nu}$. Then (1.47) becomes

$$\begin{cases} \delta p_L \left[\frac{\rho_C}{\rho_L} - 1\right] = \tilde{\gamma}_{\mu\nu} C_{\mu\nu}, \\ \tilde{\gamma}_{\mu\nu} = \gamma \delta_{\mu\nu} + \frac{\partial^2 \gamma}{\partial\theta_\mu \partial\theta_\nu}. \end{cases}$$

(1.49)

The origin of surface stiffness $\tilde{\gamma}$ is best understood by introducing fictitious 'chemical stresses' whose work is the change in surface energy as the solid grows (no relation whatsoever with usual mechanical stresses). These stresses, depicted in Fig. 5a, involve a tangential 'stretching' stress $\gamma(\theta)$, and a 'torque' $\gamma'(\theta)$ which tends to bring the interface back to the orientation of minimal energy.

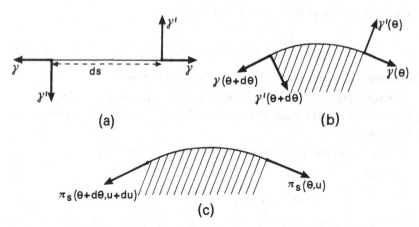

Figure 5: (a) and (b): "Chemical stresses" on an interface element, not to be confused with *mechanical* stresses (c).

For the curved interface of Fig. 5b, we see that the resultant of *tangential* forces is

$$-\gamma(\theta + \mathrm{d}\theta) + \gamma(\theta) + \gamma'\mathrm{d}\theta.$$

It vanishes, as it should (thanks to the torque). The resultant of *normal* forces is

$$-\gamma'(\theta + \mathrm{d}\theta) + \gamma'(\theta) - \gamma\mathrm{d}\theta = -\tilde{\gamma}\mathrm{d}\theta.$$

It balances the capillary pressure, in agreement with (1.47). Such a concept of 'chemical stresses,' although somewhat misleading, is often useful in assessing equilibrium shapes.

If a solid bulge grows further, its curvature increases. If $\tilde{\gamma} > 0$, the nominal equilibrium δp_L should increase. In practice δp_L is fixed: the bulge tends to melt, i.e. to regress — local equilibrium is *stable*. The reverse holds if $\tilde{\gamma} < 0$. The *local stability condition* is therefore

$$\tilde{\gamma} > 0.$$

Any orientation θ with $\tilde{\gamma} < 0$ must disappear from the equilibrium shape — leading to *angular points*.

In practice, *local* stability is one thing, *global* stability is another. For a spherical liquid drop, for instance, it is well known that the drop is stable only if the interface covers less than a half sphere — a full sphere is necessarily unstable, corresponding to the critical germ in nucleation theory.

Force equilibrium: surface stress

Once again, we consider first the simplest one dimensional geometry, namely a cylindrical interface with a single curvature $R = \mathrm{d}s/\mathrm{d}\theta$. The surface energy $\gamma(\theta, u)$ depends on the nominal surface orientation θ (before strain) and on a *tangential strain* $u_{ss} = u$. We know from (1.40) that the surface *mechanical stress* is strictly *tangential*, equal to

$$\pi_s = \gamma + \frac{\partial\gamma}{\partial u}.$$

More precisely

$$\pi_s = \frac{\partial}{\partial u}[\gamma(1 + u)], \qquad (1.50)$$

a subtlety that does not affect π_s in first order, but which is relevant for $\partial \pi_s / \partial u$:

$$\frac{\partial \pi_s}{\partial u} = \frac{\partial^2 \gamma}{\partial u^2} + 2 \frac{\partial \gamma}{\partial u}. \qquad (1.51)$$

Let n be the local orientation of the normal, t the tangential direction along the profile. A surface element ds on a bent interface is subject to *bulk* stress π_{ij} on either side — and to the surface stress π_s. Let R be the actual radius of curvature (due to *both* growth and strain). Equating to zero the total forces along the n- and t-axes, we find (see Fig. 5c):

$$\pi_{nn} + p_L = -\frac{\pi_s}{R}, \qquad (1.52)$$

$$\pi_{nt} = \frac{d\pi_s}{ds} = \frac{\partial \pi_s}{\partial \theta} \frac{d\theta}{ds} + \frac{\partial \pi_s}{\partial u} \frac{du}{ds}. \qquad (1.53)$$

In contrast to a planar interface, π_{iz} is *not* conserved. Equation (1.52) is the usual Laplace's law. According to (1.53), changes of θ and u along the interface induce *shear* in the solid. In a more general 3D geometry, (1.52) and (1.53) would be replaced by

$$\pi_{nn} + p_L = -\pi_{s,\mu\nu} C_{\mu\nu},$$

$$\pi_{n\mu} = \frac{\partial \pi_{s,\mu\nu}}{\partial s_\nu},$$

where $C_{\mu\nu}$ is the local curvature tensor.

These results should be contrasted with (1.47) and (1.49): despite the similarity in form, they describe completely different properties: mechanical equilibrium has nothing to do with phase equilibrium, surface stress π_s has nothing to do with surface stiffness $\tilde{\gamma}$.

Influence of elastic deformations on phase equilibrium

These corrections turn out to be small, typically of order a/R in which a is an atomic length and R the radius of the interface. The same order of magnitude appears if we compare the surface energy to the bulk elastic energy stored in response to capillary stresses. Let R be the size of the sample. The capillary overpressure is $\sim \gamma/R$, yielding a solid strain

$$u_{ij} \sim \frac{\kappa \gamma}{R},$$

where κ is some compressibility of the solid. The relevant quantity is the free enthalpy $g = f + p_L$ (per unit volume). The additional contribution due to u_{ij} is

$$\Delta g = \frac{u_{ij}^2}{\kappa} \sim \frac{\kappa \gamma^2}{R^2}.$$

For a total volume $\sim R^3$, we find $\Delta G \sim \kappa\gamma^2 R$, to be compared to the usual surface energy $\sim \gamma R^2$: it is clear that elastic effects are down by a factor $1/R$ — the same conclusion holds for phase equilibrium.

At such a level of accuracy, everything breaks down: the very concept of a surface energy is ambiguous! The issue is therefore somewhat semantic. Nevertheless, it is of some interest, as it may clarify some paradoxical results. For instance, an impinging acoustic wave may monitor a.c. melting at the solid-superfluid interface of ^{4}He — thereby affecting acoustic transmission. The response of the interface to the local pressure modulation δp_L is controlled by mechanical and chemical equilibrium requirements. If one uses (1.47), (1.52) and (1.53), one finds that acoustic energy is *not* conserved upon reflection and transmission — an embarrassing result, even though the missing energy is small (precisely of order $\gamma\kappa/R$). The paradox disappears if we treat *consistently* the effect of capillary elastic stresses on phase equilibrium — a feature which we have ignored until now.

This was done in [8]: since the issue is mostly pedagogic, we only sketch the method and the result. We start from a uniform solid with a planar interface $z = 0$. We then proceed in two steps:

(i) Grow some more solid at *constant strain*, raising the interface by an amount $h(x)$.

(ii) Add an additional strain by displacing the new interface by an amount $\xi(x)$. (Note that the deformation is characterized by a surface displacement, not by a bulk strain field.) In *first order* $\partial\xi_x/\partial x$ is just the tangential strain u defined previously.

The total free energy contains a surface part and an elastic part. Chemical equilibrium is achieved if it is stationary with respect to h, mechanical equilibrium if it is stationary with respect to ξ. Since we want the equilibrium condition to first order in ξ and h, we must calculate the energy to *second order* — which requires some care. The results are the following:

(i) Minimization with respect to ξ yields (1.52) and (1.53) with

$$\frac{1}{R} = -\frac{\mathrm{d}^2}{\mathrm{d}x^2}\,[h + \xi_z].$$

We recover the result of our naïve balance of forces argument.

(ii) Minimization with respect to h yields

$$\delta p_L\left[1 - \frac{\rho_C}{\rho_L}\right] = (\gamma + \gamma'')\frac{\mathrm{d}^2 h}{\mathrm{d}x^2} + \pi_s\frac{\mathrm{d}^2\xi_z}{\mathrm{d}x^2} + \frac{\partial\pi_s}{\partial\theta}\frac{\mathrm{d}^2\xi_x}{\mathrm{d}x^2}. \tag{1.54}$$

The first term is our former result (1.47), with the curvature evaluated
before strain. The last two terms express the shift of phase equilibrium
due to strain inhomogeneities. When used in the acoustic transmission
problem, (1.53) and (1.54) ensure *strict* energy conservation, showing
that the above formulation is indeed consistent.

The central issue is to compare the *growth* displacement h with the
strain displacement ξ, for a given pressure fluctuation δp_L, with typical
wave vector κ. Qualitatively, we have

$$k\xi \sim \kappa \delta p_L, \quad k^2 h \sim \left[\frac{\rho_C}{\rho_L} - 1\right]\frac{\delta p_L}{\tilde{\gamma}}.$$

The ratio ξ/h is of order k, very small unless ρ_C is very close to ρ_L.
Usually, elastic corrections to the Gibbs equilibrium condition will be
negligible.

The Grinfeld bulk instability in uniaxially stressed materials
Until now, we were concerned only with solids which in their reference
state were in a state of hydrostatic equilibrium

$$\pi_{ij}^{(0)} = -p\delta_{ij}.$$

We now consider a situation in which the solid has been prestrained —
for instance a planar interface $z = 0$ between a liquid phase (pressure p_L)
and a *uniaxially stressed* solid:

$$\pi_{zz}^{(0)} = -p_L, \quad \pi_{zx}^{(0)} = 0 \quad \text{(mechanical equilibrium)},$$
$$\pi_{xx}^{(0)} \neq \pi_{zz}^{(0)}.$$

Put another way, we *press* the crystal along the x-direction. We ignore
the y coordinate: depending on the problem at hand, we may either
set $u_{yy} = 0$ (fixed walls) or $\pi_{yy} = -p_L$ (free ends). The two phases
are initially in melting equilibrium: they have equal enthalpies per unit
mass, $g/\rho = (f + p_L)/\rho$ (see (1.42)).

Surprisingly, this state is *unstable* against 'melting crystallization
waves': corrugations of the interface develop spontaneously, monitored
by the difference $(\pi_{xx} - \pi_{zz})$. This effect, predicted recently by Grinfeld [9], is a *bulk* effect (the relevant energy is the bulk elastic energy):
it has nothing to do with capillarity, which actually counters the instability at short wavelength. Such a phenomenon may be responsible for
the irregular shapes often observed in inhomogeneous materials.

The qualitative origin of the instability is fairly obvious. Assume that the solid grows locally, the interface being displaced by an amount $h(x)$. If we ignore capillarity, the total enthalpy is unchanged as long as the growth occurs at *constant stress* π_{ij} (g_c remains the same). But then the mechanical matching conditions at the interface are violated. The normal **n** turns in an *anisotropic* stress field, by an angle dh/dx. As a result, the normal and tangential components of π are modified — in first order, a *shear* component appears

$$\pi_{nt}^{(0)} = \pi_{ij} n_i t_j = (\pi_{zz} - \pi_{xx}) \frac{dh}{dx}. \tag{1.55}$$

If we want to maintain equilibrium, we must apply an extra tangential force to the interface

$$dF_t = \pi_{nt}^{(0)} ds.$$

If we do *not* apply F_t, the surface relaxes, thereby *lowering* the elastic energy: the net cost in enthalpy of the deformation $h(x)$ is *negative* — hence the instability.

In order to render the argument more quantitative, we consider a periodic deformation

$$h(x) = h \cos kx.$$

Relaxation results in an additional displacement u_i and strain u_{ij}, which may be viewed as the response to a force $(-F_t)$ applied at the interface. The change in elastic enthalpy due to an infinitesimal *deformation* (at constant mass) is

$$\delta G = \int d\tau \left[\pi_{ij} \delta u_{ij} + p_L \delta u_{\ell\ell} \right].$$

Since $p_L = -\pi_{zz}^{(0)}$ we may expand δG as

$$\delta G = \int d\tau \left[(\pi_{xx}^{(0)} - \pi_{zz}^{(0)}) \delta u_{xx} + \Delta \pi_{ij} \delta u_{ij} \right], \tag{1.56}$$

where $\Delta \pi_{ij}$ is the additional stress due to relaxation. Integrating (1.56) for a finite deformation, we obtain

$$\Delta G = \int d\tau \left[(\pi_{xx}^{(0)} - \pi_{zz}^{(0)}) u_{xx} + \tfrac{1}{2} \Delta \pi_{ij} u_{ij} \right]. \tag{1.57}$$

In first order in h, only the first term survives. The local change in effective chemical potential μ_c (see (1.20)) is

$$\Delta\mu_c = \frac{1}{\rho}\left[\pi_{xx}^{(0)} - \pi_{zz}^{(0)}\right]u_{xx} = \lambda h. \qquad (1.58)$$

If λ is > 0, a local bulge in the solid ($h > 0$) tends to melt back ($\mu_c > \mu_L$) : equilibrium is stable. The reverse holds if $\lambda < 0$.

ΔG can be calculated directly by looking at the work of the compensating force $\mathrm{d}F_t$ as it is reduced progressively to zero

$$\Delta G = \int_0^1 [u_t\,\mathrm{d}\varepsilon]\,[\mathrm{d}F_t(1-\varepsilon)] = \tfrac{1}{2}\int \pi_{nt}^{(0)} u_x\,\mathrm{d}x. \qquad (1.59)$$

In contrast, the second term of (1.57) would be the work of u_x against a force $(-\mathrm{d}F_t)$ applied to an unstrained interface: thus $\Delta G_2 = -\Delta G$: the first term of (1.57) must yield a contribution $\Delta G_1 = -2\Delta G_2$. This is indeed true, as can be seen by carrying the integration over x first: we get a non-zero result only if z *cuts* the interface, in which case the integration is trivial. We are left with a surface integral:

$$\Delta G_1 = -\left[\pi_{xx}^{(0)} - \pi_{zz}^{(0)}\right]\int \mathrm{d}x\,\frac{\mathrm{d}h}{\mathrm{d}x}u_x = 2\Delta G.$$

In lowest order we can apply $(-\mathrm{d}F_t)$ to the original $z = 0$ plane. The only missing piece is the calculation of u_x, which is a standard problem in elasticity. As an example, consider an isotropic elastic medium, in which

$$Eu_{ij} = (1+\sigma)\pi_{ij} - \sigma\delta_{ij}\pi_{\ell\ell}$$

(E is the Young modulus, σ the Poisson coefficient). We assume $u_{yy} = 0$ (which implies $\pi_{yy} = \sigma\,[\pi_{xx} + \pi_{zz}]$). For this 2D problem, the condition $\partial\pi_{ij}/\partial x_j = 0$ is automatically satisfied if we write

$$\pi_{xx} = \frac{\partial^2\chi}{\partial z^2}, \quad \pi_{zz} = \frac{\partial^2\chi}{\partial x^2}, \quad \pi_{xz} = -\frac{\partial^2\chi}{\partial x\partial z}, \qquad (1.60)$$

where $\chi(x,z)$ is an unknown *scalar stress function*. It is known that $\Delta u_{\ell\ell} = \Delta\pi_{\ell\ell} = 0$, from which we infer $\Delta^2\chi = 0$. It follows that χ has the form

$$\chi = (az + b)\cos kx\,\mathrm{e}^{kz}.$$

The constants a and b are adjusted in such a way that π cancels out the spurious $\pi_{nn}^{(0)}$ and $\pi_{nt}^{(0)}$ at the interface — hence in first order

$$\pi_{zz}(z=0) = 0, \quad \pi_{xz}(z=0) = -\left[\pi_{zz}^{(0)} - \pi_{xx}^{(0)}\right] kh \sin kx.$$

From χ, we infer π_{ij}, then u_{ij}, then the displacement u_i and finally ΔG. The result is

$$\Delta G = \frac{1-\sigma^2}{E} \int dx\, kh^2 \sin^2 kx \left[\pi_{xx}^{(0)} - \pi_{zz}^{(0)}\right]^2. \tag{1.61}$$

ΔG is monitored by the *slope* $[-kh \sin kx]$ — and the elastic deformation extends to a depth $1/k$ in the bulk: hence the overall factor k in ΔG.

Equation (1.61) is only the *elastic* contribution to the free enthalpy, which is destabilizing.[†] If we introduce capillarity, a new term appears, $\frac{1}{2}\gamma(\partial h/\partial x)^2$, which stabilizes the interface at large k. Similarly, gravity would be stabilizing at small k: the Grinfeld instability is thus sandwiched in an intermediate window, a usual feature in bulk phenomena (see e.g. the Kelvin-Helmholtz instability). It appears if $|\pi_{xx}^{(0)} - \pi_{zz}^{(0)}|$ exceeds a threshold — which is quite modest — typically 0.1 to 1 bar.

1.4 Equilibrium shapes of crystals

We first ignore gravity: the liquid pressure p_L is the same everywhere. Equilibrium is achieved if the condition (1.49) is satisfied at every point on the interface: this determines both the *shape* of the crystal (through anisotropies of $\tilde{\gamma}$), and its *size* (determined by the supercooling δp_L). As usual, the equilibrium size is unstable unless the interface is attached to some wall, spanning less than a half sphere.

One might also look for the shape only, minimizing the surface energy at constant total solid volume V_C. One is led again to (1.49), δp_L being now a Lagrange multiplier which must be adjusted in order to have the right V_C.

It is clear that the resulting shape depends on boundary conditions. Only a completely *isolated* crystal has a well defined equilibrium shape.

[†] The same calculation yields the coefficient λ in (1.58)

$$\lambda = -2k\frac{1-\sigma^2}{E}\left[\pi_{xx}^{(0)} - \pi_{zz}^{(0)}\right]^2.$$

The interface is clearly unstable.

If it touches walls, with appropriate contact angles, the interplay of curvatures will change, and the shape will be different. As an example, we may consider soap bubbles: if they are isolated, they are spheres — but if they are attached to some wiring they can take complicated constant curvature shapes — for instance a cylinder if they are attached to two parallel wires.

In this respect, the *cylindrical* geometry, with one curvature only, is much simpler than the general 3D case. We no longer have the freedom to play with two curvatures, and as a result any small section of the interface is a piece of the same overall equilibrium shape. This is evident on the simpler equation (1.47), which is a plain first order differential equation for the abscissa $s(\theta)$ along the interface

$$\frac{\mathrm{d}s}{\mathrm{d}\theta} = \tilde{\gamma}(\theta) \qquad (1.62)$$

(the constant factors are absorbed in the scale of s). Equation (1.62) has a unique solution, whatever boundary conditions. In contrast, the 3D Euler equations are partial differential equations, with a broad variety of solutions.

We thus will discuss first the cylindrical case — as met for instance for an interface which is sandwiched *between two planar parallel walls*. Although somewhat artificial, this example displays all important physical features with a minimum of complexity.

The Wulff construction for cylindrical crystals

The solution of (1.62) can be obtained by a very simple geometrical construction due to Wulff:

(i) From some origin O, draw the line of orientation θ. On that line define M such that $OM = \gamma(\theta)$ (γ is the bare *free energy*, not the stiffness $\tilde{\gamma}$). The locus of M is the *polar plot* of $\gamma(\theta)$.

(ii) Within a change of scale, the equilibrium shape is the *pedal* of the polar plot, i.e. the envelope of the perpendicular to OM drawn through M.

The construction is illustrated in Fig. 6. In order to prove the equivalence with (1.62), we write the equation of MP in polar coordinates

$$r = \frac{\gamma(\theta)}{\cos(\varphi - \theta)}. \qquad (1.63)$$

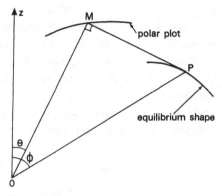

Figure 6: The Wulff construction for cylindrical interfaces.

The envelope is obtained by combining (1.63) with its derivative with respect to θ: we thus obtain the parametric set of equations

$$\varphi = \theta + \arctan \frac{\gamma'}{\gamma}, \quad r = \sqrt{\gamma^2 + \gamma'^2}. \qquad (1.64)$$

From this we infer the radius of curvature of the envelope:

$$R = \frac{\mathrm{d}s}{\mathrm{d}\theta} = \sqrt{r'^2 + r^2\varphi'^2} = \gamma + \gamma''.$$

This is just what we want!

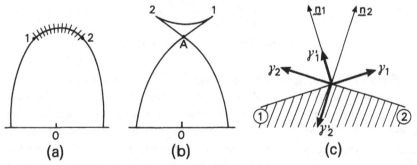

Figure 7: The formation of angular points on a cylindrical surface. (a) The polar plot. (b) The Wulff shape. (c) Chemical stresses at the angular point A.

If $\tilde{\gamma}$ is everywhere positive, the pedal is regular: the crystal is rounded. If instead there are unstable regions $\tilde{\gamma} < 0$ (Fig. 7a), the pedal develops cusps at the change of stability. The resulting shape is shown in Fig. 7b:

the actual crystal has *angular points* A and B, at which the orientation changes discontinuously. The unstable orientations have disappeared, as they should — carrying along with them some stable orientations as well. The situation is similar to a first order phase transition: the cusps are the spinodal limits, while A and B play the role of a 'Maxwell plateau' for 'orientation equilibrium.' The orientations θ_1 and θ_2 of angular points may be obtained by balancing chemical stresses on either side (Fig. 7c): projecting these stresses on the z- and x-axes, we find that the two combinations

$$\psi = \gamma \cos\theta - \gamma' \sin\theta,$$
$$\phi = \gamma \sin\theta + \gamma' \cos\theta, \tag{1.65}$$

must be continuous on going from '1' to '2' — enough to fix θ_1 and θ_2. The solution is particularly obvious when θ_1 and θ_2 are symmetric with respect to some symmetry axis — say $\theta = \frac{1}{2}\pi$. Then $\gamma(\theta) = \gamma(\pi - \theta)$: the condition $\phi_1 = \phi_2$ is ensured by symmetry, and (1.65) reduces to $\psi(\theta_1) = -\psi(\theta_2) = 0$.

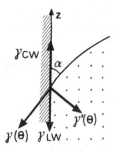

Figure 8: The contact angle with a wall.

Equation (1.62) gives the *local* crystal shape, within an arbitrary scale which is fixed by the supercooling δp_L (see (1.47)). We also need boundary conditions if the interface touches a wall. Consider the geometry of Fig. 8: the contact angle is α. If the wall is along the z-direction, the normal to the interface makes an angle $\theta = \frac{1}{2}\pi - \alpha$ with the z-axis. Let γ_{LW} and γ_{CW} be the interfacial energies of the wall with respectively the liquid and the crystal phases.[†] The contact angle is obtained by minimizing the total energy: this is Young's condition, with a slight complication due to the anisotropy of $\gamma(\theta)$. Equivalently,

---

[†] γ_{CW} depends on the crystal orientation — but for a given wall this is fixed once and for all.

one may balance the chemical stresses *along the wall* at the contact line, which yields at once

$$\gamma_{CW} - \gamma_{LW} = \gamma \sin\theta + \gamma' \cos\theta = \phi(\theta). \tag{1.66}$$

In practice $\phi' = (\gamma + \gamma'')\cos\theta$: if all orientations are stable, ϕ is monotonous, varying from $\phi_{\min}$ to $\phi_{\max}$ as θ goes from $-\frac{1}{2}\pi$ to $+\frac{1}{2}\pi$. If $(\gamma_{CW} - \gamma_{LW})$ falls in that range, there is *partial wetting*, with a well defined α. If not, the wetting is complete.

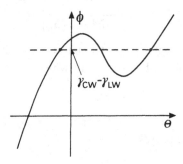

Figure 9: Hysteresis at the contact with a wall.

An interesting situation occurs if some orientations are unstable. $\phi(\theta)$ then has the shape depicted in Fig. 9. There may exist walls for which the matching equation has several locally stable solutions: the contact angle should then display *hysteresis*—a somewhat unusual feature.

Three dimensional geometries: the Landau-Andreev formulation [11]
The interface P is described in cartesian coordinates, $z(x,y)$. For an infinitesimal displacement PP'

$$\mathrm{d}z = h_\mu \mathrm{d}x_\mu \quad \text{with} \quad h_\mu = \frac{\partial z}{\partial x_\mu}.$$

The normal **n** to the interface is normal to **PP'** . It has coordinates

$$\frac{(-h_\mu, 1)}{\sqrt{1 + h^2}}$$

($h^2 = h_x^2 + h_y^2$). Let H be the projection of the origin O on the plane tangent to the interface at P :

$$|OH| = \mathbf{n} \cdot \mathbf{OP} = \frac{\zeta}{\sqrt{1 + h^2}} \tag{1.67}$$

in which we have set

$$\zeta = z - h_\mu x_\mu \implies d\zeta = -x_\mu dh_\mu \qquad (1.68)$$

(the projection is a Legendre transformation).

The surface free energy is

$$F_s = \iint \gamma \sqrt{1 + h^2}\, dx dy$$

in which γ depends on the normal orientation h_μ. Let us set

$$\gamma \sqrt{1 + h^2} = f(h_\mu)$$

Equilibrium is achieved if F_s is extremal at constant volume, i.e. for an unconditional extremum of $F - 2\lambda V$, in which V is the solid volume

$$V = z\, dx dy,$$

while λ is a Lagrange multiplier (in fact, $2\lambda = \delta p_L(\rho_C - \rho_L)/\rho_L$). The resulting Euler equation is

$$\frac{\partial}{\partial x_\mu}\left[\frac{\partial f}{\partial h_\mu}\right] = -2\lambda. \qquad (1.69)$$

Equation (1.69) is a partial differential equation which is valid for any equilibrium shape, whether isolated or between walls. This equation has a variety of solutions, among which is a simple first integral

$$f(h_\mu) = \lambda \zeta(h_\mu). \qquad (1.70)$$

The proof relies on (1.68):

$$\frac{\partial f}{\partial h_\mu} = \lambda \frac{\partial \zeta}{\partial h_\mu} = -\lambda x_\mu, \qquad (1.71)$$

from which (1.69) follows at once. Let us emphasize again that (1.70) is one particular solution among others: it is the one appropriate to a *global isolated crystal*.

Equation (1.70) is equivalent to a three dimensional Wulff construction. We construct a 3D polar plot $\gamma(\mathbf{n})$, by drawing a point M in the $\mathbf{n}$-direction such that $OM = \gamma(\mathbf{n})$. Within a change of scale, the equilibrium shape is the pedal of that polar plot, i.e. the envelope of the *plane* normal to OM at M (see Fig. 10). In order to prove it, we use (1.67)

$$|OH| = \frac{1}{\lambda}\frac{f}{\sqrt{1 + h^2}} = \frac{\gamma(\mathbf{n})}{\lambda}.$$

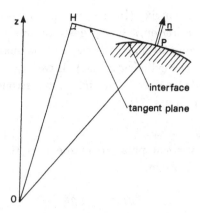

Figure 10: Three dimensional Wulff construction.

Within a constant, the locus of H is the polar plot — and conversely the locus of M (i.e. the crystal surface) is the pedal of that polar plot. The Wulff construction works equally well in two and three dimensions.

In order to specify the shape $z(x, y)$ explicitly, we push the Legendre transformation to its end. Let us set

$$\left. \begin{aligned} \eta_\mu &= \frac{\partial f}{\partial h_\mu} \\ g &= f - \eta_\mu h_\mu \end{aligned} \right\} \implies dg = -h_\mu \, d\eta_\mu. \qquad (1.72)$$

The anisotropy of γ embedded in $f(h_\mu)$ is now contained in a new function $g(\eta_\mu)$ — the calculation of which usually requires messy algebra! Using (1.68) and (1.71), we see that

$$\eta_\mu = -\lambda x_\mu,$$
$$g = \lambda\zeta + \lambda x_\mu h_\mu = \lambda z.$$

The shape of the interface is thus explicitly parametrized by the function $g(\eta_\mu)$:

$$\lambda z = g[-\lambda x_\mu]. \qquad (1.73)$$

Equation (1.73) is mathematically the most convenient formulation of the Wulff problem.

The $h = 0$ orientation is *stable* if the free energy is a minimum, i.e. if the second derivative

$$\lambda_{\mu\nu} = \frac{\partial^2 f}{\partial h_\mu \partial h_\nu} = \frac{\partial \eta_\mu}{\partial h_\nu}$$

is a *positive definite* matrix. (In the cylindrical case, $f'' = \gamma + \gamma''$: we recover our former criterion.) If $\gamma_{\mu\nu}$ has only one negative eigenvalue, the shape displays *angular lines* (the situation is qualitatively similar to that encountered in the cylindrical case). If two eigenvalues are negative — and eventually degenerate — one may have stronger singularities — for instance *conical points* [11].

Influence of gravity

Gravity creates a vertical pressure gradient in the liquid. With an appropriate choice of origin,

$$\delta p_L = -\rho_L g z.$$

We know that equilibrium requires

$$\delta p_L = \frac{\tilde{\gamma}}{R}\frac{\rho_L}{\rho_C - \rho_L}.$$

Comparison of these two terms defines a characteristic *capillary length*:

$$\ell_c = \left[\frac{\tilde{\gamma}}{g(\rho_C - \rho_L)}\right]^{1/2}. \tag{1.74}$$

Gravity is important on a scale larger than ℓ_c.

Consider an interface between two vertical plates a distance 2ℓ apart. The matching angle α with the walls is fixed.

(i) If $\ell \ll \ell_c$, gravity is negligible. The interface has its standard equilibrium shape, with a size fixed by ℓ (for a constant γ, it would be a circle with radius $R = \ell/\cos\alpha$). δp_L must adjust to produce that radius, fixed by geometry — hence a *capillary rise* of the interface (Jurin's law).

(ii) If $\ell \gg \ell_c$, most of the interface is flat, corresponding to $\delta p_L = 0$. A meniscus appears near the walls, ensuring the transition to $\theta = \frac{1}{2}\pi - \alpha$ over a distance $\sim \ell_c$.

The shape of the meniscus is governed by the second order differential equation

$$\frac{d\theta}{dx}\tilde{\gamma}\cos\theta = (\rho_C - \rho_L)gz, \tag{1.75}$$

with $\tan\theta = dz/dx$. Equation (1.75) has an obvious first integral

$$\gamma'\sin\theta - \gamma\cos\theta = \frac{1}{2}(\rho_C - \rho_L)gz^2 + \text{const.} \tag{1.76}$$

The last integration is straightforward.

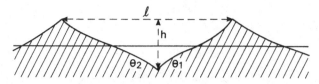

Figure 11: Formation of grooves on an unstable horizontal interface.

One pending question remains: what happens if the horizontal orientation $\theta = 0$ is unstable? In a small crystal it would disappear (angular point). Here it is forced by gravity — so what? The answer, given by Herring, is that a *groove structure* should appear on the interface, with a succession of angular points as shown in Fig. 11. The angles θ_1 and θ_2 at each angular point are bound to be those fixed by (1.65) (the 'coexistence' of two orientations is a local thermodynamic constraint). In between, the interface bends, with a characteristic length $\sim \ell_c$ (note that it explores *metastable* surface orientations). Let ℓ be the period of the structure: if $\ell \ll \ell_c$, the sides of the grooves are essentially straight lines and the depth of the grooves is

$$h = \frac{\ell}{\cot\theta_1 - \cot\theta_2}.$$

The cost in gravity energy per unit length of the interface is

$$E_g = \tfrac{1}{6}|\rho_C - \rho_L|gh^2. \tag{1.77}$$

E_g is countered by the energy cost of angular points

$$E_c = \frac{\varepsilon}{\ell},$$

where ε is the energy of an individual pair (upper and lower). We obtain ℓ by minimizing $(E_g + E_c)$.

In practice, $\varepsilon \sim \tilde{\gamma}\xi$, where ξ is the 'width' of the angular point (fixed by terms $\sim (d\theta/dx)^2$ in the surface energy, which we have been neglecting throughout). ξ is presumably of atomic size (except if θ_1 and θ_2 are very close). Minimization yields

$$\ell = \left[3\xi\ell_c^2(\cot\theta_1 - \cot\theta_2)^2\right]^{1/3}. \tag{1.78}$$

For intermediate θ_1, θ_2, the period is $\sim (\xi\ell_c^2)^{1/3}$. It is small (thereby justifying our assumption $\ell \ll \ell_c$) — but it is not ridiculous: with $\ell_c \sim 1$mm and $\xi \sim 10$Å, we find $\ell \sim 10\mu$. (ℓ is even larger if θ_1 and θ_2 are small.) The effect might thus be observable.

A detailed calculation of ε is easy to set up — but it is not worth the effort as long as we do not have a specific model.

2

Crystalline surfaces: facets, steps and kinks

Consider first a crystal surface parallel to some high symmetry crystal plane — say a (100) surface in a cubic crystal. The regular bulk structure is certainly disturbed near the interface ('surface reconstruction') — but that does not affect the *global* periodicity. If we add one atomic layer, the transition profile is *rigidly shifted* by one lattice spacing a, without any distortion.

For a *solid-vacuum* interface, the transition is abrupt. Except for a small change of lattice spacing in the last planes, the crystal atoms are stacked: the interface has no width — or say a width $\ell \sim a$ if the last layer is incompletely filled.[†] In contrast, a *solid-liquid* interface is certainly broad. The crystal periodicity penetrates into the liquid over a distance ℓ which is typically a few atomic spacings (this can be formulated precisely by looking at the decay of a density Fourier component ρ_G where G is a reciprocal lattice vector). Similarly, the change of density from ρ_C to ρ_L is progressive. As a result, *the pinning of the interface to crystal planes is rather weak*. In order to shift the whole profile by a, one must go through intermediate states where the transition region is distorted — hence an energy barrier V (per unit area) which rapidly becomes very small when the width ℓ increases (V decays *exponentially* for large ℓ). To put it short, a solid-vacuum interface is locked to crystal planes, while a solid-liquid one is weakly pinned.[‡]

Assume that the equilibrium surface is shifted by a over half its area. In order to minimize the cost in pinning energy, the transition region between the two half planes must be *finite*, with a width ξ: a localized *crystal step* develops, which may be viewed as a 'soliton' on the surface.

[†] At the moment, we ignore thermal fluctuations which act to broaden ℓ: they will be discussed later.

[‡] The situation may be different for a solid-vacuum interface if *surface melting* occurs, i.e. if the liquid wets the solid. Close to the triple point, the liquid layer thickens: the *global* interface looks very much like a solid-liquid interface!

Note that the step is an *asymptotic* structure, which is well defined even if the surface is broad. In order to estimate ξ, we ignore in a first approximation elastic strains created by the step in the bulk. The slope in the intermediate region is $\sim a/\xi$ — hence a lengthening of the interface $\sim a^2/\xi$. The net cost in energy of the step is

$$\frac{\gamma a^2}{\xi} + V\xi.$$

It will be minimum if

$$\xi \sim a\sqrt{\frac{\gamma}{V}}.$$

The weaker the pinning, the broader the step (compare with the width of a Bloch wall in ferromagnetism). The *energy of the step* (per unit length) is

$$\beta \sim a\sqrt{\gamma V}.$$

We note the important relation, independent of V,

$$\beta\xi \sim \gamma a^2. \tag{2.1}$$

Precise coefficients depend on the shape of V, which is unknown. The important conclusion is (2.1): a broad step has a small energy. At a solid-vacuum interface, $\xi \sim a$ and $\beta \sim \gamma a$ (qualitatively, β is the energy of the flank of the step, with width a). At a solid-liquid interface, ξ should be large: a width ~ 3 to 5 atomic spacings is not unreasonable.[†] An experiment which would give direct access to β is desirable.

Consider now a surface which is inclined by an angle θ with respect to crystal planes. A regular array of steps appears. Their distance along the crystal planes (x-axis) is

$$d = a \cot \theta \tag{2.2}$$

(see Fig. 12). Let $E(n)$ be the surface energy per unit dx, expressed in terms of the step density $n = 1/d$: we have

$$E(n) = \frac{\gamma(\theta)}{\cos\theta}. \tag{2.3}$$

[†] At finite temperature, the step meanders on the surface: its effective width increases. This will be described later: the divergence of ξ signals the so-called 'roughening transition.'

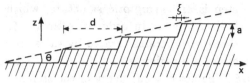

Figure 12: Steps on a vicinal surface.

The anisotropy of γ reflects the behaviour of $E(n)$: *the surface may be viewed as a distribution of steps.* In practice, such a description only makes sense if the individual steps are well defined, i.e. when

$$\xi \ll d \implies \theta \ll \frac{a}{\xi}. \qquad (2.4)$$

Such a tilted interface is called a *vicinal surface*. (If (2.4) is violated, the concept of step becomes meaningless: (2.3) is at most a convenient mathematical representation.) For small n, $E(n)$ can be expanded

$$E(n) = \gamma_0 + \beta n + E_{int}.$$

γ_0 is the energy of the (100) original surface. The linear term βn creates a *cusp* in $\gamma(\theta)$: we shall see that it is responsible for the appearance of *finite flat facets* in the equilibrium shape. The last term describes the *interactions* between steps, whatever their origin.

Consider now a curved interface. If it is *cylindrical* (with an axis in the y-direction), the steps remain parallel straight lines: only their density n is a function of x (the curvature is $\sim dn/dx$). The equilibrium shape is a balance between *step repulsion* on one hand, *supercooling* force on the other hand. In the more general 3D situations, steps are also *bent*: in addition to step interaction, we must include a step *line tension*, which tends to shrink the step length. We shall see that all our previous results are easily interpreted as an equilibrium of steps subject to these various forces: the physical picture becomes quite transparent.

Steps are due to the periodicity *perpendicular* to the crystal plane (here the z-axis). We should also worry about periodicity *within* the crystal planes. One obvious effect is to make the step energy *anisotropic*: β depends on the azimuthal angle ϕ of the normal to the step in the xy-plane. (More generally, γ is a function of both θ and ϕ.) In addition, the steps want to line up on the crystal *rows* in the xy-plane — say the x- or y-rows in a (100) cubic face: the periodicity along x or y plays for the steps the same role as the periodicity along z for the interface. Such a *locking* of steps on the in plane periodicity has important physical consequences.

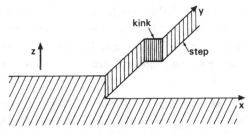

Figure 13: Steps and kinks on a crystal surface.

Consider a single step which makes on the average a small angle ϕ with a high symmetry axis, say the x-axis. In order to minimize the cost in locking energy, the passage from one row to the next must be localized: the step develops *kinks* (see Fig. 13), which are to steps what steps are to the interface. Let b be the lattice spacing along the x-axis: the kink density p is such that

$$\tan \phi = pb.$$

The energy $E(p)$ per unit length along the y-direction may be written as

$$E(p) = \frac{\beta(\varphi)}{\cos \varphi} = \beta_0 + p\varepsilon + \cdots$$

where ε is the energy of a single kink. We can repeat everything we said for steps. The kinks have a width ζ which is a compromise between step locking and stretching: as a result

$$\varepsilon\zeta \sim \beta b^2.$$

The kink picture only makes sense if $p\zeta \ll 1$ (if this condition is violated, we can practically neglect locking to the lattice rows). For a solid-vacuum interface, ζ is $\sim b$ ('atomic kinks'), and hence $\varepsilon \sim \beta b \sim \gamma ab$. For a solid-liquid interface, step broadening makes the locking extremely small: ζ is large and ε very small — so small indeed that we should worry about *thermal fluctuations*. A kink is an *atomic* entity: even for a step along the x-axis, there will always be a finite concentration $p^*(T)$ of positive and negative *thermal kinks*: our picture of a 'vicinal step' is meaningless if $p < p^*(T)$ (thermal fluctuations round off the singularity of $\beta(\phi)$ at $\phi = 0$). The kink concept is thus sandwiched between two conflicting limits:

$$p^*(\tau) \ll p \ll \frac{1}{\zeta}. \tag{2.5}$$

If $p^*\zeta > 1$, kinks 'dissolve' in thermal fluctuations: we can then ignore in-plane periodicity and assume *that the step is a continuous freely moving line* in the surface plane. This is usually the case for a solid-liquid interface: the smaller ε, the larger p^* and ζ — the inequalities (2.5) are never satisfied.

Let us return to steps along the y-axis ($\phi = 0$), and consider a *vicinal surface* in which the average step distance d is fixed by the tilt angle θ. How should we reconcile that d with the fact that steps prefer to be locked on lattice rows? If $d = Nb$, where N is an integer, there is no conflict: steps are regularly arranged on the lattice — a configuration which is nothing but a $(1, 0, N)$ facet. More generally, facets with higher order Miller indices correspond to rational values of the ratio d/b: we encounter the whole problem of *commensurability* of the step structure with the underlying in-plane periodicity. This is a whole field in itself: 'steps' in the step structure, thermal fluctuations, etc. [14].

Finally, consider a vicinal surface for $\phi \neq 0$: we have a regular array of tilted steps, each of them with a kink density p. We now face the commensurability of this kink structure with the crystal periodicity c along the y-axis. If $1/p = Mc$, we construct a commensurate structure which is a $(M, 1, MN)$ facet. The analysis rapidly becomes very complicated.

In practice, thermal fluctuations quickly wash out all higher order commensurate transitions. Thermal kinks make the steps smooth, and only the lowest order crystal faces are relevant at any realistic temperature. It is thus reasonable, as a first approximation, to ignore the in-plane periodicity completely: the crystal is a *stack of structureless plates*, steps are free, isotropic lines. The model becomes very simple, while displaying all important physical concepts. In this chapter, we will work only in that framework — which is in fact very close to reality for a solid-liquid interface. In-plane periodicity will be restored in the chapter on thermal fluctuations and roughening transitions, in order to discuss briefly the commensurability problem and the *hierarchy* of transitions.

2.1 Interaction between steps

This arises from a number of different mechanisms — and to a large extent it is still an unsettled issue.

Elastic interaction

Consider a single step located at $x = x_0$. In its vicinity surface reconstruction is shifted by one lattice spacing — hence an elastic deformation of the substrate which decays slowly in the bulk. In order to cancel such a long range strain field, we must apply extra forces $\mathbf{F}_i$ on surface atoms near the step, such as to restore the original homogeneous stress $\pi_{ij}^{(0)}$ in the bulk. If the step were a broad structure on an otherwise continuous material, these forces $\mathbf{F}_i$ would just balance the capillary forces (1.52) and (1.53). A surface element ds would be subject to

$$\mathrm{d}F_x = -\mathrm{d}[\pi_s \cos\theta], \quad \mathrm{d}F_z = +\mathrm{d}[\pi_s \sin\theta],$$

where $\theta(s)$ is the orientation profile with respect to the crystal z-axis. In this form, it is clear that *the resultant of* $\mathbf{F}_i$ *vanishes*. In practice, using a macroscopic concept such as surface stress on an atomic scale is doubtful — nevertheless, the statement $\sum \mathbf{F}_i = 0$ remains valid, as it results from global, asymptotic considerations. To see that, draw a plane P slightly below the surface, and consider forces acting on the small volume V shown in Fig. 14. The homogeneous zeroth order stress $p_{ij}^{(0)}$ gives no contribution to the sum (this is particularly obvious if $\pi_{ij}^{(0)}$ is a hydrostatic pressure). The only contributions are:

(i) the forces $\mathbf{F}_i$ applied to the surface;

(ii) the surface stress π_s on either side, which is well defined if we are far enough from the step.

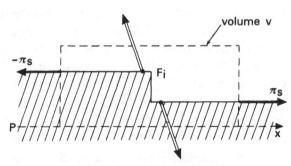

Figure 14: Forces exerted by a step
on the underlying elastic substrate.

Since the surface is the same on left and right, the stresses π_s are opposite: mechanical equilibrium implies that the resultant of $\mathbf{F}_i$ is zero.[†] Note that the argument is asymptotic: it does not rely on what happens near the step. Altogether, the forces $\mathbf{F}_i$ reduce to a localized *force doublet* applied near the step, which we can represent on a gross scale as

$$\mathbf{F}_i(x) = \mathbf{f}\delta'(x). \qquad (2.6)$$

f_x describes a local *stretch* of the surface: it is the same for steps and 'antisteps' (going up instead of down as x increases). f_z is a local *torque* which tends to twist the crystal: it changes sign as the step reverses (note that angular momentum conservation implies $f_z = \pi_s a$).

If we do *not* apply the force doublet $\mathbf{F}_i$, the solid relaxes elastically (it responds to the opposite force $-\mathbf{F}_i$). A strain field $u_{ij} \sim \kappa f/d^2$ develops at large distance d, corresponding to a displacement $u \sim \kappa f/d$ (κ is a typical compressibility of the solid). The bulk stress is accordingly shifted by $\Delta\pi_{ij}$, and the gain in elastic energy is[‡]

$$\Delta E = -\tfrac{1}{2} \int d\tau\, u_{ij}\Delta\pi_{ij}. \qquad (2.7)$$

(Note the minus sign: energy is *lowered* upon relaxation of the substrate. ΔE includes both the cost in bulk elastic energy and the work of the force (2.6) that counters the deformation: see the discussion following (1.59).)

For a single step of width ξ, the integral in (2.7) is dominated by short distances, for which macroscopic elasticity is hardly applicable.

[†] One might consider situations in which the left and right are not equivalent — for instance a *half step* on an 'a' facet parallel to the *c*-axis of an hcp crystal. Then π_s is different on left and right, and the 'step forces' $\mathbf{F}_i$ have a non-zero resultant. In such a case, however, the surface energy is also different on both sides: two half steps a distance d apart produce a 'confinement' energy $\Delta\gamma \cdot d$ which forces them to stay close to each other. The elastic interaction, although larger in such a case, is completely negligible compared to such a confining term: the cost in exposing an unfavourable surface dominates everything else.

[‡] Strictly speaking, the interface *rotates* by a small angle $\theta \sim \kappa f/d^2$, thereby affecting the balance of surface stress far away. These corrections are however negligible if d is large: the extra surface force $\pi_s \partial\theta/\partial x \sim \pi_s \kappa f/d^3$ should be compared to the primary stress $\Delta\pi \sim f/d^2$. It becomes negligible if d exceeds a characteristic length $d^* \sim \pi_s \kappa$, of atomic scale — the force doublet remains localized near the step.

Qualitatively, the elastic energy of the step is *negative*, of order

$$E_{el} \sim -\frac{\kappa f^2}{\xi^2} \sim -\kappa \pi_s^2 \frac{a^2}{\xi^2}.$$

This energy enters in the balance that determines the optimal width ξ. E_{el} should be compared with the stretching energy $\sim \gamma a^2/\xi$. Since the width ξ is necessarily $\geq a$, elastic effects will be negligible if $\pi_s^2/\gamma < a/\kappa$. This condition is usually fulfilled ($1/\kappa$ is large). If it were violated, elastic effects might act to reduce the step width ξ.

Let us now consider two steps 1 and 2 a distance d apart. The strain field produced by step 1 is picked up by step 2, thereby producing an *elastic interaction* between the steps. Quantitatively, the two steps add up to produce a total strain

$$u_{ij} = u_{ij}^{(1)} + u_{ij}^{(2)}.$$

The *cross term* in the elastic energy (2.7) is the interaction energy we are looking for. Equation (2.7) may also be written as

$$\Delta E = -\tfrac{1}{2} \int dx\, \mathbf{F}(x) \cdot \mathbf{u}(x) \qquad (2.8)$$

(the gain in elastic energy is the work of surface forces). Since the integral in (2.7) is dominated by the region *between* the steps, we can use the simple form (2.6) for $\mathbf{F}(x)$. The interaction energy of steps 1 and 2 is simply:

$$E_{12} = \mathbf{f}_2 \cdot \mathbf{u}_1'(x_2) = \mathbf{f}_1 \cdot \mathbf{u}_2'(x_1)$$

(where $\mathbf{u}_1(x_2)$ is the displacement produced by step 1 at the position of step 2).

Explicit calculations can be carried out explicitly for an isotropic half space elastic material [12]. From the force distribution (2.6) we infer the elastic displacement $\mathbf{u}$ a distance x from the step *along the surface*

$$\mathbf{u}(x) = -\frac{2(1-\sigma^2)}{\pi E} \frac{\mathbf{f}}{x}. \qquad (2.9)$$

(E and σ are respectively Young's modulus and Poisson's coefficient.) Note the minus sign: a positive f_x, which tends to *shrink* the solid near $x = 0$, produces a positive $\partial u_x/\partial x$ at large distances, i.e. an expansion. Using (2.9), we find

$$E_{12} = \frac{2(1-\sigma^2)}{\pi E} \frac{\mathbf{f}_1 \cdot \mathbf{f}_2}{d^2}. \qquad (2.10)$$

For identical steps, the interaction is *repulsive*.[†] For opposite steps, the interaction becomes attractive if $f_z > f_x$.

According to (2.10), the interaction decays as $1/d^2$, characteristic of the coupling between two force *doublets*. That can be seen directly: a point force on the surface gives a displacement $\sim 1/r$, a strain $\sim 1/r^2$. We add one power $1/r$ for a force *doublet*, which we recuperate on integrating along the step. The volume integration for the cross term is of order d^2 (per unit length along the step) — hence an interaction $(d^{-2})^2 \times d^2 \sim 1/d^2$.

This result has been questioned, some authors claiming that the interaction behaves as $\log d$ instead of $1/d^2$. The argument goes as follows:

▷ The interaction between two parallel atomic *rows* added on a flat surface goes as $1/d^2$ (reflecting the interaction $\sim 1/d^3$ between two isolated adatoms [13]). This follows from an argument similar to the one given above.

▷ Summing over *all the rows* between a step and an antistep a distance d apart yields an interaction $\sim \log d$!

The flaw in this argument is subtle. Basically, it amounts to looking at the response to the surface stress π_s at both ends of the terrace comprised between the steps — but these simply replace the stresses that existed one plane below in the absence of any terrace. The *change* due to the steps is a *force doublet* at both ends, not a plain force π_s — hence $1/d^2$ instead of $\log d$. Put another way, the force pattern exerted by a given row on the substrate, which was a dipole for a single row, becomes a *quadrupole* when the row is in the middle of a flat terrace. As usual, a continuous distribution of quadrupoles produces dipoles at both ends.

Statistical interaction of steps

The coupling is now of an entropic nature: thermal fluctuations of a step are limited by its neighbours, which means a loss of entropy and an increase in free energy — hence a *repulsion* between identical steps. Strangely enough, this repulsion also goes as $1/d^2$, despite a completely different mechanism.

[†] In contrast, the elastic mediated interaction between two point impurities is attractive (cf. the phonon interaction in the theory of superconductivity). The change of sign is traced to the reversal of strain mentioned earlier.

The central point is that two identical steps on a vicinal surface *cannot cross*: if they did, they would simply interchange their ends in order to eliminate the crossing: this is the same configuration with a different labelling of the steps (see Fig. 15). As a result, a step is confined between its neighbours.

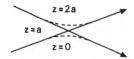

Figure 15: Elimination of crossing steps.

An accurate treatment of thermal fluctuations in an array of coupled steps is a beautiful problem in statistical mechanics, on which there exists a large literature [14]. The theories are rather elaborate — here we will content ourselves with a simple qualitative argument that displays most of the physics. Consider a single straight step along the y-axis with an energy β per unit length. Due to thermal fluctuations, an element dy is displaced by an amount $x(y)$, which we can expand in a Fourier series

$$x(y) = \sum_k x_k e^{iky}$$

(k is a multiple of $2\pi/L$, where L is the length of the step). The stretching energy is

$$E = \sum_k \tfrac{1}{2} L\beta k^2 |x_k|^2.$$

Each mode carries a thermal energy $\tfrac{1}{2}T$, which implies

$$\langle |x_k|^2 \rangle = \frac{T}{L\beta k^2}.$$

The mean square displacement of the step is

$$\langle x^2 \rangle = \sum_k \langle |x_k|^2 \rangle = \frac{1}{2\pi} \int_{-\infty}^{+\infty} dk \, \frac{T}{\beta k^2}. \qquad (2.11)$$

The integral diverges for small k: it is cut off at $k \sim 1/L$, leading to the well known result that a stretched string of length L fluctuates with an amplitude $\sim \sqrt{L}$.

We now enforce the no crossing requirement: on the average, the displacement $\langle x^2 \rangle$ is bound to be $\leq d^2$. That means that *all the long wavelength modes are cut off*, up to a wave vector k_c such that

$$\frac{1}{\pi} \int_{k_c}^{\infty} \frac{T \, dk}{\beta k^2} \sim d^2.$$

We thus find

$$k_c \sim \frac{T}{\beta d^2}. \tag{2.12}$$

For each suppressed mode, we loose a free energy $\sim -T \log T$. The free energy of the step thus *increases* by an amount of order

$$k_c L T.$$

(We discard the log for simplicity.) An *entropic repulsion* between steps appears, of order $T^2/\beta d^2$ per unit length. Comparison with the elastic interaction is a matter of specific cases.

Experimental observation of step interaction

The interaction between steps determines $\gamma(\theta)$ — and thus in the end the crystal shape: this will be discussed extensively in the next section. Here we focus on a very simple experiment, which provides direct evidence for this interaction — and which apparently contradicts the above theoretical prediction!

Consider a vicinal surface between solid ^{4}He and its superfluid melt. Because the liquid is superfluid, there is no resistance to melting: the interface responds instantaneously to any pressure fluctuation. As a result, there exist melting crystallization waves, similar to waves on a liquid surface except that only the liquid moves: the solid *grows* while staying at rest. Consider such a wave, with wave vector k, which propagates *perpendicular* to the steps of a vicinal surface. The interface remains cylindrical: the wave is simply a *compression wave of the steps*. The restoring force is the compressibility of the steps, $E''(n)$. Inertia is provided by liquid flow over a depth $\sim k^{-1}$. For a small tilt of the vicinal surface (step density n small), the dispersion relation is

$$\omega^2 = \frac{\rho_L}{(\rho_C - \rho_L)^2} \frac{E''}{a^2} k^3.$$

It gives direct access to E''. If the interaction between steps is of order $1/d^2$, the energy should behave as

$$E(n) = \gamma_0 + \beta n + \phi n^3 + \cdots. \tag{2.13}$$

Thus $E'' = 6\phi n$ should go to *zero* as the tilt angle θ vanishes.

The experiment has been done very recently by Keshishev *et al.* Apparently, E'' does *not* vanish as $\theta \to 0$ — it would rather increase! Hence a paradox which is still pending: if the experiment is confirmed, it implies other interaction mechanisms between steps, with an energy at least of order n^2. One such mechanism might be the free energy associated with step oscillations, i.e. with the *fluctuations* of melting crystallization waves. We already invoked these fluctuations when describing the statistical interaction between steps (due to freezing of the long wavelength modes). More generally, all the modes with wave vector $k \leq n$ (whether along x or y) are perturbed by the neighbouring steps: their frequency ω_k is different and the classical free energy

$$F = -\sum_k T \log \frac{T}{\hbar \omega_k}$$

is accordingly modified. Since the number of 'sensitive' modes is $\sim n^2$, one might in this way construct a term of order n^2 in F: the issue is still unsettled.

2.2 Crystal shape viewed as an equilibrium of steps

We consider a *cylindrical* vicinal interface with a profile $\theta(x)$. The characteristic size is much smaller than the capillary length ℓ_c: we neglect gravity altogether. We may view the interface as an array of parallel steps with a density $n(x)$ and energy density $E(n)$ (per unit length dx), given by

$$na = \tan\theta, \quad E(n) = \frac{\gamma}{\cos\theta}.$$

We define a step 'chemical potential' ξ and a step 'pressure' ϖ as

$$\xi = \frac{dE}{dn} = E', \quad \varpi = nE' - E. \tag{2.14}$$

ξ controls the *creation* of steps, ϖ their *spreading*. It is easily verified that

$$\xi = a[\gamma \sin\theta + \gamma' \cos\theta],$$
$$\varpi = \gamma' \sin\theta - \gamma \cos\theta \tag{2.15}$$

(compare with (1.65)). Returning to the chemical stresses of Fig. 5a, we see that ϖ measures the chemical stress along the crystal plane while ξ

measures the stress perpendicular to that plane. The *compressibility* of steps is

$$n\frac{\mathrm{d}\varpi}{\mathrm{d}n} = n^2 E''.$$

The step configuration is *locally stable* if $E'' > 0$.

For small values of n, we may expand $E(n)$ in the form (2.13): stability implies a step repulsion, $\phi > 0$. As n increases, the step picture is no longer physical (steps overlap if $n\xi \geq 1$). Nevertheless, the representation $E(n)$ remains mathematically convenient in describing the behaviour of $\gamma(\theta)$ in the range $0 < \theta < \frac{1}{2}\pi$. Outside this range, we should introduce steps of different signature, as shown in Fig. 16. If z

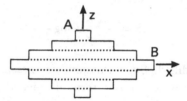

Figure 16: Four kinds of steps along x crystal planes.

and x are symmetry axes, $E(n)$ is the same for all four kinds of steps. As $\theta \to \frac{1}{2}\pi$, we expect that $\gamma(\theta)$ should behave as

$$\gamma(\theta) = \gamma_\infty \sin\theta + \frac{\lambda \cos^2\theta}{\sin\theta} + \cdots$$

(there is no cusp at $\theta = \frac{1}{2}\pi$ since we ignore periodicity along the x-axis). It is easily verified that the corresponding behaviour of $E(n)$ is

$$E = \gamma_\infty na + \frac{\lambda}{na} + \cdots.$$

The orientation $\theta = \frac{1}{2}\pi$ is stable if $\lambda > 0$.

Equilibrium of straight steps

Let δp_L be the departure of the liquid pressure from its nominal equilibrium value p^*. A step is subject to a *supercooling force* F. If δp_L is > 0, F tends to grow the solid phase: it is negative in the geometry of Fig. 12. F is such that its work is the gain of free enthalpy when the step moves by $\mathrm{d}x$: $\mathrm{d}G = -F\,\mathrm{d}x$. Using (1.41) we find

$$F = a[f_C + p_L - \rho_C\mu_L] = -a\delta p_L \frac{\rho_C - \rho_L}{\rho_L}. \qquad (2.17)$$

This 'external' force is balanced by the *repulsion of neighbouring steps*

$$F' = -\frac{d\xi}{dx} = -E''\frac{dn}{dx}.\tag{2.18}$$

Equilibrium is achieved if

$$F + F' = 0.\tag{2.19}$$

Using the definitions (2.13), it is easily verified that

$$E'' = a^2 \cos^3\theta\,(\gamma + \gamma'').\tag{2.20}$$

Moreover, the curvature of the interface is

$$\frac{1}{R} = -\frac{d\theta}{ds} = -a\cos^3\theta\,\frac{dn}{dx}$$

(R is counted positive if the convexity is in the liquid). The equilibrium condition (2.19) thus reduces to

$$\delta p_L \frac{\rho_C - \rho_L}{\rho_L} = \frac{\gamma + \gamma''}{R}.$$

We recover the equilibrium condition (1.47). The Wulff construction thus acquires a very transparent interpretation: it simply expresses equilibrium between forces acting on individual steps. If this balance is broken, steps move one way or the other: the crystal grows or melts. Steps are considered as 'particles' along the x-axis, to which we can apply all familiar concepts.

Formation of angular points

Assume that the curve $E(n)$ has the shape depicted in Fig. 17: the region of negative E'' is unstable. Then the steps will undergo a 'liquid-gas'

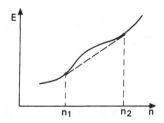

Figure 17: Formation of angular points on the interface.

phase separation: their density jumps abruptly from n_1 to n_2 — this is nothing but the *angular point* described in the preceding chapter, which in the step language appears quite familiar. The densities n_1 and n_2 'at coexistence' are obtained by a standard *double tangent construction* which minimizes the energy: as the slope $\xi = E'$ grows, one jumps discontinuously from n_1 to n_2. Such a construction implies equal values of the step chemical potential ξ and pressure ϖ on both sides: we recover the matching conditions (1.65).

Angular points may also appear between steps of different signature. An angular point between θ and $-\theta$ corresponds to A in Fig. 16. The matching conditions then reduce to $\xi(n) = 0$. $E(n)$ must have a minimum, as shown in Fig. 18a. Indeed, A must be stable against production

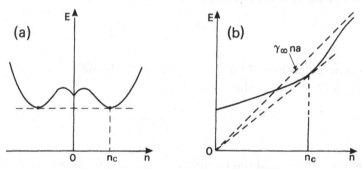

Figure 18: Angular points around a symmetry axis. (a) Around $\theta = 0$. (b) Around $\theta = \frac{1}{2}\pi$.

of $\pm$ step pairs, which implies $(\xi_+ + \xi_-) = 0$. Since $\xi_+ = \xi_-$ by symmetry, the condition $\xi = 0$ follows at once. Similarly, an angular point between θ and $(\pi - \theta)$ implies a zero *pressure* $\varpi(n)$ — an obvious condition since steps can expand freely as point B moves sideways in Fig. 16. A zero pressure means a tangent to $E(n)$ which goes through the origin (Fig. 18b) — hence $\lambda < 0$ in the expression (2.16) ($\theta = \frac{1}{2}\pi$ is unstable).

Boundary conditions

Equation (2.19) makes the determination of the profile $\theta(x)$ extremely easy. We may write it explicitly as

$$\frac{\partial \xi}{\partial x} = F \implies \xi = Fx + \text{const.} \tag{2.21}$$

From ξ we infer n — then θ: the problem is solved — we only need boundary conditions. At one end $\xi(n)$ will hit its *minimum* limit, which usually corresponds to $n = 0$, $\xi = \beta$: this corresponds to a 'vacuum of steps', i.e. to the *crystal facet* extensively discussed in the next section. On the facet the equilibrium condition (2.19) is irrelevant since there is no step there to equilibrate. At the contact between the facet and the round part, the supercooling force F tends to push the steps away from the facet (the facet wants to broaden in such a way as to make more solid). This trend is countered by the repulsion of steps further away: the balance of these forces determines the shape at facet edge.

In the opposite limit of large n and ξ we may eventually reach another high symmetry crystal orientation: the appropriate description is then in terms of steps on that new facet. We may also hit a wall, which will require a given contact angle: *the step density is fixed at the wall*. We consider only the simplest geometry shown in Fig. 19: two parallel walls,

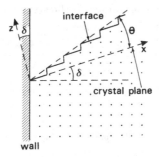

Figure 19: Contact of a vicinal surface with a wall.

parallel to the steps of our cylindrical interface. The crystal z-axis makes an angle δ with the walls, the normal to the interface makes an angle θ with the z-axis, e.g. an angle $\tilde{\theta} = \theta + \delta$ with the walls. The matching condition (1.66) becomes

$$\gamma_{CW} - \gamma_{LW} = \gamma \sin \tilde{\theta} + \gamma' \cos \tilde{\theta} = \frac{\xi}{a} \cos \delta - \varpi \sin \delta. \qquad (2.22)$$

When $\delta = 0$, (2.22) fixes the value of ξ at the wall

$$\xi = \xi_W = a[\gamma_{CW} - \gamma_{LW}]. \qquad (2.23)$$

The boundary condition on (2.21) is then trivial. Physically, the wall is a *source of steps*, which fixes their chemical potential at contact: ξ_W is just the energy cost of an extra step as it nucleates at the wall. The situation is slightly more complicated when $\delta \neq 0$: in such a case growth at the wall

implies both a *nucleation* of steps and a *compression* of the preceding
ones which are pushed away along the crystal planes. The boundary
condition involves both the chemical potential ξ and the pressure ϖ, as
shown on (2.22) (which may be derived by a simple argument of energy
conservation).

Bending of steps

If the interface is not cylindrical, the steps are bent: in addition to
supercooling and step interaction, we must worry about curvature effects.
A single step is a stretched string with line tension β: if it is bent with
a radius ρ, it feels an additional force β/ρ (similar to the Laplace force
for the interface). This force reacts on the equilibrium shape.[†]

As an illustration, we return first to the vicinal surface discussed
before. The steps are originally along the y-axis, with a density $n_0 =
(\tan\theta_0)/a$ along the x-crystal planes. If the steps move by u, the interface
grows by $\Delta z = n_0 u \cdot a$, i.e. by an amount

$$\ell = u \sin\theta_0$$

in the direction normal to the interface. Assume that ℓ is modulated
along the surface

$$\ell(x,y) = \ell e^{i\mathbf{k}\cdot\mathbf{r}}.$$

The change of surface energy per unit interfacial area is by definition

$$\Delta E = \tfrac{1}{2}\tilde{\gamma}_{\mu\nu}\bar{k}_\mu\bar{k}_\nu\ell^2, \tag{2.24}$$

where $\bar{\mathbf{k}}$ is the wave vector in the interface plane: ($\bar{k}_x = k_x\cos\theta$,
$\bar{k}_y = k_y$). $\tilde{\gamma}_{\mu\nu}$ is the surface stiffness *tensor*, with components $\tilde{\gamma}_\parallel$ and $\tilde{\gamma}_\perp$
for k respectively along the x- and y-directions. A modulation along x
amounts to a *compression* of steps (which remain straight). We have seen
that $\tilde{\gamma}_\parallel$ was related to the step *compressibility* (in (2.20)). In contrast, a
modulation along y results in a *wiggling* of the steps: the restoring force
is due to line tension: for a small step density it is basically a one step
affair. The physics is very different in the two cases: we expect $\tilde{\gamma}_\perp \neq \tilde{\gamma}_\parallel$.

[†] In the line tension picture, the change of energy is due only to *stretching* of the step. In principle, one might also envisage a *bending* energy,
which should be negligible unless the radius ρ is comparable to the *width*
ξ of the step: the step is 'stiff' only on a scale $\leq \xi$.

The formal calculation of $\tilde{\gamma}_\parallel$ and $\tilde{\gamma}_\perp$ is straightforward. In the present isotropic model, $\gamma(\theta, \phi)$ depends only on the angle θ between the crystal z-axis and the interface normal $\mathbf{n}$. In order to calculate θ, we use new coordinates $(\bar{x}, y, \bar{z})$ such that the equilibrium interface is $\bar{z} = 0$. From the displacements $\bar{z}(\bar{x}, y)$ we construct the Landau-Andreev vector $h_\mu = \partial \bar{z} / \partial x_\mu$. $\mathbf{n}$ has coordinates $[-h_x, -h_y, 1 - \frac{1}{2}(h_x^2 + h_y^2)]$. From the product $\mathbf{n} \cdot \hat{\mathbf{z}} = \cos \theta$ we infer the expansion of θ:

$$\theta = \theta_0 + h_x + \tfrac{1}{2} \cot \theta_0 \, h_y^2 + \mathrm{O}(h^3).$$

$\tilde{\gamma}_\parallel$ and $\tilde{\gamma}_\perp$ are respectively the coefficients of h_x^2 and h_y^2 in the expansion of

$$E = \gamma(h) \sqrt{1 + h_x^2 + h_y^2}.$$

A simple algebra yields

$$\tilde{\gamma}_\parallel = \gamma + \gamma'', \quad \tilde{\gamma}_\perp = \gamma + \gamma' \cot \theta_0 = \frac{\xi}{a \sin \theta_0}. \qquad (2.25)$$

The expression of $\tilde{\gamma}_\parallel$ is familiar — the one of $\tilde{\gamma}_\perp$ is new. We see that $\tilde{\gamma}_\perp$ should diverge as $\theta_0 \to 0$, a feature which could be verified in a capillary wave experiment.

It is instructive to rederive (2.25) in a step picture, which displays the underlying physics more clearly. Consider first a modulation k_x in a direction perpendicular to the steps. The step displacement u implies a change in step density in the x-plane

$$\delta n = -nku = -\frac{n\bar{k}u}{\cos \theta_0} = -\frac{\bar{k}\ell}{a \cos^2 \theta_0}.$$

The second order change in energy per unit $\mathrm{d}x$ is

$$\Delta E = \tfrac{1}{2} E'' \delta n^2.$$

The energy change per unit $\mathrm{d}s$ is

$$\Delta E \cos \theta_0 = \tfrac{1}{2} \tilde{\gamma}_\parallel \bar{k}^2 \ell^2,$$

which is equivalent to (2.25) and (2.20).

Consider now a modulation k_y along the steps. The displacement results in a rotation of the steps by an angle

$$\phi = ku.$$

As a result the step distance decreases — another way of stating that the steps lengthen. Their density increases by

$$\delta n = \frac{n\varphi^2}{2} = \frac{nk^2\ell^2}{2\sin^2\theta_0}.$$

The change of energy per unit $\mathrm{d}s$ is

$$\cos\theta_0 \, E' \delta n = \frac{E'k^2\ell^2}{2a\sin\theta_0}.$$

We recover the expression (2.25) of $\tilde{\gamma}_\perp$. For small angles θ, we can use the expansion (2.13):

$$\tilde{\gamma}_\parallel = \frac{6\phi\theta}{a^3}, \quad \tilde{\gamma}_\perp = \frac{\beta}{a\theta}. \tag{2.26}$$

The behaviour of $\gamma_\parallel$ and $\gamma_\perp$ is completely different. Note that the product $\tilde{\gamma}_\parallel\tilde{\gamma}_\perp$ is regular as $\theta \to 0$:

$$\tilde{\gamma}_\parallel\tilde{\gamma}_\perp = \frac{6\beta\phi}{a^4}.$$

If ϕ is controlled by the entropic step interaction, it is $\sim T^2/\beta$: then $\tilde{\gamma}_\parallel\tilde{\gamma}_\perp \sim T^2/a^4$ does not depend on step parameters.

Circular geometry

We now turn to the more realistic situation of an interface which has circular symmetry around the crystal z-axis, normal to the facet. (Remember that up to now we ignore any anisotropy of steps: $\gamma(\theta,\phi)$ depends only on θ.) The whole system is placed in a cylindrical vessel with the same z-axis and with radius d. The interface is made up of *circular steps*, with radius r. The density $n(r)$ fixes the profile of the meridian. Let again $\xi = E'(n)$ be the local chemical potential of a step: ξ acts as an effective *line tension* which pulls on the ends of any step element $\mathrm{d}\ell$. As a result, the step is subject to an extra radial force ξ/r, directed *inward*, which tries to reduce the perimeter. The net balance of forces on a given step reads

$$\frac{\mathrm{d}\xi}{\mathrm{d}r} + \frac{\xi}{r} = F. \tag{2.27}$$

Equation (2.27) replaces (2.21): the only new feature is the bending force $-\xi/r$.

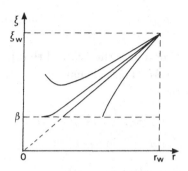

Figure 20: Determination of the profile of a circular facet.

The integration of (2.27) is easy:

$$\xi = \frac{Fr}{2} + \frac{\text{const}}{r}.$$ (2.28)

If the crystal is in a cylindrical vessel with its axis in the z-direction, the contact with the wall at $r = r_w$ fixes the local value $\xi = \xi_w$ — hence the integration constant in (2.28). Depending on the supercooling force F, the profile can have the shapes shown in Fig. 20. At small radius r, ξ eventually reaches its minimum value β, corresponding to the *edge of a circular facet* (on which there is no step). For small enough values of F, ξ bends up before reaching β. In such a case there is no equilibrium profile: we shall see in the next paragraph that the *facet collapses*. The corresponding threshold yields

$$r = r_c = \frac{\beta}{F}.$$ (2.29)

It is such that the bending force of a *single* line, β/r_c, exactly compensates the supercooling force F.

These simple results break down when the step energy β is *anisotropic*. Then a step element is subject to both a *tension force* $\beta(\phi)$ and to a *torque* $\beta'(\phi)$ in close analogy with the chemical stresses shown in Fig. 5a for the interface itself. For a single step, without any interaction, the equilibrium condition is obtained by balancing these forces against the supercooling force F. We thus find

$$F = \frac{\beta + \beta''}{\rho} = \frac{\tilde{\beta}}{\rho},$$ (2.30)

in which ρ is the local radius of curvature of the step. Here again, β is replaced by a *step stiffness* $\tilde{\beta}$. Integration of (2.30) yields the *equilibrium contour* of a single step, which is also given by a Wulff construction:

(i) Draw a *polar plot* of the step energy such that $OM = \beta(\phi)$ in the ϕ-direction.

(ii) The equilibrium contour is the pedal of that plot, i.e. the envelope of MP normal to OM at M. This envelope is conveniently parametrized in cartesian coordinates:

$$Fx = \beta \cos \phi - \beta' \sin \phi,$$
$$Fy = \beta \sin \phi + \beta' \cos \phi. \qquad (2.31)$$

(Equations (2.31) are equivalent to (1.64).)

We see that the concepts applied to the interface apply equally well to a step. Unfortunately, the description is more complicated at finite density, due to an interplay of *interaction* and *anisotropy* effects. One may still define effective forces acting on each step, but their interpretation is more delicate: this point is discussed in [15].

2.3 Crystal facets

These are due to the finite step free energy β: steps are macroscopic objects, and thermal agitation cannot produce them spontaneously. A supercooling force F tends to push the steps away toward the periphery: in the middle remains a vacuum — a *flat facet*.

While round parts grow fairly easily (we need only slide the existing steps sideways in order to enlarge the solid phase), facets grow only with great difficulty (we will discuss later possible mechanisms). As a result, a facet is often *trapped* at a given height z. While it can *enlarge* easily by adjusting the surrounding curved parts, it cannot nucleate new terraces: the facet is then *metastable*. There exists an equilibrium facet radius R^* which minimizes the energy — but in order to reach it, the facet must grow! This should be kept in mind: *facet sizes do not reflect equilibrium* unless great care is taken to allow their vertical growth.

This situation is quite clear if we consider a cylindrical interface sandwiched between two walls, a distance $2d$ apart, parallel to the vertical z-axis (Fig. 21). For a given supercooling F, the curvature of the interface is known. Starting from the contact angle θ_W at the wall, the interface reaches $\theta = 0$ a distance ℓ from the wall: this is the edge of the facet. The facet width $2L$ is just what remains once curved parts have reached equilibrium: $L = d - \ell$. If the supercooling is too small, ℓ is $> d$: there is no equilibrium configuration for that value of F (the

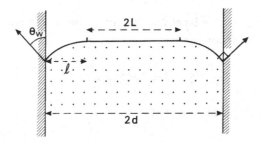

Figure 21: Determination of a metastable facet radius.

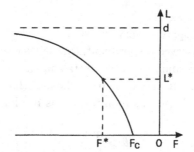

Figure 22: Width L of a strip facet as a function of supercooling (in the case $\theta_W > 0$). The facet is metastable up to the collapse threshold F_c. The equilibrium size corresponds to (L^*, F^*).

solid will melt until gravity has increased F enough to reach $\ell \leq d$). If on the other hand $\ell < d$, the facet is metastable as long as we do not allow it to grow. The behaviour of $L(F)$ is shown in Fig. 22, in the case $\theta_W > 0$. In order to reach its *equilibrium width* $2L^*$, the facet should move *vertically* until F is adjusted to produce the appropriate value of L. What we describe here is the mechanism of *capillary rise* for a faceted crystal. It goes in two steps:

▷ quick adjustment of the curved parts;

▷ slow relaxation of the facet height.

If a facet refuses to move, a large F will enhance its width for purely geometrical reasons — a trick often used in order to *reveal* facets in experimental shapes.

Strip facets in cylindrical geometries

We first look at the matching of the facet to the round part. If all crystal orientations are stable, the compressibility $E''(n)$ is everywhere > 0: $\xi(n)$ is a monotonously increasing function of n. As ξ goes down according to (2.21), n decreases regularly, down to zero when $\xi = \beta$.

58   *P. Nozières*

The round part joins the facet *tangentially*. The profile $z(x)$ close to the contact is inferred from $E(n)$:

$$\xi - \beta = 3\phi n^2 = Fx. \tag{2.32}$$

Since $na = \mathrm{d}z/\mathrm{d}x$, it follows that

$$\frac{\mathrm{d}z}{\mathrm{d}x} \sim x^{1/2}.$$

The profile should display a very unusual $x^{3/2}$ shape. In most cases this shape is not observed experimentally, neither at the liquid-solid interface of ^{4}He nor in small metallic crystals in vacuum. The reason for this failure is not clear. It may arise from the fact that the experimental window in which the shape was monitored was too large:

$$\theta > \frac{a}{\xi}.$$

In this case, the vicinal picture breaks down and $\gamma(\theta)$ has a regular θ^2 dependence. It may also signal the existence of an n^2 term in $E(n)$, as suggested independently by the results on capillary waves.

It may also happen that $E(n)$ and $\xi(n)$ behave as shown in Fig. 23: there exists an unstable range of orientations. Then a *phase separation* occurs between the facet ($n = 0$) and a finite orientation n_c, which is obtained by a standard double tangent construction. The matching of the round part to the facet is *angular* (corresponding to a Maxwell plateau in $\xi(n)$). Note that the chemical potential at contact, ξ_c, is *lower* than β (the double tangent lies *below* the tangent at $n = 0$). Such a situation is often encountered in practice.

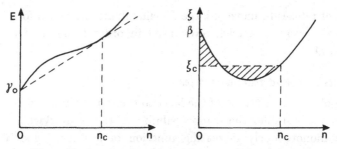

Figure 23: Angular matching of facets to the round parts.

The next question is the equilibrium *width* of the facet, $2L$. Formally, it can be inferred from the general equilibrium equation (1.47), noting that γ'' has a δ-function contribution at $\theta = 0$ due to the break of slope:

$$\tilde{\gamma} \approx \frac{2\beta}{a}\delta(\theta).$$

Integrating the equilibrium condition, we find

$$\tilde{\gamma}\frac{d\theta}{ds} \approx \tilde{\gamma}\frac{d\theta}{dx} = \frac{F}{a},$$

from which we infer

$$2L^* = x(\theta = +0) - x(\theta = -0) = \frac{2\beta}{F}. \qquad (2.33)$$

$L^* = \beta/F$ is the equilibrium width for *tangential* matching, when θ is equal to zero at step edge.

The width for *angular* matching can be obtained from the Wulff construction. The resulting shapes are shown in Fig. 24, both for tangential and for angular matching. In the latter case (Fig. 24b), the shape displays a cusp at the onset of instability ($E'' = 0$): the real shape stops at the double points A and B — hence an *angular* matching and a width $2L_c$ *smaller* than $2L^*$.

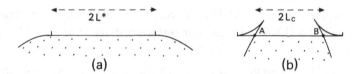

Figure 24: The Wulff shape of facets.
(a) For tangential matching.
(b) For angular matching.

All these results are easily interpreted in terms of steps. A facet is in equilibrium if the energy is stationary upon *addition of a terrace* — a statement that emphasizes the fact that the equilibrium width L^* is controlled by facet *growth*. The cost of this terrace is (per unit length along the steps):

▷ two more steps, with an energy ξ_c;

▷ a gain in bulk energy $(-2LF)$.

Equilibrium is achieved if the balance is zero:

$$L = L^* = \frac{\xi_c}{F}. \qquad (2.34)$$

For tangential matching, $\xi_c = \beta$: we recover (2.33). For angular matching, $\xi_c < \beta$: the facet is smaller.[†] In this language the physical picture is very transparent. Note that in a given vessel (2.34) implies a specific supercooling F^*: this is Jurin's law! (see Fig. 22).

A question arises: can a facet touch a wall? The answer is provided by (2.22). If all orientations are stable (tangential matching), the function

$$\tilde{\phi}(\theta) = \gamma \sin \tilde{\theta} + \gamma' \cos \tilde{\theta}$$

is monotonous, with a *discontinuity* at $\theta = 0$ (due to the break of slope of γ):

$$\Delta \phi_0 = \frac{2\beta}{a} \cos \delta. \qquad (2.35)$$

If $(\gamma_{CW} - \gamma_{LW})$ falls within that discontinuity, *the facet can touch the wall*. Such a configuration is stable for a *finite* range $\delta_1 < \delta < \delta_2$, in which δ_1 and δ_2 are given by

$$\gamma_{CW} - \gamma_{LW} - \gamma_0 \sin \delta = \pm \frac{\beta}{a} \cos \delta. \qquad (2.36)$$

(The angular rigidity of facets releases somewhat the constraint at the wall.) If the matching is angular, the range (2.36) is reduced.

Circular facets

Their equilibrium radius R^* follows from the same energy conservation argument, applied here to a circular terrace instead of a strip. The cost of this terrace is:

▷ an additional step, with energy $2\pi R \xi_c$, where ξ_c eventually incorporates a finite matching angle θ_c;

▷ a gain in bulk energy $[-\pi R^2 F]$.

The balance is zero if

$$R = R^* = \frac{2\xi_c}{F}. \qquad (2.37)$$

For tangential matching, $\xi_c = \beta$: R^* is *twice* the collapse radius r_c defined in (2.29). For angular matching, ξ_c and R^* are *reduced*.

Let us focus on the tangential matching case. The energy cost of a terrace with radius R is

$$\Delta E(R) = 2\pi R \beta - \pi R^2 F. \qquad (2.38)$$

[†] The facet may even disappear completely if the minimum of $E(n)$ is below γ_0: see Fig. 18a.

ΔE is maximum for $R = r_c$, which thus represents the critical nucleation barrier which must be overcome in order to create a new terrace. The corresponding energy cost is

$$\Delta E_c = \frac{\pi \beta^2}{F}. \qquad (2.39)$$

When $R < r_c$, the line tension takes over: the terrace *collapses*. (We saw that the solution (2.27) accordingly disappeared.) When $R > r_c$, the step grows until it reaches the edge of the facet; it is then pressed upon the round part: step repulsion balances F. The equilibrium radius $R^* = 2r_c$ only makes sense if we allow nucleation of an additionnal terrace — otherwise, any facet with $R > r_c$ is metastable.[†] The behaviour of R as a function of F for a given vessel radius d is shown in Fig. 25.

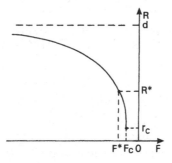

Figure 25: The radius R of a circular facet as a function of supercooling in a vessel with radius d. r_c is the collapse threshold, R^* the equilibrium radius.

In practice, the quantity which is most accessible experimentally is the collapse radius r_c: it gives direct access to the step energy β, which is a fundamental parameter of the problem.

The situation becomes more complicated when $\beta(\phi)$ is anisotropic. The geometry of a *critical nucleus* remains simple. It is just a single closed step, subject only to the supercooling force F and to its own tension: the equilibrium shape is given by (2.31). In contrast, the *equilibrium facet* cannot ignore the surrounding curved parts. The steps at the edge are *pressed* against their neighbours: in addition to F and line tension, they feel an interaction force. As a result, the equilibrium facet is *not* the critical nucleus twice as big: its determination is a difficult global problem with no obvious answer.

[†] Note that the equilibrium radius R^* corresponds to a vanishing constant in (2.28).

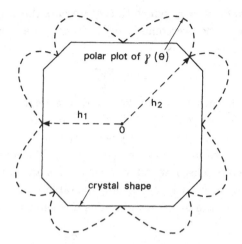

Figure 26: Total facetting, illustrated for a
square lattice with (10) and (11) facets.

Total facetting

If the cusps in $\gamma(\theta)$ are deep enough, the round parts may disappear
altogether. When carrying a double tangent construction on $E(n)$, the
absolute minimum is obtained by jumping directly from cusp to cusp.
The crystal equilibrium shape becomes a *polyhedron* with sharp edges.
This result is particularly clear within a Wulff construction, illustrated
in Fig. 26 for a simple square lattice: the polyhedral shape holds when
the pedal is completely *outside* the Wulff planes constructed on the
leading cusps of γ. It is clear that only a few *high symmetry* crystal faces
will be relevant. Those with higher Miller indices have very weak cusps:
they lie outside the leading polyhedron. Put another way, dominant
facets kill the weaker ones.

Let h_i be the distance to the origin of the i-th facet (see Fig. 26). In
view of the Wulff construction, it is clear that

$$\frac{h_i}{\gamma_i} = \text{const.} \tag{2.40}$$

Equation (2.40) is actually the original formulation of the Wulff criterion.
It can be derived directly by minimizing the energy at constant volume.
Let A_i be the area of the i-th facet: the volume is

$$V = \tfrac{1}{3}A_i h_i \implies dV = \tfrac{1}{3}[A_i\,dh_i + h_i\,dA_i].$$

But it is clear that $dV = A_i \, dh_i$: it follows that

$$dV = \tfrac{1}{2} h_i \, dA_i.$$

Equilibrium minimizes $(E - \lambda V)$, where λ is a Lagrange multiplier. Since $dE = \gamma_i \, dA_i$, (2.40) follows at once. The construction of the polyhedral shape is then trivial: it shows immediately which planes disappear from the final shape.

Isolated steps on a facet

Suppose that a screw dislocation pierces the facet: we move by one lattice spacing up or down when turning around the dislocation. The trace of that dislocation on the surface is thus the *source of a step*. It is convenient to *orient* the step in such a way that the upper side is always on the right. Positive and negative screw dislocations are respectively a *source* or a *sink* of steps (Fig. 27). *Isolated* steps may either join two

Figure 27: Formation of steps on a screw dislocation.

dislocations of opposite signs ('Frank-Read sources'), or they can go from a dislocation to the edge of the facet.

Consider first the contact line L between the facet and a curved part (Fig. 28): when reaching L, the step merges in the general step pattern that provides interface curvature. The contact angle η is such that step tension balances along L:

$$\cos \eta = \frac{\xi_c}{\beta}. \qquad (2.41)$$

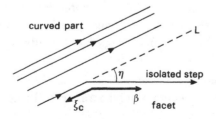

Figure 28: An isolated step merges with the curved interface.

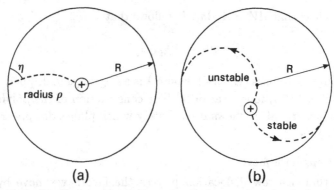

Figure 29: The growth of a facet due to the rotation of a single step.

If the facet matches the curved region *tangentially*, $\xi_c = \beta$ and $\eta = 0$: the step joins the edge of the facet tangentially. If the facet makes an angle with the curved part, ξ_c is $< \beta$ and the contact angle η is non-zero.

Equation (2.41) allows a very simple determination of the facet equilibrium radius R^*. Consider a circular facet with radius R, and a dislocation at its centre (Fig. 29a). In equilibrium, the step is a circle with radius ρ, which must reach the facet edge with an angle η — hence

$$2\rho \cos \eta = R \implies \rho = R \frac{\beta}{2\xi_c}. \tag{2.42}$$

The step will be in equilibrium if that radius is the nucleation radius $r_c = \beta/F$ for which line tension balances supercooling — which in turn implies

$$R = R^* = \frac{2\xi_c}{F},$$

in agreement with (2.37). (The result $R^* = 2r_c$ when $\eta = 0$ is obvious from Fig. 29a.) If $R \neq R^*$, the step feels a net force: it *rotates* around the centre, thereby making the facet grow or melt. If the dislocation is at the centre, such a rotation is completely free. If the dislocation is off centre, the step must overcome a *potential barrier* (as the step turns, it encounters two equilibrium configurations with radius $\rho = r_c$, shown in Fig. 29b) — but the net energy balance after one full turn still gives (2.37).

In much the same way, one may describe the 'refraction' of steps at the edge between two facets in a polyhedral crystal (Fig. 30):

$$\beta_1 \cos \eta_1 = \beta_2 \cos \eta_2.$$

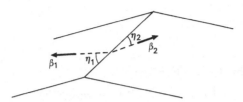

Figure 30: "Refraction" of steps between two facets.

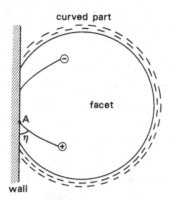

Figure 31: Contact of a facet with a wall parallel to the crystal z-axis.

We may also consider the step contact angle when the facet touches a wall. For simplicity, we assume that the wall is parallel to the crystal z-axis ($\delta = 0$ in (2.36)). The geometry is shown in Fig. 31, drawn in the xy-plane. η is obtained by requesting that the energy be stationary upon a displacement of A:

$$a[\gamma_{CW} - \gamma_{LW}] = \beta \cos \eta. \qquad (2.43)$$

η goes from $-\frac{1}{2}\pi$ to $+\frac{1}{2}\pi$ as ($\gamma_{CW} - \gamma_{LW}$) spans the stability range of the facet-wall contact given by (2.36).

We now turn to the *Frank-Read source*: a step is anchored on two opposite dislocations, a distance d apart. If a supercooling force F is applied, the step bulges with a radius $\rho = \beta/F$. As long as $\rho > \frac{1}{2}d$, the step spans less than a half circle: it is *stable*. If $\rho < \frac{1}{2}d$, there is no equilibrium configuration: the step extends irreversibly, creating a new terrace — hence a *growth mechanism* for the facet. This mechanism will be discussed extensively in the next chapter — here we note only the

threshold for growth

$$F_c = \frac{2\beta}{d}. \tag{2.44}$$

If d is small, this threshold is quite high.

It is instructive to look at the transition between Frank-Read sources and vicinal surfaces. Consider a vicinal surface with a small tilt θ: the step density $n = \theta/a$ is small. Assume now that the surface is pierced by a distribution of $\pm$ dislocations a distance $\sim d$ apart (the dislocation density is d^{-2}). The situation is very different depending on the value of nd:

(i) If $nd \ll 1$, the surface has basically independent Frank-Read sources. The step length is $\sim d$, and the tilt is achieved by a slight *polarization* of the steps, which tend to line up in the y-direction in such a way as to create an *average* tilt in the x-direction. The energy $E(\theta)$ is due to a small lengthening of the steps: it has no cusp when $\theta \to 0$.

(ii) If $nd \gg 1$, Frank-Read sources can no longer provide the average step density n: steps must *lengthen*. They are nearly parallel segments: eventually, they meet a dislocation and stop. The step pattern is a stack of long, overlapping needles, with a typical length ℓ, as shown in Fig. 32.

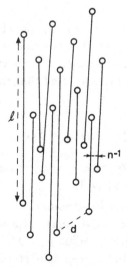

Figure 32: Step polarization on a vicinal surface.

In order to estimate ℓ, we note that a given segment corresponds to an area $n^{-1}\ell$, which must contain typically one dislocation — hence

$$\ell \sim nd^2. \tag{2.45}$$

The corresponding step energy $E(n)$ per unit area is $\sim \beta n$ (the breaking into finite segments does not matter).

The behaviour of $E(n)$ is sketched in Fig. 33.[†]

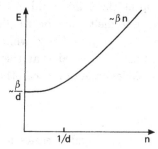

Figure 33: The step energy on a vicinal surface with dislocations.

Assume now that a supercooling force F is applied: the step acquires a radius $\rho = \beta/F$. A given step with length ℓ bulges by an amount

$$a = \frac{\ell^2}{\rho}.$$

It thereby sweeps an area $a\ell = \ell^3/\rho$. If that area contains a dislocation, the step *exchanges* with its colleague anchored on that dislocation. As this exchange process repeats, the steps slowly drift sideways: the crystal grows! The condition for growth of our vicinal surface is therefore that such an exchange occurs, which implies that the swept area contains at least one dislocation:

$$\frac{\ell^3}{\rho}\frac{1}{d^2} > 1.$$

The threshold for growth of a vicinal surface is consequently

$$\overline{F}_c = \frac{\beta d^2}{\ell^3} = \frac{\beta}{d}\frac{1}{(nd)^3}. \tag{2.46}$$

It is considerably reduced compared to the threshold (2.44) for a single Frank-Read source — but still it is finite!

The above analysis was very brief and qualitative. It can be formulated more precisely: this is done in [16].

[†] The *degeneracy* of the step configuration for a given distribution of dislocations is a beautiful counting problem, completely open.

2.4 Surface melting and crystal shape

Consider a planar solid-vapour interface close to its triple point T_p. If the liquid 'wets' the solid, a thin liquid layer may appear between solid and vapour, with a thickness ℓ that diverges when T approaches T_p. This is 'surface melting,' not to be confused with the roughening transition due to proliferation of steps on the surface. A surface molten solid may be facetted (steps develop at the solid-liquid interface) — conversely, a solid surface which does not melt may be rough. (Roughening is a direct consequence of crystalline periodicity, while surface melting is a purely thermodynamic affair.) Granted that they are different, these two phenomena can nevertheless react on each other: this is the subject of this brief section.

A survey of surface melting

Let γ_{CV}, γ_{CL}, γ_{LV} be the interfacial energies for solid-vapour, solid-liquid, liquid-vapour, respectively. We set

$$S = \gamma_{CV} - \gamma_{CL} - \gamma_{LV}.$$

In a first approximation, a liquid layer of thickness ℓ costs an extra energy

$$E(\ell) = -S + H\ell, \qquad (2.47)$$

where $H = \rho_L(\mu_L - \mu_C)$ is the cost in bulk energy per unit liquid volume upon melting. (H vanishes at the triple point, and it is positive below T_p.) The liquid 'wets' the solid if $S > 0$ [17]: in such a case, $E(\ell)$ displays a minimum when $\ell = +0$ (see Fig. 34). Such a liquid layer of vanishing thickness is clearly meaningless: we must refine (2.47), either through a more accurate treatment of the various interfaces, or by including other terms in $E(\ell)$.

Qualitatively, the structure of interfaces can be described in a Landau-Ginzburg formulation. Various phases are described by an order parameter ψ (for instance the density), and the appropriate free energy[†] is written as

$$F = \int dz \left[\frac{\lambda}{2}\left(\frac{\partial\psi}{\partial z}\right)^2 + f(\psi)\right]. \qquad (2.48)$$

f has *equal* minima $f = f^*$ for Ψ_C and Ψ_V (solid and vapour are in equilibrium). It also has a secondary minimum $f = f_L$ at Ψ_L, which

---

[†] If we fix the chemical potential, that energy is $(F - \mu N)$.

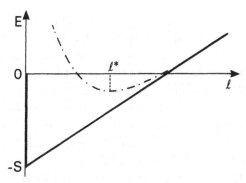

Figure 34: The extra energy $E(\ell)$ of a liquid layer of width ℓ. (i) Full curve: neglecting Van der Waals interaction and interface broadening. (ii) Dashed curve: with Van der Waals attraction.

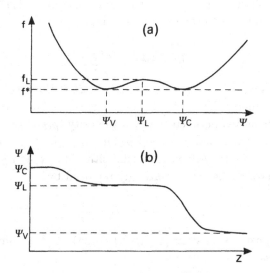

Figure 35: Surface melting in a Landau-Ginzburg description: (a) Free energy $f(\psi)$: the secondary minimum f_L is very close to f^*. (b) The interface profile $\psi(z)$.

is barely above the other two if T is close to T_p — see Fig. 35a. The 'stiffness term' in (2.48) fixes the unit of length, i.e. the width of the interface.

A standard minimization of F yields the profile $\psi(z)$ of the solid-vapour interface, given by

$$\frac{\lambda}{2}\left[\frac{\partial \psi}{\partial z}\right]^2 = f(\psi) - f^*. \tag{2.49}$$

$\psi(z)$ relaxes *exponentially* to ψ_C (resp. ψ_V) when $z \to -\infty$ (resp. $+\infty$). In between, the profile depends on the shape of $(f - f^*)$. It is clear from (2.49) that $\partial\psi/\partial z$ will be very small when ψ is close to ψ_L — hence the profile will 'linger' close to ψ_L over a fairly large range of z, as shown in Fig. 35b. Solid and liquid are nearly in equilibrium, and $\psi(z)$ nearly stops at ψ_L before finally deciding that it will feel better at ψ_V. The global solid-vapour interface is best described as the succession of two SL and LV interfaces, separated by a thickness ℓ of 'liquid.' ℓ is controlled by the overlap of exponential tails on either side [18]: as a result $\ell \sim \log(1/H)$ when $H \to 0$ — the liquid layer is very narrow.[†] Nevertheless, the phenomenon of surface melting is clearly apparent, with a well defined thickness ℓ.

The situation is even clearer if we include long range Van der Waals forces in the picture. Two atoms feel an attraction $\sim 1/r^6$, yielding a negative contribution to the energy of order

$$-\iint \mathrm{d}z\,\mathrm{d}z' \frac{\rho(z)\rho(z')}{(z-z')^4}$$

(the integral is cut off at short distances). Most of this contribution corrects the bulk energy H and the surface energy S. The only non-trivial part is the interaction between the *excess* density $(\rho_C - \rho_L)$ in the crystal and the corresponding quantity $(\rho_V - \rho_L)$ in the vapour. The corresponding energy is of order $1/\ell^2$; in the usual situation $\rho_V \ll \rho_L \ll \rho_C$, it is *positive*. Equation (2.47) is accordingly replaced by

$$E(\ell) = -S + H\ell + \frac{A}{\ell^2}, \qquad (2.50)$$

where A is a positive constant [17] (see Fig. 34). $E(\ell)$ is minimum for a finite thickness

$$\ell^* = \left[\frac{2A}{H}\right]^{1/3}. \qquad (2.51)$$

If H is large, ℓ^* is small and a detailed description of interfaces remains essential. Close to T_p, on the other hand, Van der Waals forces become dominant: *they make the liquid layer quite thick.*

[†] The breaking of an interface into two pieces close to a first order transition is a familiar phenomenon. A well known example is a *domain wall* in ferroelectrics close to a first order transition to paraelectric state: the asymptotic regions with polarizations $+P$ and $-P$ are separated by a non-polarized region $P = 0$.

Behaviour of non-wetting facets

Very recently, Heyraud *et al.* [19] were able to observe equilibrium shapes of small lead crystallites close to T_p. They found that the (111) facet persisted until melting, but with a clear change of behaviour some 20° below T_p. The matching with round parts, which was tangential well below T_p, becomes *angular*, with a limit angle θ_c that increases as melting is approached. This increase in θ_c is accompanied by an *increase* of the facet equilibrium radius R — a somewhat surprising result since usually $\theta_c \neq 0$ means a smaller R.

Very close to T_p, RHEED measurements provide unambiguous evidence for surface melting : the *thick* liquid layer screens the electron beam for the underlying substrate (away from T_p, ℓ^* is too small for the experiment to be conclusive). Heyraud *et al.* thus prove that the facet is *not* surface molten, while the round parts are (the liquid does not wet the facet : $S_0 < 0$). We now argue that such a behaviour may explain the observed shape. For simplicity, we consider only cylindrical crystals with one curvature.

Let θ be the crystal orientation of the interface. We compare two extreme situations:

(i) A plain solid-vacuum interface, without surface melting, with a surface energy $\gamma_{CV}(\theta)$.

(ii) A surface molten strucure, with a *global* surface energy $\gamma_{CLV}(\theta)$ (including *all* contributions for the equilibrium thickness ℓ^*).

Since the facet does not wet,

$$S_0 = \gamma_{CV}(0) - \gamma_{CLV}(0) < 0.$$

We now argue that the step energy is much smaller for a solid-liquid interface than it would be in the solid-vapour case:

$$\beta_{CLV} \ll \beta_{CV}.$$

Physically, the solid-liquid transition is gradual, while the solid-vapour one is sharp — hence less pinning and a smaller β. Since $\beta = a\, d\gamma / d\theta$, it follows that $S(\theta)$ *increases* rapidly with θ: it may well become positive when θ exceeds some reasonably small angle θ_m: the two surface energies cross. Put another way, the *larger step energy* β_{CV} *may induce wetting* if $\theta > \theta_m$.

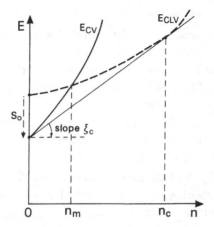

Figure 36: Matching of a non-wetting facet with a wetting round part.

Assume this situation holds: the $E(n)$ curves for the two configurations (CV) and (CLV) behave as shown in Fig. 36. The break of slope at $n = n_m$ is actually rounded off, since at that point the width of the liquid layer is very small ($S_m = 0$): (CV) and (CLV) are hardly distinguishable. However, an obvious feature remains: a double tangent can be drawn from the $n = 0$ cusp. *The matching of a non-wetting facet with a wetting round part is necessarily angular* [20]. (If the facet were also wetting ($S_0 > 0$), the whole E_{CV} curve would be above E_{CLV}: the matching would usually be tangential.)

According to (2.34), the slope of the double tangent gives the equilibrium size L^* of the facet. The observed temperature dependence of θ_c and L^* may be explained if we assume that $\gamma_{CV}(\theta)$ and $\gamma_{CLV}(\theta)$ move *rigidly* with T, the only temperature dependence being that of $S_0(T)$. Assume that S_0 *decreases* with increasing temperature, becoming negative when T reaches T_0, some 20K below T_p.[†] As S_0 goes down, it is clear from Fig. 36 that the slope of the double tangent grows, together with n_c: this is just what we want.

Unfortunately, the assumption of a constant β_{CLV} is highly questionable. As the temperature goes down below T_p, the liquid thickness ℓ^* shrinks, and a broad step can no longer exist — in the end, β_{CLV} should return to β_{CV} when ℓ^* is atomic. More realistically, the step energy $\beta(\ell)\theta/a$ should be included in the minimization that fixes the optimal

---

[†] $dS/dT < 0$ implies that the solid-vapour interface has more entropy than the CLV configuration.

liquid thickness ℓ^*. For $\theta \neq 0$, it seems that such a step energy might dominate the Van der Waals contribution: ℓ^* would then be locked to the value for which β starts increasing appreciably — our assumption of a rigid $\gamma_{CLV}(\theta)$ may not be that stupid! If this view holds, *wetting of round parts is induced by the steps* : it acts to reduce β.

At this stage, the discussion remains very speculative. Any serious argument implies a detailed description of thermal fluctuations in the (CLV) configuration — an elaborate theoretical problem [18].

Possible superheating of a solid phase

Another remarkable finding of Heyraud *et al.* [19] is the existence of *superheating*, i.e. the persistence of a metastable solid phase several degrees above T_p. This is very unusual for melting — the standard argument being that surface melting precludes metastability above T_p. Past the triple point, H is < 0 in (2.50). The slope reverses in Fig. 34, and ℓ^* increases spontaneously — there is no potential barrier to overcome, since the liquid layer existed beforehand.

The situation is different if a wetting round part is sandwiched between two non-wetting facets. Then a finite lens of liquid may appear as shown in Fig. 37. (Above T_p, the curvature of the inner interface is nega-

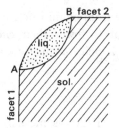

Figure 37: Formation of a liquid lens above
T_p between two non-wetting facets.

tive, since the liquid is more stable than the solid.) Such a structure may be locally stable, essentially when the meniscus spans less than a half circle (detailed calculations are given in [20]). This may explain superheating, with one proviso: we should understand why a liquid layer does not 'invade' the facets from the corners A and B in Fig. 37. Returning to the notations of Fig. 34, H and S are now *negative* ($T > T_p$, non-wetting facet). $E(\ell)$ has a potential barrier which must be overcome in order to melt the solid. This will not happen in the midst of facets — but it seems that 'creeping' from the edges should be straightforward.

The very fact that superheating is observed shows that it does not happen — why? (the fact that round parts are very small, typically 50Å, may be relevant).

3

Mobility of the interface

Consider a planar interface between a solid and its melt: it is in equilibrium if the two phases have equal chemical potentials,[†] $\mu_L = \mu_C$. More generally, the equilibrium condition for a curved interface is (1.47), which may be rewritten as

$$\Delta\mu = \mu_L - \mu_C - \frac{\tilde{\gamma}}{\rho_C R} = 0. \qquad (3.1)$$

In terms of the liquid pressure p_L, $\Delta\mu$ is given by

$$\Delta\mu = \delta p_L \frac{\rho_C - \rho_L}{\rho_C \rho_L} - \frac{\tilde{\gamma}}{\rho_C R}. \qquad (3.2)$$

$\Delta\mu$ measures the departure from equilibrium : a positive $\Delta\mu$ will drive crystal *growth* (atoms want to lower their chemical potential). In a linear regime, the interface velocity u (measured in the solid frame) may be written as

$$u = k\Delta\mu. \qquad (3.3)$$

k is the *interface mobility*. (If $\Delta\mu$ is measured per unit mass, k has the dimension of an inverse velocity.) Note that the irreversible entropy production at the interface is

$$T\dot{S} = f_C u \Delta\mu = \rho_C k \Delta\mu^2.$$

It corresponds to a genuine *surface dissipation.*

In practice, such a surface dissipation lies in series with another *bulk* dissipation, due to diffusive evacuation of the latent heat (and possibly solutes) released when the crystal grows. In usual situations, this bulk

---

[†] We disregard complications due to elastic strains in the crystal: see Section 1.

dissipation is dominant — one may then safely assume that the interface is locally in equilibrium ($\Delta\mu = 0$). Bulk diffusion leads to all sorts of growth instabilities, which are described in the lectures of C. Caroli and Y. Pomeau in this volume. Here we are concerned only with the *surface dissipation problem*: given a local $\Delta\mu$ at the surface, what are the physical mechanisms that control the mobility k? The question is relevant in at least two situations:

▷ if there is no bulk dissipation, as occurs at the liquid-solid interface of ^{4}He: heat is evacuated without resistance through the superfluid;

▷ if the surface mobility is very small, as occurs for instance at a crystal facet where growth is limited either by nucleation or by defects.

More generally, surface dissipation may be a significant correction in bulk controlled situations.

Actually, (3.3) is an oversimplified kinetic equation. It is written for a *pure* system, with a single chemical species[†] — but even in this restricted case it is too naïve a description! Difficulties stem from the release of heat upon melting: as a result, a difference ΔT is bound to occur across the interface, together with $\Delta\mu$. The *mass* flow through the interface $J = \rho_C u$ must be treated jointly with the *energy* flow J_E (flows are defined from liquid to solid). More precisely we define

$$\bar{J}_E = J_E - \mu J = J_Q + TSJ,$$

where J_Q is the *conduction* heat current and TSJ the *convective* heat flow. $\bar{J}_E$ and J are coupled to ΔT and $\Delta\mu$ by linear Onsager relationships, which are conveniently written as

$$J = \alpha\left[\Delta\mu + \lambda\frac{\Delta T}{T}\right], \quad \Delta T = R[\bar{J}_E - \lambda J]. \tag{3.4}$$

Equation (3.4) correspond to an irreversible entropy production

$$\dot{S} = \frac{R}{T^2}\left[\bar{J}_E - \lambda J\right]^2 + \frac{J^2}{\alpha T}. \tag{3.5}$$

The three dissipative coefficients α, R, λ have simple physical meaning. $\alpha = \rho_C k$ is, within a factor, our former mobility, defined for isothermal processes ($\Delta T = 0$). R is the usual Kapitza boundary thermal resistance. The cross term λ tells us how the released latent heat

$$LJ = T[S_L - S_C]J$$

[†] The extension to mixtures is discussed in the lectures of C. Caroli.

is distributed between the two phases. If $\Delta T = 0$, the conduction heat currents on either sides are

$$J_{QL} = J[\lambda - TS_L], \quad J_{QC} = J[\lambda - TS_C]. \tag{3.6}$$

The net heat release is $J_{QC} - J_{QL} = LJ$, as expected: (2.6) indicates how that heat is shared between liquid and crystal. Altogether, the mass flow and the heat flow are intricately connected!

In order to exploit (3.4), we must know ΔT — which in turn is controlled by the bulk thermal impedances. A full description is complicated [21]. Since bulk dissipation is not within the realm of these lectures, we shall neglect these complications altogether: instead of the full equations (3.4), we shall use the truncated form (3.3). (Somehow, we assume that the temperature discontinuity ΔT at the interface is zero.) This is enough to discuss *structural* aspects of surface kinetics, which are our main concern — but a word of warning on the underlying approximations was useful.

It should be emphasized that (3.3) is specific of a *liquid-solid interface*. The liquid phase is a reservoir of atoms on which growth can draw at any time. The liquid has a well defined chemical potential μ_L which acts as a reference: the growth responds to *the local* $\Delta\mu$, according to (3.3). The situation is completely different for an isolated crystal, bounded by a *solid-vacuum* interface. Then, there is no reservoir available: any relaxation of crystal shape must occur via *diffusion* of atoms along the surface. μ_L is undefined (it is a Lagrange multiplier which guarantees mass conservation): only the *gradient* of μ makes sense — and indeed the flow of atoms along the surface obeys a two dimensional Fick's law

$$J = -D\frac{dn}{d\mu}\,\mathrm{grad}[\Delta\mu], \tag{3.7}$$

where D is an appropriate diffusion coefficient. The motion of a solid-vacuum interface responds to grad $\Delta\mu$, while a solid-liquid interface responds to $\Delta\mu$ itself. We shall discuss later an example of such a surface diffusion.

3.1 Growth of a vicinal liquid-solid interface

Consider a planar surface with tilt θ. Assume first that θ is $\gg a/\xi$ where ξ is the step width. The steps are largely overlapping: they lose any reality. The surface is essentially *free* (it is not pinned to the periodic lattice): it responds to a departure of equilibrium $\Delta\mu$ according to (3.3), with a *free mobility* k_0. In principle, k_0 can depend on the crystal orientation (reflecting the anisotropy of the solid) — but this dependence is smooth: for a small range of θ, k_0 is constant. For future discussion, it is convenient to relate $\Delta\mu$ to the supercooling force F defined earlier (see (2.17)):

$$\Delta\mu = \delta p_L \frac{\rho_C - \rho_L}{\rho_C \rho_L} = \frac{F}{\rho_C a}. \tag{3.8}$$

The kinetic equation (3.3) may be written as

$$\eta u = \frac{F}{a},$$

in which we have set

$$\eta = \frac{\rho_C}{k_0}. \tag{3.9}$$

F/a is the 'driving force' on the interface per unit area (for a step, the force per unit length is $F/a \times a = F$). Its work is the gain in bulk energy. η is a *friction coefficient*, inversely proportional to the mobility. For this large tilt region, $\theta \gg a/\xi$, the response is:

(i) linear in F;

(ii) isotropic in θ (on a scale $a/\xi \ll 1$).

If the interface departs from a plane, with a displacement $z(x, y, t)$, it feels a capillary force $(-\tilde{\gamma}/R)$ in addition to the supercooling force F/a. For small displacements the curvature $1/R$ is equal to $-\nabla^2 z$ — hence a *diffusive* equation of motion for the interface:

$$\eta u = \eta \dot{z} = \tilde{\gamma} \nabla^2 z + \frac{F}{a}. \tag{3.10}$$

When $F = 0$, a deformation with wave vector q regresses with a time constant $\tilde{\gamma} q^2 / \eta$.

Let us now turn to a *vicinal* surface, for which $\theta \ll a/\xi$: it is made up of flat parts separated by localized steps. For a weak enough force F, the flat parts cannot grow: they are *trapped* by the periodic lattice potential. The only way to let the crystal grow is to *slide* the steps sideways, with a velocity u_S along the x-crystal planes. Whenever a step goes by, the interface climbs by a. The interface velocity in the z-direction is therefore

$$u_z = n u_S \cdot a = u_S \tan \theta.$$

The velocity *normal to the interface* (which is the one defined earlier) is

$$u = u_z \cos\theta = u_S \sin\theta. \tag{3.11}$$

If we assume that the step responds to $\Delta\mu$ with some characteristic *step mobility*

$$u_S = k_S \,\Delta\mu,$$

then the *interface* mobility is

$$k = k_S \sin\theta.$$

$k(\theta)$ should vanish linearly as $\theta \to 0$ — a signature of the vicinal regime.

Before calculating k_S, let us look more carefully at the trapping of flat parts by the lattice. Let $V(z)$ be the pinning periodic potential, with typical amplitude V: we saw that the width of the facet was

$$\xi \sim a\sqrt{\frac{\tilde{\gamma}}{V}}. \tag{3.12}$$

Let us now assume that a supercooling force F/a is applied: it is equivalent to an "external" potential $(-Fz/a)$ added to $V(z)$. Clearly, two regimes are possible:

▷ If $F \gg V$, the net energy is a monotonous function of z: the minima of V have been washed out by F. The interface will simply spill down the slope, like a particle on a tilted washboard: it moves essentially freely, and the periodic pinning potential $V(z)$ is mostly irrelevant except close to threshold. The interface velocity is $k_0\Delta\mu$, with a small modulation due to $V(z)$. The *net mobility* is k_0, the same as when $\theta \gg a/\xi$.

▷ If instead $F \ll V$, the minima of $V(z)$ are hardly affected. The interface remains *pinned*, and it does not move unless some nucleation mechanism is at work — which is unlikely except very close to a roughening transition.

In between, there exists a characteristic supercooling force $F_m \sim V$ for which the potential barrier that opposes the growth of facets vanishes. F_m is the *limit of metastabilty* of the facet, similar to the spinodal limit in a liquid-gas phase transition. At zero temperature, facet growth begins sharply at F_m — at finite temperature, the transition is blurred by thermal activation (i.e. homogeneous nucleation, discussed in the next section). Using (3.12), we may estimate F_m as

$$F_m \sim \frac{\gamma a^2}{\xi^2} \sim \frac{\beta}{\xi}, \tag{3.13}$$

where β is the step energy.

In order to interpret this result, we note that the characteristic size of a facet was its collapse radius

$$r_c = \frac{\beta}{F}$$

(r_c is the radius of a critical nucleus for terrace creation, $2r_c$ is the equilibrium radius of the facet.) The threshold (3.13) is consequently such that

$$r_c \sim \xi. \tag{3.14}$$

Surface pinning disappears when the facet size r_c is smaller than the step thickness ξ — an important result which clarifies the physical nature of the transition:

▷ If $r_c < \xi$, the very idea of a facet becomes meaningless. The bordering steps 'spread in' and suppress completely the flat part. One cannot define a facet radius to better than ξ!

▷ If on the other hand $r_c > \xi$, step blurring is only a minor correction, and the flat parts remain well defined.

We thus arrive at an important conclusion: a strong enough supercooling 'roughens' the interface — the so-called *dynamical roughening*. The relevant criterion is a comparison of the two length scales, r_c and ξ. If $r_c < \xi$, the interface grows freely, as it would if θ were $\gg a/\xi$. If $r_c > \xi$, it grows only by sideways motion of the steps. For a given θ, this will show up as a *non-linear* relation $u(F)$: for $F \ll F_m$, the slope is $k_S \sin\theta$, while it becomes k_0 for $F \gg F_m$.

In order to complete the discussion, we must calculate k_S. This can be done using (3.10), with an additional *pinning force* $-\partial V/\partial z$. The general equation of motion reads

$$\eta \dot{z} = \tilde{\gamma} z'' - \frac{\partial V}{\partial z} + \frac{F}{a}. \tag{3.15}$$

In equilibrium ($F = 0$), integration of (3.15) yields the step profile $z_0(x)$

$$\tilde{\gamma} z'' - \frac{\partial V}{\partial z} = 0 \implies \tfrac{1}{2}\tilde{\gamma} z'^2 = V - V_{\min}.$$

The step energy β is equal to

$$\beta = \tilde{\gamma} \int_{-\infty}^{+\infty} z_0'^2 \, dx. \tag{3.16}$$

Suppose now that a force F is applied, the step moving at a *constant* velocity u_S. Equation (3.16) becomes

$$-\eta u_S z' = \tilde{\gamma} z'' - \frac{\partial V}{\partial z} + \frac{F}{a}.$$

Multiplying by z' and integrating from $x = -\infty$ to $+\infty$, we find

$$\eta u_S \int_{-\infty}^{+\infty} z'^2 \, dx = F. \tag{3.17}$$

(We took a step that goes *down* as x increases — and which therefore moves to the *right* if F is positive: $u_S > 0$.) Equation (3.17) is nothing but a statement of energy conservation: the power dissipated by friction is

$$\eta \int_{-\infty}^{+\infty} u_z^2 \, dx = \eta u_S^2 \int_{-\infty}^{+\infty} z'^2 \, dx.$$

It is equal to the power $F u_S$ drawn from the force F that drives the step at velocity u_S. In lowest order, we can replace z by z_0 in (3.17). We thus find

$$u_S = \frac{F \tilde{\gamma}}{\beta \eta}. \tag{3.18}$$

Equation (3.18) should be compared to the velocity for a *free* interface (no pinning)

$$u = \frac{F}{\eta a}.$$

Using (3.11), we see that the vicinal structure reduces the interface mobility by a factor

$$\frac{k}{k_0} = \frac{\tilde{\gamma} a}{\beta} \sin \theta \sim \frac{\xi}{a} \sin \theta. \tag{3.19}$$

When the steps merge together for $\theta \sim a/\xi$, the mobility $k(\theta)$ goes smoothly into the mobility k_0 characteristic of a free interface. The behaviour of $k(\theta)$ is sketched in Fig. 38.

In conclusion, we emphasize the two main results of this discussion:

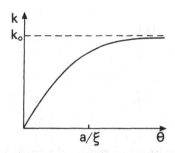

Figure 38: The mobility of a vicinal surface as a
function of tilt angle θ for a small supercooling F.

(i) For a small F, there is a crossover from step sideways drift to
free interface growth when $\theta \sim a/\xi$: hence a *static roughening* due to
interface tilt.

(ii) For small θ, *dynamical roughening* occurs when F reaches F_m for
which ξ equals the nucleation radius r_c.

We shall recover the same conclusions when studying the roughening
transition in Section 4.

3.2 Facet growth at a liquid-solid interface: homogeneous nucleation

Consider a circular facet with radius R (Fig. 39). For simplicity, we

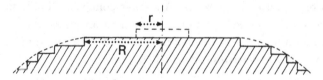

Figure 39: Geometry of a circular facet.

assume that it matches the curved part tangentially: the *equilibrium*
radius of the facet is

$$R^* = 2r_c = \frac{2\beta}{F}. \tag{3.20}$$

The formation of a terrace with radius r means a cost in energy

$$\Delta E(r) = 2\pi r\beta - \pi r^2 F \tag{3.21}$$

(see (2.38)). R^* is such that $\Delta E(R^*) = 0$: the energy is stationary
with respect to facet growth. If $R > R^*$, the formation of a new terrace

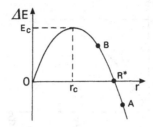

Figure 40: Nucleation barrier for creation of a terrace on a circular facet.

is energetically favourable: if we can overcome the potential barrier, we go from the origin to A in Fig. 40: the facet *grows* by one lattice spacing, and consequently it *shrinks*: R decreases and we return toward equilibrium.

This idea of facet shrinking is in fact more subtle. Consider for instance a capillary rise experiment, in which the solid grows in a vertical cylindrical tube. The contact angle at the walls is fixed, the curvature of round parts is known for a given supercooling F: the facet radius R follows by difference. When we add one terrace, R would be unchanged were it not for *gravity*: as the facet goes up, supercooling goes down, the curved parts are less curved — and the resulting R does shrink! Clearly, what matters is not R itself — but its value compared to round parts. The quantity which does shrink upon growth is R/R^* — either because R decreases or because R^* goes up! A similar conclusion holds for shape relaxation of an isolated crystallite. F is then an adjustable Lagrange multiplier. Adding one terrace enhances the radius of round parts — and thus it shrinks the *relative* facet width R.

If $R < R^*$ (point B in Fig. 40), the reverse process occurs: a step bordering the facet shrinks "in" until a terrace annihilates in the centre: we go from B to O — the facet melts.[†] Note that *there remains a potential barrier* as long as $R > r_c$: the edge step must first fight supercooling before line tension takes over.

[†] Facet melting could also proceed via nucleation of 'negative' terraces, dug into the facet, which would expand and disappear when two opposite steps annihilate at the edge. The resulting potential barrier is higher than for step shrinking. If the negative terrace nucleates at the centre, that barrier is $4\pi R\beta$. If it nucleates near the edge, the negative terrace is simply the first stage in step shrinking.

The nucleation barriers for growth and melting are respectively

$$E_{c\uparrow} = \pi \frac{\beta^2}{F}, \quad E_{c\downarrow} = \pi \frac{(\beta - FR)^2}{F}, \tag{3.22}$$

(see Fig. 40). We define a *nucleation rate* ν as the number of critical nuclei produced per unit time and per unit area on the facet. We have a $\nu_\uparrow$ and a $\nu_\downarrow$ respectively for facet growth and facet melting : equilibrium corresponds to $\nu_\uparrow = \nu_\downarrow$. A precise calculation of ν is difficult. While everybody agrees about the *Arrhenius factors*

$$\nu_\uparrow \sim \exp[-E_{c\uparrow}/T], \quad \nu_\downarrow \sim \exp[-E_{c\downarrow}/T]$$

the prefactor is a much more controversial issue. The difficulty is to assess the *density* of potential nucleation centres on the facet, i.e. their *entropy*. This issue is addressed in the lectures of J.S. Langer (Chapter 3). We will skip it: ν is an input of our description.

If R is very close to the equilibrium radius R^*, it is clear that facet growth and melting behave *symmetrically*. The net growth rate is proportional to

$$\nu_\uparrow - \nu_\downarrow \sim F e^{-E_c/T}, \quad E_c = \frac{\pi \beta^2}{F} = \frac{\pi \beta R}{2}. \tag{3.23}$$

For large R, E_c is $\gg T$: the growth rate is very small — albeit linear in F.

Such a nice symmetry disappears away from equilibrium: growth and melting behave very differently for large supercooling F. Let $F^* = 2\beta/R$ be the value of F for which the initial facet is in equilibrium.

▷ If $F \gg F^*$, point A goes down the slope in Fig. 40. $\nu_\downarrow$ quickly disappears: growth is controlled by $\nu_\uparrow$ (terrace creation), with a potential barrier $E_{c\uparrow}$ that goes *down* as F grows. The response $u(F)$ is highly *nonlinear*.

▷ If F approaches $\frac{1}{2}F^*$, point B in Fig. 40 approaches the maximum: $E_{c\downarrow}$ vanishes and the facet *collapses*, at a rate limited by the step mobility. This collapse threshold is somewhat rounded by thermal fluctuations — but for a large facet it remains quite sharp.

The corresponding variation of $\nu = \nu_\uparrow - \nu_\downarrow$ as a function of F is sketched in Fig. 41.

When F approaches the metastability limit F_m given by (3.13), the barrier $E_{c\uparrow}$ is of order

$$\frac{\beta^2 \xi^2}{\gamma a^2} \sim \gamma a^2.$$

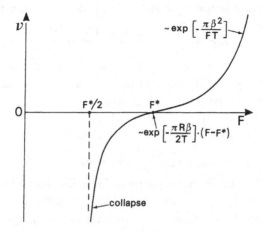

Figure 41: Nucleation rate of a facet as a function of the supercooling force F.

It becomes an atomic energy, easily gathered at any reasonable temperature. In fact, (3.21) and (3.22) are no longer valid close to F_m, since the barrier profile is modified by the bias F. The barrier lowers, and $E_{c\uparrow}$ accordingly decreases, going to zero when F reaches the threshold F_m (at which free growth begins). The calculation of $E_{c\uparrow}$ in this transition region was done long ago by Cahn and Hilliard [22]: it amounts to a steadily decreasing β in (3.22).

In usual situations, the temperature T is comparable to γa^2. As F approaches F_m critical nuclei proliferate, and the facet "roughens" with a large density of thermal steps. It will eventually move as if it were free, with a mobility k_0. We recover, in another guise, the *dynamical roughening* mentioned before. Indeed, the characteristic "unlocking" force F_m is precisely that for which critical nuclei become meaningless, since their radius r_c is comparable to the width ξ of the bordering step. F_m also represents the passage from a nucleation regime to free growth: it is just the range around which the growth law $u(F)$ returns to linear behaviour. Above F_m, facets disappear altogether.

In order to make this discussion more quantitative, we address three points:

(i) The subsequent growth of a single supercritical terrace: the point is to show that it remains circular.

(ii) The coalescence of germs, which affects the growth rate of the interface.

(iii) The onset of collapse when $R \sim r_c$, which may shed some light on step interaction.

Shape of a growing terrace

The growth law for a single step may be written as[†]

$$\eta_S v_n = F\left[1 - \frac{r_c}{\rho}\right], \qquad (3.24)$$

in which v_n is the velocity normal to the step, ρ the radius of curvature, η_S a 'step friction coefficient' related to k_S. (The second term in the bracket is the bending force.) For a circular facet with radius r, (3.24) reduces to

$$\eta_S \dot{r} = F\left[1 - \frac{r_c}{r}\right]. \qquad (3.25)$$

Close to r_c the growth is slow, but it quickly becomes linear as r increases.

In order to study the stability of this circular shape, we consider a small deformation, described in polar coordinates:

$$r(\phi) = r_0 + \sum_{m \neq 0} \varepsilon_m e^{im\phi}.$$

The interface rotates by a small angle $\alpha = -r'/r$: the normal velocity becomes $v_n = \dot{r}\cos\alpha$. Using the expression of curvature in polar coordinates

$$\frac{1}{R} = \frac{r^2 + 2r'^2 - rr''}{(r^2 + r'^2)^{3/2}},$$

we obtain the equation of motion for r. In zeroth order in ε_m, we recover (3.25). In first order, we find

$$\eta_S \dot{\varepsilon}_m = F(1 - m^2)\frac{r_c \varepsilon_m}{r_0^2}. \qquad (3.26)$$

(Note that $\dot{\varepsilon}_1 = 0$: a deformation with $m = 1$ is a plain translation for which there is no restoring force.) Dividing (3.26) by (3.25), we obtain the *relative* growth law

$$\frac{d\varepsilon_m}{dr_0} = (1 - m^2)\frac{r_c \varepsilon_m}{r_0(r_0 - r_c)}. \qquad (3.27)$$

[†] The step responds to the *local* $\Delta\mu$, which implies a reservoir of atoms, provided by the liquid phase.

For $m > 1$, any deformation relaxes to zero: *the circular shape is stable as the terrace grows.* Qualitatively, such a conclusion is fairly obvious: a bulge in the circle has an increased curvature — it consequently grows more slowly, thereby regressing.[†]

Coalescence of germs

Consider a facet with area $S \gg r_c^2$. In our former language, that means a large supercooling, $F \gg F^*$: reverse nucleation is negligible and only the rate $\nu_\uparrow = \nu$ is relevant. The number of germs produced per unit time is $S\nu$. After a time t, there are $S\nu t$ of them, and their average distance is

$$\delta = \frac{1}{\sqrt{\nu t}}.$$

These germs subsequently grow, with an asymptotic step velocity $v = F/\eta_S$. After a time t, they have a radius $r = vt$. Coalescence[‡] occurs when $\delta = r$, i.e. for

$$t \sim t_m = (\nu v^2)^{-1/3}, \quad r \sim r_m = \left(\frac{v}{\nu}\right)^{1/3}. \tag{3.28}$$

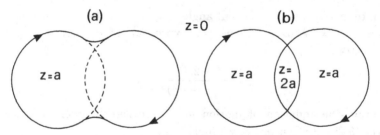

Figure 42: Coalescence of two terraces.

Then the terraces merge together and the interface goes up by one lattice spacing.

[†] The conclusion would be different if growth were limited by *diffusion* along the surface. Then we expect 2D growth instabilities, as discussed by C. Caroli. The terrace shape may become quite erratic.

[‡] We assume a plain coalescence, as shown in Fig. 42. One might also envisage a situation in which the two colliding terraces would generate a terrace at height $2a$ (Fig. 42) — a process which might make sense if steps have inertia. Then an avalanche process sets up.

The whole issue is to compare r_m with the facet size $S^{1/2}$. If the nucleation rate ν is small, $Sr_m^2 > 1$: the germs reach the edge of the facet before coalescing. Facet growth is a 'one germ process': a terrace appears and fills the whole facet — then we wait a long time until another terrace does it again. The facet velocity is

$$u = a\nu S. \qquad (3.29)$$

It depends on the facet area.

If ν is larger, Sr_m^2 becomes < 1. Then coalescence is crucial. The net interface velocity is

$$u = \frac{a}{t_m} \sim a(\nu v^2)^{1/3}. \qquad (3.30)$$

It involves a power $\frac{1}{3}$ of the nucleation rate — the so-called 'Kolmogorov-Avrami' law — and as a result the Arrhenius factor of u is $\exp[-\frac{1}{3}E_c/T]$ instead of the usual $\exp[-E_c/T]$. This point is essential in interpreting experimental data.

Experimental observation at the solid-superfluid interface of ^{4}He

In practice, homogeneous nucleation is observed only when the barrier is low, which means either a large supercooling F or a small step energy β — i.e. the vicinity of a roughening transition (see Section 4). Such a measurement was carried out by Wolf *et al.* [23] on the (0001) facet of ^{4}He (perpendicular to the hexagonal c-axis). The corresponding roughening temperature is $T_R = 1.28$K: homogeneous nucleation is clearly observed above 1.1K.

The experimental set up is sketched in Fig. 43: a crystal with a horizontal facet grows between two parallel vertical plates. Outside

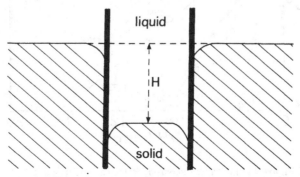

Figure 43: Experimental set up for facet growth measurements in ^{4}He.

the plates, the large horizontal solid-liquid interface sits at the nominal equilibrium pressure p_L^*: it fixes the origin of the supercooling δp_L. The value of δp_L on the facet is just the hydrostatic pressure difference

$$\delta p_L = \rho_L g H$$

(see Fig. 43). The resulting supercooling force F follows from (3.8):

$$F = (\rho_C - \rho_L) a g H.$$

Within a constant factor, F is just the level difference H. The experiment consists in measuring the interface velocity $u = -\dot{H}$ as a function of H for various temperatures.

Far from equilibrium, u should behave as

$$u \sim \exp\left[-\frac{E_{c\uparrow}}{3T}\right] = \exp\left[-\frac{\pi\beta^2}{3FT}\right].$$

The plot of $\log u$ as a function of $1/H$ should thus be *linear*, the slope yielding the step energy $\beta(T)$. Results are shown in Fig. 44. Instead

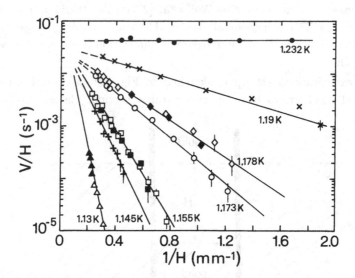

Figure 44: $\mathrm{Log}(u/H)$ versus $1/H$ for the (0001) facet of ^{4}He, for various values of the temperature T.

of log u, we plot $\log[u/H]$, partly because more elaborate theories suggest a prefactor $\sim H$ to the Arrhenius exponential — partly also because such a prefactor interpolates with the linear behaviour $u \sim H$ expected in the rough state. (Anyhow, this prefactor plays a very minor role.) The exponential behaviour of u is clearly observed over five decades, yielding unambiguous evidence of homogeneous nucleation (this is probably the neatest experimental demonstration). From the slope, we infer $\beta(T)$, which is shown in Fig. 45. We see that $\beta(T)$ decays very fast as T approaches T_R. Indeed, it is impossible to locate T_R accurately from such a measurement alone — a detailed analysis must use other experimental input, as discussed in Section 4.

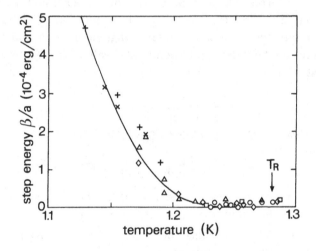

Figure 45: The step energy β as a function of temperature for the (0001) facet, as extracted from Fig. 44. Note that the roughening transition is well into the tail of $\beta(T)$.

At lower temperatures, the growth rate is no longer measurable: homogeneous nucleation practically disappears. Eventually, it will be relayed by growth on crystal defects.

Facet collapse upon melting

Consider a circular facet with radius R, which matches the surrounding curved parts tangentially. If the supercooling force F goes below $F_c = \beta/R$, the facet *collapses*: we want to understand how it reaches a steady state. For simplicity, we ignore nucleation, which acts to blur the transition at $F = F_c$: there is no thermal fluctuation. The facet is strictly metastable for $F > F_c$ — we are only concerned with the onset of collapse when $F < F_c$.

Technically, we consider a crystal in a cylindrical vessel, with an axis parallel to the crystal z-axis. The vessel radius is d. If $F > F_c$, the facet is locked: nothing moves. If $F < F_c$, steps 'pour in' the facet and annihilate at the centre: the crystal *melts*. We are interested in a *steady state*, in which the whole pattern moves down at a constant vertical velocity u: we want to find the steady pattern $n(r)$ for a given u, and the growth law $u(F)$ in the vicinity of $F = F_c$. We shall see that this growth law at threshold depends on the nature of step interactions: it may be an interesting experimental information.

Let $\xi(r)$ be the step chemical potential a distance r from the axis. ξ is itself a function of the step density $n(r)$. In equilibrium, ξ obeys (2.27), which expresses the balance of supercooling, interaction and bending forces. Using the boundary condition at the wall, $\xi(d) = \xi_W$, we obtain the explicit solution

$$\xi = F\left[\frac{r^2 - d^2}{2r}\right] + \xi_W \frac{d}{r}. \tag{3.31}$$

The facet edge $r = R$ corresponds to $\xi = \beta$. It exists if

$$F > F_c = \frac{1}{d}\left[\xi_W + \sqrt{\xi_W^2 - \beta^2}\right].$$

Below F_c, ξ never reaches β (see Fig. 18): there is no metastable equilibrium. At threshold, the radius $R = r_c = \beta/F_c$.

Assume now that F is slightly below F_c: steps start moving, with a radial velocity v which we write as

$$\eta_S v = F - \frac{\partial \xi}{\partial r} - \frac{\xi}{r}. \tag{3.32}$$

Compared to (3.24), we have included the effect of *step interactions*. (Note that v is negative since the steps move inward.) If needed, the step friction coefficient η_S may be obtained from (3.18):

$$\eta_S = \frac{\eta \beta}{\gamma}. \tag{3.33}$$

As steps drift by, the interface *melts*, with a local velocity

$$u(r) = -n(r)v(r)a$$

(for convenience, we take u positive for *melting* — hence the minus sign). In a *steady* state, we want $u(r)$ to be constant (the profile moves rigidly) — hence a constraint between step density and velocity:

$$v(r) = -\frac{u}{an[\xi(r)]}.$$

Expression (2.32) becomes a non-linear differential equation for $\xi(r)$:

$$\frac{d\xi}{dr} = F - \frac{\xi}{r} + \frac{\eta_S u}{an(\xi)} \tag{3.34}$$

($n(\xi)$ is obtained by inverting $\xi(n)$). Only one solution of (3.34) is regular at the origin. Since that solution must also obey the wall boundary conditions $\xi(d) = \xi_W$, an *eigenvalue* condition ensues, that fixes u for a given F — hence the growth law. The problem is mathematically well posed.

It can be solved explicitly very close to threshold, when $F_c - F \ll F_c$. Then the *steps are injected into the facet very slowly* (they are nearly in equilibrium near the edge). Once they have moved in, they quickly accelerate toward the centre, and consequently their density $n(r)$ is very small, everywhere except close to the edge (where they are 'balancing on whether to go or not'). n is obtained by setting ξ equal to β in (3.34):

$$n = \frac{\eta_S u}{aF} \frac{r}{r_c - r}, \qquad (r < r_c), \tag{3.35}$$

(remember that $F \approx F_c = \beta/r_c$). n grows close to r_c (where steps go slowly), and it vanishes at the origin (where the tension force is infinite). From n we infer the value ξ_{in} *inside* the facet.

Conversely, the profile well *outside* the facet is very close to the equilibrium profile at threshold, given by (3.31) (the small velocity u is only a correction). We thus know ξ_{out} for $r > r_c$. We are left with a *matching* problem of these two asymptotic solutions, depicted in Fig. 46, which will yield the desired eigenvalue condition. As long as $F_c - F \ll F_c$, the matching occurs in a very *narrow* region near $r = r_c$, where the step density n is still small, allowing an expansion of $\xi(n)$. The matching depends on $(\xi - \beta)$, i.e. on the *interaction* between the steps: physically, it is that interaction that pushes the outer steps inward, thereby controlling the initial stages of step shrinking. Consequently, the results will depend crucially on the nature of step interactions. If $E(n)$ has the form (2.13), we have

$$\xi - \beta = 3\phi n^2 \implies n \sim \sqrt{\xi - \beta}.$$

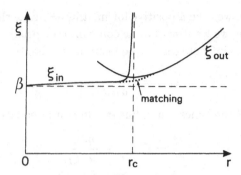

Figure 46: The matching of inner and outer
solutions in the problem of terrace collapse.

In contrast, if $E(n)$ had a term $\sim n^2$, we would have

$$n \sim (\xi - \beta).$$

The nature of matching changes completely.

In principle, matching requires a numerical solution of (3.34). Actually, using appropriately scaled variables one may reduce the differential equation to a universal dimensionless form: the answer is known within numerical factors of order 1. Detailed calculations are given in appendix B of [15]: we only state the results:

(i) For a step interaction $\sim 1/d^2$, the growth characteristic is linear:

$$u \sim F_c - F \quad \text{if} \quad F < F_c. \tag{3.36}$$

(ii) If $E(n)$ has a leading n^2 interaction term, u grows more slowly

$$u \sim (F_c - F)^{3/2} \quad \text{if} \quad F < F_c \tag{3.37}$$

(u is always zero when $F > F_c$). Homogeneous nucleation will round off the singularity at F_c — but the change of regime should be clearly visible experimentally. Observation of facet collapse provides important information:

▷ the threshold F_c yields β;

▷ the dependence of u on $(F_c - F)$ yields information about step interaction.

Such a measurement is highly desirable.

We may add one final remark: as the facet melts, *it is no longer planar*. The small density of shrinking steps creates a *bulge*, which is initially small, but which becomes dramatic if F is well below F_c — then the facet has completely disappeared.

3.3 Frank-Read sources

We consider first the simpler case of a single step anchored on a dislocation. The facet radius is R, the dislocation lies a distance d away from the centre (see Fig. 29b). We assume that the step joins the edge tangentially ($\eta = 0$ in Fig. 29). The step is in equilibrium if its radius is equal to r_c. This is possible if

$$R - d < 2r_c < R + d.$$

In such a case the facet does not move in the absence of nucleation (the step must overcome a potential barrier). It follows that the facet is *metastable* for all supercooling forces F between F_1 and F_2, with

$$F_1 = \frac{2\beta}{R + d}, \quad F_2 = \frac{2\beta}{R - d}.$$

Below F_1, the facet melts and eventually collapses at $F_c = \beta/R$. Above F_2, it grows. The interface velocity u as a function of F is sketched in Fig. 47.

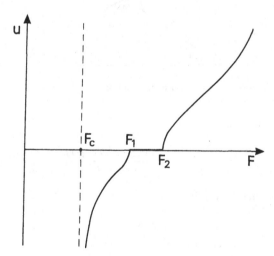

Figure 47: The facet growth rate u as a function of supercooling force F when a step is anchored on a single dislocation.

The interesting case is $F \gg F_2$. Then the step winds up around the dislocation, forming a *spiral* (hence the so-called 'spiral growth'). The asymptotic spacing ℓ of that spiral is $\sim r_c$ (in fact about $20r_c$). The spiral expands with a normal velocity v_n given by (3.24), resulting in a vertical facet velocity

$$u = \frac{a v_n}{\ell}.\tag{3.38}$$

Since v_n and $1/\ell$ are both proportional to F, u goes as F^2, an unexpected behaviour characteristic of spiral growth. That result holds if the spiral is well developed, i.e. if $r_c \ll R - d$.

In order to make that argument more quantitative, we describe the step in polar coordinates centred on the dislocation : the azimuth ϕ is a function of radius r. Locally the step makes an angle δ with the tangential direction (Fig. 48), such that

$$\sin \delta = \frac{1}{\sqrt{1 + r^2 \phi'^2}}.$$

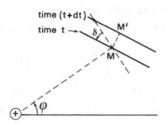

Figure 48: Rotation of the step in spiral growth.

The local curvature is

$$\frac{1}{\rho} = \frac{\mathrm{d}(\phi + \delta)}{\mathrm{d}s} = \frac{2\phi' + r^2 \phi'^3 + r\phi''}{(1 + r^2 \phi'^2)^{3/2}}.$$

The normal velocity is finally

$$v_n = \frac{MM'}{\mathrm{d}t} = r\dot{\phi}\sin\delta.$$

Putting all these results into (3.24), we obtain the equation of motion of the step. We look for a *steady state* in which the whole pattern *rotates regularly* with a constant angular frequency $\dot{\phi} = \omega$ — hence the equation

$$\eta_S \frac{\omega r}{\sqrt{1 + r^2 \phi'^2}} = F\left[1 - r_c \frac{2\phi' + r^2 \phi'^3 + r\phi''}{(1 + r^2 \phi'^2)^{3/2}}\right].\tag{3.39}$$

Equation (3.39) is a (complicated!) first order differential equation for $\phi'(r)$. Its solution must be regular when $r = 0$ and $r = \infty$ — hence an eigenvalue constraint that fixes $\omega(F)$.

In their famous 1951 paper, Burton, Cabrera and Frank [24] devised an approximate solution of (3.39), based on an interpolation between small and large r. More generally, (3.39) can be solved numerically. It is found that

$$\omega \sim 0.32 \frac{F}{\eta_s r_c}. \tag{3.40}$$

The resulting facet velocity is

$$u = \frac{a\omega}{2\pi} \sim 0.05 \frac{aF^2}{\beta \eta_s} \tag{3.41}$$

(the interface moves up by a whenever the spiral has turned by 2π). Note that at large distance we recover an Archimedean spiral, with a spacing $\sim 20 r_c$ (compare (3.38) and (3.41)). One point should be emphasized: the result (3.41) is valid only at a *liquid-solid interface*, when the supply of atoms for growth is not a problem (hence the use of step mobility η_s). At a *solid-vacuum* interface, diffusion along the surface becomes the dominant factor: the law (3.41) may be considerably altered, as discussed in [24].

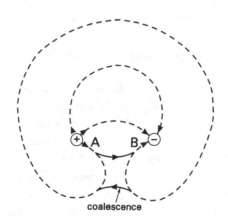

Figure 49: Terrace generation by a Frank-Read source.

Let us now turn to the Frank-Read source anchored on two opposite dislocations a distance d apart. When F exceeds the threshold

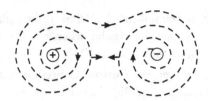

Figure 50: Frank-Read growth for large supercooling $F \gg F_0$.

$F_0 = 2\beta/d$, the step grows, turning around the dislocations. Eventually, the two 'lips' coalesce on the back side, as shown in Fig. 49: a terrace forms, while a new step AB appears, which repeats the process. The interface growth rate is again $u = a\omega/(2\pi)$, ω being *the average angular velocity* of the step around a given dislocation. For very large supercooling, $F \gg F_0$, the pattern becomes that of Fig. 50: each dislocation is surrounded by a spiral, and terraces are nucleated when the two growing spirals merge in the middle. In that limit the growth rate is that of a single spiral, given by (3.41).

Assume that the facet has *two* Frank-Read sources (Fig. 51): common sense would suggest that the facet grows twice as fast! This is *wrong*:

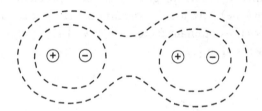

Figure 51: Two Frank-Read sources do not make the crystal grow faster.

when the terraces nucleated by each source coalesce, they produce a single terrace with the *same height a*. Of course, the time it takes for a terrace to cover the whole facet is shorter if there are more Frank-Read sources — but the *nucleation rate* of terraces remains controlled by ω, according to (3.40): it is a *one source* affair. Adding more Frank-Read sources makes no difference to the facet velocity u. Put another way, *Frank-Read sources are not cumulative* as long as they are well separated. This rather paradoxical conclusion is very important: spiral growth does not accelerate if we add more dislocations — the interface is just more messy!

Altogether, the growth law of a single Frank-Read source behaves as shown in Fig. 52 (disregarding facet collapse when $F < F_c$). One

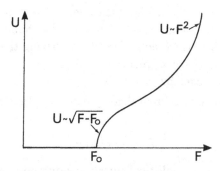

Figure 52: Facet growth law for a single Frank-Read source.

may show that $u \sim \sqrt{F - F_0}$ at threshold — but this is a somewhat semantic point since in practice there is a *distribution* of distances d and thresholds F_0 on a given facet.

3.4 Diffusion limited growth at a solid-vacuum interface

The physics is now completely different, since any change of shape is due to *atomic diffusion along the surface*, instead of transfer from a liquid reservoir. Such a situation is often met experimentally: a crystallite is prepared with some initial shape. One studies the *relaxation* towards equilibrium at constant total mass. The initial shape may result from previous fast growth — or it may be prepared on purpose (like grooves on a flat surface). The final question is always the same: how long does it take to reach equilibrium?

We shall discuss this problem very cursorily, since the main object of these lectures is the liquid-solid interface. We shall however give a few examples that stress the peculiarities of this diffusion limited regime.

Let us first adopt a thermodynamic point of view. For a liquid-solid interface, the departure from equilibrium was measured by $\Delta\mu$, defined in (3.1). One way to view it is to define an *effective chemical potential* for solid atoms at the surface

$$\mu_{\text{eff}} = \mu_C + \frac{\tilde{\gamma}}{\rho_C R}. \qquad (3.42)$$

μ_{eff} incorporates capillary effects that make parts of the interface preferable to others for solid growth. Equilibrium is achieved if μ_{eff} is equal to μ_L. For a solid-vacuum interface, μ_L becomes physically meaningless: it is only a Lagrange multiplier that guarantees mass conservation. The effective chemical potential μ_{eff} is thus defined within an arbitrary constant — as such, it is unphysical. On the other hand, its *gradient* is a significant quantity, which acts like a *force* driving the motion of atoms along the solid interface: as mentioned in (3.7), physics is controlled by grad μ_{eff}, and no longer by μ_{eff} itself.

Let J be the surface current density of atoms. In analogy with Ohm's law, we write

$$J = -\sigma \, \text{grad} \, \mu_{\text{eff}}, \qquad (3.43)$$

in which σ is a phenomenological *surface conductivity*, whose mechanism should be clarified. The divergence of J produces accumulation of surface atoms — i.e. crystal growth with a normal interface velocity u such that

$$u \rho_C = -\,\text{div}\, J.$$

Hence the net growth law in this diffusive regime:

$$u = \frac{\sigma}{\rho_C} \Delta \mu_{\text{eff}}. \qquad (3.44)$$

Crystal growth responds to $\Delta \mu_{\text{eff}}$ instead of μ_{eff}. Equation (3.44) was established long ago by Mullins [25]. It is traced back to the fact that the density of surface atoms is a *conserved* quantity at a solid-vacuum interface, while it is *not* conserved in the solid-liquid case (the liquid easily provides new material!).

Within a constant, μ_{eff} reduces to the capillary term in (3.42). We can thus write (for a cylindrical geometry with one curvature only: the generalization is evident):

$$u = \frac{\sigma}{\rho_C^2} \Delta\Big(\frac{\tilde{\gamma}}{R}\Big). \qquad (3.45)$$

Consider for instance a small distortion $z(x, y, t)$ of a planar interface, with wave vector q. If we assume an isotropic surface tension

$$\tilde{\gamma} = \gamma = \text{const.}$$

then (3.45) predicts a relaxation rate

$$s = -\frac{\gamma\sigma}{\rho_C^2} q^4 \qquad (3.46)$$

(the curvature is $1/R = q^2 z$). For a liquid-solid interface, s was $\sim q^2$ — the difference is evident.

In practice, $\tilde{\gamma}$ is not constant, and we should reconsider the whole section in the light of a diffusion limited growth: what happens to homogeneous nucleation, to dislocation mediated growth? How does a vicinal surface grow when atoms are injected at the edge of a surface element and diffuse slowly toward the centre? We shall only comment very briefly on these questions, referring to the large literature on the subject.

The first problem is to reconsider the definition of μ_{eff}. It is clear that (3.42) does not make sense at the level of one step, since curvature is then meaningless: we must return to a more microscopic picture. Let $\mu(r)$ be the actual chemical potential of surface adatoms, defined everywhere on each terrace. Consider first a *single straight* step. If the step moves by dx, a number of atoms

$$dN = -\rho_C a\, dx \qquad (3.47)$$

(per unit step length) is attached to the solid (Fig. 53). Since the step

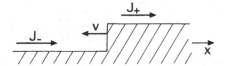

Figure 53: Motion of a step at a crystal-vacuum interface.

is merely translated, only the bulk energy is affected: equilibrium is achieved if

$$\mu(r) = \mu_C \quad \text{at the step.} \qquad (3.48)$$

Equation (3.48) is a *boundary condition* for $\mu(r)$. More generally, assume that the step moves with velocity v. Atoms are brought in from either side with a current density $J_\pm$ (see Fig. 53). Mass conservation implies

$$J_+ - J_- = \rho_C av = \frac{av}{\Omega} \qquad (3.49)$$

(in which we have introduced the atomic volume of the crystal, $\Omega = 1/\rho_C$). If the attachment of atoms to the steps meets some resistance, (3.48) should be modified according to

$$\alpha_- J_- = \mu(-0) - \mu_C, \quad \alpha_+ J_+ = \mu_C - \mu(+0). \qquad (3.50)$$

$\mu(\pm 0)$ is the atomic chemical potential immediately on right (left) of the step. $\alpha_\pm$ are 'attachment resistances'. Equations (3.50) are the *dynamic* version of (3.48). The contact resistances $\alpha_{\pm}$ will act in series with the bulk resistance that drives J in the middle of terraces,

$$J = -\sigma \operatorname{grad} \mu.$$

Depending on the situation, one or the other may dominate.

We now return to equilibrium, and we consider a *vicinal surface* with a distribution of steps $n(x)$. (For simplicity we consider only a cylindrical surface with straight steps parallel to the y-axis: the generalization to circular steps is straightforward.) When a step moves by dx, atoms are once again attached to the solid, according to (3.47) — but now the energy of the step has changed, because of *interaction with its neighbours*. ξ has been shifted by

$$\frac{\mathrm{d}\xi}{\mathrm{d}x}\,\mathrm{d}x = E''\frac{\mathrm{d}n}{\mathrm{d}x}\,\mathrm{d}x.$$

Equilibrium is achieved if the total energy is stationary — hence the condition

$$\mu(r) - \mu_C = -\frac{E''}{\rho_C a}\frac{\mathrm{d}n}{\mathrm{d}x}. \tag{3.51}$$

We know from (2.20) that $E''\mathrm{d}n/\mathrm{d}x = -a\tilde{\gamma}/R$: the new equilibrium condition (3.51) reduces to our original macroscopic result (3.42). The physical meaning of that result is now clear: the chemical potential of an atom at a given step is shifted due to *interaction between crystal steps* (attachment of the atom moves the step and changes its energy[†]).

In conclusion, the chemical potential of atoms at the surface is fixed at each step. For a single straight step in equilibrium, it obeys (3.48). Step interactions shift this equilibrium condition to (3.51). In the more general dissipative regime, the matching conditions (3.50) should be written as

$$\begin{aligned}
\alpha_+ J_+ &= \mu_C - \mu(+0) - \frac{1}{\rho_C a}\frac{\mathrm{d}\xi}{\mathrm{d}x}, \\
\alpha_- J_- &= \mu(-0) - \mu_C + \frac{1}{\rho_C a}\frac{\mathrm{d}\xi}{\mathrm{d}x}.
\end{aligned} \tag{3.52}$$

(It is the departure from *local* equilibrium which drives attachment to the steps.)

[†] If the steps were circular, attachment of atoms would increase their radius r and thus their energy $2\pi r\xi$: in this way we generate the *bending* term ξ/r as well as the interaction term $\mathrm{d}\xi/\mathrm{d}r$.

Motion of a single step on a facet

Consider for instance a Frank-Read source: atoms must diffuse along the surface in order to reach the growing spiral. This problem is discussed in detail in the original reference [24]. Such a diffusive regime can kill the growth law $u \sim F^2$ found at a solid-liquid interface. Since there are many different cases, we do not explore it in further detail: see [24].

Similarly, homogeneous nucleation is deeply affected. Once a critical nucleus has passed its potential barrier, its subsequent growth is completely different in a diffusive regime. Consider for instance a germ with radius $r_g \gg r_c$ in the middle of a facet with radius R. We neglect attachment dissipation: $\mu(r) = \mu_C$ on the germ, while $\mu = \mu_R$ is fixed at the facet edge. Let $v = \dot{r}_g$ be the growth rate of the germ. In a quasistatic approximation, the adatom current on the facet is

$$J = -\frac{av}{\Omega}\frac{r_g}{r}.$$

The μ profile is therefore logarithmic, and the growth law is

$$v = \frac{\Omega\sigma}{ar_g}\frac{\mu_R - \mu_C}{\log[R/r_g]}.$$

The growth rate of the germ is *not* constant for large r_g, in contrast with (3.25). Moreover, the circular shape is highly unstable. A detailed treatment of nucleation in this regime is difficult, involving an analysis 'à la Lifshitz-Slyozov' [26]: it has been carried out by Villain [27].

Steady shape of curved parts between two parallel facets

We consider the situation of Fig. 54, in which a curved part is sandwiched between two semi-infinite parallel facets, a distance $2h$ apart.

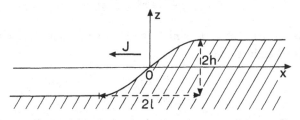

Figure 54: Steady shape of a groove between two facets.

Eventually, such a structure should relax toward a uniform plane. If there is no nucleation on the facets, h is fixed. Steps can only spill down the slope from the upper facet to the lower one. The width 2ℓ of the curved region slowly increases, and facets shrink *at constant height*. There is no stationary state, and eventually ℓ will go to infinity, thereby 'eating up' the facets on either side. In the end, we recover a planar interface $h = 0$ (grooves drawn on a surface slowly disappear).

A detailed description of this transient regime is not difficult. Nevertheless, in order to avoid mathematical complexities, we consider a simpler problem — admittedly somewhat artificial — which displays the main physical ideas. We assume that a *constant* adatom current J flows from the upper facet to the lower one, and we look for a *steady* profile of the intermediate curved region. It is clear that such a configuration is strictly impossible if the solid is in contact with a liquid. Mathematically, a given supercooling F cannot equilibrate regions of both positive and negative curvature. Physically, step repulsion will conspire with a positive F to push steps in the lower part toward the left, thereby filling the lower facet. In contrast, Fig. 54 is perfectly eligible for a solid-vacuum interface. In the intermediate region, J implies a gradient of chemical potential. If σ is angle independent, we must have

$$\Delta\mu = \mu(r) - \mu_C = \frac{Jx}{\sigma}. \tag{3.53}$$

$\Delta\mu$ changes sign at $x = 0$ — and thus it can accommodate opposite curvatures on both sides.

The simplest model is to take a constant $\tilde{\gamma}$. Then

$$\Delta\mu = -\frac{\tilde{\gamma}}{\rho_C} z'',$$

from which we infer the intermediate profile

$$z = -\frac{\rho_C J}{\sigma\tilde{\gamma}}\left[\tfrac{1}{3}x^3 - \ell^2 x\right]. \tag{3.54}$$

The slope is zero for $x = \ell$, which is chosen in such a way that the corresponding height is $\pm h$

$$\ell^3 = \frac{3}{2}\frac{\sigma\tilde{\gamma}}{\rho_C J} h. \tag{3.55}$$

Equation (3.55) yields the width of the transition region, which goes as $[h/J]^{1/3}$: the smaller J, the broader the curved part. (If $J = 0$, the only steady state is $\ell = \infty$, as surmised at the beginning.)

In practice, a constant $\tilde{\gamma}$ is completely unrealistic close to a facet orientation: we should treat the interface as a vicinal surface, characterized

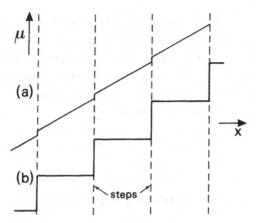

Figure 55: Chemical potential profile at a vicinal surface with a fixed current J. (a) Dissipation dominated by terrace diffusion. (b) Dissipation dominated by step attachment.

by a step density $n(x)$. The chemical potential profile is then a compromise between transport on the terraces and attachment at the steps — hence two limiting situations illustrated in Fig. 55.

(i) In case (a), dissipation occurs entirely on the terraces: this is the view adopted by Rettori and Villain [28]. μ is continuous at the steps, (3.53) remains valid. In view of (3.51),

$$\Delta\mu = -\frac{1}{\rho_C a}\frac{d\xi}{dx}.$$

It follows that

$$\xi = \beta - \frac{\rho_C aJ}{\sigma}\frac{(x^2 - \ell^2)}{2}. \tag{3.56}$$

If we assume $\xi - \beta = 3\phi n^2$, we find

$$\frac{dz}{dx} = na = \left[\frac{\rho_C a^3 J}{6\sigma\phi}(\ell^2 - x^2)\right]^{1/2}. \tag{3.57}$$

The integration of (3.57) is straightforward: the width ℓ is of order $h^{1/2}$.

(ii) In case (b), transport through the terrace is easy, and dissipation is dominated by attachment to the steps: we should use (3.52). Such a situation is analyzed in [29]. Assuming equal attachment resistances on both sides ($\alpha_+ = \alpha_- = \alpha$), one finds

$$\frac{d\xi}{dx} = -2\alpha J\rho_C z. \tag{3.58}$$

The integration of (3.58) is again straightforward: ℓ now behaves as $h^{1/3}$.

Detailed results are not essential: the only reason for this discussion was to stress the salient features of the problem, and to show how it should be formulated. The important question that must be asked beforehand is 'What is the dominant dissipation mechanism? Diffusion on the terraces or attachment to the steps?'

4

Thermal fluctuations: the roughening transition

4.1 Fluctuations of a single step

Consider a single straight step along the y-axis, in the absence of any locking periodic potential. If the step is displaced by an amount $x(y)$, the corresponding stretching energy is

$$E = \beta_0 \int \mathrm{d}y \sqrt{1 + x'^2}, \qquad (4.1)$$

where β_0 is the original step energy, in the absence of any fluctuation. If the slope x' is small, we may expand (4.1) into

$$\Delta E = E - E_0 = \int \tfrac{1}{2} \beta_0 x'^2 \, \mathrm{d}y. \qquad (4.2)$$

The corresponding thermal fluctuations were already mentioned when we discussed the statistical interaction due to step confinement: for convenience, we repeat the argument. We expand x in a Fourier series

$$x = \sum_k x_k e^{iky} \implies \Delta E = \sum_k \tfrac{1}{2} \beta_0 k^2 L \langle |x_k^2| \rangle \qquad (4.3)$$

(L is the length of the chain). Assigning an energy $\tfrac{1}{2}T$ to each mode, we find

$$\langle |x_k^2| \rangle = \frac{T}{\beta_0 k^2 L} \implies \langle x^2 \rangle = \sum_k \langle |x_k|^2 \rangle.$$

The allowed k are multiples of $2\pi/L$: usually, we would replace the sum by an integral

$$\langle x^2 \rangle = \frac{1}{\pi} \int_0^\infty \mathrm{d}k \, \frac{T}{\beta_0 k^2}. \qquad (4.4)$$

Here the integral (4.4) diverges at small k: we must stick to the discrete summation. We thus find that the fluctuations of the step are

$$\langle x^2 \rangle = \frac{LT}{12\beta},$$

a well known result for a stretched string.[†]

The same argument gives the fluctuations of *slope* of the step

$$\langle x'^2 \rangle = \frac{1}{\pi} \int_0^{k_{\max}} dk \frac{T}{\beta_0}. \tag{4.5}$$

Here the divergence occurs at *large* k: we must cut off the integral at some $k_{\max}$. Physically, the cut off will arise from the natural *width* ξ_0 of the step. When the wavelength is comparable to ξ_0, the cost in energy is due not only to *stretching*, but also to *bending* of the step. The step behaves like a rod whose structure is distorted on a scale $\lesssim \xi_0$. Such a distortion yields an additional contribution to ΔE, which should ensure the convergence of (4.5). We thus set $k_{\max} \sim 1/\xi_0$, thereby obtaining

$$\langle x'^2 \rangle \sim \frac{T}{\beta_0 \xi_0}. \tag{4.6}$$

Remember that β_0 and ξ_0 are *zero temperature* quantities, which reflect the internal structure of the step. We conclude that slope fluctuations become large when

$$T \sim T_R \sim \beta_0 \xi_0 \sim \gamma a^2. \tag{4.7}$$

When $T > T_R$, the expansion (4.2) breaks down: the step starts meandering in the surface plane, with large slope fluctuations. At this stage, we do not know yet what happens really — but it is already clear that T_R is a characteristic temperature where 'something' happens.

The same characteristic temperature can be inferred from another argument — which sheds some more light on the underlying physics. Consider a step of length L with free ends: can we calculate its *configuration entropy*? Since the step cannot cross itself, it is a standard case of self avoiding random walk, a well known problem in statistical mechanics. For our purpose, a very crude oversimplified estimate is sufficient. Since

---

[†] Remember that the statistical interaction between steps was precisely due to freezing of those modes that are responsible for the divergence in (4.4).

the step cannot bend on a scale $< \xi_0$, we divide the step into L/ξ_0 pieces, each of length ξ_0, and we assume that an additional piece can freely rotate around the preceding one, with p choices. On a square lattice, p would be equal to 3. If we ignore the no crossing constraint, all pieces are statistically independent, the resulting number of configurations being

$$W = p^{L/\xi_0}. \tag{4.8}$$

The self avoiding constraint will modify (4.8) — the dominant effect being a reduction in p. Since we do not specify p accurately, we ignore these complications, and we take (4.8) at face value. The entropy is then

$$S = \log W = \frac{L}{\xi_0} \log p. \tag{4.9}$$

The crucial result is that S is *proportional to* L. As a result, the step *free energy* may be written as

$$E - TS = \left[\beta_0 - \frac{T}{\xi_0} \log p\right] L = \beta L. \tag{4.10}$$

It is clear that β becomes negative if T is large enough — actually if

$$T > \frac{\beta_0 \xi_0}{\log p} \sim T_R. \tag{4.11}$$

We recover the same characteristic temperature, in another guise. If $T > T_R$, the step free energy becomes negative, *steps proliferate spontaneously*, with completely erratic shapes — a feature which was already suggested by the large value of $\langle x'^2 \rangle$.

The thermal fluctuations that control the free energy β are short wavelength fluctuations, with a scale $\sim \xi_0$. The quantity β is thus usable at all scales $\gg \xi_0$. Let us for instance look at the scale ξ on which slope fluctuations are of order 1. ξ is an important parameter of the problem, which separates two distinct regimes:

(i) On a scale $\ll \xi$, the step meanders like a polymeric chain, as shown in Fig. 56. The concept of a 'straight' step is washed out by fluctuations.

Figure 56: The effective width of a meandering step.

(ii) On a scale $\gg \xi$, the step behaves like a *stretched string*, the only difference being the replacement of β_0 by a free energy β which incorporates the effect of short wavelengths fluctuations. (This is a first example of 'mode-mode coupling,' a concept which will recur at all stages of this section.)

ξ may be viewed as a *statistical width* of the step, which is to ξ_0 what β is to β_0. Technically, $1/\xi$ is the value of $k_{\max}$ in (4.5) for which $\langle x'^2 \rangle \sim 1$, the only change being the replacement of β_0 by β. We thus find

$$\xi \sim \frac{T}{\beta}. \qquad (4.12)$$

Close to T_R, β is very small: ξ becomes large and (4.12) is valid. The statistical width ξ thus diverges when $\beta \to 0$ — but the product $\beta\xi$ is not affected much

$$\beta\xi \sim T \sim T_R \sim \beta_0\xi_0. \qquad (4.13)$$

Formulae (4.13) will be essential in understanding the roughening transition.

Equation (4.12) may be obtained using another route, which stresses the weak points of the present argument. If the step has a free energy β, *terraces* appear spontaneously on the surface, with a typical size ℓ such that $\beta\ell \sim T$ — hence $\ell \sim \xi$. The surface is paved with random terraces of size $\sim \xi$. A single straight step cuts through these terraces. Since crossings are not allowed, all intersections will rearrange: the final step will meander from one terrace to the next, with typical excursions $\sim \xi$, as expected. Where is the weak point? In assuming that terraces develop next to each other, in a *single* crystal plane! In reality, the crystal can grow indefinitely, with one small terrace on top of a bigger one, etc. As a result, the statistics of steps is deeply affected as soon as they begin proliferating (in a somewhat pompous language, the universality class of the transition is different). We shall remedy these weaknesses later. Here, we only want to provide simple qualitative arguments for two fundamental results:

▷ The step free energy vanishes at some critical temperature.

▷ The step has a *statistical width* ξ below which its shape is an erratic meandering curve. The product $\beta\xi$ is roughly constant as temperature varies.

The behaviour of $\xi(T)$ is sketched in Fig. 57. While at $T = 0$ the width is natural, due to the internal structure of the step, it diverges near T_R, due

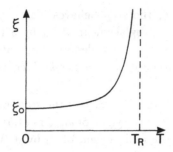

Figure 57: Step width as a function of temperature.

to increasing thermal fluctuations. Equations (4.10) and (4.12) suggest $\xi \sim 1/(T_R - T)$: we shall see that such a guess is *wrong* — a consequence of the 'weak point' mentioned above.

Step locking: statistics of kinks

Assume now that the step is subject to an in-plane lattice potential periodic in x. Any deformation tends to localize into discrete kinks, positive or negative (i.e., the step shifts to right or left as y increases). Each kink has an energy ε and a width ζ such that

$$\varepsilon \zeta \sim \beta b^2. \tag{4.14}$$

Let p_+ and p_- be the densities of $\pm$ kinks per unit length. An average tilt ϕ of the step (with respect to the y-axis) implies a 'charge' of kinks

$$p_+ - p_- = \frac{\phi}{b} \tag{4.15}$$

(we suppose ϕ small). Moreover, thermal fluctuations can *create pairs of opposite kinks* (i.e. a bulge in the step): the 'kink chemical potentials' μ_+ and μ_- are opposite. If the kink densities are small, we may neglect their interaction in first approximation. The densities $p_\pm$ at temperature T are then of order

$$p_\pm \sim \frac{1}{\zeta} \exp\left[-\frac{\varepsilon - \mu_\pm}{T}\right]$$

(a length L of step can accommodate at most L/ζ non-overlapping kinks). If $\phi = 0$, (4.15) implies $\mu_+ = \mu_- = 0$, hence

$$p_+ = p_- = p_0 \sim \frac{1}{\zeta} e^{-\varepsilon/T}. \tag{4.16}$$

More generally, p_+ and p_- obey the equilibrium relation

$$p_+ p_- = p_0^2 \qquad (4.17)$$

(compare with the distribution of electrons and holes in a semi-conductor). Equations (4.15) and (4.17) determine p_+ and p_-. They are meaningful only if

$$T < \varepsilon, \quad \phi < \frac{b}{\zeta}. \qquad (4.18)$$

If one of these conditions is violated, the kinks overlap, thereby losing any physical reality: the *step is free*.

Consider first the case $\phi = 0$. If we sit somewhere in the middle of the step, the net charge Q on one side fluctuates, with an amplitude

$$\langle \Delta Q^2 \rangle \sim p_0 L.$$

ΔQ in turn implies an excursion in the x-direction $\Delta x = b \Delta Q$, hence a position fluctuation

$$\langle \Delta x^2 \rangle \sim b^2 p_0 L. \qquad (4.19)$$

Fluctuations are again $\sim \sqrt{L}$, but $\langle \Delta x^2 \rangle$ is reduced by a factor

$$p_0 \frac{\beta b^2}{T} \sim p_0 \zeta \frac{\varepsilon}{T}$$

compared to a free step. This factor is very small if $T \ll \varepsilon$. As T approaches ε, p_0 becomes $\sim 1/\zeta$: kink fluctuations join the free step regime smoothly.

Assume now that the step is tilted by ϕ: *minority kinks are negligible* if $\phi > p_0 b$. The argument that led to (4.19) is still valid, except that p_0 is replaced by the density of majority kinks, $p = \phi/b$:

$$\langle \Delta x^2 \rangle \sim b \phi L. \qquad (4.20)$$

When approaching overlap, $\phi \sim b/\zeta$, we should worry about kink interactions. We may write the free energy as

$$F \sim L f\left(\frac{Q}{L}\right) \implies \langle \Delta x^2 \rangle \sim \frac{b^2 L T}{f''} \qquad (4.21)$$

(for free kinks, $f \sim T p \log p$ yields (4.19)). When $p\zeta \sim 1$, we expect

$$f'' = \frac{\partial \varepsilon}{\partial p} \sim \varepsilon \zeta \sim \beta b^2.$$

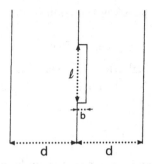

Figure 58: Confinement of kinks due to neighbouring steps.

We recover our earlier result for a free step.

In brief, the kink structure *reduces* step fluctuations — which nevertheless remain $\sim \sqrt{L}$. It becomes irrelevant when either of the two conditions (4.18) is violated: it seems that we could forget about kinks completely. *This is not so in practice, due to the presence of other steps.* To see that, consider a straight step along the y-axis, sandwiched between two parallel neighbours a distance d apart (Fig. 58). The interaction energy with these neighbours is $2V(d)$. We next create a kink and an antikink a distance ℓ apart, thereby shifting a length ℓ of step by an amount b to the right. The interaction energy is corrected by

$$\Delta V = \left[V(d+b) + V(d-b) - 2V(d)\right] \approx b^2 V''. \qquad (4.22)$$

If the surface is stable, V'' is > 0: (4.22) is a *confining* energy, growing with ℓ, which prevents the kinks from spreading too far apart. ΔV is not a real interaction between the kinks, but rather an effect of the step drift that ensues when the kinks separate — a concept familiar in particle physics. It follows that the idea of *independent kinks* is highly suspect in the presence of *step interactions*: it will only hold if some mechanism kills the confinement term (4.22). Such a mechanism might be the proliferation of $\pm$ thermal kinks, which would 'screen out' the confinement potential (think of a metal insulator transition in a semiconductor). At this stage, we are not equipped to treat this difficult problem — but it is important to be aware of it: *kink separation may be inhibited in the presence of other steps*. We shall see that this very concept is at the root of roughening transitions for higher order facets.

At this stage of the game, we ignore these difficulties: we return to our stack of plates with no in-plane periodicity. Then kinks never arise to start with.

4.2 Fluctuations of the interface

Consider first a planar *free* interface $z = 0$. Due to thermal fluctuations it is slightly displaced, by an amount $z(x, y)$. The corresponding energy cost is[†]

$$E = \tfrac{1}{2} \int \mathrm{d}x \, \mathrm{d}y \, \tilde{\gamma} \operatorname{grad} z^2. \tag{4.23}$$

We can repeat everything we said for a step, the only difference being that the problem is 2D instead of 1D. We Fourier expand z and we assign $\tfrac{1}{2}T$ to each mode. Instead of (4.4), we find

$$\langle z^2 \rangle = \frac{1}{4\pi^2} \int 2\pi k \, \mathrm{d}k \, \frac{T}{\tilde{\gamma} k^2}. \tag{4.24}$$

The integral is divergent both at large and small k — a typical 2D feature. We cut if off when $k \sim 1/L$, where L is the size of the interface.[‡] At large k, (4.23) ceases to be correct when $k = \Lambda \sim \ell^{-1}$, where ℓ is the *thickness of the interface* — hence a natural physical cut off. Altogether

$$\langle z^2 \rangle = \frac{T}{2\pi\tilde{\gamma}} \log(\Lambda L). \tag{4.25}$$

Fluctuations diverge *logarithmically* as a function of L (in contrast to the $\sqrt{L}$ behaviour found for a step).

In the same way we calculate *slope* fluctuations

$$\langle \operatorname{grad} z^2 \rangle = \frac{T}{2\pi\tilde{\gamma}} \int k \, \mathrm{d}k \sim \frac{T\Lambda^2}{\tilde{\gamma}}.$$

They are controlled by *large* k's. If $T < \tilde{\gamma}/\Lambda^2$, this result calls for no comment. If T happens to be $> \tilde{\gamma}/\lambda^2$, the surface has short scale *meandering*, with a typical size

$$\ell_{\mathrm{eff}} \sim \sqrt{\frac{T}{\tilde{\gamma}}}$$

(ℓ_{eff} if the size of 'thermal bubbles' for which the surface energy is $\sim T$). The summation on k should then be cut off at $\Lambda_{\mathrm{eff}} = 1/\ell_{\mathrm{eff}}$, instead of Λ (beyond that wave vector, the slope is large and the expansion (4.23) is

---

[†] $\tilde{\gamma}$ is the surface stiffness, $\gamma + \gamma''$, due both to *stretching* and to *rotation* of the surface when $\operatorname{grad} z \neq 0$. Strictly speaking, $\tilde{\gamma}$ should be a tensor.
[‡] The interface is a square $L \times L$, with the constraint $z = 0$ at the edge.

not valid). Put another way, the effective thickness of the interface has two origins:

▷ an intrinsic thickness ℓ_0 due to decay of the crystal order parameter;

▷ a statistical thickness ℓ_T due to short range domains and surface meandering.

ℓ_{eff} is the biggest of these two lengths. The situation is the same as for steps — with one major difference: the surface configurational entropy is *not* proportional to L^2 — hence the surface free energy does *not* vanish as T increases (which would indicate a second order transition) — hence ℓ_{eff} does *not* diverge. *Statistical broadening* will just correct Λ, with no qualitative consequence. We assume this has been done and we forget about the whole issue.

We now restore *pinning* of the interface by the lattice periodic potential. We assume that it can be represented by a potential energy

$$E_p = \int dx \, dy \, V(z), \qquad (4.26)$$

where V is some periodic function of z, with period a, the lattice spacing along z. The interplay of (4.23) and (4.26) is the subject of this whole section. Here, we begin with a very naïve argument, due to J. Lajzerowicz, which displays most of the physics without any fancy mathematics. Let us assume for a moment that V is *small*: we shall argue later that this is indeed a sensible assumption. We can then calculate the thermal average $\langle E_p \rangle$, integrated over the surface. When $L \to \infty$, two situations may occur:

(i) Either $\langle E_p \rangle$ diverges: the larger the surface, the more relevant the pinning to lattice planes. In fact, there should exist a characteristic length $L = \xi$ for which $\langle E_p \rangle \sim T$: ξ will turn out to be the *correlation length* of the problem. $\langle E_p \rangle$ is negligible on scales $L \ll \xi$ (free behaviour), while it is dominant when $L \gg \xi$ (locking of facets).

(ii) Or $\langle E_p \rangle \to 0$ as $L \to \infty$: the interface is *free on a macroscopic scale*. The potential energy averages to zero: it is irrelevant.

A simple *first order* calculation of $\langle E_p \rangle$ is thus very instructive.

In order to obtain $\langle E_p \rangle$ in first order we use the distribution $P(z)$ in zeroth order, i.e. a Boltzmann distribution with energy $E(z)$. Since that energy is quadratic, $P(z)$ is a gaussian:

$$P(z) \sim \exp\left[-\frac{z^2}{2\langle z^2 \rangle}\right].$$

We expand $V(z)$ in a Fourier series, and we pick one particular term, $V_n \cos(2\pi n z/a)$. The average of the cosine is trivial[†]

$$\left\langle \cos \frac{2\pi n z}{a} \right\rangle = \exp\left[-2\pi^2 n^2 \frac{\langle z^2 \rangle}{a^2}\right].$$

The fundamental $n = 1$ is clearly dominant (harmonics are exponentially smaller). We thus find

$$\langle E_p \rangle \sim \tfrac{1}{2} V_1 L^2 \exp\left[-2\pi^2 \frac{\langle z^2 \rangle}{a^2}\right] \qquad (4.27)$$

(the factor L^2 is the area of the sample). Using (4.25), we finally obtain

$$\langle E_p \rangle \sim \tfrac{1}{2} V_1 L^2 \frac{1}{L^{\pi T/\tilde\gamma a^2}}. \qquad (4.28)$$

Obvious as it may look, this result is very important:

(i) If $T > T_R = 2\tilde\gamma a^2/\pi$, then $\langle E_p \rangle$ goes to zero as $L \to \infty$: the interface is free.

(ii) If $T < T_R$, then $\langle E_p \rangle$ goes to infinity as $L \to \infty$: the interface is locked on a macroscopic scale.

T_R thus marks the transition from low temperature facetting to high temperature free fluctuations: this is the *roughening transition*.

According to (4.28), there exists a *universal relationship* between T_R and the surface stiffness $\tilde\gamma$:

$$T_R = \frac{2\tilde\gamma a^2}{\pi}. \qquad (4.29)$$

A question arises: is that relationship a peculiarity of our first order calculation, or is it really universal? The answer is clear if we realize that the L dependence in (4.28) is entirely due to *long wavelength fluctuations*, with wave vectors $k \sim 1/L$. On that macroscopic scale, all short wavelength modes only act to *renormalize*[‡] the surface stiffness $\tilde\gamma$ and the periodic pinning potential V. As long as $T > T_R$, this renormalized V goes to zero as $L \to \infty$: we thereby justify *a posteriori* our first order

---

[†] Write the cosine as a sum of exponentials and close the square with the gaussian exponents.

[‡] Here again, we see mode mode coupling in action: the energy $\tfrac{1}{2}\tilde\gamma z_k^2$ of the kth mode is corrected due to interaction with the short wavelength degrees of freedom.

calculation. Even if V was large to start with, it becomes very small once we have averaged over short wavelength modes: our first order treatment of the remaining low k modes is then correct. Equation (4.28) and (4.29) are *bona fide* results as long as they involve *macroscopic* quantities, appropriately renormalized by microscopic fluctuations. (Indeed, this is just what we want: $\tilde{\gamma}$ in (4.29) is what we measure in the lab!)

In conclusion, we can trust (4.29) because it relies only on *macroscopic* properties of the system $(k \to 0)$. Admittedly, the 'proof' is somewhat handwaving: the systematic 'renormalization program' which we now set up will make it rigorous.

4.3 Static renormalization: the Kosterlitz-Thouless transition

We start from a *microscopic* surface energy

$$E[z(r)] = \iint d^2r \left[\gamma_0 + \tfrac{1}{2}\tilde{\gamma}\,\mathrm{grad}\,z^2 + V(z)\right]. \qquad (4.30)$$

γ_0 is the *energy* of the original planar interface. The *stiffness* $\tilde{\gamma}$ measures the energy cost of surface distortions. The periodic pinning potential $V(z)$ will usually be approximated by its fundamental

$$V(z) = -V\cos\frac{2\pi z}{a}. \qquad (4.31)$$

(Harmonics are less and less relevant as renormalization proceeds.) With that definition the interface is locked at $z = 0$ at zero temperature.

The partition function of the interface is calculated in the usual way

$$\mathcal{Z} = \int \mathcal{D}z(r) \exp\left[-\frac{E[z(r)]}{T}\right].$$

The sum over configurations is conveniently carried out in terms of the Fourier components z_k:

$$\mathcal{Z} = \prod_{k=0}^{\Lambda} \int dz_k \exp\left[-\frac{E(z_k)}{T}\right]. \qquad (4.32)$$

At this stage, (4.32) is exact — albeit intractable! The cut off Λ is fixed by the interface — whatever its origin. The energy $E(z_k)$ may be written as[†]

$$E(z_k) = \sum_k \tfrac{1}{2}\tilde{\gamma}k^2|z_k|^2 + V(z_k).$$

[†] We set $L = 1$ in order to simplify writing.

The periodic potential introduces a *mode-mode coupling* which is the key to our theory. From the partition function $\mathcal{Z}$, we infer the interface free energy

$$F = -T \log \mathcal{Z}.$$

F incorporates the entropy of surface fluctuations.

The philosophy of renormalization is then very simple. We divide fluctuations in two classes:

(i) 'Macroscopic' ones, with $k < \bar{\Lambda}$ where $\bar{\Lambda}$ is a new cut off $< \Lambda$: these we want to keep explicitly in our description.

(ii) 'Microscopic' ones with $\bar{\Lambda} < k < \Lambda$ on which we average at the outset. More precisely, we set

$$\bar{z}(r) = \sum_{k < \bar{\Lambda}} z_k e^{i\mathbf{k} \cdot \mathbf{r}}, \quad \delta z(r) = \sum_{\bar{\Lambda} < q < \Lambda} z_q e^{i\mathbf{q} \cdot \mathbf{r}}.$$

The microscopic energy E becomes a function $E(\bar{z}, \delta z)$: we want to average over δz, thereby defining an effective energy $\bar{E}(\bar{z})$ for the remaining modes. Technically, we break the summation in (4.32) in two parts, integrating first on δz and subsequently on $\bar{z}$

$$\prod_{\bar{\Lambda} < q < \Lambda} \int d\delta z_q \exp\left[-\frac{E(\bar{z}, \delta z)}{T}\right] = \exp\left[-\frac{\bar{E}(\bar{z})}{T}\right],$$

$$\mathcal{Z} = \prod_{k < \bar{\Lambda}} \int d\bar{z}_k \exp\left[-\frac{\bar{E}(\bar{z})}{T}\right].$$

(4.33)

The first equation (4.33) *defines* an effective energy $\bar{E}(\bar{z})$ for the surviving modes, appropriately modified by coupling[†] to the short wavelength part δz. Put another way $\bar{E}$ is still a microscopic energy as regards the components $\bar{z}$, but it is a *statistical free energy* as regards the components δz. Equilibrium thermal fluctuations of δz act to modify the effective energy for $\bar{z}$: $\bar{E}(\bar{z})$ has been 'renormalized' by short wavelength fluctuations.

Equation (4.33) is still exact: if we were able to carry the algebra accurately, we could choose $\bar{\Lambda}$ as we want — going for instance directly from Λ to a very small 'macroscopic' cut off. This is not possible in

---

[†] If V is zero, all modes are decoupled: averaging over δz does not affect $\bar{z}$, and $\bar{E}(\bar{z})$ is just the original streching energy $E(\bar{z})$.

practice: we are forced to make approximations — specifically, we will *truncate* the effective energy $\bar{E}(\bar{z})$, retaining only terms which we suspect to be the leading ones. The minute we do such a truncation, the result depends on how we proceed: going from Λ to Λ_1 and then from Λ_1 to $\bar{\Lambda}$ will *not* give the same result as going directly from Λ to $\bar{\Lambda}$. (Note that the discrepancy is an artefact of approximations: in itself, the renormalization machinery has no guilt!) In an attempt to optimize the process, we shall carry the renormalization *progressively*, reducing the cut off Λ by successive infinitesimal amounts. Each infinitesimal step incorporates *the effect of previous renormalizations*: in this way, we hope to account for the cumulative effect of mode-mode coupling. As Λ will go to zero, this coupling will decrease (when $T > T_R$): our approximations should become better and better. The asymptotic macroscopic behaviour will be essentially exact, earlier errors being absorbed into parameters which can be measured directly *e.g.* the macroscopic $\bar{\gamma}$. We thereby justify the naïve first order calculation given above.

Altogether, this progressive elimination of degrees of freedom generates a 'renormalization group.' As the cut off Λ is *scaled down* (corresponding to a coarse graining of our physical description), the effective energy $\bar{E}$ slowly changes. As $\Lambda \to 0$, it should evolve toward a 'fixed point' which embodies the macroscopic behaviour of the interface. Approximations arise in the *truncation* of $\bar{E}$.

Derivation of scaling equations

The literature on this subject is very large [30]. It is often plagued with unwarranted approximations that make the results quantitatively inaccurate. Here we follow the approach of Knops and den Ouden [31] which carefully avoids all pitfalls. Some further developments are given in [32]. We stick to our sharp cut off picture, which is much easier to handle[†] (the summation over k stops abruptly at $k = \Lambda$).

We start from (4.33). When carrying the average over δz, we can extract the zeroth order part out of $E(\bar{z}, \delta z)$: it generates a *gaussian*

[†] For practical purposes, it is often preferable to use a *smooth* cut off function $f(k/\Lambda)$ that goes from 1 to 0 as k increases: one thereby avoids unpleasant oscillations. Technicalities of this smooth cut off are described in [26].

probability

$$P_0(\delta z_q) \sim \exp\left[-\frac{\tilde{\gamma} q^2 \delta z_q^2}{T}\right].$$

This probability weighs the remaining exponential $\exp[-E_p/T]$. The effective energy $\bar{E}(\bar{z})$ thus takes the form

$$\bar{E}(\bar{z}) = -T \log\left\langle \exp\left[-\frac{E_p(\bar{z}+\delta z)}{T}\right]\right\rangle_0, \tag{4.34}$$

where $\langle\ \rangle_0$ means an average with weight P_0.

Now come the approximations! First we expand in powers of V — the so-called 'cumulant expansion':

$$-T \log\langle e^{-E_p/T}\rangle = \langle E_p\rangle - \frac{1}{2T}\left[\langle E_p^2\rangle - \langle E_p\rangle^2\right] + \cdots \tag{4.35}$$

(in order to prove (4.35), expand the exponential and the log in a power series). We stop in *second order*: our theory is a *weak coupling* theory, possibly acceptable at a solid-liquid interface where pinning is small, but certainly very bad in the solid-vacuum case, when the interface is rigidly locked to crystal planes. Our excuse is that V will eventually become small as scaling proceeds: if this is so, the first stages of renormalization are not to be trusted — but the weak coupling approximation should become better and better as we go on.

Using (4.31), we expand the pinning energy as

$$E_p = -\int \mathrm{d}^2\mathbf{r}\, V\left[\cos\left(2\pi\frac{\bar{z}}{a}\right)\cos\left(2\pi\frac{\delta z}{a}\right) - \sin\left(2\pi\frac{\bar{z}}{a}\right)\sin\left(2\pi\frac{\delta z}{a}\right)\right].$$

In *first order*, everything is simple: the modulation $\cos(2\pi\bar{z}/a)$ is just reduced by a constant factor

$$\left\langle\cos\left[2\pi\frac{\delta z}{a}\right]\right\rangle_0 = \exp\left[-\frac{2\pi^2}{a^2}\langle\delta z^2\rangle\right]$$

(the distribution of $\delta z(\mathbf{r})$ is gaussian). $\langle\delta z^2\rangle$ is in turn given by (4.24), except that the integral runs from $\bar{\Lambda}$ to Λ, thereby avoiding any divergence:

$$\langle\delta z^2\rangle = \frac{T}{2\pi\tilde{\gamma}} \log\frac{\Lambda}{\bar{\Lambda}}. \tag{4.36}$$

In practice, we are interested in an *infinitesimal* renormalization, $\bar{\Lambda} = \Lambda(1-\varepsilon)$, for which δz is small:

$$\langle\delta z^2\rangle = \frac{T}{2\pi\tilde{\gamma}}\varepsilon.$$

In such a case

$$\left\langle \cos\left(2\pi\frac{\delta z}{a}\right)\right\rangle_0 \approx 1 - \frac{2\pi^2}{a^2}\langle\delta z^2\rangle = 1 - \frac{\pi T}{\tilde\gamma a^2}\varepsilon. \qquad (4.37)$$

(Equation (4.37) could be obtained directly by expanding the cosine: the gaussian average can be bypassed.) The amplitude V of the periodic potential is corrected by a small amount δV such that

$$\frac{\delta V}{V} = -\frac{\pi T}{\tilde\gamma a^2}\varepsilon. \qquad (4.38)$$

As surmised, renormalization *reduces* V.

Life is more complicated in second order: we give the algebra in some detail in order to avoid any misunderstanding. According to (4.35), the corresponding contribution to $\bar E$ is

$$\delta E^{(2)} = -\frac{V^2}{2T}\int d^2r\, d^2r'\Big[\Big\langle\cos\frac{2\pi}{a}(\bar z+\delta z)\cos\frac{2\pi}{a}(\bar z'+\delta z')\Big\rangle_0$$
$$- \Big\langle\cos\frac{2\pi}{a}(\bar z+\delta z)\Big\rangle_0\Big\langle\cos\frac{2\pi}{a}(\bar z'+\delta z')\Big\rangle_0\Big].$$

z, z' are respectively shorthands for $z(\mathbf r)$, $z(\mathbf r')$. Since δz is small (ε is infinitesimal), we expand directly up to order δz^2. Only the cross term survives, yielding

$$\delta E^{(2)} = -\frac{V^2}{2T}\frac{4\pi^2}{a^2}\int d^2r\, d^2r'\sin\frac{2\pi}{a}\bar z\sin\frac{2\pi}{a}\bar z'\langle\delta z\,\delta z'\rangle_0. \qquad (4.39)$$

The correlation function $\langle\delta z\,\delta z'\rangle_0$ is easily calculated:

$$\langle\delta z\,\delta z'\rangle_0 = \sum_{k=\bar\Lambda}^{\Lambda}\exp\big[i\mathbf k\cdot(\mathbf r-\mathbf r')\big]\frac{T}{\tilde\gamma k^2}.$$

If we set $\mathbf r'-\mathbf r=\rho$, we find for an infinitesimal ε

$$\langle\delta z\,\delta z'\rangle_0 = \frac{T}{2\pi\tilde\gamma}\varepsilon J_0(\Lambda\rho)$$

(J_0 arises from the angular average on $\mathbf k$). Equation (4.39) may thus be written explicitly as

$$\delta E^{(2)} = \frac{\pi V^2}{2\tilde\gamma a^2}\varepsilon\int d^2r\, d^2r'\, J_0(\Lambda\rho) \qquad (4.40)$$
$$\times\Big\{\cos\Big[\frac{2\pi}{a}(\bar z+\bar z')\Big] - \cos\Big[\frac{2\pi}{a}(\bar z-\bar z')\Big]\Big\}.$$

To first order in ε and to second order in V, (4.40) is still exact.

Now comes the *truncation*!

(i) We discard the term $\cos[2\pi(\bar{z} + \bar{z}')/a]$, which becomes the *second harmonic* of the main oscillation when $\mathbf{r} = \mathbf{r}'$: as scaling proceeds, it will decrease faster than V, and thus it is 'irrelevant' on a macroscopic scale.

(ii) We expand the other term, assuming that $\boldsymbol{\rho} = \mathbf{r}' - \mathbf{r}$ is small

$$\cos\left[\frac{2\pi}{a}(\bar{z} - \bar{z}')\right] \approx 1 - \frac{2\pi^2}{a^2}(\bar{z} - \bar{z}')^2$$
$$\approx 1 - \frac{2\pi^2}{a^2}\left[\boldsymbol{\rho} \cdot \operatorname{grad}\bar{z}\right]^2. \tag{4.41}$$

The first term in (4.41) yields a constant in $\delta E^{(2)}$: it renormalizes the facet energy γ_0 by an amount

$$\delta\gamma_0 = -\frac{\pi V^2}{2\tilde{\gamma}a^2}\varepsilon\int \mathrm{d}^2\rho\, J_0(\Lambda\rho). \tag{4.42}$$

The second term corrects the coefficient of $\operatorname{grad}\bar{z}^2$: it changes slightly the stiffness $\tilde{\gamma}$:

$$\delta\tilde{\gamma} = \frac{\pi^3 V^2}{\tilde{\gamma}a^4}\varepsilon\int \rho^2\, \mathrm{d}^2\rho\, J_0(\Lambda\rho). \tag{4.43}$$

$\delta\gamma_0$ and $\delta\tilde{\gamma}$ arise from coupling of long wavelength modes to those that were eliminated ($\bar{\Lambda} < k < \Lambda$). Hopefully, the other terms in the expansion (4.41) are irrelevant as scaling proceeds.

Equation (4.43) is commonly found in the literature — unfortunately it is *wrong*! Clearly, the ρ integration raises difficult convergence problems. Even if we sort them out, the result is necessarily *zero* as can be seen from (4.39). Our calculation to order $(\bar{z} - \bar{z}')^2$ is tantamount to an expansion of the sines, yielding an integral

$$\int \mathrm{d}^2\mathbf{r}\, \mathrm{d}^2\mathbf{r}'\, \bar{z}\,\bar{z}'\langle\delta z\,\delta z'\rangle_0.$$

Since $\bar{z}$ and δz correspond to *different wave vectors*, the integral is zero. The result (4.43) is phony!

The correct procedure was given by Knops and den Ouden [31]: the expansion (4.41) is too naïve as it ignores completely coupling between long wavelength modes. What we should do instead is to consider $\bar{z}$ as the sum of two terms:

▷ a particular long wavelength component that we single out, for which we want to extract the corrective term $\sim \mathrm{grad}\, \bar{z}^2$;

▷ the rest which is in thermal equilibrium, with a distribution $P_{eq}(\bar{z})$ over which we average.

Expansion (4.41) is thus replaced by

$$\cos\left[\frac{2\pi}{a}(\bar{z} - \bar{z}')\right] \implies$$

$$\left[1 - \frac{2\pi^2}{a^2}(\rho \cdot \mathrm{grad}\,\bar{z})^2\right] \cdot \left\langle \cos\left[\frac{2\pi}{a}(\bar{z} - \bar{z}')\right]\right\rangle_0.$$

The only new feature is the last factor, which expresses *the effect of thermal fluctuations of $\bar{z}$ on that particular component that we singled out*. Since the distribution of $(\bar{z} - \bar{z}')$ is gaussian, the average is easily carried out[†]:

$$\left\langle\cos\left[\frac{2\pi}{a}(\bar{z} - \bar{z}')\right]\right\rangle_0 = \exp\left[-\frac{2\pi^2}{a^2}\langle(\bar{z} - \bar{z}')^2\rangle_0\right],$$

$$\langle(\bar{z} - \bar{z}')^2\rangle_0 = \frac{T}{\pi\tilde{\gamma}}\int_0^\Lambda \frac{dk}{k}[1 - J_0(k\rho)] = \frac{T}{\pi\tilde{\gamma}}h(\rho). \tag{4.44}$$

The integrands in (4.42) and (4.43) just contain an additional factor $\exp[-2nh(\rho)]$, in which we have set

$$n = \frac{\pi T}{\tilde{\gamma}a^2}. \tag{4.45}$$

We should write for instance

$$\delta\tilde{\gamma} = \frac{\pi^3 V^2}{\tilde{\gamma}a^4}\varepsilon\int \rho^2\, d^2\rho\, J_0(\Lambda\rho)e^{-2nh(\rho)}. \tag{4.46}$$

The last factor ensures convergence if $n > 2$.[‡]

We are now equipped to write the *differential scaling equations* we are looking for: we define

$$\ell = \log\frac{\Lambda}{\tilde{\Lambda}} \implies d\ell = \varepsilon.$$

[†] Note that there is no divergence at $k = 0$, in contrast to (4.24): we study fluctuations of a *difference* $(\bar{z} - \bar{z}')$, instead of z itself.

[‡] In the dynamic version to be discussed shortly, convergence will be ensured for any n.

We also introduce the pinning energy per area $\Lambda^{-1} \times \Lambda^{-1}$:

$$U = \frac{V}{\Lambda^2}.$$

Putting together (4.38) and (4.46), noting also that h depends only on the 'reduced distance' $\tilde{\rho} = \Lambda \rho$, we obtain finally

$$
\boxed{
\begin{aligned}
\frac{dU}{d\ell} &= U[2 - n], \\
\frac{d\tilde{\gamma}}{d\ell} &= \frac{2\pi^4}{\tilde{\gamma} a^4} U^2 A(n),
\end{aligned}
}
\tag{4.47}
$$

in which we have set

$$n = \frac{\pi T}{\tilde{\gamma} a^2}, \quad A(n) = \int_0^\infty \tilde{\rho}^3 \, d\tilde{\rho} \, J_0(\tilde{\rho}) e^{-2nh(\tilde{\rho})}. \tag{4.48}$$

$A(n)$ is defined unambiguously for $n \geq 2$. For $n = 2$,

$$A(2) = 4e^{-4\Gamma} = 0.398 \tag{4.49}$$

(Γ is Euler' constant).[†]

Equations (4.47) give the evolution of U and $\tilde{\gamma}$ as we progressively eliminate degrees of freedom. In *first order*, $\tilde{\gamma}$ and n would remain constant: this is the naïve calculation given earlier (compare with (4.28)). The new feature is the change of $\tilde{\gamma}$ to second order in U, which reacts on n and thus on U: the resulting *coupled equations* describe a new kind of transition, first analyzed by Kosterlitz and Thouless in a different context [33].

Let us emphasize that (4.47) is an *expansion* in powers of U. In order to assess its validity we note that the cut off Λ corresponds to a number of modes per unit area

$$\frac{1}{(2\pi)^2} \pi \Lambda^2.$$

[†] In order to prove (4.49) we note that

$$h = \int_0^{\tilde{\rho}} \frac{dk}{k} [1 - J_0(k)] \implies J_0(\tilde{\rho}) = 1 - \tilde{\rho} \frac{dh}{d\tilde{\rho}}.$$

We may integrate (4.48) by parts, which gives

$$A(n) = \frac{1}{2n} \int_0^\infty \tilde{\rho}^{4-2n} \, d\tilde{\rho} \frac{d}{d\tilde{\rho}} [\tilde{\rho}^{2n} e^{-2nh}].$$

$4A(2)$ is the limit of $\tilde{\rho}^4 e^{-4h}$ as $\tilde{\rho} \to \infty$.

It follows that the effective 'unit cell' on the interface has an area $4\pi/\Lambda^2$. The expansion should be valid if the *pinning energy per unit cell* is smaller than T, i.e. if

$$4\pi U < T. \tag{4.50}$$

It is convenient to introduce reduced variables x and y defined as

$$x = \frac{2\tilde{\gamma}a^2}{\pi T} = \frac{2}{n}, \quad y = \frac{4\pi U}{T}. \tag{4.51}$$

Weak coupling means $y < 1$, while the roughening transition corresponds to $x = 1$ (see 4.29). In terms of these variables, (4.47) become

$$\frac{dy}{d\ell} = 2y\left[1 - \frac{1}{x}\right], \quad \frac{dx}{d\ell} = \frac{y^2}{2x}A\left[\frac{2}{x}\right]. \tag{4.52}$$

The only difficulty is to obtain A.

Scaling trajectories

The scaling equations (4.52) no longer depend on temperature (T has been absorbed in the reduced variables x and y). Dividing one by the other, we obtain a *separable* differential equation for the trajectory $y(x)$

$$\frac{dy}{dx} = \frac{x-1}{y}\frac{4}{A(2/x)},$$

with an obvious integral

$$y^2 = 8\int\frac{(x-1)\,dx}{A(2/x)} + C. \tag{4.53}$$

Here C is an integration constant fixed by initial conditions at $\ell = 0$. In practice, we will study only the vicinity of the fixed point $x = 1$, for which we set

$$A(2/x) \approx A(2) = A = 0.398.$$

(Away from that fixed point, y is not small, and the expansion implicit in (4.53) is highly doubtful.[†]) Equation (4.53) then reduces to

$$y^2 = \frac{4}{A}(x-1)^2 + C. \tag{4.54}$$

[†] Nevertheless, using the exact kernel $A(2/X)$ improves the fit with experiments: see [32].

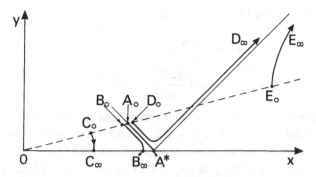

Figure 59: Scaling trajectories in the plane (x, y). The dashed line is the locus of initial points $(\ell = 0)$ as a function of temperature for a given interaction strength U_0. The critical trajectory is $A_0 A^*$.

The trajectories $y(x)$ are hyperbolae, depicted in Fig. 59.

As ℓ goes to infinity, trajectories evolve toward *fixed points*, obtained by setting the right hand side of (4.52) equal to 0. Clearly, there exists a line of fixed points, $y = 0$ for any value of x. A linear stability analysis shows that these fixed points are attractive for $x < 1$, repulsive for $x > 1$ (an obvious statement in the first equation (4.52)). *The roughening transition corresponds to the change of stability of the fixed point $y = 0$,* i.e. to $x = 1$: we recover the prediction of our naïve first order argument (which becomes exact as $y \to 0$).

The constant C is fixed by initial conditions. When $\ell = 0$ the surface stiffness and pinning potential have their original values, $\tilde{\gamma}_0$ and U_0. It follows that

$$x_0 = \frac{2\tilde{\gamma}_0 a^2}{\pi T}, \quad y_0 = \frac{4\pi U_0}{T}. \tag{4.55}$$

The ratio y_0/x_0 is constant

$$\frac{y_0}{x_0} = \frac{2\pi^2 U_0}{\tilde{\gamma}_0 a^2}.$$

Consequently, the locus of initial points as temperature varies is a straight line, shown dashed in Fig. 59. (High temperatures are close to the origin, low temperatures far to the right.) The physical properties for various temperatures can be read off the figure directly.

The critical temperature T_R corresponds to the trajectory $A_0 A^*$ in Fig. 59, i.e. to a constant $C = 0$ in (4.54). The corresponding initial conditions must be such that

$$\frac{\sqrt{A}}{2} y_0 = 1 - x_0,$$

from which we infer the values of T_R:

$$T_R = \frac{2\tilde{\gamma}_0 a^2}{\pi} \left[1 + \tfrac{1}{2} t_c \right], \tag{4.56}$$

$$t_c = \sqrt{A} \, \frac{2\pi^2 U_0}{\tilde{\gamma}_0 a^2} = \sqrt{A} \, \frac{y_0}{x_0}. \tag{4.57}$$

t_c is a dimensionless parameter which measures the *strength of pinning*. Weak coupling means $t_c < \sqrt{A} \sim 0,6$. It should be emphasized that $\tilde{\gamma}_0$ in (4.56) is the *bare* stiffness, before pinning has been taken into account: it corresponds to point A_0 in Fig. 59. The real, physical *macroscopic* stiffness $\tilde{\gamma}^*$ corresponds to the fixed point A^*, reached when the scale has been extended to infinity ($\ell \to \infty$). Since $x^* = 1$, we have

$$\tilde{\gamma}^* = \frac{\pi T_R}{2a^2},$$

in agreement with (4.29). The result (4.56) does not contradict the universal relationship between surface stiffness and roughening temperature T_R: it simply points to the fact that such a relationship involves the *macroscopic* $\tilde{\gamma}^*$, not the original $\tilde{\gamma}_0$.

At high temperatures, $T > T_R$, we start from points B_0, C_0 in Fig. 59. y scales to zero, while x goes to a limit x_∞ as $\ell \to \infty$. That limit gives the *macroscopic surface stiffness* at temperature T,

$$\tilde{\gamma}_\infty = \frac{\pi T}{2a^2} x_\infty.$$

The ratio $x_\infty / x_0 = \tilde{\gamma}_\infty / \tilde{\gamma}_0$ depends on the trajectory — and as a result $\tilde{\gamma}_\infty$ depends on T. Physically, this dependence is an effect of *mode-mode coupling* (due to the pinning potential): thermal short wavelength fluctuations affect the long wavelength restoring force. Results are conveniently expressed in terms of a reduced temperature

$$t = \frac{T - T_R}{T}.$$

Inserting the initial conditions (4.55) into (4.54), we easily find (in leading order in t, t_c)

$$C = -\frac{4}{A} t(t + t_c). \tag{4.58}$$

From this we infer the fixed point value of x

$$x_\infty = 1 - \tfrac{1}{2}\sqrt{AC} = 1 - \sqrt{t(t + t_c)}.$$

The macroscopic surface stiffness is

$$\tilde{\gamma}_{\infty} = \tilde{\gamma}^* \frac{T}{T_R} x_{\infty} = \tilde{\gamma}_0 \left[1 + t + \tfrac{1}{2} t_c\right] x_{\infty}.$$

We finally obtain

$$\tilde{\gamma}_{\infty}(T) = \tilde{\gamma}_0 \left[1 + t + \tfrac{1}{2} t_c - \sqrt{t(t + t_c)}\right]. \tag{4.59}$$

The corresponding behaviour is shown in Fig. 60. We can distinguish two ranges of temperatures:

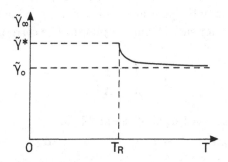

Figure 60: The macroscopic surface stiffness γ_{∞} as a function of temperature.

(i) A 'critical' regime $t \ll t_c$, for which

$$\tilde{\gamma}_{\infty} \approx \tilde{\gamma}^* \left[1 - \sqrt{tt_c}\right].$$

A typical trajectory is shown as $B_0 B_{\infty}$ in Fig. 59. It follows the separatrix closely, departing only when very close to the fixed point.

(ii) A 'normal' regime $t \gg t_c$, for which

$$\tilde{\gamma}_{\infty} \approx \tilde{\gamma}_0 \left[1 + \frac{t_c^2}{8t}\right].$$

The corresponding trajectory is shown as $C_0 C_{\infty}$ in Fig. 59: $\tilde{\gamma}$ is hardly renormalized. In a first approximation, we could take it as constant — which is nothing but the naïve first order calculation described earlier.

Two points are worth emphasizing. First the result (4.59) for $\tilde{\gamma}_{\infty}(T)$ refers to an interface *parallel* to crystal planes. Since any measurement of surface curvature necessarily implies a finite angular window, an experimental check of (4.59) is problematic — we return to this point later. Second, the calculation is only valid for *weak coupling*, $t_c < 1$: in that limit, the variations of $\tilde{\gamma}$ are relatively small ($\tilde{\gamma}^*$ is close to $\tilde{\gamma}_0$).

When $T < T_R$, the free surface fixed point is repulsive: y will eventually go to infinity, either directly (trajectory $E_0 E_\infty$ in Fig. 59, which corresponds to a 'normal' regime $|t| > t_c$), or after a long hesitation close to the critical point A^* (trajectory $D_0 D_\infty$, typical of the 'critical' regime $|t| \ll t_c$). In practice, our calculation breaks down when $y \geq 1$: the surface then '*locks*' to the crystal planes — facets appear on a macroscopic scale.

$T < T_R$: *the correlation length*

Let ℓ_c be the value of our scaling parameter ℓ for which $y \sim 1$ (the estimate of ℓ_c can only be qualitative, since our second order calculation is at best indicative for such strong coupling). The corresponding length scale is

$$\xi \sim \frac{1}{\bar{\Lambda}} \sim \frac{1}{\Lambda} e^{\ell_c}. \tag{4.60}$$

ξ acts as the *correlation length* of the problem:

▷ On scales shorter than ξ, the pinning potential has no time to build up: y remains $\ll 1$ and the interface fluctuates freely.

▷ On scales larger than ξ, fluctuations are blocked. The cost in pinning energy becomes too large to allow translation from one *average* crystal plane to the next: localized steps appear, which are the signature of a facetted state (localized step width means a *finite step energy* β, hence a cusp in $\gamma(\theta)$).

ξ thus marks the onset of facetting on large scales.[†] The crossover is of course progressive — but ξ remains the characteristic length of the problem. Note that ξ is infinite at all temperatures above T_R, since in that case the interface is always macroscopically free.

The correlation length ξ defined in (4.60) is also the width ξ_S of a step (roughly speaking, of course, since all these quantities are not sharply defined). Let $\tilde{\gamma}_{\text{eff}}$, V_{eff} be the effective parameters once scaling has reached the cut off $\bar{\Lambda} \sim 1/\xi$ (renormalized by all 'microscopic'

[†] The notion of a *free* interface on scales $\ll \xi$ is only meaningful if $\xi \gg a$, the lattice spacing. Such a condition is met at any temperature if the initial pinning is weak: $t_c \ll 1$. In a strong coupling material, $t_c \geq 1$, the interface is locked on all scales except if T is very close to T_R, in which case ξ diverges.

fluctuations, $k\xi \gg 1$). Qualitatively, we expect

$$\tilde{\gamma}_{\text{eff}} \sim \tilde{\gamma}_0, \quad V_{\text{eff}} \sim \frac{U_{\text{eff}}}{\xi^2} \sim \frac{T}{\xi^2}$$

(it is clear in Fig. 59 that $\tilde{\gamma}$ retains the same order of magnitude throughout). In cases of interest, T is of order $T_R \sim \tilde{\gamma}_0 a^2$. The resulting step width is

$$\xi_S \sim a \sqrt{\frac{\tilde{\gamma}_{\text{eff}}}{V_{\text{eff}}}}.$$

As expected, $\xi_S \sim \xi$. (The step width is the 'last attempt' at having a free surface — if ξ_S were $> \xi$, the cost in pinning energy would be unacceptable.) Once we know ξ_S, we infer the step energy

$$\beta \sim \frac{\tilde{\gamma} a^2}{\xi}.$$

Above T_R, ξ is infinite and $\beta = 0$: thermal fluctuations have killed the cusp in $\gamma(\theta)$ — facets disappear.

In order to estimate ℓ_c, we must integrate the first equation (4.52), $x(y)$ being extracted from (4.54). Since everything is dominated by the vicinity of the critical point, $x \simeq 1$, we simplify the equation somewhat:

$$\frac{dy}{d\ell} \approx 2y [x - 1] = \pm \sqrt{A}\, y \sqrt{y^2 - C}, \qquad (4.61)$$

where the sign $\pm$ depends on whether we are on the ascending or descending branch of the hyperbola in Fig. 59. The constant C is given by (4.58): it is positive in the critical regime $|t| < t_c$ (trajectory D in Fig. 59), becoming negative again when $|t| > t_c$ ('normal' regime, trajectory E).

The integration of (4.61) is straightforward. Results are simple in limiting cases:

(i) In the critical regime, $|t| \ll t_c$, the integration over y runs from y_0 down to the minimum $y_m = \sqrt{C}$, and then up again from y_m to 1. The integral is dominated by the vicinity of y_m, yielding

$$\ell_c = \frac{2}{\sqrt{A}} \int_{\sqrt{C}}^{\infty} \frac{dy}{y\sqrt{y^2 - C}} = \frac{\pi}{2\sqrt{|t| t_c}}. \qquad (4.62)$$

The corresponding expression of ξ is

$$\xi = \frac{1}{\Lambda_0} \exp\left[\frac{\pi}{2\sqrt{|t| t_c}}\right]. \qquad (4.63)$$

ξ diverges *very fast* close to T_R — and accordingly the step energy β becomes very small — albeit finite. Such a behaviour was clearly apparent in Fig. 45: it makes the determination of T_R very delicate. The result (4.63) is very unusual: it was derived for the first time by Kosterlitz and Thouless [33] in a different context.

(ii) In the normal regime, $|t| \gg t_c$, the integral runs directly from y_0 to 1:

$$\ell_c \approx \frac{1}{\sqrt{A}} \int_{y_0}^{1} \frac{dy}{y\sqrt{y^2 - C}}.$$

It is dominated by the vicinity of y_0, itself $\approx t_c/\sqrt{A}$. We thus find

$$\ell_c \approx \frac{1}{2|t|} \log \frac{4|t|}{t_c}. \qquad (4.64)$$

Altogether, ξ increases with temperature, ultimately diverging at T_R: we recover, in another guise, the thermal 'step wandering' discussed at the beginning of this chapter — but with a *completely different critical behaviour* in the vicinity of T_R: a naïve Landau description in which $\xi^{-1} \sim (T_R - T)$ is completely wrong. Looking at a single step shows the existence of a transition, but it grossly distorts the critical behaviour.

The Kosterlitz-Thouless length (4.63) results from a delicate balance in the hyperbolic trajectories of Fig. 59. As y decreases, x gets closer to 1 — and as a result y decreases more slowly — hence the very slow evolution near the minimum. The result would be quite different in a *first order* calculation in which x is not renormalized at all. Then the equation (4.62) reads

$$\frac{dy}{d\ell} = 2y \frac{x_0 - 1}{x_0}.$$

The roughening transition corresponds to $x_0 = 1$, i.e.

$$T_R = \frac{2\tilde{\gamma}_0 a^2}{\pi}.$$

Close to T_R we may expand $(x_0 - 1)/x_0 \simeq -2t$, from which we infer

$$\ell_c = \frac{1}{2|t|} \log \frac{1}{y_0}. \qquad (4.65)$$

Here again the critical behaviour near $t = 0$ is wrong (although less so than in the single step description) — basically, we find the 'normal' behaviour (4.64) at all temperatures, while we should have (4.62) when $|t| \ll t_c$.

The physical meaning of ξ is apparent if we look at *fluctuations* of the interface height z. For a free interface, we found

$$\langle z^2 \rangle = \sum_k \langle |z_k|^2 \rangle = \frac{T}{2\pi} \int \frac{dk}{\tilde{\gamma}k}. \qquad (4.66)$$

In the spirit of a renormalization scheme, each wave vector k should have its own $\tilde{\gamma}(k)$, renormalized by all shorter wavelength fluctuations. Equation (4.66) is accordingly replaced by the more refined expression

$$\langle z^2 \rangle = \frac{T}{2\pi} \int_0^{\ell_{\max}} \frac{d\ell}{\tilde{\gamma}(\ell)}, \qquad (4.67)$$

where $\ell = \log(\Lambda_0/k)$ is our usual logarithmic coordinate.

(i) Above T_R, scaling proceeds to $\ell = \infty$: $\ell_{\max}$ is limited only by the size L of the sample. We recover the logarithmically divergent fluctuations of a free surface

$$\langle z^2 \rangle = \frac{T}{2\pi\tilde{\gamma}_\infty} \log[L\Lambda_{\text{eff}}]. \qquad (4.68)$$

Note that *macroscopic* fluctuations dominate: the coefficient in (4.68) is controlled by the *macroscopic* stiffness $\tilde{\gamma}_\infty$. The cut off Λ_{eff} depends on the early stages of scaling.

(ii) Below T_R, scaling stops on a scale ξ: $\ell_{\max}$ is roughly the value ℓ_c for which the interface locks ($y \sim 1$). We thus find

$$\langle z^2 \rangle \sim \frac{T}{2\pi\tilde{\gamma}} \log[\Lambda_0\xi]. \qquad (4.69)$$

Fluctuations are finite — albeit large if ξ is large.

The fact that fluctuations are *finite* below T_R is crucial: it allows a non-zero average periodic potential V — hence locking of the interface on a macroscopic scale, with all its physical consequences: steps, facetting, etc.

Vicinal surfaces: interrupted renormalization and blurring of the transition

Consider an interface tilted by a small angle θ with respect to crystal planes. We change our notations somewhat, and we choose the z-axis normal to the *interface*, not to the crystal planes, as shown in Fig. 61. The periodic pinning potential is

$$V \cos\left[\frac{2\pi}{a}(z + \theta x)\right].$$

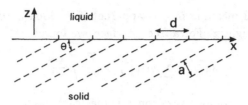

Figure 61: Renormalization on a vicinal surface.

It oscillates *along* the interface, with a period $d = a/\theta$ which was the step distance in our former language. On any scale larger than d, V averages to zero for purely geometric reasons, without invoking thermal fluctuations of z.

Let us again carry our renormalization program: degrees of freedom are eliminated progressively, down to a momentum cut off $\overline{\Lambda}$. Put another way, the interface is *coarse grained* on a length scale $\sim 1/\overline{\Lambda}$. This length scale should be compared to d: the physics is very different in the two opposite limits.

(i) If $\overline{\Lambda}d \gg 1$, the interface tilt does not matter much. On a scale $\overline{\Lambda}^{-1}$, it simply acts to shift the *local* origin of z. The renormalization of V and $\tilde{\gamma}$ proceeds as before, and the scaling equations (4.47) are essentially unaffected by the small angle θ.

(ii) If $\overline{\Lambda}d \ll 1$, the situation is completely different. The periodic potential averages to zero automatically, and consequently the effective V quickly drops to zero. As a result, the renormalization of $\tilde{\gamma}$ stops: $\tilde{\gamma}$ is more or less *frozen* at the value it reached when $\overline{\Lambda} \sim 1/d$.

In between, there is a *smooth* crossover — which is nevertheless fairly rapid on the logarithmic scales usual in renormalization. A few periods d are enough to kill the *average* pinning potential V: in terms of our logarithmic variable $\ell = \log(\Lambda/\overline{\Lambda})$, scaling stops in a range $\Delta\ell \sim 1$. For a very small tilt θ, that range is narrow compared to ℓ itself.

Consequently, it is not unreasonable to try a naïve 'yes or no' approach in order to estimate $\tilde{\gamma}$:

▷ We carry renormalization ignoring tilt, up to

$$\ell_m = \log(\Lambda_0 d) \sim \log(1/\theta).$$

▷ We stop it abruptly at ℓ_m: $\tilde{\gamma}$ remains constant.

Using the scaling equations (4.47), we obtain in this way an 'effective' stiffness $\tilde{\gamma}(\ell_m)$: $\tilde{\gamma}$ is a function of $|\log\theta|$. In principle, the calculation has

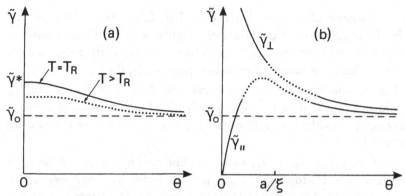

Figure 62: The surface stiffness as a function of angle. (a) When $T \geq T_R$: the evolution is logarithmic. (b) When $T < T_R$: the dotted region is an interpolation between an interrupted renormalization approach at large angles, a step description at small angles. The crossover corresponds to $\theta - a/\xi$ (step overlap).

logarithmic accuracy: the leading dependence on $\log \theta$ should be properly accounted for. An obvious failure is the absence of any anisotropy: such an *interrupted renormalization scheme* gives a single $\tilde{\gamma}$, in contradiction with (2.25) which predicted different values for the longitudinal and transverse stiffnesses $\tilde{\gamma}_\parallel$ and $\tilde{\gamma}_\perp$ (modulations parallel or perpendicular to the x-axis). More generally, the thermodynamic relationships (2.25) between $\tilde{\gamma}_\parallel$, $\tilde{\gamma}_\perp$ and the surface *energy* γ are not obeyed. Hopefully, the corresponding discrepancies are of higher order whenever scaling is valid — indeed they are!

Before discussing more technical aspects of this calculation, let us qualitatively examine the physical consequences of such an approach. We first consider the case $T \geq T_R$: scaling is free to proceed up to ℓ_m. It is clear from Fig. 59 that the larger ℓ_m, the larger $\tilde{\gamma}$: the surface stiffness γ *increases* as the tilt angle θ gets smaller. The exact form of $\tilde{\gamma}(\ell_m)$ is obtained by integrating (4.52). The calculation is simple just at the transition:

$$T = T_R \implies \tilde{\gamma}(\ell_m) = \tilde{\gamma}_0 \left[1 + \frac{t_c^2 \ell_m}{2(1 + t_c \ell_m)} \right]. \qquad (4.70)$$

As surmised, $\tilde{\gamma}$ depends on $\log \theta$. It goes from the bare $\tilde{\gamma}_0$ at large angle θ to the universal $\tilde{\gamma}^*$ at $\theta = 0$. When $T > T_R$, a similar calculation shows that

$$\tilde{\gamma}_\infty - \tilde{\gamma} \sim \theta^{4\sigma} \quad \text{with } \sigma \sim \begin{cases} \sqrt{t t_c} & \text{if } t \ll t_c, \\ t & \text{if } t \gg t_c. \end{cases}$$

The behaviour of $\tilde{\gamma}(\theta)$ is sketched in Fig. 62a. From $\tilde{\gamma}(\theta)$ we may infer the equilibrium shape in a *finite* angular window, thereby allowing comparison with experiment. The detailed result (4.70) is necessary in order to check the universal curvature at $T = T_R$, $\theta = 0$.

Let us now consider low temperatures, $T < T_R$: the cut off ℓ_m should be compared to ℓ_c — or equivalently the step distance d should be compared to the correlation length ξ — hence two clear cut limiting cases:

(i) $d \ll \xi$ (i.e. $\ell_m < \ell_c$, corresponding to *large* angles $\theta \gg a/\xi$). The growth of U stops at ℓ_m: the pinning potential never reaches the value $\sim T$ needed in order to lock the interface. The latter is therefore free, despite the fact that $T < T_R$. The interrupted renormalization scheme remains valid (U is always small): we can use it to obtain the surface stiffness $\tilde{\gamma}(\theta)$, with the same provisos that we had when $T > T_R$. $\tilde{\gamma}(\theta)$ grows *logarithmically* as θ goes down. As θ approaches a/ξ (i.e. as ℓ_m approaches ℓ_c), $\tilde{\gamma}$ increases appreciably, as seen from Fig. 59: the separatrix reaches $y = 1$ for $x \sim 2$, corresponding to $\tilde{\gamma} \sim 2\tilde{\gamma}^*$.

(ii) $d \gg \xi$ (i.e. $\ell_m > \ell_c$, corresponding to small angles θ). Then the interface is *locked* before it has a chance to feel the tilt angle θ: the renormalization approach breaks down. Physically, we enter the *vicinal regime*, characterized by *well separated steps*, with a distance d much larger than their width (in contrast, steps largely overlap when $d \ll \xi$: they practically disappear). The appropriate description is in terms of *step equilibrium*, as discussed in Section 2. According to (2.26), the longitudinal and transverse stiffnesses should behave very differently:

$$\tilde{\gamma}_\| = \frac{6\phi\theta}{a^3}, \quad \tilde{\gamma}_\perp = \frac{\beta}{a\theta}.$$

The anisotropy of $\tilde{\gamma}$, which was small in the renormalization regime, becomes dramatic when $\theta \ll a/\xi$. (Note that $\tilde{\gamma}_\|$ and $\tilde{\gamma}_\perp$ both return to a value $\sim \tilde{\gamma}_0$ when $\theta \sim a/\xi$.)

In between there is a crossover region where neither the renormalization group nor the phenomenological step picture work: in that region we have no theory at all. We know that $\tilde{\gamma}$ should be a *universal* function of a single variable $\theta\xi/a$ (at least when $\xi \gg a$) — but this function can only be approached from both sides. In the middle we are left with an 'artist's touch interpolation,' which is reasonable but not quantitative. The corresponding behaviour of $\tilde{\gamma}(\theta)$ is sketched in Fig. 62b.

The important physical result to be remembered is the gradual crossover at $\theta \sim a/\xi$: *the interface roughens at large angles*. For a given value of θ, it remains rough below T_R, down to a temperature T such that $\xi(T) \sim a/\theta$. The corresponding transition is *blurred*, in much the same way as a ferromagnetic transition is blurred in a non-zero external magnetic field. In a sense, *tilting the interface kills the roughening transition*, which becomes a mere crossover somewhere below T_R.

We now add a few more technical comments on renormalization at finite θ, without going into any detail: we only want to give a rough idea of how the calculation goes on (see [32] for a more complete discussion).

(i) First of all, we must care for anisotropy of $\tilde{\gamma}$: even if there was none to start with, it will be generated by scaling. The gaussian fluctuations of z become

$$\langle z^2 \rangle = \sum_{\mathbf{k}} \frac{T}{\tilde{\gamma}_{\parallel} k_x^2 + \tilde{\gamma}_{\perp} k_y^2}.$$

Carrying an angular average, we see that $\tilde{\gamma}$ is replaced by $\sqrt{\tilde{\gamma}_{\parallel} \tilde{\gamma}_{\perp}}$ in (4.38). This is the only change in first order in V.

(ii) In second order, scaling used to generate a term proportional to $\cos[2\pi(z'-z)/a]$, from which we extracted the renormalization of $\tilde{\gamma}$. Here this term is transformed into

$$\cos\left[\frac{2\pi}{a}(z' - z + \theta\rho_x)\right],$$

where $\rho = \mathbf{r}' - \mathbf{r}$ as before. As a result, an additional factor $\cos(2\pi\theta\rho_x/a)$ enters the integral in (4.46). Since that factor breaks rotational invariance in the xy-plane, we should write the scaling in a more general tensor form:

$$\delta\tilde{\gamma}_{ij} = \frac{2\pi^3 V^2}{\tilde{\gamma}a^4} \varepsilon \int \rho_i \rho_j \, \mathrm{d}^2\rho \, J_0(\Lambda\rho) \mathrm{e}^{-2nh(\rho)} \cos\left[\frac{2\pi\theta\rho_x}{a}\right]. \qquad (4.71)$$

$\delta\tilde{\gamma}_{\parallel}$ involves ρ_x^2, $\delta\tilde{\gamma}_{\perp}$ involves ρ_y^2. We may carry the angular average over ρ. We thus recover the former scaling equation (4.46) with an additional factor $K[2\pi\theta\rho/a]$ due to the tilt, the function K being

$$K(u) = \begin{cases} J_1'(u) & \text{for } \tilde{\gamma}_{\parallel}, \\ J_1(u)/u & \text{for } \tilde{\gamma}_{\perp}. \end{cases} \qquad (4.72)$$

This factor oscillates for $u \gg 1$: it kills the integral if

$$\rho \sim \frac{1}{\Lambda} \gg \frac{a}{\theta} = d.$$

The renormalization of $\tilde{\gamma}$ then stops: this is just the mechanism described qualitatively before.[†]

(iii) We may carry the same machinery for the surface *energy* γ (i.e. the *constant* term in the free energy $\bar{F}(\bar{z})$. The change $\delta\gamma$ in an infinitesimal scaling is given by an improved form of (4.42). It is easily verified [32] that

$$\delta\tilde{\gamma}_{\parallel} = \frac{\partial^2}{\partial\theta^2}[\delta\gamma], \quad \delta\tilde{\gamma}_{\perp} = \frac{1}{\theta}\frac{\partial}{\partial\theta}[\delta\gamma]. \tag{4.73}$$

We recover for the increments $\delta\tilde{\gamma}$ the general thermodynamic relationships (2.25) — with one error: the 'stretching term' γ in the stiffnesses has been lost on the way! The error stems from the approximation made at the very beginning of the calculation:

$$\sqrt{1+h^2} \implies 1 + \tfrac{1}{2}h^2.$$

Doing that, we neglect terms of order h^4, which also yield *mode-mode coupling*. They are precisely responsible for the 'stretching term' γ in (2.25). In practice, this term is small except at very early stages in scaling. Comparing (4.42) and (4.43), we see that

$$\frac{\delta\gamma}{\delta\tilde{\gamma}} \sim \frac{1}{(\bar{\Lambda}a)^2}.$$

Once $\bar{\Lambda} \ll \Lambda_0$, $\delta\gamma$ is only a very small correction. We neglect it and we write

$$\tilde{\gamma}_{\parallel} = \gamma + \frac{\partial}{\partial\theta}[\theta(\tilde{\gamma}_{\perp} - \gamma)] \approx \tilde{\gamma}_{\perp} + \theta\frac{\partial\tilde{\gamma}_{\perp}}{\partial\theta}. \tag{4.74}$$

In terms of our logarithmic variable ℓ_m, (4.74) reads

$$\tilde{\gamma}_{\parallel} = \tilde{\gamma}_{\perp} - \frac{\partial\tilde{\gamma}_{\perp}}{\partial\ell}. \tag{4.75}$$

It is clear from (4.75) that $\tilde{\gamma}_{\parallel} < \tilde{\gamma}_{\perp}$ — a feature we incorporated in Fig. 62b. Also $\partial\tilde{\gamma}_{\perp}/\partial\ell_m \sim \tilde{\gamma}_{\perp}/\ell_m$: as predicted, the *anisotropy* $(\tilde{\gamma}_{\perp} - \tilde{\gamma}_{\parallel})$ is of higher order in the logarithmic variable ℓ_m.

[†] Note that renormalization also generates a term of order dz/dx in the free energy (the x-symmetry is broken by θ). Such a term reflects the first derivative $d\tilde{\gamma}/d\theta$. It is irrelevant since it integrates to zero when the ends of the interface are fixed.

If needed, we could calculate the integral (4.71) more accurately. It would not make much sense in view of the many approximations made throughout our renormalization program. We should rather stick to a *simple* formulation:

(i) use interrupted scaling in order to calculate *one* stiffness, say $\tilde{\gamma}_\perp$;

(ii) infer the other $\tilde{\gamma}_\parallel$ from (4.75), thereby ensuring the internal consistency of the calculation.

In this way, we obtain a reasonable qualitative description of *rough vicinal surfaces*.

4.4 Dynamic renormalization

The preceding calculation was *static*, referring only to thermal equilibrium. We now extend it to dynamical phenomena, involving *motion* of the interface. Our motivation is two fold:

▷ Dynamic scaling is in a sense more natural — and it ensures fully *convergent* equations, even below T_R.

▷ We want to show that an applied supercooling F *blurs* the transition, in much the same way as an interface tilt θ. Such a *dynamic roughening* of the interface shows up as a non-linear growth law $u(F)$, which eventually will evolve smoothly into homogeneous nucleation.

Our starting point is the equation of motion of the interface,

$$\eta \dot{z} = \tilde{\gamma} \nabla^2 z + \Phi + R + \frac{F}{a}. \tag{4.76}$$

η is the *bare* friction coefficient which describes microscopic surface dissipation. The various 'chemical forces' acting on the interface are

▷ a bending force $\tilde{\gamma} \nabla^2 z$;

▷ a pinning force $\Phi = -\dfrac{\partial V}{\partial z} = -\dfrac{2\pi V}{a} \sin\left(\dfrac{2\pi z}{a}\right)$;

▷ a thermal noise force $R(t)$ which we characterize by a white noise spectrum

$$\langle R_k(t) R_k(t') \rangle = G_k \delta(t - t'); \tag{4.77}$$

▷ a supercooling force per unit area F/a (our definition of F is the same as in Section 3). In a first stage, we set $F = 0$.

Since friction implies a finite memory, (4.76) yields the *retarded* displacement $z(t)$ in terms of all forces at earlier times $t' < t$.

Let us first ignore pinning and supercooling. The solution of (4.76) is then straightforward:

$$z(\mathbf{r}, t) = \int_{-\infty}^{t} \mathrm{d}t' \int \mathrm{d}^2 r' \chi_0(\mathbf{r} - \mathbf{r}', t - t') R(\mathbf{r}', t'). \qquad (4.78)$$

$\chi_0(\rho, \tau)$ is the usual response function of a diffusion equation,

$$\begin{aligned}
\chi_0(\rho, \tau) &= \frac{1}{4\pi\tilde{\gamma}\tau} \exp\left[-\frac{\eta\rho^2}{4\tilde{\gamma}\tau}\right], \\
\chi_0(k, \tau) &= \frac{1}{\eta} \exp\left[-\frac{\tilde{\gamma}k^2\tau}{\eta}\right].
\end{aligned} \qquad (4.79)$$

For a given *spatial* scale ρ, the relaxation *time scale* is

$$\tau \sim \frac{\eta\rho^2}{\tilde{\gamma}}. \qquad (4.80)$$

Equation (4.78) must produce the equilibrium fluctuations (4.24) at temperature T: it follows that the noise amplitude G_k must obey the fluctuation-dissipation theorem

$$G_k = 2\eta T. \qquad (4.81)$$

The only dynamical parameter is η.

In order to treat mode-mode coupling, we need a *cut off* Λ in momentum space. In the static formulation, this was done on the displacement z itself: Fourier components z_k with $k > \Lambda$ were simply dropped out. Mathematically, that may be viewed as an extreme non-linearity in the stretching energy

$$\sum_{\mathbf{k}} \tfrac{1}{2}\tilde{\gamma}_k \langle|z_k|^2\rangle.$$

We set $\tilde{\gamma}_k$ infinite for $k > \Lambda$: fluctuations are frozen. Here we take a different course: we *cut off the random force* R_k. Accordingly, G_k is given by (4.81) if $k < \Lambda$, and it is zero when $k > \Lambda$. Qualitatively, it does not matter much whether we cut off z_k or R_k — still, the two approaches are different, and they will yield slightly different scaling trajectories.

The renormalization program is now straightforward. The displacement $z[\mathbf{r}, t]$ is a given functional of the random force $R(\mathbf{r}', t')$ at previous times. In zeroth order in V, it is given by (4.78). In higher orders, it becomes very complicated — still it is well defined:

$$z(\mathbf{r}, t) = z\big[R(\mathbf{r}', t')\big].$$

We separate the random force into two parts

$$R = \bar{R} + \delta R.$$

The long wavelength part $\bar{R}$ contains all wave vectors below a new cut off $\bar{\Lambda}$, while δR corresponds to $\bar{\Lambda} < k < \Lambda$. We then average over the noise δR, thereby defining

$$\bar{z}(r, t) = \langle z[\bar{R} + \delta R] \rangle_{\delta R}. \tag{4.82}$$

Due to mode-mode coupling, $\bar{z}$ contains terms of order δR^2, δR^4 which give non-trivial contributions. We want to construct an equation of motion for $\bar{z}$, which will be cast in the form (4.76) with new coefficients $\bar{\eta}$, $\bar{\gamma}$, $\bar{V}$: the problem has been *renormalized* through the elimination of δR.

We will encounter the same difficulties as in the static case, and the structure of the calculation will be the same: the only new feature is the presence of a *time* variable together with position **r**.

Scaling equations

These were derived for the first time by Chui and Weeks [34] who gave all important qualitative answers. We only sketch the argument, referring to [32] for a more detailed discussion. Let us set

$$\delta z = z - \bar{z}.$$

Averaging (4.76) over δR, we find

$$\eta \dot{\bar{z}} = \tilde{\gamma} \nabla^2 \bar{z} + \bar{R} + \langle \Phi \rangle_{\delta R}. \tag{4.83}$$

The equation for δz is obtained by difference with (4.76):

$$\eta \delta \dot{z} = \tilde{\gamma} \nabla^2 \delta z + \delta R + \Phi - \langle \Phi \rangle_{\delta R}. \tag{4.84}$$

We are concerned with *infinitesimal* scaling ($\bar{\Lambda}$ very close to Λ). δz is small and we can expand (4.83) and (4.84):

$$\delta \Phi = \Phi - \langle \Phi \rangle_{\delta R} = -\frac{4\pi^2 V}{a^2} \cos\left(2\pi \frac{\bar{z}}{a}\right) \delta z + \cdots, \tag{4.85}$$

$$\langle \Phi \rangle_{\delta R} = -\frac{2\pi V}{a} \sin\left(\frac{2\pi \bar{z}}{a}\right) \left[1 - \frac{2\pi^2}{a^2} \langle \delta z^2 \rangle + \cdots \right]. \tag{4.86}$$

At this stage, we have made no approximation.

We now expand in powers of V — which amounts to solving (4.84) by iteration. We write

$$\delta z(\mathbf{r},t) = \int d^2\mathbf{r}' \int_{-\infty}^{t} dt' \chi_0(\mathbf{r}-\mathbf{r}', t-t')$$
$$\times [\delta R(\mathbf{r}',t') + \delta\Phi(\mathbf{r}',t')]. \tag{4.87}$$

The zeroth order $\delta z^{(0)}$ is the response to δR. The first order $\delta z^{(1)}$ is the response to $\delta\Phi$ with δz replaced by $\delta z^{(0)}$ in (4.85). We carry the resulting δz into $\langle\Phi\rangle$ and we average over δR: (4.83) thus appears as a power series in V. The algebra is a little heavy — but completely straightforward.

In first order in V, we set $\delta z^2 = \delta z^{(0)2}$ in (4.86). V is reduced by a small amount δV: we recover (4.38) with no change. In next order we set $\delta z^2 = 2\delta z^{(0)}\delta z^{(1)}$ — then the algebra is different. Collecting all factors, we find

$$\langle\Phi^{(2)}(\mathbf{r},t)\rangle = -\frac{32\pi^5 V^2}{a^5} \int d^2\mathbf{r}' \int_{-\infty}^{t} dt' \chi_0(\mathbf{r}-\mathbf{r}', t-t')$$
$$\times \sin\left[\frac{2\pi}{a}\bar{z}(\mathbf{r},t)\right] \cos\left[\frac{2\pi}{a}\bar{z}(\mathbf{r}',t')\right] \langle\delta z^{(0)}(\mathbf{r},t)\delta z^{(0)}(\mathbf{r}',t')\rangle. \tag{4.88}$$

The *retarded* correlation function of $\delta z^{(0)}$ is easily evaluated:

$$\langle\delta z^{(0)}(\mathbf{r},t)\delta z^{(0)}(\mathbf{r}',t')\rangle = \frac{T}{2\pi\tilde{\gamma}}\varepsilon J_0(\Lambda\rho)e^{-\tilde{\gamma}\Lambda^2\tau/\eta}. \tag{4.89}$$

Equation (4.88) is still exact to order V^2.

We now carry the machinery set up in the static case:

(i) We transform the product of sin and cos into terms

$$\sin\left[\frac{2\pi}{a}(\bar{z}' \pm \bar{z})\right].$$

We drop the combination $(\bar{z} + \bar{z}')$ which is an irrelevant harmonic.

(ii) In the remaining term, we single out one *slowly varying* component of $\bar{z}$, and we average over the remaining equilibrium distribution:

$$\sin\left[\frac{2\pi}{a}(\bar{z}' - \bar{z})\right] \implies \frac{2\pi}{a}\left[\frac{\rho^2}{4}\nabla^2\bar{z} + \tau\dot{\bar{z}}\right]\left\langle\cos\left[\frac{2\pi}{a}(\bar{z}' - \bar{z})\right]\right\rangle. \tag{4.90}$$

The term $\sim \nabla^2\bar{z}$ renormalizes $\tilde{\gamma}$, the term $\sim \dot{\bar{z}}$ renormalizes η. The average in (4.90) may be written as

$$\exp[-2n\,h(\rho,\tau)],$$

where $h(\rho, \tau)$ is a retarded version of (4.44):

$$
\begin{aligned}
h(\rho, \tau) &= \frac{\pi\tilde{\gamma}}{T}\Big\langle \big[\bar{z}(\mathbf{r}, t) - \bar{z}(\mathbf{r} - \boldsymbol{\rho}, t - \tau)\big]^2 \Big\rangle \\
&= \int_0^\Lambda \frac{\mathrm{d}k}{k}\,[1 - J_0(k\rho)]\,\exp\Big[-\frac{\tilde{\gamma}k^2\tau}{\eta}\Big].
\end{aligned}
\tag{4.91}
$$

The rest is plain algebra. We obtain the same scaling equations (4.47), with a kernel $A(n)$ which differs from (4.48). Setting $\tilde{\rho} = \Lambda\rho$ and $x = \gamma\tau/\eta\rho^2$,

$$
A(n) = n\int_0^\infty \tilde{\rho}^3\,\mathrm{d}\tilde{\rho}\int_0^\infty \frac{\mathrm{d}x}{x}\,\mathrm{e}^{-1/4x}\mathrm{e}^{-2nh}J_0(\tilde{\rho})\mathrm{e}^{-\tilde{\rho}^2 x}.
\tag{4.92}
$$

A similar scaling equation holds for the friction coefficient η:

$$
\frac{\mathrm{d}\eta}{\mathrm{d}\ell} = \frac{8\pi^4}{\tilde{\gamma}a^4}U^2 B(n)\frac{\eta}{\tilde{\gamma}},
\tag{4.93}
$$

where $B(n)$ is given by (4.92) with an additional factor x in the integral. Our dynamic renormalization program is thus completed.

The kernels $A(n)$ given by (4.48) and (4.93) are different, a fact which is not surprising: static and dynamic renormalization involve different approaches. One may show that $A(2)$ is the same in both methods: the behaviour at fixed point is *universal*. Away from $n = 2$, we lose universality. Equation (4.92) remains however a *consistent* expression of $A(n)$, which helps in fitting experiment. (In contrast to the static case, integrals converge for any n.) The resulting behaviour of $A(n)$ is shown in Fig. 63: it is quite different from the usual 'logarithmic approximations' found in the literature.

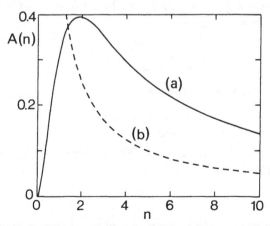

Figure 63: The kernel $A(n)$ in (4.47). (a) From the 'exact' equation (4.92). (b) From a logarithmic approximation [34].

Dynamical scaling does not change the physics. Trajectories in the $(\tilde{\gamma}, U)$-plane are the same — the only new feature is the scaling of η. Comparing (4.47) and (4.93), we see that

$$\frac{d\eta}{d\tilde{\gamma}} = \frac{\eta}{\tilde{\gamma}} \frac{4B(n)}{A(n)}. \tag{4.94}$$

Above T_R, we can scale ℓ to infinity. The *macroscopic* friction η_∞ is a function of temperature, which remains *finite* at the transition T_R: the dynamical behaviour is 'normal.' Below T_R, η is meaningless: a facet forms which does not grow except by homogeneous nucleation — the *linear* mobility is zero (i.e. $\eta = \infty$). The behaviour of $\eta(T)$ is sketched in Fig. 64.

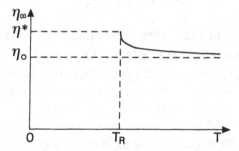

Figure 64: The microscopic friction coefficient η_∞ as a function of temperature.

Effect of an applied force: crossover to nucleation

We now restore the supercooling force F. In the absence of pinning, the interface moves with a velocity

$$u = \frac{F}{a\eta}.$$

The periodic potential is accordingly

$$V \cos\left[\frac{2\pi}{a}(z + ut)\right]. \tag{4.95}$$

Crystal planes *sweep* by the interface. The potential (4.95) oscillates in time, with a period

$$t_0 = \frac{a}{u}$$

which should be compared to the 'memory' of relaxation kernels. The situation is similar to that encountered for vicinal surfaces: there the pinning potential was modulated by x, with a characteristic scale d. Here it is modulated by t, with a scale t_0. The consequences are the same: *blurring* of the transition at T_R, and *roughening* of the interface below T_R.

We could formulate the problem mathematically as we did for vicinal surfaces. We will not do it, since in the end the 'interrupted scaling' approach relies only on qualitative arguments. When the cut off is $\overline{\Lambda}$, the spatial scale is $\rho \sim 1/\overline{\Lambda}$, and the corresponding time scale is

$$\tau = \frac{\eta}{\tilde{\gamma}}\overline{\Lambda}^2.$$

τ is the range of time integrations in (4.92). It should be compared to the sweeping time t_0:

(i) If $\tau \ll t_0$, the motion of the interface does not matter much: scaling proceeds as if F were zero.

(ii) If $\tau \gg t_0$, sweeping of crystal planes during a correlation time τ washes out any pinning potential: V averages to zero.

Just as we did for vicinal surfaces, we use an extreme 'yes or no' approximation. Scaling is carried out ignoring F up to a maximum ℓ_m for which $\tau = t_0$:

$$\Lambda_m = \sqrt{\frac{\eta u}{\tilde{\gamma} a}} \implies \ell_m = \log \frac{\Lambda_0}{\Lambda_m}. \tag{4.96}$$

Beyond ℓ_m, it stops *abruptly*: $\tilde{\gamma}$ and η are *frozen* to the value they had at ℓ_m.

Let us first consider the case $T > T_R$. Λ_m depends on the applied supercooling F,

$$\Lambda_m \approx \sqrt{\frac{F}{\tilde{\gamma}a^2}}. \tag{4.97}$$

As a result, the effective friction coefficient η_{eff} depends on F (via ℓ_m): the response $u(F)$ is *non-linear*. Detailed calculations are straightforward using the dynamical scaling equations (4.47) and (4.94). Qualitatively, the larger F, the smaller ℓ_m, the smaller η: the ratio u/F increases for large F.

Consider now the case $T < T_R$: the cut off ℓ_m should be compared to ℓ_c — i.e. the length $1/\Lambda_m$ should be compared to the correlation length ξ.

(i) Close to T_R, $\Lambda_m \xi \gg 1$: scaling stops *before* the onset of pinning. (V is still small when $\overline{\Lambda} = \Lambda_m$.) The interface remains rough: this is the so-called '*dynamical roughening*' due to interface drift (in contrast to static roughening due to interface tilt). The interrupted scaling calculation of $\eta(\ell_m)$ is still reliable: we can thus extend the non-linear characteristics $u(F)$ *below* T_R.

(ii) At lower temperatures, $\Lambda_m \xi \ll 1$: the interface ceases to grow freely well before the scale Λ_m. In such a case, growth can only occur through *homogeneous nucleation*: in order to maintain the velocity u which was assumed at the outset, we need a very large force F. Scaling is useless — we should resort instead to a phenomenological description of nucleation, in much the same way as we used a 'step picture' for vicinal surfaces.

The crossover between the two regimes is *smooth* (there is no sharp transition as soon as $F \neq 0$), and it occurs when $\Lambda_m \xi \sim 1$, i.e.

$$F_c \sim \frac{\tilde{\gamma} a^2}{\xi^2}. \tag{4.98}$$

Expression (4.98) is identical to our former result (3.13). This is hardly surprising since (3.13) characterized *unlocking* of the facet due to the bias F. F_c marks the transition between *free* behaviour for $F \gg F_c$ and *facet* formation for $F \ll F_c$. In (3.13) we approached it from the low F side, while in (4.98) we start from the free side. It is a comforting result that we do find the same F_c from both ends. (Note that F_c is such that ξ is comparable to the nucleation radius r_c: the crossover also corresponds to the disappearance of well defined critical nucleation germs.)

To sum it up, a surface subject to a finite supercooling F remains rough below T_R, down to a temperature T such that $\xi(T) \sim [\tilde{\gamma} a^2 / F]^{1/2}$. The corresponding 'mobility' u/F as a function of temperature T is sketched in Fig. 65 for various forces F: non-linearities are clearly apparent. We have a reliable theory close to T_R (interrupted scaling), or well below (homogeneous nucleation). In between, we have nothing: the dotted curve interpolations are pure guess work! Nevertheless, the picture is consistent: high temperature non-linearities are the premises of homogeneous nucleation.

A non-linear behaviour similar to Fig. 65 has been observed in ^{4}He, close to the roughening transition of the (0001) facet [35]. It allowed detailed comparison between experiment and theory. The theoretical

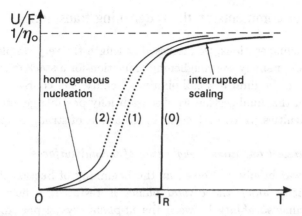

Figure 65: The interface mobility u/F as a function of temperature T for various supercoolings F. The curve (0) corresponds to an infinitesimal F, the curves (1) and (2) to increasing values of F. The dotted part interpolates between calculations at low and high T.

parameters were T_R, t_c and the original cut off Λ_0 (which enters in the expression of ξ). The experimental data were three fold:

▷ curvature measurements near $\theta = 0$;

▷ homogeneous nucleation growth, shown in Figs. 44 and 45;

▷ non-linearities of u/F.

Using our interrupted scaling description, one may fit all these data with the choice

$$T_R = 1.28\text{K}, \quad t_c = 0.63, \quad \Lambda_0^{-1} = 2.6\text{Å}.$$

The value of Λ_0 is reasonable. The measured t_c is surprisingly large: it means that the original pinning of the interface is of *medium strength*, while we expected it to be small for a solid-liquid interface. This result needs more thought — and experimental confirmation on other roughening transitions (there is some evidence that the roughening transition observed around 0.1K in ^{3}He corresponds to a much weaker coupling). Here we only want to emphasize one point: a *single* experiment does not allow a reliable analysis of roughening transitions. It is only the *combination* of several different phenomena that leads to a consistent theoretical picture.

4.5 Further comments on the roughening transition

In the preceding sections, we discussed at length the very simplest physical situation, namely the roughening transition for a stack of structureless plates, in the limit of weak pinning. Reality is of course more complicated. In this final section we discuss briefly possible generalizations
— and difficulties which may occur in situations of strong coupling.

Commensurate transitions: roughening of vicinal surfaces

The issue was briefly addressed at the beginning of Section 2. When
the periodicity *along the crystal planes* is restored, a new problem
arises: *commensurability* between the *in-plane* crystal periodicity and
the step average distance d (fixed by the tilt angle θ of the interface). A
commensurate structure is nothing but a facet with higher Miller indices.
There is an infinite number of such facets: the problem is mathematically
very complicated, leading to a hierarchy of transitions known as the
'devil's staircase' in the context of commensurate transitions.

The simplest situation is the $(1,0,N)$ facet, in which steps are parallel
to the crystal y-symmetry axis, separated by a distance $d = Nb$ where b
is the crystal x-period and N an integer. Starting from this 'primary
structure,' we may build localized 'solitons' by changing the distance
of two consecutive steps to $(N \pm 1)b$. These solitons can in turn form
a periodic lattice on the surface, etc. In this way, one generates any
facet $(p,0,q)$. One may also tilt the steps by introducing a finite
density of kinks: if these kinks form a commensurate structure in the
y-direction, they produce a (p,q,r) facet. Deciphering the hierarchy of
these transitions is a field in itself!

Here, we only look at the very first stage: can we estimate the
roughening temperature of a $(1,0,N)$ facet? The issue is of experimental
interest, as roughening of vicinal surfaces has been extensively studied by
ion diffraction techniques [36]. A detailed theory has been put forward
by Villain *et al.* [37]. We only give a brief qualitative sketch of the
argument, as it raises a fundamental question — see [37] for details.

Start from a regular array of steps. In order to 'roughen' the corresponding facet, we need infinite fluctuations of the interface — i.e. *infinite fluctuations of the step* in the x-direction. This in turn implies
free fluctuations of the 'kink charge density' (4.15). If steps were independent, such fluctuations would always exist (see (4.19)): the vicinal
surface would always be rough. Facetting is only possible because of *step*

interaction, which produces the *confinement potential* (4.22) between $+$ and $-$ kinks. The situation is reminiscent of a *metal-insulator transition*:

▷ At very low temperature, thermal agitation cannot overcome the confining potential. The $\pm$ kink gas is a 'dielectric,' in which $\pm$ pairs remain attached. There is no global *charge* fluctuation — hence no appreciable excursion of the step.

▷ If the temperature increases, $\pm$ kink pairs ionize: they form a 'conducting plasma' which allows large charge fluctuations.

The roughening temperature T_R must therefore arise from step interaction. It will be small if the Miller index N is large.

Following Villain *et al.* [37], we consider the geometry of Fig. 58, in which a length ℓ of step has moved by one lattice spacing b between two rigid parallel steps. The cost of this displacement is

$$2\varepsilon + W\ell,$$

where ε is the kink energy and $W = b^2 V''$ the confining potential (4.22) per unit length. It is convenient to use an atomic picture: the lattice spacing in the y-direction is c, the number of displaced atoms is $n = \ell/c$. We may calculate the probability P that a *given* site on the step has been shifted: it then belongs to the segment of Fig. 58, with any number of shifted sites above (n_1) or below (n_2). In equilibrium, the liquid is a reservoir of atoms at no cost: the probability is a plain Boltzmann factor, and

$$P \sim \sum_{n_1,n_2=0}^{\infty} e^{-\beta(2\varepsilon + Wc(n_1+n_2+1))} = \frac{e^{-2\beta\varepsilon - \beta Wc}}{\left[1 - e^{-\beta Wc}\right]^2}. \qquad (4.99)$$

The roughening transition occurs when the step 'unlocks,' i.e. when $P \sim 1$ — hence an estimate of T_R.

W is small for high index facets (steps are far away). We then have $Wc \ll T \ll \varepsilon$. We may expand (4.99), which yields a condition for T_R:

$$T_R \sim Wc \exp\left[\frac{\varepsilon}{T_R}\right]. \qquad (4.100)$$

Qualitatively, we find

$$T_R \sim \frac{\varepsilon}{\log[\varepsilon/Wc]}. \qquad (4.101)$$

Expression (4.101) clearly shows the ingredients needed in order to produce a vicinal facet:

▷ kinks must have a finite (free) energy $\varepsilon \neq 0$;

▷ steps must interact $(W \neq 0)$ — but once it exists this interaction only enters logarithmically.

The argument is admittedly very crude — still it has the right physics! A more elaborate theory is given in [37].

Second order or first order roughening transitions?

Our weak coupling approach leads to a second order roughening transition. A question immediately comes up: could a more realistic strong pinning model lead to a *first* order transition? The question is not specific of roughening: for any phase transition, the same possibility exists once we go beyond a small amplitude Landau-Ginzburg expansion. In principle, only exact solutions for specific models can provide unambiguous answers. (Such solutions were considered by a number of authors [38, 39].)

Such elaborate theories are beyond the realm of our qualitative standpoint. We will content ourselves with a very naïve argument which shows *how* a first order transition might occur. Specifically, we calculate the free energy $F^{(1)}$ to *first order* in V, and we show that for large enough V this energy $F^{(1)}$ leads to a first order transition. In a sense, the argument is completely self contradicting: if V is large, a first order calculation is not valid. Nevertheless, such an estimate is instructive: it sheds a different light on roughening — and it shows that any first order transition must occur at a temperature T_c *larger* than T_R. Strong pinning can *lock* an *a priori* rough interface, not the reverse — a fairly obvious statement. Let us strongly emphasize that the present paragraph is *not* a theory — it just has suggestive value.

In order to calculate F to first order in V, we use a zeroth order statistical distribution of z_k, i.e. a gaussian. The net pinning energy is given by (4.27):

$$E_p = -VL^2 \exp\left[-\frac{2\pi^2}{a^2}\langle z \rangle^2\right].$$

Since E_p depends only on $\langle z^2 \rangle$, it is convenient to perform the statistical average in two steps:

(i) Carry a partial average over all $|z_k|^2$ at *constant* $\langle z^2 \rangle$. This is achieved by introducing a Lagrange multiplier into (4.23): the statistical distribution corresponds to an effective energy

$$\bar{E} = \int d^2 r \frac{\tilde{\gamma}}{2} [\text{grad } z^2 + \varepsilon z^2].$$

The average stretching free energy F_0 can be calculated as a function of the Lagrange multiplier ε — or equivalently as a function of $\langle z^2 \rangle$.

(ii) Then average over $\langle z^2 \rangle$. Since the latter is a macroscopic variable, that second step reduces to a *minimization* of the total free energy $F = F_0 + E_p$ with respect to ε (put another way, we use a saddle point approximation in carrying the average over $\langle z^2 \rangle$).

The algebra is straightforward. From the thermal average of z_k

$$\langle |z_k|^2 \rangle = \frac{T}{\tilde{\gamma} L^2 (k^2 + \varepsilon)}, \tag{4.102}$$

we infer the interface fluctuations:

$$\langle z^2 \rangle = \sum_k \langle |z_k|^2 \rangle = \frac{T}{4\pi\tilde{\gamma}} \log \left[\frac{\Lambda_0^2 + \varepsilon}{\varepsilon} \right] \tag{4.103}$$

(Λ_0 is the initial momentum cut off that we have been using throughout). If $\varepsilon \neq 0$, $\langle z^2 \rangle$ is finite and the pinning energy E_p is dominant on macroscopic scales: the interface is *faceted*. If on the contrary $\varepsilon = 0$, (4.103) diverges and $E_p \to 0$: the interface is *free*. The issue 'rough or faceted?' reduces to a single question: is the minimum of $F(\varepsilon)$ achieved for $\varepsilon = 0$ or $\varepsilon \neq 0$?

Within an irrelevant constant, the stretching free energy F_0 is

$$F_0 = \sum_k \left[L^2 \tilde{\gamma} \frac{k^2}{2} \langle |z_k|^2 \rangle - \frac{T}{2} \log\langle |z_k|^2 \rangle \right]. \tag{4.104}$$

The two terms of (4.104) correspond respectively to the energy E and entropy $(-TS)$. Using (4.102), we find[†]

$$F_0 = \frac{T}{2} \sum_k \left[\frac{k^2}{k^2 + \varepsilon} + \log(k^2 + \varepsilon) \right]. \tag{4.105}$$

[†] We could also include the Lagrange multiplier in the energy: the corresponding partition function Z_ε yields a free energy

$$F_\varepsilon = -T \log Z_\varepsilon = -\frac{T}{2} \sum_k \log \left[\frac{T}{\tilde{\gamma}(k^2 + \varepsilon)} \right].$$

The actual free energy is $F_0 = F_\varepsilon - \frac{1}{2}\varepsilon\tilde{\gamma}L^2\langle z^2 \rangle$: within a constant, we recover (4.105).

The integration is elementary, yielding

$$F_0(\varepsilon) = F_0(0) + \frac{TL^2\Lambda_0^2}{8\pi}\log\frac{\Lambda_0^2 + \varepsilon}{\Lambda_0^2}. \qquad (4.106)$$

This result should be compared to the pinning energy

$$E_p = -VL^2\left[\frac{\varepsilon}{\Lambda_0^2 + \varepsilon}\right]^{\pi T/2\tilde{\gamma}a^2}. \qquad (4.107)$$

We are looking for the *minima* of $(F_0 + E_p)$.

The behaviour near $\varepsilon = 0$ is obvious. $\Delta F_0 = F_0(\varepsilon) - F_0(0)$ is *positive* and $\sim \varepsilon$, while E_p is *negative* and $\sim \varepsilon^m$: ΔF is minimum at $\varepsilon = 0$ if $m > 1$, while it is maximum if $m < 1$ (Fig. 66a). We recover the result of our naïve first order calculation[†]: the transition occurs at $T_R = 2\tilde{\gamma}a^2/\pi$ — the interface is rough when $T > T_R$ (i.e. $m > 1$).

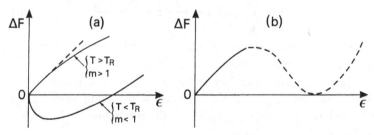

Figure 66: The free energy as a function of the Lagrange multiplier ε. (a) For weak coupling: the change of slope at the origin yields the usual second order roughening transition. (b) For strong coupling: a first order transition might appear.

We may also extend the discussion to larger ε. It is clear from (4.106) and (4.107) that F_0 will ultimately take over, $F(\varepsilon)$ increasing for large ε. If V is strong enough, $F(\varepsilon)$ may display a secondary minimum, as shown in Fig. 66b. The interface then undergoes a *first order transition* from a facetted state $\varepsilon \neq 0$ to a rough state $\varepsilon = 0$.

Such a conclusion may well be an artefact of our first order calculation applied in a region where obviously it does not hold. If V is large, the

---

[†] When $T < T_R$, we easily obtain the value of ε_m for which $F = F_0 + E_p$ is minimum. ε_m yields the *correlation length* of the problem, i.e. the scale ξ at which the interface locks: $\xi \sim \varepsilon^{-1/2}$. For weak coupling V, we obtain (4.65) which we know is wrong in the critical region: a first order calculation is not enough.

statistical distribution $P(z^2)$ is certainly not a gaussian: it is strongly modulated by the lattice potential. Indeed, the extreme case of a discrete lattice (the so-called SOS model) seems to give also a *second order transition*. Thus the above discussion should not be taken seriously. We do not argue that a first order transition does happen — we only claim that it *might* happen in specific situations. If it does, Fig. 66 strongly suggests that T_c will be $> T_R$ (for large V, we need a large initial slope in order to counter the downward trend of E_p).

One or several transitions?

The original description of roughening transitions by Burton, Cabrera and Frank [24] referred to a *single* crystal layer at the surface. All layers are filled, except the last one which has fractional occupation $n = xN_L$, where N_L is the number of sites per layer. The free energy is a function $F(x)$. According to [24], two situations may occur:

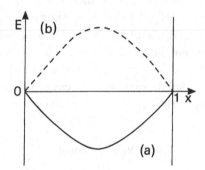

Figure 67: Energy as a function of filling ratio x of the last layer. (a) is supposedly rough, (b) is supposedly faceted.

▷ If $F(x)$ behaves as curve (a) in Fig. 67, all filling ratios x are stable. We go smoothly from one plane to the next (x going from 0 to 1): the interface is rough.

▷ If $F(x)$ behaves as curve (b) in Fig. 67, a 'liquid-gas' phase separation occurs in the last layer. Part of it is $x = 0$, part of it is $x = 1$, i.e. a *localized step* (the 'domain wall') separates filled layers. This is the faceted state.

In this approach, the roughening transition is equivalent to a two dimensional Ising model — with a critical behaviour which is completely different from the Kosterlitz-Thouless behaviour described above.

Such a picture has been historically very important, as it emphasized the existence of *surface transitions*. It is however quite misleading, as

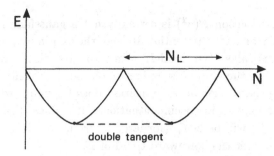

Figure 68: Step formation on a supposedly rough interface.

actually *both* curves (a) and (b) in Fig. 67 lead to a *facetted* state: the transition described by Burton, Cabrera and Frank is a genuine surface transition, but it is *not* a roughening transition in the sense that we have been using. To see that, we plot the energy $E(N)$ of the crystal as a function of the number of atoms N. E has period N_L. In the supposedly 'rough' state of Fig. 67, $E(N)$ behaves as shown in Fig. 68. In such a case, a double tangent can still be drawn between successive minima: *steps will be formed*, not between $x = 0$ and $x = 1$ as in Fig. 67b, but between $x = \frac{1}{2}$ and $x = \frac{3}{2}$ — it does not make much difference! If we tilt the interface by a small angle θ, the *average* density N must drift linearly with x (the position along the interface):

$$N \sim N_L \frac{\theta x}{a} + \text{const.}$$

In order to minimize the energy, N will jump from one minimum of $E(N)$ to the next one, wherever it is, the transition being localized in a *finite* region (the step). The resulting 'soliton' energy is finite — hence a surface energy proportional to the soliton density, i.e. $\gamma(\theta)$ proportional to θ. *Any* modulation of $E(n)$ results in a cusp of $\gamma(\theta)$, i.e. in a *faceted* state. Both regimes in Fig. 67 are equivalent in this respect.

The only way to kill the steps is to have *no modulation at all*, a condition which is achieved if the fluctuation amplitude $\langle z^2 \rangle$ is infinite. We conclude that a *bona fide* roughening transition has nothing to do with a liquid-gas separation in the last crystal layer. *Roughening is due to infinite fluctuations.* This important point is often overlooked. Among other things, it precludes *quantum roughening* due to zero point fluctuations in crystals like ^{3}He and ^{4}He. Let ω_k be the spectrum of surface fluctuations. Assigning half the zero point energy $\frac{1}{2}\hbar\omega_k$ to the *potential* stretching energy, we find

$$\langle z^2 \rangle = \sum_k \frac{\hbar \omega_k}{4L^2 \tilde{\gamma} k^2}.$$ (4.108)

Since $\omega_k \to 0$ as $k \to 0$, the integral (4.108) is convergent: fluctuations are finite and the interface is *faceted*. This result was emphasized by Fisher and Weeks [40]: it contradicts views of the Soviet school.

Granted that the Burton-Cabrera-Frank transition is not roughening, it may exist independently in specific models — hence the possibility of *several* surface transitions — one being usual roughening, the other one a rearrangement of atomic layers of the Ising type. This possibility has been discussed by a number of authors [39, 41]. We refer to the literature for details — here we only stress what we believe is the underlying physical argument behind these extra transitions.

These very superficial and qualitative remarks show that the statistical mechanics of crystal surfaces is a very rich problem. We only scratched it: many features need to be explored — but the qualitative physical picture is reasonably well understood.

Acknowledgments

These lectures originate in a course given at Collège de France during the Fall of 1983. Since that time, many features have been clarified, and new problems have emerged, mostly through discussions with my colleagues in Grenoble and Paris. I would like to express my gratitude to all of them, and especially to Makio Uwaha and Dietrich Wolf with whom I worked at ILL, to Sébastien Balibar, Étienne Wolf and Francois Gallet with whom I shared the joy of discovery in the attic of their laboratory at École Normale Supérieure — to Bernard Castaing, also, who introduced me to the physics of melting ten years ago.

In the end, manuscripts are always typed in a rush. This one was no exception, and F. Parisot and F. Giraud did not spare their effort in order to complete it in time. I want to thank them very warmly: their enthusiasm is one of the features that make the Theory College of ILL such a lively place.

References

The literature on this subject is extremely large: an extensive bibliography was out of reach. We list only a few references which are either the background or direct extensions of the subjects treated in these lectures.

[1] See for instance the review of G. Navascues, *Rep. Prog. Phys.*, **42**, 1131, 1979.

[2] See for instance F.C. Larche, J.W. Cahn, *Acta Metall.*, **26**, 1579, 1978;
W.W. Mullins, R.F. Sekerka, *J. Chem. Phys.*, **82**, 5192, 1985.

[3] A.F. Andreev, Yu. A. Kosevich, *Sov. Phys. JETP*, **54**, 761, 1981.

[4] P. Nozières, D.E. Wolf, *Z. Phys.*, **B70**, 399, 1988.

[5] C. Herring, *Structure and properties of solid surfaces*, R. Gomer, C.S. Smith Ed., Univ. of Chicago Press, 1953.

[6] R. Shuttleworth, *Proc. Phys. Soc. London*, **163**, 444, 1950; (See also [5].).

[7] See for instance V.I. Marchenko, A. Ya. Parshin, *JETP Lett.*, **31**, 724, 1980.

[8] D.E. Wolf, P. Nozières, *Z. Phys.*, **B70**, 507, 1988.

[9] M.A. Grinfeld, *Sov. Phys. Dokl.*, **31**, 831, 1986.

[10] See for instance L.D. Landau, E.M. Lifschitz, *Theory of Elasticity*, Pergamon, pp. 17–18.

[11] L.D. Landau, E.M. Lifschitz, *Statistical Physics*, Pergamon, Vol. 1, p. 520;
A.F. Andreev, *Sov. Phys. JETP*, **53**, 1063, 1981.

[12] V.I. Marchenko, A. Ya. Parshin, *Sov. Phys. JETP*, **52**, 129, 1980; A detailed calculation of the elastic response for a half space is given in [10], p. 22.

[13] K.H. Lau, W. Kohn, *Surface Science*, **65**, 607, 1977.

[14] See for instance J. Villain, Two dimensional solids and their interaction with substrates, in *"Ordering in strongly fluctuating condensed matter systems"*, T. Riste Ed., 1979;
Applications to crystal shapes are treated by H.J. Schulz, *J. Physique*, **46**, 157, 1985.

[15] M. Uwaha, P. Nozières, *Proceedings of the 1985 OJI Symposium on Morphology and Growth unit of Crystals*, Terra Scientific, Tokyo.

[16] M. Uwaha, P. Nozières, *J. Physique*, **48**, 407, 1987.

[17] For an extensive review on wetting, see P.G. de Gennes, *Rev. Mod. Phys.*, **57**, 827, 1985.

[18] R. Lipowski, W. Speth, *Phys. Rev.*, **B28**, 3983, 1983;
A detailed treatment of thermal fluctuations is given by R. Lipowski, *Phys. Rev.*, **B32**, 1731, 1985.

[19] J.C. Heyraud, J.J. Métois, J.M. Bermond, *J. Cryst. Growth*, (in press).

[20] P. Nozières, *J. Physique*, **50**, 2541, 1989.

[21] See for instance P. Nozières, M. Uwaha, *J. Physique*, **48**, 389, 1987.

[22] J.W. Cahn, J.E. Hilliard, *J. Chem. Phys.*, **31**, 688, 1959.

[23] P.E. Wolf, F. Gallet, S. Balibar, E. Rolley, P. Nozières, *J. Physique*, **46**, 1987, 1985.

[24] W.K. Burton, N. Cabrera, F.C. Frank, *Phil. Trans. Roy. Soc.*, **243**, 299, 1951.

[25] W.W. Mullins, *J. Appl. Phys.*, **28**, 333, 1957.

[26] I.M. Lifschitz, V.V. Slyozov, *J. Phys. Chem. Solids*, **19**, 35, 1963.

[27] J. Villain, *Europhys. Lett.*, **2**, 531, 1986.

[28] A. Rettori, J. Villain, *J. Physique*, **49**, 257, 1988.

[29] P. Nozières, *J. Physique*, **48**, 1605, 1987.

[30] See for instance:
S.T. Chui, J.D. Weeks, *Phys. Rev.*, **B14**, 4978, 1978;
H.J.F. Knops, *Phys. Rev. Lett.*, **49**, 776, 1977;
J.V. Jose, L.P. Kadanoff, S. Kirkpatrick, D.R. Nelson, *Phys. Rev.*, **B16**, 1217, 1977;
H. Van Beijeren, *Phys. Rev. Lett.*, **38**, 993, 1977;
T. Ohta, D. Jasnow, *Phys. Rev.*, **B20**, 139, 1979.

[31] H.J.F. Knops, L.W.J. Den Ouden, *Physica*, **A103**, 579, 1980.

[32] P. Nozières, F. Gallet, *J. Physique*, **48**, 353, 1987.

[33] J.M. Kosterlitz, D.J. Thouless, *J. Phys.*, **C6**, 1181, 1973; *J. Phys.*, **C7**, 1046, 1974.

[34] S.T. Chui, J.D. Weeks, *Phys. Rev. Lett.*, **40**, 733, 1978.

[35] F. Gallet, S. Balibar, E. Rolley, *J. Physique*, **48**, 369, 1987.

[36] For recent results, see B. Salanon, F. Fabre, J. Lapujoulade, *Phys. Rev.*, **B38**, 7385, 1988.

[37] J. Villain, D.R. Grempel, J. Lapujoulade, *J. Phys.*, **F15**, 809, 1985.

[38] H. Van Beijeren, *Phys. Rev. Lett.*, **38**, 993, 1977.

[39] H.J.F. Knops, *Phys. Rev.*, **B20**, 4670, 1979.

[40] D.S. Fisher, J.D. Weeks, *Phys. Rev. Lett.*, **50**, 1077, 1983.

[41] K. Rommelse, M. den Nijs, *Phys. Rev. Lett.*, **59**, 2578, 1987.

CHAPTER 2

INSTABILITIES OF PLANAR
SOLIDIFICATION FRONTS

B. Caroli, C. Caroli and B. Roulet

1

Introduction

Solidification i.e. the growth of a solid from a liquid phase — has been
studied for a long time for practical reasons. In a solid, atomic motion
is extremely slow diffusion coefficients in solids at room temperature
are typically of order 10^{-11}–10^{-13} cm^2/sec. So one can consider, at
least roughly speaking, that the microstructures of solids — e.g. of metals
as obtained by casting or by progressive cooling of liquid ingots — are
frozen in the state in which the solid has grown. It is the microstructures,
i.e. the nature and density of defects, as well as the chemical homogeneity,
which are responsible not only for the mechanical properties of solids but,
also, for the electronic performances of many materials used in modern
technologies (e.g. optics or semiconducting devices).

For the same reason, only crystals of microscopic dimensions[†] — in
the micron range — can reach on reasonable time scales, of the order of a
few hours, their equilibrium shape, which minimizes the thermodynamic
potential. The shapes of naturally occurring crystals are growth shapes
of dynamical origin. One may think, for example, of the so-called
dendritic shapes of ice or snow crystals (Fig. 1). These patterns, with

---

[†] And helium crystals, in which atomic motions are extremely fast, due
to the importance of zero point motion (quantum fluctuations).

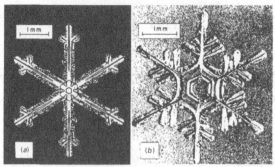

Figure 1: Snow crystals photographed a) in reflected light, b) in oblique illumination (U. Nakaya, *Snow crystals*, Harvard University Press, Cambridge, 1954).

quasi-periodic branches which, as they grow, themselves emit secondary branches, etc. are typical of free growth, i.e. growth from an isolated germ from the supercooled or supersaturated melt. Metallurgical studies have identified and classified other types of regular structures, for example:

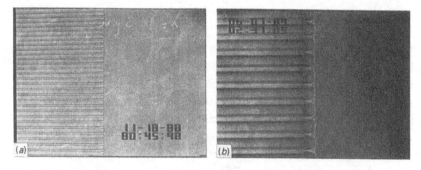

Figure 2: CBr_4–C_2Cl_6 eutectic alloy in directional solidification: a) $V \simeq 2\ \mu m/s$; $\lambda = 12.3\ \mu m$; b) $V \simeq 0.8\ \mu m/s$; $\lambda = 18.5\ \mu m$; (courtesy of G. Faivre).

▷ Lamellar eutectics [1, 2] (Fig. 2). They are obtained by direction-ally solidifying alloys at (or close to) the eutectic concentration, i.e. by pulling at an imposed velocity V in an imposed external temperature gradient. It is observed that, in such systems, the two solid phases into which the eutectic separates are organized in planar periodic lamellae, the period of which depends on the growth velocity. Periodic rod ar-rangments may also occur.

▷ Cellular growth of dilute alloys [3, 4]: when these are grown by directional solidification, it is found that, at low growth velocities (typically, $V < 1 \ \mu$m/sec) the solute concentration in the solid alloy is homogeneous. When V increases above a threshold velocity V_c (which depends on the thermal gradient and on the impurity content), the solid alloy becomes inhomogeneous; it exhibits periodic striations perpendicular to the growth front, due to solute segregation. Typical scales are in the 50–100 μm range.

▷ Let us also mention the existence of growth modes which give rise to solid composition striations parallel to the growth front, observed in electrodeposition [5, 6], in rapid solidification [7, 8, 9], and possibly responsible for the structure of some striated rocks [10, 11].

That all these structures originate from the dynamics of solidification is proven by the fact that they depend in a systematic and reproducible fashion on the growth velocity.

It was thus natural for metallurgists and for specialists of crystal growth to try to understand how one should choose growth methods and conditions in order to avoid as much as possible deformations of the liquid-solid front during growth (and thus, as we shall see, the inhomogeneities which come with them) or in order to control them for particular practical purposes — e.g. mechanical hardening.

These brief remarks put forward the fact that metallurgists and material physicists would like to be able to predict how the dynamical growth conditions influence the structure of the resulting solids. Let us mention that this is also of interest for geophysicists in a different perspective: in this case, one of course does not control growth conditions, and the problem is to know whether it is possible to obtain, from some rock structures — e.g. composition striations — at least qualitative informations about the conditions which prevailed when they grew.

Finally, much more recently, solidification has become a subject of interest for specialists of statistical physics from still a different point of view [12]: it is by nature an out-of-equilibrium phenomenon. So, growth front morphologies are a subclass of the general problem of pattern formation in dissipative systems, other well studied examples of which are met in hydrodynamics (Rayleigh-Bénard convection, Taylor-Couette vortices...), chemistry (Belousov-Zhabotinsky reactions...), laser physics, superconducting Josephson junctions...

One of the central problems in this field is to understand what determines the selection, if any, of observed structures. In particular, it was hoped for a long time that it would be possible to find a general selection principle which would play for out-of-equilibrium systems the role played, at equilibrium, by the free-energy minimization principle. As will appear in Y. Pomeau and Ben Amar's chapter on dendritic growth and as we will discuss here on the case of cellular growth of mixtures, the recent experimental and theoretical advances about solidification have contributed to prove, in agreement with hydrodynamical studies, that the situation is, from that point of view, unfortunately less simple than could have been hoped (non-validity of the 'principle of marginal statiblity' [13] in the case of dendrites, non-unique selection of growth structures in cellular growth or for lamellar eutectics [14]).

This does not mean, as we shall see, that no general schemes of behavior of out-of-equilibrium systems emerge, but that we are still far from understanding completely which, among the phenomena taking place on the mesoscopic scale of interest here, may be considered 'generic', and which are controlled by specific microscopic properties of the systems under study. An obvious example, in the case of solidification, is that of the role of crystal structure, which comes into play in several ways, e.g. capillary anisotropy — now believed to play a generic role in the selection of velocity of freely growing dendrites — and plastic properties, which possible influence on solidification patterns is still essentially unexplored.

From this point of view, it would be, in our opinion, quite artificial to try and separate completely, when studying solidification as well as in many other fields, a 'generic-fundamental' approach from a 'materials science-empirical' one. As will appear in this chapter, many fundamental problems in solidification dynamics remain, at least partly, unsolved, and the physicists interested in them have certainly much to learn from the experimental results of metallurgical studies. Conversely, we believe that progress in some problems set by high-tech industrial processes can only come from the identification of relevant parameters and criteria through model laboratory experiments and their theoretical analysis.

All the types of growth patterns mentioned above exhibit characteristic length scales which lie, typically, in the 10–100 μm range, and are responsible, at least grossly, for a given growth mode, for the details of the microstructure of the resulting solid. It is thus reasonable to think that, in order to describe these phenomena, and at least in a first stage, a 'mesoscopic' description — i.e. a description of the concerned liquids and solids as continuous media — of solidification is appropriate.

On this scale, which forgets about the details of atomic organization and motion, it is sufficient to describe solidification as a *first-order transition*, i.e. a process giving rise to *release of heat* (the latent heat of melting) and, if one is dealing with a mixture — or, more generally, an impure system — to *rejection or incorporation of solute*, since the solid and liquid phases of a given mixture have, at equilibrium, different concentrations.

The moving solid-liquid interface can therefore be viewed as a source (or sink) of heat and solute which, once produced, are evacuated by diffusive transport in the adjacent volume phases. If transport did not take place, heat (solute) would accumulate close to the front, temperature (concentration) would rise so that the liquid would become locally more stable than the solid, and solidification would stop. So, it is the *dynamical balance* between production and transport of heat — and, for impure materials, solute — which is responsible, for given external constraints, for the growth mode, i.e. which determines the position and shape of the growth front and, consequently, the microstructure of the solid.

The model of solidification based on this description, which we will treat in detail, can be considered as a *minimal model*. We will see that it is sufficient to describe the basic instability mechanisms and the main types of front patterns. However, it neglects physical effects which may be important in practical situations:

(i) There may be fluid flows in the liquid phase. Such flows may be externally imposed, e.g. because the solid is being rotated while pulled out (Czochralski's method [15], electrochemical growth) or because the cooling liquid is being stirred electromagnetically in order to accelerate transport and/or homogenize the melt. One may also be dealing with natural flows: heat and solute transport create, ahead of the front in the melt, temperature and concentration gradients which may give rise to natural convection. Convective flows contribute to volume transport and at the same time create temperature and concentration gradients along the front, which they consequently tend to deform.

On the other hand, a first order phase change gives rise to a density change. The solidification of a given volume of liquid thus implies a change of mass which must be fed by a flow in the liquid.

(ii) Solids are deformable — elastically at low stress level, plastically at high stresses. Solidification may give rise to dynamically induced stresses. Such is obviously the case when, for example:

▷ a liquid drop becomes for some reason trapped inside the solid. When it cools down, it solidifies and thus contracts; the corresponding stresses may be large enough to create plastic defects, or to induce the formation of semi-macroscopic voids.

▷ an alloy solidifies with an inhomogeneous concentration (solute striations). The difference between the solute and solvent atomic volumes may result, along the striation lines or planes, in the generation of dislocations or sub-grain boundaries.

Finally, such mesoscopic models should also describe the dynamics of other kinds of first order phase transitions, e.g. solid/solid, or liquid/liquid crystal. It is clear that stress effects are very important in solid/solid transitions. On the other hand it is well known that, when liquid crystals are generated from the isotropic liquid phase, the dynamics of phase change have a strong influence on the optical quality, i.e. on the defect content, of the liquid crystal.

These remarks point to the fact that a detailed realistic model of solidification (and, more generally, of the dynamics of first order transitions) must be able to take into account hydrodynamic, elastic — and, hopefully,[†] plastic — effects.

Since we are not aware of the existence of any textbook dealing in a systematic way with the inclusion of these effects into the formalism of solidification, we will organize this chapter as follows.

In Section 2, we show how, starting from the formalism of out-of-equilibrium thermodynamics, one can write down the set of phenomenological equations which describe solidification while taking into account the possibility of fluid flow in the liquid phase. For simplicity, we will assume that the stresses in the solid reduce to a hydrostatic pressure. An example of how elastic effects can be included into the formalism may be found in reference [16].

We then analyze, on the minimal model, the Mullins-Sekerka instability (Section. 3) which is of central importance in the formation of front patterns. Sections 4 to 6 are devoted to cellular growth of mixtures — position and nature of the bifurcation, weakly non-linear regime, deep cells structures, and the important question of pattern selection.

[†] This is considerably more difficult, since it is necessary to specify in each case the nature of the important defects, the mechanisms by which they are created and, finally, their dynamics.

Finally, we touch, in Sections 7 and 8, on two less well known — and much less explored, problems, namely the coupling between the Mullins-Sekerka and convective instabilities, and the destabilization of a facet in directional solidification of impure solids. We will not consider dendritic growth, which is the subject of Chapter 4 by Pomeau and Ben Amar.

2

Thermo-hydrodynamic formalism

2.1 The one-phase system: a brief summary

Let us briefly recall the main steps of the construction of non-equilibrium thermodynamics for a single-phase system. We follow here closely the approach of de Groot and Mazur [17].

The system is considered as a continuous medium, assumed to be large enough to be divided into 'cells', of dimensions d_{cell} such that:

▷ The macroscopic properties (quantities) are quasi-constant within a cell.

▷ The cell is large enough to be considered itself as a thermodynamical system — so that thermodynamic fluctuations can be neglected.

This *coarse-graining* approximation:

(i) Makes it possible to define continuous fields of intensive quantities (e.g. mass density $\rho(\mathbf{r}, t)$; temperature $T(\mathbf{r}, t)$, densities of momentum $\mathbf{g}(\mathbf{r}, t)$, internal energy $u(\mathbf{r}, t)$, entropy $s(\mathbf{r}, t), \ldots$).

(ii) Implies — and this is the *fundamental hypothesis* of out-of-equilibrium thermodynamics — that each cell is at *local equilibrium*,[†] and therefore that the usual fundamental thermodynamic relations are valid *locally*. In particular:

▷ the equilibrium equation of state remains valid locally without any modification (without any additional dependence on gradients of thermodynamic quantities);

[†] Which is necessary to define local thermodynamic quantities, e.g. a local temperature.

▷ the thermodynamic identity[†]

$$T\,\mathrm{d}s = \mathrm{d}u - \sum_k \mu_k\,\mathrm{d}\rho_k \tag{2.1}$$

(where μ_k is the chemical potential per unit mass of species k, s, u, ρ_k the entropy, internal energy and mass of species k per unit volume) is valid locally, so that it reads, for the non-equilibrium system:

$$T(\mathbf{r}, t)\frac{\mathrm{d}s(\mathbf{r}, t)}{\mathrm{d}t} = \frac{\mathrm{d}u(\mathbf{r}, t)}{\mathrm{d}t} - \sum_k \mu_k(\mathbf{r}, t)\frac{\mathrm{d}\rho_k(\mathbf{r}, t)}{\mathrm{d}t}. \tag{2.2}$$

This local equilibrium assumption is valid only in the *hydrodynamic regime*, i.e. when the frequencies and wavevectors ω, q which characterize the space and time variations in the system verify:

$$\omega\tau_c \ll 1, \qquad q\ell \ll 1,$$

where τ_c and ℓ are the collision time and mean free path of particles in the system — i.e. the typical time and length scales for reaching equilibrium.

The procedure is then as follows:

2.1.A *Conservation laws*

One first formulates in this continuous description all the conservation laws obeyed by the system (mass, momentum, energy, angular momentum...).

Examples:

a) *Mass conservation:*

We consider a binary isotropic fluid, composed of a mixture of the two chemically inert species $k = 1, 2$. Conservation of the mass of species k is then expressed by:

$$\frac{\partial\rho_k(\mathbf{r}, t)}{\partial t} + \mathrm{div}\big[\rho_k(\mathbf{r}, t)\mathbf{v}_k(\mathbf{r}, t)\big] = 0, \tag{2.3}$$

[†] This expression involving quantities per unit volume is easily deduced from the more usual one, valid for quantities per unit mass:

$$T\,\mathrm{d}\tilde{s} = \mathrm{d}\tilde{u} + p\,\mathrm{d}\widetilde{\Omega} - \sum_k \mu_k\,\mathrm{d}c_k$$

by using $s = \rho\tilde{s}$, $u = \rho\tilde{u}$, $\widetilde{\Omega} = 1/\rho$.

where $\mathbf{v}_k(\mathbf{r}, t)$ is the velocity of species k. Equation (2.3) expresses the fact that variations of the mass of (k) contained in a volume element dV can only be due to the flux through the surface limiting dV.

We define the total density:

$$\rho(\mathbf{r}, t) = \sum_k \rho_k(\mathbf{r}, t); \tag{2.4}$$

the (barycentric) velocity of the mixture

$$\mathbf{v} = \frac{1}{\rho} \sum_k \rho_k \mathbf{v}_k \tag{2.5}$$

and the total mass current

$$\mathbf{J} = \rho \mathbf{v} = \sum_k \rho_k \mathbf{v}_k. \tag{2.6}$$

One is then led to define the *diffusion current* of species k :

$$\mathbf{J}_k = \rho_k (\mathbf{v}_k - \mathbf{v}), \tag{2.7}$$

which measures how much the motion of species k differs from that of the center of mass, in a given volume element; i.e. it characterizes the *relative motion* of the two species. Note that, from (2.5):

$$\mathbf{J}_1 = -\mathbf{J}_2. \tag{2.8}$$

The two equations (2.3) $(k = 1, 2)$ can then be rewritten as:

$$\frac{\partial \rho}{\partial t} = -\operatorname{div}(\rho \mathbf{v}) = -\operatorname{div} \mathbf{J}, \tag{2.9}$$

$$\rho \frac{\partial c_k}{\partial t} + \rho \mathbf{v} \cdot \boldsymbol{\nabla} c_k = -\operatorname{div} \mathbf{J}_k, \tag{2.10}$$

where we have defined the mass concentration:

$$c_k = \frac{\rho_k}{\rho}. \tag{2.11}$$

b) *Equation of motion (momentum balance):*

The motion of a given element of matter is described by Newton's equation. That is, if δV is the volume element under consideration:

$$\delta F = \rho \delta V \frac{d\mathbf{v}}{dt} = \delta V \cdot (\text{sum of forces/unit volume}).$$

The 'material derivative' d/dt defines time variations affecting a 'labelled' element of matter. So:

$$\frac{d}{dt} = \frac{\partial}{\partial t} + \frac{d\mathbf{r}}{dt} \cdot \nabla + \mathbf{v}(\mathbf{r}, t) \cdot \nabla. \qquad (2.12)$$

For the sake of physical interpretation it is useful to decompose forces into:

▷ external forces, the expression of which can be written explicitly (e.g. gravity, electric, magnetic, centrifugal... forces). We denote them by $\sum_k \rho_k \mathbf{F_k}$, where $\mathbf{F}_k$ is the external force per unit mass of species k.

▷ internal forces, originating from molecular interactions. Due to the short range of these interactions, they reduce to a set of forces acting on the *surface* of the volume element. This entails that the internal force on volume δV can be written as[†]

$$\delta \mathbf{F}_{\text{int}} = -\delta V \operatorname{Div} \overline{\overline{P}}(\mathbf{r}, t) \qquad (2.13)$$

(i.e. $\delta F_{\text{int}}^{(\alpha)} = -\delta V \sum_{\beta=1}^{3} \partial P_{\beta\alpha}(\mathbf{r}, t)/\partial x_\beta$, where α, $\beta = 1, 2, 3$ denote cartesian coordinates).

One shows [17] that $\overline{\overline{P}}$ is a symmetric tensor (except if the constitutive molecules have a chirality or a spin). It is called the *generalized pressure tensor* or *stress tensor*.

One can separate out, from the internal forces, the pressure term (the only one which is present in the fluid at equilibrium). By definition, for an isotropic fluid it does not depend on the orientation of the limiting surface δS and is perpendicular to it, so that the corresponding contribution to the pressure tensor reads:

$$p\delta_{\alpha\beta} = p\left(\overline{\overline{I}}\right)_{\alpha\beta} \qquad (2.14)$$

[†] This can be seen by noticing, for example, that the sum of the forces acting on the two surfaces $\perp Ox$, of area $\delta y \delta z$, of abscissae x and $(x + \delta x)$, is:

$$\varphi_x(x + \delta x) - \varphi(x) = \delta x \frac{\partial}{\partial x} \varphi_x(x),$$

while φ_x is proportional to the area: $\varphi_x = -\mathbf{P}_x \delta y \delta z$, so that the corresponding contribution to δF_{int} is $(-\delta V \partial \mathbf{P}_x/\partial x)$.

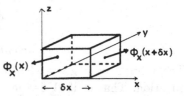

where $\bar{I}$ is the unit tensor. One then decomposes the pressure tensor $\bar{\bar{P}}$ of the non-equilibrium system into:

$$P_{\alpha\beta} = p\delta_{\alpha\beta} + \Pi_{\alpha\beta}. \qquad (2.15)$$

The tensor $\bar{\bar{\Pi}}$, which is zero at equilibrium, is the *viscous pressure tensor*; it is associated with the flow-induced viscous stresses.

Newton's equation can then be written, with the help of (2.9), as

$$\frac{\partial(\rho,\mathbf{v})}{\partial t} = -\operatorname{Div}\left(\rho\mathbf{v}\mathbf{v} + \bar{\bar{P}}\right) + \sum_k \rho_k \mathbf{F}_k. \qquad (2.16)$$

It appears as a momentum balance equation. In the absence of external forces ('sources' of momentum) it expresses momentum conservation.

The tensor $(\rho\mathbf{v}\mathbf{v} + \bar{\bar{P}})$ — where $(\mathbf{v}\mathbf{v})_{\alpha\beta} = v_\alpha v_\beta$ — appears as a momentum flux, $\rho\mathbf{v}\mathbf{v}$ being its *convective* part (matter moving at velocity $\mathbf{v}$ carries a momentum $\rho\mathbf{v}$); $\bar{\bar{P}}$ describes internal transfers. As is well known, the pressure part $p\bar{I}$ corresponds to a reversible exchange of work; as we will see, $\bar{\bar{\Pi}}$ corresponds to dissipative (heat producing) exchanges.

One can write as well the equation expressing total energy conservation and, from the set of conservation laws, the equation describing the internal energy balance, which is precisely the explicit expression of the first principle, written here locally for a moving binary fluid.

2.1.B *Entropy production*

Let $s(\mathbf{r}, t)$ be the entropy per unit volume. The entropy balance can then be written formally as:

$$\frac{\partial s}{\partial t} = -\operatorname{div}(\mathbf{J}_{s,\text{tot}}) + \sigma, \qquad (2.17)$$

i.e., by definition, that part of the time variation of s which is not carried in by the entropy current $\mathbf{J}_{s,\text{tot}}$ (i.e. provided by the world outside the considered volume element) is due to *entropy production* within the (infinitesimal) system. That is, σ accounts for dissipative (irreversible) phenomena.

The second principle of thermodynamics (here expressed locally) therefore imposes:

$$\sigma \geq 0. \qquad (2.18)$$

On the other hand, one writes down the entropy balance equation with the help of the local expression of the thermodynamic identity

(Eq.(2.2)) into which one inserts the expressions of du/dt, $d\rho/dt$,...
provided by the conservation laws. One then obtains:

$$\rho \frac{d}{dt}\left(\frac{s}{\rho}\right) = \rho \left(\frac{\partial}{\partial t} + \mathbf{v} \cdot \boldsymbol{\nabla}\right)\left(\frac{s}{\rho}\right) \tag{2.19}$$

$$= -\operatorname{div}\left(\frac{\mathbf{J}_q - \sum_k \mu_k \mathbf{J}_k}{T}\right) - \frac{\boldsymbol{\nabla} T}{T^2} \cdot \mathbf{J}_q$$

$$- \frac{1}{T}\sum_k \mathbf{J}_k \cdot \left\{T\boldsymbol{\nabla}\left(\frac{\mu_k}{T}\right) - \mathbf{F}_k\right\} - \frac{1}{T}\overline{\overline{\Pi}} : \boldsymbol{\nabla}\mathbf{v}.$$

$\mathbf{J}_q$ is the heat current, defined as that part of the energy current
which is neither convective energy transport nor transport of mechanical
work [17]. $\overline{\overline{\Pi}} : \boldsymbol{\nabla}\mathbf{v} \equiv \sum_{\alpha\beta} \Pi_{\alpha\beta} \partial v_\alpha/\partial x_\beta$.

Identification with expression (2.17) then yields the expression of the
non-convective part of the entropy current:

$$\mathbf{J}_s = \mathbf{J}_{s,\text{tot}} - \mathbf{v}s = \frac{1}{T}\left(\mathbf{J}_q - \sum_k \mu_k \mathbf{J}_k\right) \tag{2.20}$$

and that of the entropy production σ. Using Eq. (2.8) and separating
the viscous tensor into a scalar part $(O = \frac{1}{3}(\operatorname{tr}\overline{\overline{\Pi}})\overline{\overline{I}})$ and a zero-trace
symmetric tensor $\overset{\overline{\overline{0}}}{\Pi}$ (i.e. separating terms of different tensorial orders):

$$\sigma = -\frac{\boldsymbol{\nabla} T}{T^2} \cdot \mathbf{J}_q - \mathbf{J}_1 \cdot \frac{1}{T}\left[T\boldsymbol{\nabla}\left(\frac{\mu_1 - \mu_2}{T}\right) - \mathbf{F}_1 + \mathbf{F}_2\right]$$

$$- \frac{\Pi}{T}(\operatorname{div}\mathbf{v}) - \frac{1}{T}\overset{\overline{\overline{0}}}{\Pi} : (\boldsymbol{\nabla}^0\mathbf{v})^s \tag{2.21a}$$

where

$$(\boldsymbol{\nabla}^0\mathbf{v})^s_{\alpha\beta} = \frac{1}{2}\left(\frac{\partial v_\beta}{\partial x_\alpha} + \frac{\partial v_\alpha}{\partial x_\beta}\right) - \frac{1}{3}\delta_{\alpha\beta}\operatorname{div}\mathbf{v}. \tag{2.21b}$$

The separation of $\overline{\overline{\Pi}}$ into two components amounts to separating shear
effects from those due to flow-induced variations of specific volume.

2.1.C *Linear laws*

Entropy production then appears as a sum of terms each of which is the
product of a (heat, diffusion, momentum) current J_i by a 'thermody-
namic force' X_i which is zero at equilibrium (where $\boldsymbol{\nabla} T = \boldsymbol{\nabla}\mu = \boldsymbol{\nabla} v = 0$).
At equilibrium, currents must as well vanish. They depend on the
forces — the specific form of the corresponding function(al) of course

depends on the physical nature of the system. For not too large[†] departures from equilibrium, one may approximate these functions by the first linear term of their Taylor expansion. This results in the *phenomenological linear transport laws* or *linear kinetic laws:*

$$J_i = \sum_j L_{ij} X_j \tag{2.22}$$

(in this symbolic expression i and j label both the type of current or force and the cartesian components).

In this linear approximation, the entropy production reads:

$$\sigma = \sum_i J_i X_i = \sum_{ij} L_{ij} X_i X_j. \tag{2.23}$$

L_{ij} is the generalized matrix of transport coefficients. Its elements must obey three sets of conditions, which follow from general physical principles.

a) Second principle
This imposes (Eq.(2.18)) that the quadratic form (2.23) be non-negative, from which:

$$L_{ii} \geq 0, \qquad L_{ii} L_{jj} \geq \tfrac{1}{4}(L_{ij} + L_{ji})^2. \tag{2.24}$$

b) Curie principle
This states that 'if the system is invariant under a set (S) of symmetry operations, the current-force relations must also be invariant under these transformations'. The conditions on the L_{ij}'s resulting from it can be stated explicitly with the help of group theory.

For an isotropic fluid, Curie's principle reduces to the simple result that forces and fluxes of different tensorial characters do not couple.

c) Onsager's relations
These result from the time-reversal invariance of the equations of motion of individual particles and, as such, are general. They impose that (in the absence of magnetic field) the matrix of transport coefficients be symmetric:

$$L_{ij} = L_{ji}. \tag{2.25}$$

[†] The precise meaning of this restriction must obviously be specified for each particular system — or class of systems — and transport property.

When all these conditions are taken into account, the phenomenological linear transport laws reduce, for a binary isotropic fluid, to:

$$\Pi = -L(\text{div }\mathbf{v}), \tag{2.26a}$$

$$\mathbf{J}_q = L_{qq}\left(-\frac{\nabla T}{T^2}\right) + L_{q1}\frac{1}{T}\left[-\nabla(\mu_1 - \mu_2) + \mathbf{F}_1 - \mathbf{F}_2\right], \tag{2.26b}$$

$$\mathbf{J}_1 = L_{q1}\left(-\frac{\nabla T}{T^2}\right) + L_{11}\frac{1}{T}\left[-\nabla(\mu_1 - \mu_2) + \mathbf{F}_1 - \mathbf{F}_2\right], \tag{2.26c}$$

$$\overset{\overset{\circ}{=}}{\Pi} = -L'\frac{1}{T}(\nabla^0\mathbf{v})^s. \tag{2.26d}$$

For a pure fluid ($\mathbf{J}_1 = 0$, $(\mu_1 - \mu_2, F_1 - F_2) \to 0$), setting

$$\mathcal{K} = \frac{L_{qq}}{T^2}, \quad \eta = \frac{L'}{2T}, \quad \eta_v = \frac{L}{T}, \tag{2.27}$$

and plugging Eqs.(26) into the momentum and internal energy balance equations, one obtains the *Navier-Stokes* equation and the equation of heat transport in a moving fluid which reduces, in the absence of fluid flow, to the *heat diffusion equation*:

$$C_p\frac{\partial T}{\partial t} = -\text{div }\mathbf{J}_q = -\text{div}(-\mathcal{K}\nabla T), \tag{2.28a}$$

or
$$\frac{\partial T}{\partial t} = D_{\text{th}}\nabla^2 T. \tag{2.28b}$$

$\mathcal{K}$ is the thermal conductivity, C_p the specific heat per unit volume, $D_{\text{th}} = \mathcal{K}/C_p$ the thermal diffusion coefficient. η and η_v (Eq.(2.27)) are the two viscosities of the isotropic fluid.

For a binary fluid at rest, equations (2.26) describe how heat transport and chemical diffusion (relative motion of the two species) are induced by temperature and concentration gradients. Plugging them into the concentration (Eq. (2.9)) and internal energy balance equations, one obtains (using the standard notations defined in detail in [17], and setting $c \equiv c_2$):

$$C_p\frac{\partial T}{\partial t} = -\text{div }\mathbf{J}_q = \text{div}\left\{\mathcal{K}\nabla T + \frac{\partial\mu_1}{\partial c}TD'\nabla c\right\}, \tag{2.29a}$$

$$\frac{\partial c}{\partial t} = -\frac{1}{\rho}\text{div }\mathbf{J}_2 = \text{div}\left\{c(1-c)D'\nabla T + D\nabla c\right\}. \tag{2.29b}$$

The cross-coefficient D' characterizes the couplings between heat and concentration transport (Soret and Dufour effects). It is called the Dufour or thermal diffusion coefficient.

In an *isothermal* system, equation (2.29b) reduces to Fick's diffusion equation:

$$\frac{\partial c}{\partial t} = D\nabla^2 c. \tag{2.30}$$

Note, finally, that it is also possible to build along the same lines an out-of-equilibrium thermodynamics for elastic or viscoelastic [18] solids. However, no such general formalism including plasticity exists — this results in particular from the fact that plastic defects are not conservative objects and that they can be created or destroyed by a variety of material-dependent mechanisms. A general phenomenology would then be rather empty, and one must resort in each case where it is possible to identify the relevant mechanisms, to a detailed specific model.

2.2 Generalization to two-phase systems. Solidification

As mentioned in Section 1, we will treat here the solid, for the sake of simplicity, as mechanically isotropic and non-deformable.

A system undergoing solidification is, by definition, composed of two volume phases L (liquid) and S (solid), which are in contact with each other along an interface I. In order to describe it, it is thus necessary to extend the formalism sketched out above to the case of heterogeneous systems which are the seat of a phase change. This can be done [19, 20] with the help of a method first developed by Bedeaux, Albano and Mazur [21] to describe non-miscible fluids.

We have seen that, for the single-phase system, non-equilibrium thermodynamics defines fields of densities of extensive quantities $y(\mathbf{r}, t)$ (e.g. $\rho(\mathbf{r}, t)$, $s(\mathbf{r}, t), \ldots$) which vary continuously in space. For a system containing two adjacent phases L, S, one extends this notion by setting:

$$y(\mathbf{r}, t) = y_L(\mathbf{r}, t)\theta^L(f) + y_s(\mathbf{r}, t)\theta^S(f) + y_I(\mathbf{r}, t)\delta^I(\mathbf{r}, t), \tag{2.31}$$

where

$$f(\mathbf{r}, t) = 0 \tag{2.32}$$

defines the position of the interface.

$$\delta^I(\mathbf{r}, t) = |\boldsymbol{\nabla} f|\delta\big(f(\mathbf{r}, t)\big) \tag{2.33}$$

is a δ-function localized on the interface. θ^L (resp. θ^S) is restricted to the region of space occupied by the liquid (resp. solid) phase:

$$\theta^L(f) = \begin{cases} 1 & f > 0, \\ 0 & f \leq 0; \end{cases} \qquad \theta^S(f) = \begin{cases} 0 & f \geq 0, \\ 1 & f < 0. \end{cases} \tag{2.34}$$

Equation (2.31) therefore implies that the interface has zero thickness — at least on the mesoscopic scale — but carries finite densities of mass, energy, entropy....

2.2.A *Thermodynamic characterization of an interface*

Let us first consider briefly the definition and meaning of these interface quantities.

An interface is not a well-defined object on the microscopic scale. When two phases at equilibrium are in contact, the properties of the system (e.g. density, concentrations...) vary continuously (see Fig. 3). Most of this variation takes place on a finite distance, $\underline{a}$, of the order of a few molecular layers[†] (typically, $\sim 10\text{Å}$).

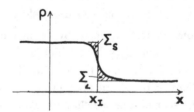

Figure 3: Schematic density profile close to an interface.

It is therefore impossible to elaborate any strictly geometric criterion about the shape of a profile such as that of Fig. 3 that would define uniquely and precisely an interface position. In order to get rid of this degree of arbitrariness, one must impose an additional condition. This should of course be chosen in such a way that the interface position thus defined lies in the region of fast variation of the physical properties. The point of view most commonly adopted in thermodynamic descriptions is that of Gibbs.

[†] Except in the vicinity of a critical point. This restriction, while unimportant for solidification, may come into play in the case of 'very weakly first-order' solid-solid or liquid-liquid crystal transitions.

a) *Gibbs definition of the interface:*

Let x_I be the — for the moment arbitrary — position assigned to the interface (here assumed planar for the sake of simplicity). It separates two volumes V_L (liquid region $x > x_I$) and V_S, with $V_L + V_S = V$, the total volume of the system. Let M be the total mass:

$$M = \int_{(\text{total volume})} \rho(x) \mathrm{d}^3 r. \qquad (2.35)$$

One then defines an *interface mass* as:

$$M_I = M - \rho_L V_L - \rho_S V_S \qquad (2.36)$$

$$= \int_{(x < x_I)} \mathrm{d}^3 r \left[\rho(x) - \rho_S \right] + \int_{(x > x_I)} \mathrm{d}^3 r \left[\rho(x) - \rho_L \right].$$

M_I is the difference between the mass of the real heterogeneous system and that of volumes V_L and V_S of liquid and solid if they where quasi-infinite (and therefore homogeneous). It is represented on Fig. 3 by the area difference $\Sigma_L - \Sigma_S$.

Gibbs's definition of the interface position is specified by the condition

$$M_I = 0, \qquad (2.37)$$

i.e. $\Sigma_L = \Sigma_S$, which entails that x_I indeed lies in the physically reasonable region.

x_I is no longer arbitrary, so that one can now define, for every extensive thermodynamic quantity Y, the corresponding *interface quantity* Y_I by:

$$Y_I = Y - V_L y_L - V_S y_S, \qquad (2.38)$$

where y_L (resp. y_S) is the density of Y in the infinite liquid (resp. solid) — e.g. concentration, specific internal energy, entropy.... From Y_I, one defines a surface density $y_I = Y_I / \mathcal{A}$, where $\mathcal{A}$ is the interface area.

Two qualitative remarks may be of interest:

▷ Since the definition of x_I involves getting rid of a single degree of arbitrariness, for a binary system one can only impose one condition of vanishing interface mass. This may be, for example, imposed on the mass of solvent. Since the density profiles of the two species are in general different, the solute interface mass remains in general finite. It may even be negative.

▷ For the same reason, it may happen that the interface entropy defined from (2.38) be negative.

These apparently paradoxical results simply express the fact that an interface in Gibbs's sense is not an independent thermodynamic phase, but only a concept which is convenient to describe on the semi-macroscopic scale fast but continuous variations within a single system.

b) *Guggenheim's point of view: the notion of surface phase*

In this approach, built from more intuitive considerations, the interface is *a priori* described as a region of *finite thickness*, possessing all the thermodynamic characteristics of a phase. This approach is different from the Gibbs one, but is of course not contradictory — most of the disputes about this, which can be met in the literature about surfaces, either are completely formal or result from a misinterpretation of the underlying definitions.

In practice, it is the microscopic physical structure of the system which should dictate which of the two approaches is best adapted. As seen above, when the physical properties vary quasi-monotonously across the interface, Gibbs's point of view is well-suited. If, on the contrary, one at least of the profiles (e.g. the density one) behaves as indicated on Fig. 4, as may for instance be the case for a mixture exhibiting selective adsorption — it is obviously appropriate to define a surface phase. In this latter case, all interface quantities are non-zero. (However, for completeness, one must then define two (Gibbs?) surfaces of contact between the surface and volume phases!)

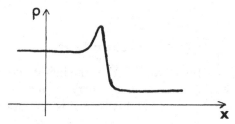

Figure 4: Typical shape of the density profile
for a mixture exhibiting selective adsorption.

In any case, and this is the important point for our concern, which is to study phenomena on at least the micron scale, the thickness of this phase is of atomic order, i.e. is negligible on this scale. So, definition (2.31), as well as the associated formalism, is valid whatever the choice of definition of the interface. The difference between both approaches only appears indirectly, via differences in the space variations of the volume kinetic coefficients in the immediate vicinity of the interface. This, again, decides on the description to be chosen. Indeed, non-equilibrium thermodynamics is only valid, within a given phase, to describe variations on distances much larger than the atomic ones. So, all fast variations must be included into interface effects, and it is this heuristic condition which should finally prevail.

2.2.B *Interface conservation and balance equations*

Let $y(\mathbf{r}, t)$ be the density of an extensive quantity in our two phase system. Its evolution is described by the general equation

$$\frac{\partial}{\partial t} y(\mathbf{r}, t) = -\nabla \cdot \mathbf{Y}(\mathbf{r}, t) + \chi_y(\mathbf{r}, t), \tag{2.39}$$

where $\mathbf{Y}$ is the flux associated with y, and χ_y a source term which vanishes if y is a conserved quantity. We formally decompose $\mathbf{Y}$ and χ as well as y, according to (2.31), into volume and interface sources and currents. The evolution equations for y_L, y_S and y_I are then obtained [20] by plugging expressions (2.31) for $y, \mathbf{Y}, \chi_y$ into (2.39) and equating separately the coefficients of θ^L, θ^S, δ^I as well as the component of $\mathbf{A}$ normal to the interface in terms of the form $\mathbf{A} \cdot \nabla \delta^I$. Moreover, one takes advantage of the following relations [21]:

$$\frac{\partial}{\partial t} \theta^{L,S} \big(f(\mathbf{r}, t) \big) = \mp \mathbf{v}_I \cdot \hat{n} \delta^I(\mathbf{r}, t), \tag{2.40a}$$

$$\nabla \theta^{L,S} \big(f(\mathbf{r}, t) \big) = \pm \hat{n} \delta^I(\mathbf{r}, t), \tag{2.40b}$$

$$\frac{\mathrm{d}^I}{\mathrm{d}t} \delta^I(\mathbf{r}, t) = 0, \qquad \text{where} \tag{2.40c}$$

$$\frac{\mathrm{d}^I}{\mathrm{d}t} \equiv \left(\frac{\partial}{\partial t} + \mathbf{v}_I \cdot \nabla \right). \tag{2.40d}$$

$\hat{n}$ is the unit vector normal to the interface pointing into the liquid. The normal component $\mathbf{v}_I \cdot \hat{n}$ of the 'interface velocity' ($\mathbf{v}_I \cdot \hat{n} = -|\nabla f|^{-1} \partial f / \partial t$) is the *growth velocity* of the solid (the only displacement of the interface which is associated with variations of volume of the two L, S phases is that along the local normal). The tangential component has a different physical meaning: it describes convective transport along the interface.

One thus obtains:

$$\frac{\partial}{\partial t} y_{L,S}(\mathbf{r}, t) + \nabla \cdot \mathbf{Y}_{L,S}(\mathbf{r}, t) - \chi_{L,S}(\mathbf{r}, t) = 0, \tag{2.41a}$$

$$\frac{\partial}{\partial t} y_I(\mathbf{r}, t) + \boldsymbol{\nabla} \cdot \mathbf{Y}_I - \chi_I \tag{2.41b}$$

$$= -\hat{n} \cdot \left[\mathbf{Y}_L - \mathbf{Y}_S - \mathbf{v}_I (y_L - y_S) \right]_{\text{int}},$$

$$(\mathbf{Y}_I - y_I \mathbf{v}_I) \cdot \hat{n} = 0. \tag{2.41c}$$

In Eq. (2.41b), the subscript 'int' means that limits on the interface of volume quantities are to be taken. Equation (2.41a) simply reproduces, as necessary, the one-phase balance or conservation equations (see § 2.A). The presence of the interface gives rise to interface balance equations (Eq. (2.41b)) and to conditions (Eq. (2.41c)) expressing the fact that, by definition, interface quantities propagate with the surface to which they are linked.

The balance equation (2.41b) can be visualized as the balance for a Gauss pillbox linked to the interface, which thickness is of atomic order in Guggenheim's description, zero if one uses Gibbs's definition, but which anyhow receives through its lateral faces a total flux $(-\boldsymbol{\nabla} \cdot \mathbf{Y}_I)$, due to interface transport.

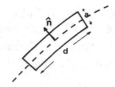

Figure 5: Gauss pillbox attached to the interface.

If the interface was motionless, the effect of the contact with the volume phases would only be to feed into the Gauss box a total flux $(\mathbf{Y}_S - \mathbf{Y}_L) \cdot \hat{n}$. The box moves at velocity $\mathbf{v}_I$, which corresponds to a rate $\mathbf{v}_I \cdot \hat{n}$ of liquid-solid transformation. In the interface (box) frame, the fluxes are thus $(\mathbf{Y}_{L,S} - y_{L,S} \mathbf{v}_I) \cdot \hat{n}$ (the box swallows liquid with a density y_L and spits out solid with density y_s).

One can thus immediately write down the interface mass and energy conservation equations and the momentum balance which complement, for the solidifying system, the corresponding volume equations.

a) *Mass conservation at the interface*

For a binary system $(k = 1, 2)$, it reads:

$$\frac{\partial \rho_{kI}}{\partial t} + \operatorname{div}(\rho_{kI} \mathbf{v}_{kI})$$

$$= \hat{n} \cdot \left[\rho_{kL}(\mathbf{v}_I - \mathbf{v}_{kL}) - \rho_{kS}(\mathbf{v}_I - \mathbf{v}_{kS}) \right]_{\text{int}}, \tag{2.42a}$$

with (cf. Eq. (2.41c)):

$$(\mathbf{v}_{kI} - \mathbf{v}_I) \cdot \hat{n} = 0. \tag{2.42b}$$

It is important to evaluate the respective orders of magnitude of the various terms appearing in the conservation equations (2.42a). Let us assume, for the sake of simplicity of the following argument, that we have chosen the 'surface phase' point of view (this is of no consequence on the conclusions). The Gauss box then has a thickness a of atomic order and lateral dimensions of 'hydrodynamic' order $(d \gg a)$ — fixed by the scale of volume space variations. Conservation in the box is then of the form: $(j_L - j_S) \simeq (a/d) j_I$, where the j's are current densities. This expression takes into account the fact that ρ_{kI}, which is a *surface* density, is of order $a \cdot \rho^{\mathrm{vol}}$, and that $\mathrm{div}(\rho_I v_I) \sim (1/d) \rho_I v_I$, since d is the lateral length scale.

Typically, $a/d \sim 10^{-3}$–10^{-4}, and the terms involving interface densities and currents in (2.42a) — and, more generally, (2.41b), are in general negligible,[†] which we will from now on assume. Conservation of mass at the interface is then simply expressed by:

$$\rho_{kL}(\mathbf{v}_{kL} - \mathbf{v}_I) \cdot \hat{n} = \rho_{kS}(\mathbf{v}_{kS} - \mathbf{v}_I) \cdot \hat{n} \equiv \mathbf{j}_k \cdot \hat{n}, \tag{2.43}$$

where the *conserved quantity* $(-\mathbf{j}_k \cdot \hat{n})$ is the *solidification rate* of species k.

One can also define a conserved total solidification rate $(-\mathbf{J} \cdot \hat{n})$ by:

$$\mathbf{J} \cdot \hat{n} \equiv \mathbf{J}_{L,S} \cdot \hat{n} = \sum_k \mathbf{j}_{kL,S} \cdot \hat{n} = \hat{n} \cdot [\rho_{L,S}(\mathbf{v}_{L,S} - \mathbf{v}_I)]_{\mathrm{int}}. \tag{2.44}$$

$\mathbf{J} \cdot \hat{n}$ is the mass flow towards the interface measured in the interface frame.

[†] Let us however insist that this approximation is no longer justified when current densities in the volume phases become extremely small. This is for example the case:

• for extremely slow growth — due either to the fact that departure from equilibrium is very small, or to a very slow attachment kinetics (see below);

• in sintering, where the flow of mass from the gas environment of the solid is negligible.

In such cases, surface mass transport (surface diffusion) becomes important, and the full formalism [19, 20] must be used.

There is no convection in a solid, and we neglect elastic motions. So:

$$\mathbf{v}_s = 0, \tag{2.45}$$

and (2.44) reduces to:

$$\mathbf{v}_L \cdot \hat{n} = \frac{\rho_L - \rho_S}{\rho_L} \mathbf{v}_I \cdot \hat{n}. \tag{2.46}$$

Equation (2.46) is the expression of the *advection* effect mentioned in Section 1: the densities ρ_L, ρ_S of the two phases are in general different. In order for a given volume of liquid to transform into solid, mass must be fed into (or rejected from) it. This, since the solid is assumed non-deformable, gives rise to a flow ($\mathbf{v}_L$) in the liquid. In practice, for many pure materials, $\delta\rho/\rho$ is of the order of a few per cent, and the advection effect is small enough to be neglected. It may however become important in some situations such as:

▷ rapid solidification where growth velocities may reach the m/sec range;

▷ some cases of growth from a solution, where $\delta\rho/\rho$ may be large (for example, $\delta\rho/\rho \sim 30\%$ for NH_4Br growing from its aqueous solution);

▷ in experiments in which one tries to get rid of natural convection (e.g. microgravity), one should keep in mind the fact that advective flows remain present;

▷ in experiments on very thin samples: the presence of confining plates may reduce considerably flows in the thin liquid layer. It may then be possible that the solid does not grow with its equilibrium density, but under a (dilation) stress depending on the growth velocity.

b) *Momentum conservation*

Since we neglect elastic deformations and assimilate the solid, from the mechanical point of view, to an isotropic medium, stresses in the solid phase reduce to a hydrostatic pressure p_s ($\overline{\overline{P}}_s = p_s \overline{\overline{I}}$). The effect of external forces (e.g. gravity) on the interface is negligible, for the same reasons as those discussed in § 2.2.B.a) above, as well as the contribution of surface inertia (terms $\partial(\rho_I \mathbf{v}_I)/\partial t$ and $\mathrm{Div}(\rho_I \mathbf{v}_I \mathbf{v}_I)$; see Eq. (2.16)). Using Eq.(2.45) and decomposing the volume momentum flux into $\rho \mathbf{v}\mathbf{v} + \overline{\overline{P}}$, one finds that the momentum equation deduced from (2.41b) reduces to:

$$\hat{n} \cdot [\overline{\overline{P}}_L - \overline{\overline{P}}_s + \rho_L(\mathbf{v}_L - \mathbf{v}_I)\mathbf{v}_L] + \mathrm{Div}\,\overline{\overline{P}}_I = 0. \tag{2.47}$$

$\overline{\overline{P}}_I$ is the interface stress tensor. From (2.41c) ($\overline{\overline{P}}_I \cdot \hat{n} = 0$) it only has components in the plane tangent to the interface. One then decomposes it into a 'surface pressure' term and a viscous one $\overline{\overline{\Pi}}_I$, which we drop since we neglect surface currents:

$$\overline{\overline{P}}_I \cong p_I(\overline{\overline{I}} - \hat{n}\hat{n}), \tag{2.48}$$

where we assume for the moment that the interface is isotropic.

Equation (2.47) is the condition of mechanical equilibrium of the interface (since we have neglected surface inertia, this equilibrium is instantaneous). In the absence of fluid flow, once projected on $\hat{n}$, it simply yields:

$$\rho_L - \rho_S = -\hat{n} \cdot \mathrm{Div}[\rho_I(\overline{\overline{I}} - \hat{n}\hat{n})]. \tag{2.49}$$

Now, by definition[†]:

$$p_I = -\gamma, \tag{2.50}$$

where γ is the *interface energy*.

On the other hand, it can be shown [20] that:

$$\hat{n} \cdot \mathrm{Div}(\overline{\overline{I}} - \hat{n}\hat{n}) = \mathcal{K} = \frac{1}{R_1} + \frac{1}{R_2}, \tag{2.51}$$

where $\mathcal{K}$ is the total curvature of the interface, defined as positive for a convex solid. R_1, R_2 are the principal radii of curvature.

When the *interface energy* can be assumed *isotropic*, equation (2.49) can then be rewritten as:

$$p_S - p_L = \gamma\mathcal{K}, \tag{2.52}$$

i.e. , it reduces to Laplace's equation.

In the presence of fluid flow, following (2.47), it generalizes into:

$$p_S - p_L - \hat{n} \cdot \overline{\overline{\Pi}}_L \cdot \hat{n} - (\mathbf{J} \cdot \hat{n})(\mathbf{v}_L \cdot \hat{n}) = \gamma\mathcal{K} \tag{2.53}$$

which contains contributions due to viscous forces and to the change in momentum which takes place when liquid with velocity $\mathbf{v}_L$ transforms into solid.

[†] γ is defined from the expression of the work received by the system when its interface area is increased, at constant V_L, V_S by $\delta\mathcal{A}$: $\delta W = \gamma\delta\mathcal{A}$, which is also given by $\delta W = -p_I\delta\mathcal{A}$.

Finally, due to the crystal structure of solids, the interface energy γ is in general anisotropic. It can be shown [22] that, in this case, the Laplace force is:

$$\left(\gamma + \frac{\partial^2 \gamma}{\partial \theta_1^2}\right)\frac{1}{R_1} + \left(\gamma + \frac{\partial^2 \gamma}{\partial \theta_2^2}\right)\frac{1}{R_2}, \qquad (2.54)$$

where θ_1, θ_2 are the polar angles, in each principal plane measured from the local normal to the interface.

One may of course, at this stage, wonder why we have not neglected Laplace's force, as far as it is much smaller than volume forces except for very large curvatures, which makes it usually irrelevant except for very small systems. The reason for retaining it is the following: besides the fact that we may be interested in the growth of small germs with radii just above the nucleation threshold (the value of which is determined by surface tension), capillarity is, as we will see, the *only* force which stabilizes growth fronts against short wavelengths deformations. So, although it is a weak force, it is essential for our concern and must be retained even in the most simplified model of solidification.

c) *Energy conservation*

When surface mass and surface heat transport are neglected — as well as the temperature dependence of the surface energy [23], it reads [19, 20]

$$(\mathbf{J}_{qL} - \mathbf{J}_{qS})\cdot\hat{n} + (\mathbf{J}\cdot\hat{n})\left[\frac{1}{2}\left((\mathbf{v}_L - \mathbf{v}_I)^2 - \mathbf{v}_I^2\right) + h_L - h_S\right]$$
$$+ \hat{n}\cdot\overline{\overline{\Pi}}_L\cdot(\mathbf{v}_L - \mathbf{v}_I) = 0, \qquad (2.55)$$

where h is the enthalpy per unit mass. Equation (2.55) means that the energy current in the interface frame, the normal component of which:

$$\mathcal{J}_e \cdot \hat{n} = (\mathbf{J}_e - \mathbf{v}_I e)_{L,S}\cdot\hat{n}$$
$$= \hat{n}.\left[\mathbf{J}_q + \mathbf{J}\left\{h + \frac{1}{2}(\mathbf{v} - \mathbf{v}_I)^2 + \overline{\overline{\Pi}}\cdot(\mathbf{v} - \mathbf{v}_I)\right\}\right]_{L,S} \qquad (2.56)$$

is conserved, can be decomposed into heat current, viscous energy transport and convective transport of kinetic energy and enthalpy.

2.2.C *Interface entropy production and kinetic laws*

In order to write the expression for entropy production in the heterogeneous system, it is of course necessary to extend to the (two-phase + interface) system the local equilibrium assumption. That is, we assume that the thermodynamic identity is valid locally, not only within the volumes (L) and (S) (Eq. (2.2)), but, as well, for interface thermodynamic quantities, i.e.:

$$\frac{d^I(s_I)}{dt} = \frac{1}{T_I}\left[\frac{d^I(u_I)}{dt} - \sum_k \mu_{kI}\frac{d^I(\rho_{kI})}{dt}\right], \qquad (2.57)$$

where T^I is an interface temperature, and d^I/dt ('interface-bound' derivative) is defined by (2.40d).

The assumption expressed by Eq. (2.57) clearly implies that interface thermodynamic quantities vary, along the interface, slowly on the scale of the local mean free path, which is determined by interactions between surface excitations. However, it should be pointed out that the existence itself of the interface implies that some at least of the relevant quantities vary rapidly (on the atomic scale) in the transverse direction — as expressed by decomposition (2.31). In other words, the Gauss pillbox of Fig. 5 has a microscopic thickness, and a large contact area along the volume phases. The assumption that it can be considered as a quasi-isolated thermodynamic system therefore appears to us to be a strong hypothesis, the conditions of validity of which remain ill-defined. While it is probably justified for rough interfaces — which, as we shall see, function practically without discontinuity of chemical potentials — the question of its validity seems less clear in the case, for example, of facets growing by 2D nucleation. In this case, is the probability of a critical nucleation fluctuation independent of the (strong) departure from equilibrium across the interface? This type of question remains, as far as we know, completely open.

Proceeding, as for the homogeneous system, with the help of Eqs. (2.2), (2.57) and of the conservation laws, one finds that the entropy production can be decomposed following (2.31). One can then identify volume production terms, given for each phase by the standard expression (2.21), and an *interface entropy production* σ_I, which must also be non-negative. When interface mass and transport are neglected, and once conservation of mass and energy currents transverse to the interface is taken into account (Eqs. (2.43) and (2.56)), one finds:

$$\sigma_I = -\sum_k (\mathbf{j}_k \cdot \hat{n})\left[\frac{1}{T}\left\{\mu_k + \tfrac{1}{2}(\mathbf{v} - \mathbf{v}_I)^2 + \hat{n}\cdot\frac{\overline{\overline{\Pi}}}{\rho}\cdot\hat{n}\right\}\right]_S^L$$

$$+ (\boldsymbol{\mathcal{J}}_e \cdot \hat{n})\left(\frac{1}{T}\right)_S^L - \frac{1}{T_L}\hat{n}\cdot\overline{\overline{\Pi}}_L\cdot(\overline{\overline{I}} - \hat{n}\hat{n})\cdot(\mathbf{v}_L^{\text{int}} - \mathbf{v}_I)_t, \qquad (2.58)$$

where $[A]_S^L = (A^L - A^S)_{\text{int}}$, and $\mathbf{v}_t$ is the tangential velocity (projection of $\mathbf{v}_L$ on the plane tangent to the interface).

This expression, which generalizes the one written, for a pure non-viscous fluid (superfluid He), by Castaing and Nozières [24], has the same structure as expression (2.21): each term is the product of a conserved transverse current by an interface thermodynamic force. In the volume the forces are gradients of continuous thermo(hydrodynamic) quantities. At the interface, the corresponding forces are the discontinuities of these same quantities across the interface.

Note that, in the presence of a flow in the liquid, mass currents are driven by variations of the *effective chemical potential* $\tilde{\mu}_k$, which contains a viscous contribution and a kinetic energy term:

$$(\tilde{\mu}_k)_{L,S} = \left(\mu_k + \hat{n} \cdot \overline{\overline{\Pi}} \cdot \hat{n} + \tfrac{1}{2}(\mathbf{v} - \mathbf{v}_I)^2\right)_{L,S}. \qquad (2.59)$$

Proceeding as for the homogeneous system, one linearizes the forces in the departures from equilibrium $\tilde{\mu}_{kL} - \tilde{\mu}_{kS}$, $T_L - T_S$, and writes the linear phenomenological laws (note that $\mathbf{j}_k \cdot \hat{n}$ and $\mathcal{J}_e \cdot \hat{n}$ are scalars, $(\mathbf{v}_L - \mathbf{v}_I)_t$ is a vector).

For an isotropic interface, these relations read:

$$\frac{\tilde{\mu}_{1S} - \tilde{\mu}_{1L}}{T} = \alpha_{11}\mathbf{j}_1 \cdot \hat{n} + \alpha_{12}\mathbf{j}_2 \cdot \hat{n} + \alpha_{13}\bar{\mathbf{J}}_q \cdot \hat{n}; \qquad (2.60a)$$

$$\frac{\tilde{\mu}_{1S} - \tilde{\mu}_{2L}}{T} = \alpha_{12}\mathbf{j}_1 \cdot \hat{n} + \alpha_{22}\mathbf{j}_2 \cdot \hat{n} + \alpha_{23}\bar{\mathbf{J}}_q \cdot \hat{n}; \qquad (2.60b)$$

$$\frac{T_S - T_L}{T^2} = \alpha_{13}\mathbf{j}_1 \cdot \hat{n} + \alpha_{23}\mathbf{j}_2 \cdot \hat{n} + \alpha_{33}\bar{\mathbf{J}}_q \cdot \hat{n}; \qquad (2.60c)$$

$$(\mathbf{v}_I - \mathbf{v}_L)_t = \alpha_{44}\hat{n} \cdot \overline{\overline{\Pi}}_L \cdot (\bar{I} - \hat{n}\hat{n}), \qquad (2.60d)$$

where

$$\bar{\mathbf{J}}_q \cdot \hat{n} = (\mathcal{J}_e - \tilde{\mu}_1\mathbf{j}_1 - \tilde{\mu}_2\mathbf{j}_2) \cdot \hat{n} \qquad (2.61)$$

is the heat current across the interface. $\tilde{\mu}_k, T$ are average values at the interface (e.g. $T = \tfrac{1}{2}(T_L + T_S)$).

The matrix of the α_{ij} coefficients is non-negative (second law) and symmetric (Onsager's relations):

$$\alpha_{ii} \geq 0, \qquad \alpha_{ii}\alpha_{jj} \geq (\alpha_{ij})^2, \qquad (2.62)$$

$$\alpha_{44} \geq 0. \qquad (2.63)$$

α_{44} relates the shear in the liquid close to the interface with the discontinuity of tangential velocities. So, α_{44}^{-1} is a *contact viscosity*. On the other hand, within our non-deformable solid assumption, $\mathbf{v}_S = \overline{\overline{\Pi}}_S = 0$. Mechanical contact therefore imposes $(\mathbf{v}_I)_t = 0$, and equation (2.60d) determines $(\mathbf{v}_L)_t$. Contact viscosities are in general very large, as far as the interface can be considered quasi-solid, and (2.60d) then reduces to the *no-slip condition*:

$$(\mathbf{v}_L)_t = 0. \tag{2.64}$$

Note, however, that while (2.64) is certainly correct for the growth of an ordinary solid, it might not apply to solidification of polymers: on the one hand, their hydrodynamic behavior close to an interface is known to be non-standard, due to reptation phenomena [25]; on the other hand, polymer melts are extremely viscous, and volume viscous stresses may be considerable.

Equations (2.60a–2.60c) relate the mass and heat currents across the interface with the temperature and chemical potential jumps. The off-diagonal coefficients α_{13}, α_{23} describe the Peltier effect for each species. $\alpha_{23} T^2$ is the *Kapitza resistance* of the interface. These effects are negligible, for the solid-liquid interface, for practically all materials, due to the smallness of the phonon mean free path at ordinary melting temperatures. Equation (2.60c) then reduces to the condition of *thermal equilibrium* at the interface:

$$T_S\big|_{\text{int}} = T_L\big|_{\text{int}}. \tag{2.65}$$

Let us insist on the fact that this is *not valid for helium*, in which the dissipative thermal interface effects are non-negligible [26].

2.2.D *Attachment kinetics*
For solidification of ordinary materials, the non-trivial part of Eqs. (2.60) therefore reduces to the so-called *attachment kinetic laws* which relate the chemical potential jumps across the interface with the solidification rates. For a pure material, this reads:

$$[\tilde{\mu}_S - \tilde{\mu}_L]_{\text{int}} = \beta \mathbf{J} \cdot \hat{n} \qquad (\beta = \alpha T), \tag{2.66}$$

where, in the absence of fluid flow, $-\rho^{-1}\mathbf{J} \cdot \hat{n} = \mathbf{v}_I \cdot \hat{n}$ is simply the local growth velocity of the solid.

The effect described by Eq. (2.66) is specific of solidification (more generally, of the dynamics of first order phase transitions). It is an *interface dissipation*, i.e. a 'resistance' of the system against the formation of the growing phase, due to the difference between the microscopic structures (the symmetries) of the two phases. Indeed, in order for the atoms or molecules of the liquid in the immediate vicinity of the interface to incorporate into the crystal lattice of the growing solid, they must adapt their positions and orientations — which means rearranging their local environment. This process may be pictured, very schematically (and this is a reasonable first approximation for rough interfaces), by analogy with self-diffusion, as a thermally activated jump over a potential barrier [27]. The higher the barrier, the smaller the rate of incorporation into the solid for a given $\tilde{\mu}_L - \tilde{\mu}_S$ (the larger β).

Clearly, equation (2.66) is only the linearized version of the real mass current — chemical potential difference relation:

$$[\tilde{\mu}_S - \tilde{\mu}_L]_{\text{int}} = -f(\mathbf{J} \cdot \hat{n}). \tag{2.67}$$

The sign of f is that of $(-\mathbf{J} \cdot \hat{n})$: when the solid grows ($\mathbf{J} \cdot \hat{n} < 0$), necessarily $\tilde{\mu}_S < \tilde{\mu}_L$.

The functional form of the kinetic law (of function f) — in particular the extension of the growth velocity range where the linear approximation (2.66) is legitimate — depends very strongly on the interface microstructure.[†] This question is discussed in detail by P. Nozières's [28], so we only sum up here the main conclusions.

As far as growth kinetics is concerned, interfaces exhibit widely different behaviors depending on whether, close to the melting point, they are rough on the macroscopic scale or not.

a) *Faceted interfaces* (quasi-perfect atomic planes)

These have *slow growth kinetics*. In the absence of steps associated with the emergence of dislocations, in order for a new solid terrace to grow, its radius must first exceed the critical value determined by the balance between the gain in volume free energy and the step energy cost associated with the terrace edge. Growth thus takes place via 2D nucleation. It is the probability for a critical terrace to appear which determines the nucleation rate, and thus the kinetic law, which has the form:

$$(-\mathbf{J} \cdot \hat{n}) = \gamma \exp - \left[\frac{A}{k_B T (\tilde{\mu}_L - \mu_S)_{\text{int}}} \right], \tag{2.68}$$

[†] The kinetic coefficients and the f-function also depend, in general, on the direction of the interface relative to the crystal lattice (anisotropic kinetics).

i.e., in practice, the linear approximation is never valid.

If screw dislocations emerge onto the facet, this results into the presence of atomic steps on which atoms may attach without having to form critical terraces. The corresponding kinetic law is of the form [28]

$$(\tilde{\mu}_L - \tilde{\mu}_S)_{\text{int}} = \beta_d (\mathbf{J} \cdot \hat{n})^2 \qquad (2.69)$$

b) *Rough interfaces*

As compared with ideal facets, these can be considered as containing a large density of defects (vacancies, terraces, steps) i.e. of sites of easy attachment. So, their kinetics is very fast — the 'jumping time' for an atom to incorporate into the solid is, roughly speaking, of the order of the inverse atomic vibration frequency ($\sim 10^{-12}$ sec). This time is much smaller than the time $(\mathbf{v}_I \cdot \hat{n}/a)^{-1}$ for growth of an atomic layer of solid as long as the growth velocity is smaller than, typically, $\sim$ 1m/sec.

So, *except in rapid solidification regimes*, one can consider that attachment onto a rough liquid-solid interface is quasi-instantaneous. This means that β in Eq. (2.66) is very small (a very small $\delta\mu$ is enough to drive a large mass current), and one may admit that the interface functions at *local thermodynamic equilibrium*:

$$(\tilde{\mu}_L - \tilde{\mu}_S)_{\text{int}} = 0. \qquad (2.70)$$

When the solid grows, there is of course a chemical potential drop in the system, but in this case it takes place entirely in the volume of the liquid and solid.

The above remarks also apply to *alloys*. One should simply keep in mind that it is necessary, in principle, to write *one kinetic law for each consistuent*, and that the linear laws (2.60a, 2.60b) *a priori* contain off-diagonal terms. These are the phenomenological expression of the fact that the attachment kinetics of an atom of type (1) is modified by interactions with neighboring (2) atoms. For instance, it may happen that (2) can be incorporated only together with a (1) atom, or by exchange with a (1) of the solid, etc. These effects, which are usually negligible for dilute mixtures, are of importance in metallurgy, since they are responsible for the 'solute trapping' phenomenon [29], which gives rise to the formation of metastable or unstable alloys.

Finally, let us mention the case of polymers. Their solidification kinetics is very slow, but also much more complex than those described by (2.68) or (2.69). Solidifying a polymer necessitates that long chain segments come into 'the right position'. This means configuration changes which may involve for example reptation of the liquid chains, the dynamics of which resorts to polymer physics. Let us only point out that polymers exhibit specific growth morphologies, such as spherulites [30] which have not been studied experimentally in a very systematic way yet, and the theoretical interpretation of which is a completely open subject [31, 32].

So, when one wants to study the solidification dynamics of a particular material, it is desirable to have at least an approximate prediction of the (rough or faceted) microstructure of its liquid-solid interface. When no experimental information about the roughening transition temperature is available, an estimate can be obtained from *Jackson's criterion* [4]. One calculates the ratio $\psi = L/k_B T_M$, where L is the latent heat of melting, T_M the melting temperature. That is, ψ compares the enthalpy difference between the two phases (roughly, their 'atomic or molecular relative affinity') with the entropy gained by creating a defect on a facet. So, if ψ is distinctly smaller than 1 (resp. larger), the interface should be rough (resp. faceted): the system gains more in entropy by disordering its interface microstructure than it would gain in energy by maximizing the number of solid-solid bonds.

It is however necessary to insist on the fact that this is only a *semi-quantitative* criterion. The value $\psi = 1$ by no means fixes an exact frontier between the two regimes.

In practice, most metal interfaces are rough, as well as those of organic materials which solidify into a plastic phase (in this case L is very small). This is the case for example of succinonitrile and CBr_4, on which many recent experimental studies have been performed. On the contrary, ice and most ionic crystals have faceted interfaces, as well as many semiconducting and organic materials.

2.2.E *The generalized Gibbs-Thomson equation*

We will derive it here for a pure material; its generalization to binary mixtures is performed in Appendix A.

Using the definition (2.59) of the effective chemical potential, the kinetic law (2.67) can be written:

$$(\mu_S - \mu_L)_{\text{int}} = \left[\hat{n} \cdot \left(\frac{\overline{\overline{\Pi}}_L}{\rho_L} \right) \cdot \hat{n} + \tfrac{1}{2} v_L^2 - \mathbf{v}_L \cdot \mathbf{v}_I \right]_{\text{int}} - f(\mathbf{J} \cdot \hat{n})$$

$$\equiv \Delta\mu. \tag{2.71}$$

Each chemical potential is in principle a known function of state variables (the equilibrium one):

$$\mu_\alpha \equiv \mu_\alpha(p_\alpha, T_\alpha) \qquad (\alpha = L, S). \tag{2.72}$$

Assuming that the interface jump remains reasonably small, one expands $\mu_S - \mu_L$ about the reference point (p_L, T_M), where $T_M \equiv T_M(p_L)$ is the melting temperature at pressure p_L, i.e.

$$\mu_L(p_L, T_M) = \mu_S(p_L, T_M). \tag{2.73}$$

Then:

$$\mu_\alpha(p_\alpha, T_\alpha) = \mu_\alpha(p_L, T_M) + V_\alpha(p_\alpha - p_L) - s_\alpha(T_\alpha - T_M), \tag{2.74}$$

where V_α, s_α are the volume and entropy per unit mass in the α phase. Taking into account the continuity of temperature at the interface (2.65) (we thus forget here about He) and the condition of mechanical equilibrium (2.53) (generalized Laplace equation), one can then rewrite (2.71), for a system with isotropic surface energy γ, as:

$$T_{\text{int}} = T_M + \frac{T_M}{L} \Big\{ - V_S \gamma \mathcal{K} - f(\mathbf{J} \cdot \hat{n}) \tag{2.75}$$

$$- \left(V_S - \frac{1}{\rho_L} \right) (\hat{n} \cdot \overline{\overline{\Pi}}_L \cdot \hat{n})$$

$$+ \rho_S V_S (\mathbf{v}_I \cdot \hat{n})(\mathbf{v}_L \cdot \hat{n}) + \left(\tfrac{1}{2} \mathbf{v}_L^2 - \mathbf{v}_L \cdot \mathbf{v}_I \right) \Big\}.$$

$\overline{\overline{\Pi}}_L$ and $\mathbf{v}_L$ are the interface values. If the interface energy is anisotropic, $\gamma \mathcal{K}$ must be replaced by expression (2.54).

Using $V_S = \rho_S^{-1}$, $L = T_M(s_L - s_S)$, $\mathbf{J} \cdot \hat{n} = -\rho_S \mathbf{v}_I \cdot \hat{n}$ (Eq. (2.44) with $\mathbf{v}_s = 0$), and the no-slip condition (2.64), we finally get:

$$T_{\text{int}} = T_M - \frac{\gamma T_M}{\rho_S L} \mathcal{K} - \frac{T_M}{L} f(\mathbf{J} \cdot \hat{n})$$

$$+ \frac{T_M}{L} \left(\tfrac{1}{2} \mathbf{v}_L^2 + \frac{\rho_S - \rho_L}{\rho_L \rho_S} \hat{n} \cdot \overline{\overline{\Pi}}_L \cdot \hat{n} \right). \tag{2.76}$$

This equation defines, for a given growth regime, the temperature of the moving interface.

When hydrodynamic effects are neglected, and when attachment is assumed instantaneous $(f = 0)$, it reduces to the *Gibbs-Thomson equation* for a pure material:

$$T_{\text{int}} = T_M - \frac{\gamma T_M}{\rho_S L} \mathcal{K}, \tag{2.77}$$

which expresses how the melting temperature is shifted by the finite curvature of the interface (by the Laplace pressure). This shift is negative for a convex solid.

The hydrodynamic term on the r.h.s. of (2.76) contains a kinetic energy contribution proportional to $(\delta\rho/\rho)^2$, which is in general negligible. The viscous pressure term, proportional to $\delta\rho/\rho$, can also usually be neglected, except for highly viscous melts [32] (polymers, glass-forming liquids).

Finally, it is seen on (2.76) that non-instantaneous attachment kinetics ($f \neq 0$) decreases the interface temperature. The corresponding shift $\delta T_{\text{kin}} = -T_M/Lf(\mathbf{J} \cdot \hat{n})$ is called *kinetic undercooling*: at given liquid pressure, and for a planar front, T must be smaller than T_M in order for the system to build up the finite $\Delta\mu$ necessary to drive growth at velocity $\mathbf{v}_I \cdot \hat{n}$.

Equation (2.76) is easily generalized to the case of binary mixtures. Since, in that case, there is one kinetic equation for each species, one obtains *two* equations relating interface temperature and concentrations. For a *dilute alloy*, they read (see Appendix A):

$$T_{\text{int}} = T_M + m_L C_L - \frac{\gamma T_M}{\rho_S L}\mathcal{K} - \frac{T_M}{L}f_1(\mathbf{J} \cdot \hat{n})$$
$$+ \frac{T_M}{L}\left(\tfrac{1}{2}\mathbf{v}_L^2 + \frac{\rho_S - \rho_L}{\rho_L \rho_S}\hat{n} \cdot \overline{\overline{\Pi}}_L \cdot \hat{n}\right) \qquad (2.78a)$$

and[†]

$$C_S = K C_L. \qquad (2.78b)$$

m_L is the slope $(\mathrm{d}T/\mathrm{d}C)_{L0}$ of the liquidus curve of the phase diagram at $C \to 0$. K is the *equilibrium segregation coefficient* (the ratio of the liquidus and solidus slopes). f_1 is the kinetic function associated with the solvent. Note that the shift of interface temperature (Eq. (2.78a)) is given by the same expression as for the pure solvent, the reference temperature being that of the alloy at the local concentration, $T_M + m_L C_L$.

2.2.F *The minimal model of solidification*

In §§ 2.2.A–E we have derived the whole set of equations needed to describe the growth of a non-deformable binary solid from its melt. On

---

[†] Eq. (2.78b) is only valid as such in the dilute limit. For concentrated alloys, there appear curvature and kinetic corrections (Appendix A).

the mesoscopic scale, this system is completely described by three fields: temperature, concentration, (liquid) velocity. In the volume of the two phases, these fields obey the usual transport equations (thermal and chemical diffusion, Navier-Stokes). At the S/L interface — the shape of which remains to be determined — they obey a set of equations which we have derived above, namely: mass conservation, heat balance, generalized Laplace law, attachment kinetic laws, hydrodynamic boundary conditions.

This set of equations, when complemented with the boundary conditions appropriate to a given experiment, define a closed mathematical problem — of the 'free-boundary' type — the solution(s) of which in principle describe(s) the shape of the growth front, its evolution, and the solute distribution in the growing solid.

As is known from experiments, a variety of growth regimes and often complex front morphologies are possible. This wealth of solutions also indicates that the associated general mathematical problem is highly non-trivial. Indeed, even when fluid flow is neglected, while the remaining volume diffusion equations are linear, the interface equations contain non-linearities coming into play via the interface curvature (Eq. (2.76 or 2.78a)) and the components of the interface normal unit vector. (For the simple case where the interface shape $z = z_I(x)$ depends on a single coordinate, $\mathcal{K} = -z_I''(1 + z_I'^2)^{-3/2}$, $n_z = (1 + z_I')^{-1/2}$, etc.)

It is therefore necessary, in order to try and identify the mechanisms responsible for instabilities (i.e. for changes in front morphologies) as well as the important parameters, to first reduce the above description to its simplest possible form, which we call the *minimal model*. It is the systematic analysis of this model, following the pioneering work of Mullins and Sekerka [33], which has brought up most of the recent advances in the theory of solidification.

The minimal model assumes that:

(i) There is no flow in the liquid phase. That is, the system is not stirred, nor rotated, and natural convection is also neglected, as well as advection — i.e. the two phases are assumed to have the same density ρ.

(ii) Attachment is instantaneous. This implies that the S/L interface is microscopically rough. It is then reasonable, at least as a first approximation, to assume that S/L surface tension is isotropic.[†]

The equations describing solidification in this model are then obtained by setting, in §§ 2.2.A–E:

$$\mathbf{v}_L = 0, \quad \rho_S = \rho_L = \rho, \quad (\mu_S - \mu_L)_{\text{int}} = 0. \tag{2.79}$$

For a *dilute* binary mixture, they read:

a) *In the volume of each phase* $(\alpha \equiv L, S)$:

$$\frac{\partial T_\alpha}{\partial t} = D_{\alpha,\text{th}} \nabla^2 T_\alpha, \tag{2.80a}$$

$$\frac{\partial C_\alpha}{\partial t} = D_\alpha \nabla^2 C_\alpha, \tag{2.80b}$$

where $D_{\alpha,\text{th}}$, D_α are the thermal and chemical diffusion coefficients.

b) *On the interface, defined by* $z = z_I(\rho, t)$ $(\rho \equiv (x, y))$:

▷ continuity of temperature:

$$T_S = T_L \ ; \tag{2.81a}$$

▷ heat balance:

$$\hat{n} \cdot (K_S \nabla T_S - K_L \nabla T_L) = L \mathbf{v}_I \cdot \hat{n} \ ; \tag{2.81b}$$

▷ solute conservation:

$$\hat{n} \cdot (D_L \nabla C_L - D_S \nabla C_S) = (C_S - C_L) \mathbf{v}_I \cdot \hat{n} \ ; \tag{2.81c}$$

▷ local equilibrium:

$$C_S = K C_L, \tag{2.81d}$$

$$T_L = T_M + m_L C_L - \frac{\gamma T_M}{\rho L} \mathcal{K}, \tag{2.81e}$$

where we have rewritten (2.55) after noticing that, since the interface is at local equilibrium $(\mu_L = \mu_S)$, $h_L - h_S = T(s_L - s_S) = L$.

[†] However, this last restriction must be released in the case of dendritic growth in which capillary anisotropy plays a fundamental role (see Pomeau and Ben Amar's Chapter 4).

Equation (2.81c), which expresses solute mass conservation, is obtained from the interface mass conservation relations (2.43, 2.44) when use is made of the fact that (cf. Eqs. (2.6, 2.7, 2.9)) the transverse diffusion currents $\mathbf{J}_c^{(\alpha)} \cdot \hat{n}$ — given in each phase by Fick's law — are defined by:

$$\mathbf{j}_2 \cdot \hat{n} = \mathbf{J}_c^{(\alpha)} \cdot \hat{n} + C_\alpha \mathbf{J} \cdot \hat{n}. \qquad (2.82)$$

It simply states that the solute rejected (or absorbed) at the moving interface must be evacuated by volume diffusion.

For a *pure material*, the above set reduces to (2.80a), (2.81a, b, e).

For a *binary mixture*, it can be simplified further by exploiting the fact that *thermal diffusion is much faster than solute diffusion*. Typically, in the temperature range of interest, $D_L \sim 10^{-5} \, \mathrm{cm}^2/\mathrm{sec}$, $D_{\alpha,\mathrm{th}}$ ranges, roughly, from $10^{-1} \, \mathrm{cm}^2/\mathrm{sec}$ (metals) to $10^{-3} \, \mathrm{cm}^2/\mathrm{sec}$ (organic materials). So, the evacuation of the latent heat released on the moving front is quasi-instantaneous as compared with solute transport, and the thermal field should adjust quasi-adiabatically to the front position. This can be formalized by using reduced variables built from the mechanism that we expect to be limiting for solid growth, namely solute diffusion.

We choose for our length scale the *solute diffusion length in the liquid*

$$\ell = \frac{D_L}{V}, \qquad (2.83)$$

where V is the front velocity in the case of stationary growth and, for non-stationary regimes, some typical velocity scale.

The time scale is the diffusion time associated with this length:

$$\tau = \frac{D_L}{V^2}. \qquad (2.84)$$

Equations (2.80a), (2.81b) then read, when written in terms of the reduced variables $\mathbf{r}' = \mathbf{r}/\ell$, $t' = t/\tau$, $\hat{n} \cdot \mathbf{v}_I' = (\hat{n} \cdot \mathbf{v}_I)/V$:

$$\nabla'^2 T_\alpha = \frac{D_L}{D_{\alpha,\mathrm{th}}} \frac{\partial T_\alpha}{\partial t'}, \qquad (2.85a)$$

$$\hat{n} \cdot \left(\nabla' T_L - \frac{K_S}{K_L} \nabla' T_S \right)_{\mathrm{int}} = \frac{D_L}{D_{L,\mathrm{th}}} \frac{L}{C_p} (\mathbf{v}_I' \cdot \hat{n}), \qquad (2.85b)$$

where we have used $D_{\mathrm{th}} = K/C_p$, C_p being the specific heat at constant pressure. To lowest order (of an asymptotic expansion in the small parameter $D_L/D_{L,\mathrm{th}}$), the r.h.s. of Eqs. (2.85) are negligible, and equations (2.80, 2.81) for the dilute mixture become:

a) *Volume:*

$$\nabla'^2 T_\alpha = 0, \tag{2.86a}$$

$$\frac{\partial C_\alpha}{\partial t} = \frac{D_\alpha}{D_L} \nabla'^2 C_\alpha \; ; \tag{2.86b}$$

b) *Interface:*

$$T_S = T_L, \tag{2.87a}$$

$$\hat{n} \cdot \left(\nabla' T_L - \frac{K_S}{K_L} \nabla' T_S \right) = 0, \tag{2.87b}$$

$$\hat{n} \cdot \left(\nabla' C_L - \frac{D_S}{D_L} \nabla' C_S \right) = (C_S - C_L) \mathbf{v}'_I \cdot \hat{n}, \tag{2.87c}$$

$$C_S = K C_L, \tag{2.87d}$$

$$T_L = T_M + m_L C_L - \frac{\gamma T_M V}{\rho L D_L} \mathcal{K}'. \tag{2.87e}$$

It is important to mention that this set of volume and interface equations can be reduced to a single non-linear integro-differential equation for the front profile in two simple limits, namely the one-sided ($D_S = 0$) and the symmetric ($D_S = D_L$) models. The corresponding integral equations (which are in some cases, e.g. for the purpose of numerical simulations, or of analytical studies of dendritic growth, more easily to handle than (2.86, 2.87)) are derived in Appendix B.

The question of course arises of whether this minimal model is sufficient to describe real experimental situations. It is difficult in practice to eliminate natural convection when solidifying a bulk sample. As already mentioned in Section 1, the first order phase transition induces density gradients ahead of the front which may be destabilizing — whether this is the case must be checked for each material and growth configuration. Moreover, when working in an external thermal gradient (directional solidification; see Section 4), it is impossible to eliminate completely lateral gradients, since the only thermal contact, is necessarily, at the sample edges. Now, horizontal gradients induce convection without threshold [34].

One may of course consider performing experiments in microgravity. However, even if one neglects the fact that convection may still be driven by the Marangoni effect [35], in view of the technical demands of space experiments, of the considerable difficulty of any modifications while they are taking place, of the specific problems (e.g. control of residual gravity level, possible effects of the large accelerations when coming back into the atmosphere...), and — last but not least — of their price, we believe that they should be considered as tools which can be used only in quite exceptional cases.

However, the use of *thin samples* (with thickness typically in the 10–50 μm range) provides a rather easily accessible method for getting to a large extent rid of liquid flow effects.[†]

The assumptions of the minimal model concerning advection and attachment kinetics are of much lesser importance. Since many materials have $\delta\rho/\rho \sim$ a few per cent, advection currents are usually negligible at ordinary growth velocities. On the other hand, the local equilibrium assumption is well justified for materials with rough interfaces (metals, plastic crystals...) except in rapid solidification.

It is in view of these remarks that most of the recent experiments aiming at precise quantitative studies of growth morphology and dynamics have been performed on transparent materials (which allow for direct continuous observation) with rough interfaces such as succinonitrile [36, 37, 38], CBr_4 [39, 40], pivalic acid [41, 42], Xe and Kr [43] — as well as liquid crystals [44] — many experiments being performed on thin samples [37–41, 45].

3

The Mullins-Sekerka instability: free growth of a spherical germ

3.1 Free growth of a pure solid

This is *a priori* the simplest situation since, in the minimal model, the solidifying system is defined by a single field, the thermal one. The growth equations (2.80–2.81) reduce to:

[†] As may be suspected, there is a price to be paid for this advantage: some of the observed effects then exhibit a non-trivial dependence on sample thickness (see Section 4), and direct observation of the front profile along the small dimension is in practice impossible.

(i) Volume:

$$\frac{\partial T_\alpha}{\partial t} = D_{\alpha,\text{th}} \nabla^2 T_\alpha \; ; \tag{3.1}$$

(ii) Interface ($z = z_I(\rho, t)$):

$$\hat{n} \cdot (K_S \nabla T_S - K_L \nabla T_L) = L\mathbf{v}_I \cdot \hat{n}, \tag{3.2}$$

$$T_S = T_L = T_M - \frac{\gamma T_M}{\rho L} \mathcal{K}, \tag{3.3}$$

to which we must add boundary conditions. *Free growth* is defined as the case where the solid grows from an initial germ, either resulting from nucleation or, preferably, introduced through a capillary [36, 43] into the quasi-infinite[†] liquid undercooled at temperature $T_0 < T_M$. The boundary condition is then that $T_L \to T_0$ in the liquid infinitely far from the front.

3.2 The planar stationary solution

We will first investigate whether it is possible to find a regime in which the solid grows at constant velocity V with a planar front, i.e. whether (3.1–3.3) have a solution which is stationary in a frame moving at velocity $V\hat{z}$ with respect to the material (laboratory) frame. In this frame $(x, y, \tilde{z} = z - Vt, t)$

$$\frac{\partial}{\partial t} \longmapsto \frac{\partial}{\partial t} - V\frac{\partial}{\partial \tilde{z}}.$$

The temperature field must depend on z (the coordinate normal to the planar interface) only. It obeys the following stationary diffusion equations:

(a) *In the liquid ($\tilde{z} > 0$):*

$$\left(\frac{\partial^2}{\partial \tilde{z}^2} + \frac{1}{\ell_L^{\text{th}}} \frac{\partial}{\partial \tilde{z}}\right) T_L(\tilde{z}) = 0, \tag{3.4a}$$

$$T_L(\tilde{z} \to \infty) = T_0. \tag{3.4b}$$

[†] That is, the liquid should be, in all space directions, much thicker than the relevant scale, which is here (see below)

$$\ell_{\text{th}} = D_{\text{th}}/V,$$

where V is the growth velocity measured in the experiment.

(b) *In the solid* ($\tilde{z} < 0$):

$$\left(\frac{\partial^2}{\partial \tilde{z}^2} + \frac{1}{\ell_S^{\text{th}}}\frac{\partial}{\partial \tilde{z}}\right)T_S(\tilde{z}) = 0, \tag{3.5}$$

where ℓ_L^{th} (resp. ℓ_S^{th}) is the *thermal diffusion length* in the liquid (resp. solid):

$$\ell_{L,S}^{\text{th}} = \frac{D_{L,S}^{\text{th}}}{V}. \tag{3.6}$$

The solution of Eqs. (3.4) reads

$$T_L^{(0)}(\tilde{z}) = T_0 + A\exp\left(-\frac{\tilde{z}}{\ell_L^{\text{th}}}\right). \tag{3.7}$$

The only physical (non-diverging at $\tilde{z} \to -\infty$) solution of (3.5) is:

$$T_S^{(0)}(\tilde{z}) = B = \text{C}^{\text{st}}. \tag{3.8}$$

The continuity and local equilibrium conditions (3.3) at $\tilde{z} = 0$ impose ($\mathcal{K} = 0$):

$$A = T_M - T_0, \quad B = T_M. \tag{3.9}$$

The corresponding temperature field is represented in Fig. 6: the latent heat released by solidification accumulates ahead of the front, giving rise to a diffusion layer of thickness ℓ_L^{th}.

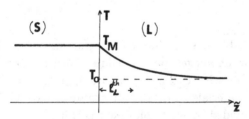

Figure 6: Planar stationary solution: temperature profile.

This solution must satisfy the heat balance condition (3.2), i.e.:

$$T_M - T_0 = L\frac{D_L^{\text{th}}}{K_L} = \frac{L}{C_p}. \tag{3.10}$$

So, we find that a planar stationary growth regime is possible only when the reduced undercooling

$$\Delta = \frac{C_p}{L}(T_M - T_0) \tag{3.11}$$

is equal to unity. In this case, nothing determines the growth velocity V, which may take any value.

This result can be understood by noticing that $\Delta = 1$ precisely corresponds to the case where solidification of a volume δV of liquid releases the heat $L\delta V = C_p(T_M - T_0)\delta V$ which is exactly that necessary to bring the temperature of the material from T_0 (liquid disappearing in the $\tilde{z} \rightarrow \infty$ region) to T_M (temperature of the solid produced). This is true whatever the rate of phase change.

One may of course ask whether planar growth is possible for $\Delta \neq 1$, i.e., with most materials, for $\Delta < 1$.[†] This question, the 1D Stefan problem has been studied extensively by mathematicians [47]. It is found that non-stationary planar growth is possible: the planar front moves following a diffusion law: $z_I = \sqrt{\mathcal{D}t}$. The front diffusion coefficient $\mathcal{D}$ is a function of the undercooling. It increases monotonously from 0 for $T_0 = T_M$ to infinity for $\Delta \rightarrow 1$.[‡]

The stationary solution calculated above therefore corresponds to a highly singular limit, and it is essentially of academic interest. However, it is worth studying its stability since, while the mathematical formulation is reduced to its maximal degree of simplicity, we will see that it nevertheless permits to identify the physical mechanism which is basically responsible for most growth front instabilities.

[†] Indeed, realizable Δ's are limited by spontaneous nucleation (either homogeneous or inhomogeneous) to values of order, typically, $C_p T_M/4L$. It takes materials with $C_p T_M > 4L$ to obtain $\Delta > 1$. This is not the case for usual materials — metals, for example, have $C_p T_M < 3L$. An exception is provided by white phosphorus [46].

Note that large reduced undercoolings can also be obtained in glass forming liquids (e.g. Se, As, polymers...). However, when they are deeply undercooled, their attachement kinetics is extremely slow, their viscosity very high, and the minimal model is insufficient to describe their solidification.

[‡] For $\Delta > 1$, in this model, the mathematical problem diverges at finite time, thereby predicting an instantaneous solidification which is of course physically irrealistic: in this regime, it becomes necessary to take into account attachment kinetics and advection.

3.3 Linear stability of this solution: the Mullins-Sekerka instability

In order for a stationary solution to be observable, it must at least be *locally stable*. That is fluctuations of infinitesimal amplitude about it must regress. If this is not the case, any inhomogeneity, (and, if this were absent, thermal noise) is amplified and the system evolves towards a different regime.

In order to decide about this, a *linear stability analysis* must be performed, i.e. we must look for the spectrum of eigenmodes of the (temperature field + front shape) system by linearizing the growth equations in the departures:

$$\delta T(\rho, \tilde{z}, t) = T(\rho, \tilde{z}, t) - T^{(0)}(\tilde{z}), \tag{3.12a}$$

$$\zeta(\rho, t) = \tilde{z}_I(\rho, t) \tag{3.12b}$$

from the stationary solution. Since the equations governing the evolution of δT, ζ are linearized, and since the zeroth order system about which fluctuations take place is translationally invariant in the (xy)-plane, the eigenmodes are necessarily of the form:

$$\zeta(\rho, t) = \zeta_k \exp(i\mathbf{k} \cdot \rho + \Omega t), \tag{3.13a}$$

$$\delta T_\alpha(\rho, \tilde{z}, t) = \delta T_{k\alpha}(\tilde{z}) \exp(i\mathbf{k} \cdot \rho + \Omega t). \tag{3.13b}$$

From the boundary condition (3.4b)

$$\lim_{\tilde{z} \to 0} \delta T_{kL}(\tilde{z}) = 0. \tag{3.13c}$$

In the $(x, y, \tilde{z})$-frame, Eqs. (3.1, 3.3) read:

$$\frac{\partial T_\alpha}{\partial t} = D_{\alpha,\text{th}} \tilde{\nabla}^2 C_\alpha + V \frac{\partial C_\alpha}{\partial \tilde{z}}, \tag{3.14}$$

from which:

$$\delta T_{kL}(\tilde{z}) = D_k e^{-q\tilde{z}}, \qquad \delta T_{kS}(\tilde{z}) = D'_k e^{q'\tilde{z}}. \tag{3.15}$$

q and q' are the positive solutions of:

$$(q\ell_L)^2 - q\ell_L - (k^2\ell_L^2 + \Omega\tau_L) = 0, \tag{3.16a}$$

$$(q'\ell_S)^2 + q'\ell_S - (k^2\ell_S^2 + \Omega\tau_S) = 0, \tag{3.16b}$$

where we have dropped the 'th' superscripts on thermal diffusion lengths. $\tau_{L,S}$ are the thermal diffusion times

$$\tau_{L,S} = \frac{D_{L,S}^{\text{th}}}{V^2}. \tag{3.17}$$

In order to write the linearized boundary conditions we need to compute the values of the functions T_α, ∇T_α on the deformed front (3.13a). For example, from (3.12a):

$$T_{L|\text{int}} = \delta T_L\big(\rho, \zeta(\rho,t), t\big) + T_L^{(0)}\big(\zeta(\rho,t)\big), \tag{3.18}$$

which yields, to first order in the fluctuation amplitudes δT_k, ζ_k :

$$T_L\big|_{\text{int}} - \big(T_L\big|_{\text{int}}\big)^{(0)} = \delta T_L(\rho,0,t) + \zeta(\rho,t)\frac{\partial T_L^{(0)}}{\partial \tilde{z}}\Big|_{\tilde{z}=0}. \tag{3.19}$$

That is, the first order variation of the interface temperature is the sum of two terms:

▷ The first one is due to the variation of the temperature field δT at the position of the undeformed front.

▷ The second one describes the fact that, when it deforms, the front explores the zeroth-order thermal field $T_L^{(0)}$, and therefore feels its gradient.

Let us choose for convenience the x-axis along the direction of $\mathbf{k}$. The front curvature is given by:

$$\mathcal{K} = -\frac{\zeta_{xx}''}{(1+\zeta_x'^2)^{3/2}} = k^2\zeta_k + \mathrm{O}(\zeta^2). \tag{3.20}$$

We then find, from Eqs. (3.3):

$$D_k' = -\frac{\gamma T_M}{\rho L}k^2\zeta_k, \tag{3.21a}$$

$$D_k = \left[\frac{T_M - T_0}{\ell_L} - \frac{\gamma T_M}{\rho L}k^2\right]\zeta_k. \tag{3.21b}$$

We linearize in the same way Eq. (3.2), noticing that

$$\begin{aligned}
n_z &= (1+\zeta_x'^2)^{-1/2} = 1 + \mathrm{O}(\zeta^2),\\
n_x &= -\zeta_x'(1+\zeta_x'^2)^{-1/2} = \mathrm{O}(\zeta),
\end{aligned} \tag{3.22}$$

and that

$$\mathbf{v}_I = \hat{z}(V + \dot{\zeta}) \qquad (\dot{\zeta} \equiv \partial\zeta/\partial t), \tag{3.23}$$

we obtain

$$K_S q' D'_k + K_L \left[q D_k + \frac{T_M - T_0}{\ell_L^2} \zeta_k \right] = \Omega L \zeta_k. \qquad (3.24)$$

Taking into account the condition of unit reduced undercooling and using (3.21), we get the dispersion equation of front deformations:

$$\Omega \tau_L = -1 + q \ell_L - d_0 \ell_L^2 k^2 (q + n q'). \qquad (3.25)$$

We have set $n = K_S / K_L$ and defined the *capillary length* of our thermal problem:

$$d_0 = \frac{\gamma T_M C_p}{\rho L^2}. \qquad (3.26)$$

Since q and q' are functions of k and Ω (Eqs. (3.16)), a complete analysis of the exact relation (3.25) is rather heavy. Note however that one can check immediately that:

$$\Omega(k = 0) = 0, \qquad (3.27)$$

as imposed by the free growth situation: a $k = 0$ deformation is a mere infinitesimal translation along Oz, and the free growth configuration is translationally invariant (no external gradient).

Equation (3.25) simplifies considerably for the modes which satisfy the *condition of quasi-stationarity*. The corresponding quasi-stationary (QS) approximation, very useful and often well justified in solidification, formally amounts to neglecting the $\partial T_\alpha / \partial t$ term in Eqs. (3.14). This means that the front moves slowly enough for the temperature field to adapt to its instantaneous shape as if it were stationary. That is, this approximation neglects retardation effects due to the finite propagation time of the thermal signal (here the latent heat release).

For a deformation of wavevector $\mathbf{k}$, the characteristic delay time is of order $\Delta t \sim (D^{th} k^2)^{-1}$, and the QS-approximation is justified if Δt is much smaller than the characteristic time of evolution of the mode, Ω^{-1}, i.e. if

$$\Omega \ll D^{th} k^2. \qquad (3.28)$$

When we are restricted to the modes which satisfy this condition, and which are moreover such that[†]:

$$k \ell_L \gg 1, \qquad (3.29)$$

[†] For growth velocities $V < 100 \ m\mu/sec$, with $D^{th} \gtrsim 10^{-3} \ cm^2/sec$, $\ell_L \gtrsim 1 \ mm$, so condition (3.29) is in fact a very weak restriction.

we get (cf. Eqs. (3.16))

$$q \cong q' \cong k \tag{3.30}$$

and

$$\Omega \cong kV\left[1 - d_0 \ell_L^2 (1+n) k^2\right]. \tag{3.31}$$

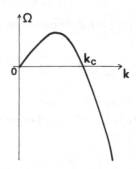

Figure 7: The Mullins-Sekerka growth rate spectrum.

The spectrum of growth rates given by (3.31) has the form shown in Fig. 7. At long wavelengths, $\Omega > 0$. The corresponding deformation modes are amplified: the *planar front is unstable* against long wavelength deformations whatever the growth velocity V and the values of the material parameters d_0 and n. This is the *Mullins-Sekerka (MS) instability*.

On the other hand, short wavelength modes are regressing ($\Omega < 0$). The k^3-term in (3.31) responsible for this is proportional to the capillary length d_0; it is the high interface energy cost which stabilizes the system against short wavelength deformations. Here appears the crucial role played by capillarity in moving front problems: however small d_0 is, the k-range corresponding to unstable modes remains finite. Capillarity provides a short-range cut-off for deformations, thus prohibiting the development of cusped front shapes. This behavior of the deformation spectrum is a manifestation of the mathematical fact — central in the problem of velocity selection of dendrites — that capillarity acts as *singular perturbation*.

Let us now try to understand the qualitative nature of the MS instability mechanism [12]. Imagine that some fluctuation in the system creates a bump of the front pointing into the liquid (Fig. 8). Ahead of the bump, the (previously planar) deformed isotherms get closer to

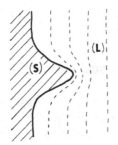

Figure 8: Isotherms (dashed lines) in the
liquid ahead of a bump of the solid front.

each other. The temperature gradient, and therefore the heat current,
increases, which — due to the heat balance condition (3.2) — increases
the rate of production of latent heat, i.e. the front velocity. The bump
gets amplified. The same mechanism is easily checked to decrease the
front velocity in a trough, i.e. to amplify it.

So, it is *diffusion* which *destabilizes the planar front*, by a mechanism
of the same nature as the point effect in electrostatics.

The unstable region of the spectrum is limited on the short wavelength
side, by the *stabilizing effect of capillarity*, to $\lambda > \lambda_c$, where the
wavelength of the *marginal (or neutral) mode*, defined by $\Omega(k_c) = 0$,
is given by

$$\lambda_c = 2\pi \sqrt{1+n} \sqrt{d_0 \ell_L}. \tag{3.32}$$

$\sqrt{1+n}$ is a number of order 1 (typically, $n \simeq 2$ can be taken as a
reasonable value in many cases).

The *most amplified* mode $(d\Omega/dk = 0)$ has wavevector $k_m = k_c/\sqrt{3}$.
So, one can reasonably guess that the deformed pattern towards which
the unstable planar front evolves should have, at least at short times
(when non-linearities are still relatively weak), a characteristic space
scale of order $\lambda_c \sim \sqrt{d_0 \ell_L}$.

The thermal capillary length d_0 (Eq. (3.26)) is typically of the order
of a few Angströms ($\gamma \sim 10^{-7}$ cal/cm^2, $L \sim C_p T_M$, $\rho L \sim 10$ cal/cm^3).
For growth velocities in the 10 μm/sec range, ℓ_L is in the cm range
at least. One therefore expects front structures on the 1–10 μm scale,
which should — in agreement with experiments — be *observable by optical methods*.

Finally, $\Omega_{\text{max}} = \Omega(k_m) \sim k_m V$. So, the effects of interest are well
described within the quasi-stationary approximation ($\Omega_{\text{max}}/D^{\text{th}} k_m^2 \sim$

$(k_m \ell_L)^{-1} \sim \sqrt{d_0/\ell_L} \ll 1$, cf. Eq. (3.28)) and take place in the strongly non-local domain defined by Eq. (3.29).

3.4 Free growth of a spherical germ [33]

This is a more realistic situation, which provides a good description of the initial evolution of a solid germ which has nucleated in the undercooled liquid. Since in this case the volume of the solid is finite, the growth regime cannot be stationary. A simple calculation of the velocity evolution of the spherical solid is however possible within the quasi-static approximation (quasi-adiabatic adaptation of the thermal field to the instantaneous front configuration) [33]. This amounts to neglecting the $\partial T/\partial t$ term in the diffusion equation (3.1), which thus reduces to the Laplace equation. Its spherically symmetric solution satisfying $T_L(r \to \infty) = T_0$ is:

$$T_L^{(0)} = T_0 + \frac{\alpha}{r}, \qquad (r > R_S), \tag{3.33a}$$

$$T_S^{(0)} = T_{S_0} = \mathrm{C^{st}}, \qquad (r < R_S), \tag{3.33b}$$

where $R_S \equiv R_S(t)$ is the instantaneous value of the solid sphere radius. The interface conditions (3.3) — where $\mathcal{K} = 2/R$ — impose that:

$$T_0 + \frac{\alpha}{R_S} = T_{S_0} = T_M - \frac{L}{C_p} \frac{2d_0}{R_S}, \tag{3.34}$$

and the conservation equation (3.2) determines the growth velocity:

$$v_R \equiv \dot{R}_S = \frac{D_L^{\mathrm{th}}}{R_S} \Big[\Delta - \frac{2d_0}{R_S} \Big], \tag{3.35}$$

where Δ is the reduced undercooling (Eq. (3.11)). The behavior of $v_R(R_S)$ is sketched in Fig. 9.

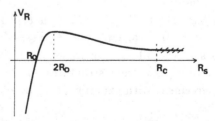

Figure 9: Growth velocity of a sphere as a function
of its radius. Above R_c the sphere become unstable.

v_R vanishes for

$$R_S = R_0 = \frac{2d_0}{\Delta}. \tag{3.36}$$

This is, as expected, the nucleation radius as obtained directly by extremalizing the energy of the liquid + germ system (see Nozières's chapter).

For $R_S < R_0$, $v_R < 0$: a subcritical germ regresses by melting. For $R_S > R_0$, the germ is supercritical, it grows. When $R_S \gg R_0$, growth becomes diffusive: $R_S \sim \sqrt{2D_L^{\text{th}}t}$.

Finally, we must check on the validity of the quasi-static approximation. It is justified when $\partial T_L/\partial t \ll D^{\text{th}}(\partial^2 T_L/\partial r^2)$, i.e. when $\dot{R}_S(\partial\alpha/\partial R_S) \ll D^{\text{th}}(\alpha/r^2)$, and thus for $r \ll r_m = R_S/\sqrt{\Delta}$. As long as r_m is much smaller than the distance $\sim D^{\text{th}}/\dot{R}_S$, beyond which retardation effects due to the propagation of the thermal signal become important, corrections to (3.35) are negligible. This entails that the quasi-static approximation is valid only for small undercoolings ($\Delta \ll 1$).

The question then arises of the stability of this spherical solution against small deformations of the growing germ. As in the planar front case, we therefore linearize the growth equations about the basic spherical solution and look for the spectrum of eigenmodes involving front deformations δR and variations δT_L, δT_S of the temperature field.

The spherical symmetry of the basic solution entails that the eigenmodes have the following form:

$$\delta R(\theta, \varphi, t) \equiv R(\theta, \varphi, t) - R_S(t) = \varepsilon_\ell Y_{\ell m}(\theta, \varphi) e^{\Omega_\ell t}, \tag{3.37a}$$

$$\delta T_{L,S}(\mathbf{r}, t) = \theta_{L,S}^{(\ell)}(r) Y_{\ell m}(\theta, \varphi) e^{\Omega_\ell t}, \tag{3.37b}$$

where $Y_{\ell m}$ is a spherical harmonic, and θ, φ polar angles on the sphere.

Solving Laplace's equation for $\delta T_{L,S}$ (with $\delta T_L(r \to \infty) = 0$, δT_S finite) yields:

$$\theta_L^{(\ell)}(r) = \frac{\alpha_\ell}{r^{\ell+1}}, \quad \theta_S^{(\ell)}(r) = \beta_\ell r^\ell. \tag{3.38}$$

We then linearize the interface equations (3.2, 3.3) (cf. Eq. (3.19)).

Taking advantage of the fact that, up to 1st order in ε_ℓ, the curvature of the surface: $R(\theta, \varphi, t) = R_S(t) + \varepsilon_\ell Y_{\ell m} \exp(\Omega_\ell t)$ is given by[†]:

$$\mathcal{K} \cong \frac{2}{R_S} + (\ell - 1)(\ell + 2)\frac{\delta R}{R_S^2}, \qquad (3.39)$$

we obtain:

$$\frac{\alpha_\ell}{R_S^{\ell+1}} - \frac{\alpha}{R_S^2}\varepsilon_\ell = \beta_\ell R_S^\ell = -\frac{L}{C_p}(\ell - 1)(\ell + 2)\frac{d_0\,\varepsilon_\ell}{R_S^2}, \qquad (3.40)$$

$$\varepsilon_\ell \Omega_\ell = -\frac{2v_R}{R_S}\varepsilon_\ell + \frac{C_p}{L}D_L^{\text{th}}\left[\frac{\ell+1}{R_S^{\ell+2}}\alpha_\ell + n\ell R_S^{\ell-1}\beta_\ell\right]. \qquad (3.41)$$

Solving (3.40) for α_ℓ, β_ℓ, we find that the growth rate of the (ℓm)-mode is given by:

$$\Omega_\ell = (\ell - 1)\frac{v_R}{R_S}\left[1 - \frac{d_0\,D_L^{\text{th}}}{v_R R_S^2}\{1 + (1 + n)\ell\}(\ell + 2)\right]. \qquad (3.42)$$

The growth rate is independent of m, as imposed by the basic symmetry (the spherical germ has no preferred axis).

From (3.42), $\Omega_1 = 0$. The $\ell = 1$ modes describe infinitesimal translations of the sphere. Since the growth dynamics is independent of the position of the germ center, these modes are neutral.

Moreover, (3.42) shows that Ω_ℓ is a function of the instantaneous radius of the spherical solid. Since $R_S(t)$ increases with t, we can define, for each value of ℓ, a radius of instability with respect to modes (ℓm), $R_C^{(\ell)}$, such that $\Omega_\ell(R_C^{(\ell)}) = 0$. Using Eqs. (3.35), (3.36):

$$R_C^{(\ell)} = R_0\left\{1 + \tfrac{1}{2}(\ell + 2)\left[1 + (1 + n)\ell\right]\right\}, \qquad (\ell \geq 2), \qquad (3.43)$$

and it is immediately checked on (3.42) that $\Omega_\ell < 0$ (resp. > 0) for $R_S < R_C^{(\ell)}$ (resp. $R_S > R_C^{(\ell)}$).

The first instability of the spherical growth mode appears when R_S reaches the smallest of the $R_C^{(\ell)}$, which corresponds to $\ell = 2$. One thus defines a critical radius of destabilization of the sphere:

$$R_C = R_0(7 + 4n). \qquad (3.44)$$

[†] This is for example found with the help of $\mathcal{K} = \operatorname{div}\hat{n}$, where $\hat{n}$ is the unit vector normal to $R(\theta, \varphi, t)$, and of

$$-\left[\frac{1}{\sin\theta}\frac{\partial}{\partial\theta}\left(\sin\theta\frac{\partial}{\partial\theta}\right) + \frac{1}{\sin^2\theta}\frac{\partial^2}{\partial\varphi^2}\right]Y_{\ell m} = \ell(\ell + 1)Y_{\ell m}.$$

If, for example, $n = 2$, the spherical nucleus begins to deform for $R \geq 15R_0$, and the spherical growth regime should be hardly observable in practice, except possibly at extremely small undercoolings (remember that $R_0 \propto \Delta^{-1}$). Free nuclei very rapidly develop bumped front patterns which later develop into dendrites.

Note finally that the planar spectrum (3.31) can be obtained by formally taking the limit $R_S \to \infty$, $\ell \to \infty$, $k = \ell/R_S$ of (3.43): in this limit, deformations with $\ell \to \infty$ are indeed equivalent to plane waves with wavevector ℓ/R_S.

3.5 Isothermal spherical growth from a supersaturated melt

In the preceding paragraphs we considered the case of a pure material growing from its undercooled metastable liquid. The analogous situation for a mixture is that of isothermal growth from the supersaturated liquid phase. Let T be the temperature of the system, $C_{L,S}^{(0)}(T)$ the concentrations of the (infinite) liquid and solid at equilibrium at this temperature ($T = T_M + m_{L,S} C_{L,S}^{(0)}$, where T_M is the melting temperature of the pure solvent, $m_{L,S}$ the liquidus and solidus slopes, and we assume the mixture to be dilute), C_∞ the concentration of the nutrient liquid. When the solid grows, heat is being produced but, as discussed in § 2.2.F, its evacuation is quasi-instantaneous on the scale of chemical diffusion — which is now the limiting transport. The corresponding heat gradients are negligible, and growth from the supersaturated melt can be considered isothermal.

The growth equations then reduce to (2.80b) and (2.81c, d, e), which have essentially the same structure as those describing free growth of a pure material (Eqs. (3.1–3.3)). The two problems become identical — up to appropriate scalings — if one makes the supplementary assumption that the concentration gap $\Delta C = C_S^{(0)} - C_L^{(0)}$ on the phase diagram (Fig. 10) can be taken as (locally) constant.

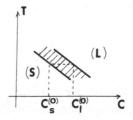

Figure 10: Schematic phase diagram for a binary mixture with $K = 1$.

This is the $K = 1$ (K is the segregation coefficient) limit. It is of course academic, but is nevertheless often useful in problems of growth of mixtures, since it simplifies the algebra without modifying qualitatively the solute rejection phenomenon. In this limit, one immediately checks that the transformations $T_\alpha \mapsto -C_\alpha/\Delta C$, $D_\alpha^{th} \mapsto D_\alpha$, $L/C_p \mapsto 1$ transform Eqs. (3.1–3.3) into (2.80b), (2.81c, d, e). So the growth velocity of the impure spherical solid germ is given by Eq. (3.35) where $D_L^{th} \mapsto D_L$, and where one has defined:

▷ a reduced supersaturation:

$$\Delta_{ch} = \frac{C_\infty - C_L^{(0)}}{\Delta C}, \tag{3.45}$$

▷ a *chemical capillary length* of the mixture:

$$d_0^{(ch)} = \frac{\gamma T_M}{\rho L m_L \Delta C}. \tag{3.46}$$

It is easily checked in the expression (3.35) for v_R that growth can occur only if $\Delta > 0$, i.e., following (3.45), when the liquid phase is metastable ($C_\infty < C_L^{(0)}$ (resp. $> C_L^{(0)}$) for $\Delta C < 0$ (resp. > 0)).

Note that $m_L \Delta C$ is always positive, whatever the sign of $m_{L,S}$. We should also insist on the fact that chemical capillary lengths are always much larger than thermal ones (Eq. (3.26)).

$$\frac{d_0^{(ch)}}{d_0} \sim \frac{L}{C_p(m_L \Delta C)}$$

is, typically, for dilute mixtures, of order 10^2, which means that the $d_0^{(ch)}$'s commonly are in the 100–1000 Å range.

The instability radius R_C of the impure sphere is given by Eq. (3.44), where a further simplification can be made. Chemical diffusion is considerably slower in the solid than in the liquid ($D_S/D_L \lesssim 10^{-6}$), so that n in Eq. (3.44), which should now be understood as D_S/D_L, can be neglected. Then

$$R_C = 7R_0^{(ch)} = \frac{14 d_0^{(ch)}}{\Delta_{ch}}. \tag{3.47}$$

These results are only slightly modified [33] when a more realistic model, with $K \neq 1$, is used for the phase diagram of the dilute mixture. The differences come from the fact that the concentration gap at the interface now varies with local liquid concentration. One finds that the growth velocity of the sphere is given by:

$$v_R = \frac{D_L}{R_S}\left(\Delta_{\text{ch}} - \frac{2d_0^{(\text{ch})}}{R_S}\right)\left[1 + (K-1)\frac{2d_0^{(\text{ch})}}{R_S}\right]^{-1}, \qquad (3.48)$$

where the concentration gap appearing in Δ_{ch} and $d_0^{(\text{ch})}$ is, as could be expected, the equilibrium gap at temperature T. The last factor on the r.h.s. of (3.48), which accounts for the local adjustment of solute rejection, only weakly modifies v_R. This same effect also induces a small shift of the instability radius, which can be neglected in first approximation [33].

The above calculation naturally raises a practical question. It is clear that, the more the material is purified, the smaller solute rejection at the interface, and heat transport, although it is faster than solute diffusion, must finally become the limiting mechanism. In other words, when is a material pure enough for its growth dynamics to be determined by thermal effects?

This may be evaluated by computing the growth velocity v_R of a solid sphere growing from a liquid at temperature T_0, concentration C_∞, from the complete set of equations (2.80, 2.81). The calculation is completely analogous to that of § 3.4. Assuming, for simplicity, that $K = 1$, one finds:

$$v_R = \frac{D_L^{\text{th}}}{R_S}\left[\frac{C_p}{L}\left(T_M + m_L C_\infty - T_0\right) - \frac{2d_0}{R_S}\right] \qquad (3.49)$$

$$\times \left[1 + \frac{D_L^{\text{th}}}{D_L}\frac{(m_L \Delta C)C_p}{L}\right]^{-1},$$

which reduces to (3.35), with $\Delta = \dfrac{C_p}{L}\left(T_{\text{eq}}(C_\infty) - T_0\right)$, provided that:

$$m_L \Delta C \ll \frac{L}{C_p}\frac{D_L}{D_L^{\text{th}}}. \qquad (3.50)$$

For a metal with, for example, $L/C_p \sim 150\text{K}$, $D_L^{\text{th}} \sim 10^{-1}$ cm^2/sec, this implies that $m_L \Delta C \ll 15 \cdot 10^{-3}$ K. For an organic material with $D_L^{\text{th}} \sim 10^{-3}$ cm^2/sec, $L/C_p \sim 20$ K (succinitrile, CBr$_4$), (3.50) gives $m_L \Delta C \ll 2 \cdot 10^{-1}$ K.

So, it takes in practice very pure materials to reach the thermal regime.

Let us finally mention that, following the seminal work of Mullins and Sekerka [33] various calculations of the growth velocity and critical radius of a sphere have been performed on more refined models of the liquid-solid system. Let us quote for example: inclusion of kinetic attachment effects [48] — which induce a decrease in the maximum growth velocity and give rise to a plateau of the $v_R(R_S)$ curve; analytical and numerical studies [49, 50] of the non-linear evolution of growth shapes for $R_S > R_C$, which point in particular to the importance of the effect, on these shapes, of capillary anisotropy; extension of the initial MS calculation to the case where a unique crystal orientation corresponds to slow kinetics [51, 52], in order to describe the destabilization of flat 'lollipop' ice crystals [53], etc.

4

Directional solidification of mixtures: linear stability of the planar front

The principle of directional solidification experiments is sketched in Fig. 11: the sample is pulled at a constant imposed velocity V (which we choose // Oz) in a temperature gradient established, schematically, with the help of two thermal contacts at temperatures T_1 and T_2 respectively larger and smaller than the melting temperature of the material. The experimental setup is such that, when the sample is at rest, the isotherms — and, therefore, the L/S interface — are planar (xy-plane).[†]

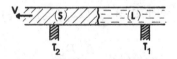

Figure 11: Sketch of a directional solidification setup.

[†] This schematic description forgets about the experimental problems met in real growth situations. In practice, lateral thermal gradients are unavoidable in bulk samples — even when they are minimized, they induce a slight curvature of the interface [54]. When using thin samples, it is difficult to eliminate completely vertical thermal gradients while allowing for optical observation.

Within the minimal model, the equations describing growth in this situation are obtained from (2.86, 2.87), which are written in a frame bound to the sample. We will first rewrite them in the laboratory frame, which moves with velocity $V\hat{z}$ with respect to the sample. Moreover, in order to simplify the algebra, we limit ourselves for the moment to the case of solidification proper, for which chemical diffusion in the cold (solid) phase is negligible ($D_s = 0$). We will discuss later how the results are modified when $\eta = D_s/D_L$ is finite. In the laboratory frame, and coming back to the physical (dimensional) length and time variables, Eqs. (2.86–2.87) become:

a) *Volume* ($\alpha \equiv L, S$)

$$\nabla^2 T_\alpha = 0, \qquad (4.1a)$$

$$\frac{\partial C}{\partial t} = D\nabla^2 C + V\frac{\partial C}{\partial z}. \qquad (4.1b)$$

b) *Interface* ($z = z_I(\rho, t)$)

$$T_s = T_L, \qquad (4.2a)$$

$$\hat{n} \cdot (\nabla T_L - n\nabla T_s) = 0, \qquad (4.2b)$$

$$D\hat{n} \cdot \nabla C = (K - 1)C(V + \dot{z}_I)n_z, \qquad (4.2c)$$

$$T = T_M + m_L C - \frac{\gamma T_M}{\rho L}\mathcal{K}, \qquad (4.2d)$$

where C and D are the concentration and solute diffusion coefficient in the liquid, K the segregation coefficient, $n = K_s/K_L$.

Equations (4.1, 4.2) must be supplemented with the following boundary conditions:

$$\lim_{z=L_1} T_L(\mathbf{r}, t) = T_1, \qquad (4.3a)$$

$$\lim_{z=-L_2} T_s(\mathbf{r}, t) = T_2, \qquad (4.3b)$$

$$\lim_{z\to\infty} C(\mathbf{r}, t) = C_\infty. \qquad (4.3c)$$

As we will check below, the thickness of the diffusion layer ahead of the front is of order

$$\ell = \frac{D}{V}. \qquad (4.4)$$

Typically, for $V \gtrsim 1$ μm/sec, $\ell \lesssim 1$ mm, i.e. the diffusive layer is much thinner than the liquid part of the sample. For this reason, one may in fact consider that, in Eq. (4.3c), the reservoir of solute is infinitely far from the front.

4.1 The planar stationary solution

Let us choose for the origin of the z-coordinate the position of this planar front. The geometry of the system (which we assume to be of infinite extent in the (xy)-directions) imposes that, for such a solution, T_α and C depend on z only. We then find immediately, from Eqs. (4.1, 4.3), (4.2a, b, c):

(i) *In the liquid* $(z > 0)$

$$T_L^{(0)}(z) = G(z - L_1) + T_1,\qquad (4.5a)$$

$$C^{(0)}(z) = C_\infty\left[1 + \frac{1-K}{K}e^{-z/\ell}\right].\qquad (4.5b)$$

(ii) *In the solid* $(z < 0)$

$$T_s^{(0)}(z) = \frac{G}{n}(z + L_2) + T_2,\qquad (4.5c)$$

where the thermal gradient in the liquid is given by

$$G = \frac{T_1 - T_2}{L_1 + L_2/n}.\qquad (4.6)$$

The Gibbs-Thomson equation then determines the front temperature

$$T_I^{(0)} = T_M + \frac{m_L C_\infty}{K}.\qquad (4.7)$$

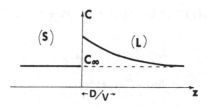

Figure 12: Planar stationary solution for a mixture: concentration profile.

The concentration profile (4.5b) is represented by Fig. 12: in the stationary regime, solidification gives rise, ahead of the front on the liquid side, to the build-up of a solute diffusion layer, of thickness $\ell = D/V$, the chemical diffusion length.

At the interface, $C = C_\infty/K$, i.e. (cf. Eq. (2.87d)) the solid formed in the stationary regime has the concentration C_∞ of the nutrient liquid. This follows immediately from solute mass conservation. The front

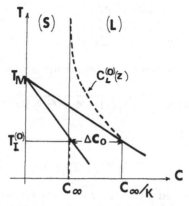

Figure 13: Phase diagram (solid lines) for a dilute alloy. Use has been made of the linear shape of the thermal profile to sketch (dashed line) the concentration profile for a planar front.

temperature is the equilibrium melting temperature for this solid (see Fig. 13) and the corresponding concentration gap $(C_S - C_L)_{\text{int}} = \Delta C_0$ is given by:

$$\Delta C_0 = \frac{K-1}{K} C_\infty. \tag{4.8}$$

4.2 Linear stability of the planar front (deformation spectrum)

The mathematical method used to derive the growth rate of deformation modes in the linear approximation is identical to the one used in § 3.3. Taking advantage of the (xy)-translational invariance of the basic planar solution, we look for solutions of the growth equations of the form:

$$T_\alpha(\mathbf{r}, t) = T_\alpha^{(0)}(z) + T_{\alpha 1}(z) \exp(i\mathbf{k} \cdot \boldsymbol{\rho} + \Omega t), \tag{4.9a}$$

$$C(\mathbf{r}, t) = C^{(0)}(z) + C_1(z) \exp(i\mathbf{k} \cdot \boldsymbol{\rho} + \Omega t), \tag{4.9b}$$

$$z_I(\rho, t) = \zeta_1 \exp(i\mathbf{k} \cdot \boldsymbol{\rho} + \Omega t), \tag{4.9c}$$

with

$$\lim_{z=L_1} T_{L_1}(z) = \lim_{z=-L_2} T_{S_1}(z) = \lim_{z \to \infty} C_1(z) = 0. \tag{4.9d}$$

We obtain, with the help of Eqs. (4.1b), (4.9d):

$$C_1(z) = C_1 e^{-qz}, \tag{4.10}$$

where q is the positive solution of:

$$q^2 \ell^2 - q\ell - (k^2 \ell^2 + \Omega\tau) = 0, \tag{4.11}$$

where $\tau = D/V^2$ is the solute diffusion time for our problem,

$$q\ell = \tfrac{1}{2}\left[1 + \sqrt{1 + 4(k^2\ell^2 + \Omega\tau)}\right]. \tag{4.12}$$

Similarly, from (4.1a), (4.9d)

$$T_{L_1}(z) = T_{L_1}\,e^{-kz}, \quad T_{S_1}(z) = T_{S_1}\,e^{kz}, \tag{4.13}$$

where we have assumed that $kL_1, kL_2 \gg 1$ (relevant wavelengths are smaller than the z-dimensions of the system, which we will check later to be a good approximation).

Linearizing the interface equations (4.2), with the help of expansions of the type of (3.19), in the fluctuation amplitudes $T_{\alpha 1}, C_1, \zeta_1$, we obtain:

$$T_{L_1} - T_{S_1} + G\left(1 - \frac{1}{n}\right)\zeta_1 = 0, \tag{4.14a}$$

$$T_{L_1} + nT_{S_1} = 0, \tag{4.14b}$$

$$C_1[q\ell + K - 1] + \frac{\Delta C_0}{\ell}(K + \Omega\tau)\zeta_1 = 0, \tag{4.14c}$$

$$m_L C_1 + \left[m_L \Delta C_0\left(\frac{1}{\ell} - d_0 k^2\right) - \frac{2G}{1+n}\right]\zeta_1 = 0, \tag{4.14d}$$

where we have dropped the 'ch' superscript on the chemical diffusion length:

$$d_0 = \frac{\gamma T_M}{\rho L m_L \Delta C_0}. \tag{4.15}$$

Equations (4.14) form a linear homogeneous system, the solvability condition of which yields the dispersion relation of front deformations, namely:

$$1 - \frac{\ell}{m_L \Delta C_0}\frac{2G}{1+n} - \frac{K + \Omega\tau}{q\ell + K - 1} - d_0 \ell k^2 = 0. \tag{4.16}$$

As in the case of free growth, the Ω-dependence of q (Eq. (4.12)) makes the analysis of the exact equation (4.16) quite heavy. We will therefore only study it in the quasi-stationary approximation, where, as discussed in § 3.3, this dependence may be neglected. We will see below that the phenomena of interest are well described by this approximation.

Equation (4.16) can then be rewritten as:

$$\Omega\tau = \left(1 - \frac{\ell}{\ell_T} - d_0\ell k^2\right)(q\ell + K - 1) - K, \qquad (4.17)$$

with

$$q\ell \cong \tfrac{1}{2}\left[1 + \sqrt{1 + 4k^2\ell^2}\right], \qquad (4.18)$$

and we have defined the *thermal length* ℓ_T associated with the externally imposed gradient by:

$$\ell_T^{-1} = \frac{2G}{(1+n)m_L\Delta C_0}. \qquad (4.19)$$

It is immediately seen from Eq. (4.17) that, at large k, $\Omega \sim -d_0 D k^3$: as was the case in free growth, the planar front is stabilized against short wavelength deformations by capillarity. On the other hand, $\Omega(k=0) = -KV/\ell_T < 0$. The homogeneous mode (translation along Oz) is now stable. This is due to the presence of the external thermal gradient $G \sim \ell_T^{-1}$, which stabilizes the system against long wavelength deformations: if the front moves translationally forward ($\delta z_I > 0$), its temperature becomes too high, it melts away. Finally, a detailed analysis of equation (4.17) [55] shows that the $\Omega(k)$ curve has at most one maximum (Fig. 14), which originates for the 'diffusive factor' $(q\ell + K - 1)$ on the r.h.s. of the equation.

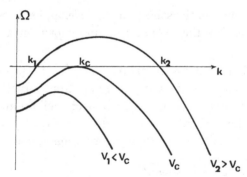

Figure 14: Directional solidification of a mixture: the Mullins-Sekerka spectrum for various velocities. $V = V_c$ corresponds to the cellular threshold.

We therefore meet again, here, with the physical mechanisms — stabilizing capillarity, destabilizing diffusion — identified in § 3.3 as being responsible for the Mullins-Sekerka instability of the freely growing system, to which we must add, in directional solidification, the stabilizing effect of the thermal gradient.

4.3 The cellular bifurcation

The condition for the planar front to be *locally stable* is that $\Omega(k)$ be negative for all values of k (all modes of infinitesimal amplitude regress).

The existence and position of the maximum of the $\Omega(k)$ curve of course depend on the values of the parameters ℓ_T, ℓ, d_0, K, etc. We will separate them into two classes:

▷ chemical (or material) parameters $(K, m_L, D, \gamma, L/C_p, n)$, which only depend on the nature of the binary mixture;

▷ control parameters: C_∞, V, G. They play in practice non-equivalent roles, since, in an experiment on a given sample, C_∞ is fixed, while V and G can be modified. Note also that it is usually more easy to vary the pulling velocity than the thermal gradient.

Let us therefore assume for the moment that C_∞ and G are fixed and that, for some $V = V_1$, $\Omega(k)$ has a maximum for $k = k_{m1}$, with $\Omega(k_{m1}) < 0$ (Fig. 14). If, when V is increased, the $\Omega(k)$ curve rises (Fig. 14), for a value V_c of the pulling velocity, its maximum reaches $\Omega = 0$ for $k = k_c$. Then, for $V > V_c$, a finite band of wavevectors (k_1, k_2) corresponding to *unstable modes* $(\Omega > 0)$ develops around k_c : the planar front has become unstable, beyond $V = V_c$, against deformations of wavevector $k \sim k_c$.

One therefore expects the solid front to develop, for $V > V_c$, a quasi-periodic structure. It is the *cellular structure* known for a long time by metallurgists [56].

Such a morphological transition (change, on the present example from a translationally invariant to a periodic front structure) is called a *bifurcation*. It is the analogous, for out-of-equilibrium systems, of phase transitions for equilibrium ones — the role of the equilibrium order parameter is played here by the front deformation amplitude.

a) *Principle of exchange of stability*

We now want to calculate the value $V_c(G, C_\infty)$ of the *critical velocity*. However, in the above discussion, we have anticipated, when performing the quasi-stationary approximation, a non-trivial result, namely the fact that, close to the bifurcation and for $k \sim k_c$, Ω is real. This is not, *a priori*, necessarily always true: a bifurcation is a point in the space of parameters where the relaxation rate of a mode changes sign, i.e. where $\mathrm{Re}\,\Omega = 0$. It may happen that, when this occurs, $\mathrm{Im}\,\Omega \neq 0$. The bifurcation is then of the *Hopf* (or oscillatory) *type*. If such was the case for our problem, the front position would oscillate in the laboratory frame, i.e. the growth velocity of the solid would oscillate around the pulling velocity V.

In order to clear up this point, one must come back to the exact expression (4.16), (4.12) of the dispersion relation, and examine whether it is possible that, when the $\mathrm{Re}\Omega(k)$ curve is at a tangent to the $\Omega = 0$ axis, $\mathrm{Im}\Omega \neq 0$. Wollkind and Segel [57] have shown that this is impossible. So, the cellular bifurcation is not oscillatory — this is expressed in technical language as 'the principle of exchange of stability is verified'. At the bifurcation, for $k = k_c$, $\mathrm{Im}\Omega = 0$ so that, by continuity, Ω is real and small in the vicinity of the bifurcation, which justifies our anticipated use of the quasi-stationary approximation (4.18).

b) *Calculation of V_c*

Since, when the $\Omega(k)$ curve becomes tangent to the $\Omega = 0$ axis, the contact point is necessarily at the maximum of $\Omega(k)$, the bifurcation occurs when the two conditions:

$$\Omega = 0, \qquad \frac{\mathrm{d}\Omega}{\mathrm{d}k} = 0, \tag{4.20}$$

are verified simultaneously. Using Eqs. (4.17), (4.18), they can be rewritten as:

$$1 - \frac{\ell}{\ell_T} = d_0 \ell k^2 + \frac{K}{K - \frac{1}{2} + \sqrt{\frac{1}{4} + k^2 \ell^2}}, \tag{4.21a}$$

$$\frac{d_0}{\ell} = K(1 + 4k^2\ell^2)^{-1/2}\left[K - \frac{1}{2} + \sqrt{\frac{1}{4} + k^2\ell^2}\right]^{-2}. \tag{4.21b}$$

The two equations (4.21a, b) determine, for given ℓ_T, d_0, one (or several) value(s) of the (ℓ, k) couple, i.e. of the critical velocity V_c and critical wavevector k_c.

For a given material, V_c and k_c depend on the other two control parameters G, C_∞, via ℓ_T, d_0. We assume from here on that C_∞ is fixed and will now discuss the shape of the *bifurcation diagram* $V_c(G)$ for a given mixture.

Let us first consider the case where Eqs. (4.21) have a solution such that $k_c\ell \gg 1$. In this case, from (4.21b),

$$k_c \cong \left(\frac{K}{2d_0\ell_c^2}\right)^{1/3}. \tag{4.22a}$$

We then get, with the help of (4.21a)

$$1 - \frac{\ell_c}{\ell_T} = 3\left(\frac{d_0 K^2}{4\ell_c}\right)^{1/3}, \tag{4.22b}$$

which may be rewritten, since $k_c \ell \gg 1$, and using the definition (4.19) of ℓ_T, as:

$$V_c = \frac{2D}{(n+1)m_L \Delta C_0} G\left[1 + O(G^{1/3})\right] \tag{4.22c}$$

The condition of strong non-locality $k_c \ell_c \gg 1$ is verified if $\ell_c/d_0 \gg 1$, i.e. for threshold velocities $V_c \ll D/d_0$. Since $D \sim 10^{-5}$ cm^2/sec, with d_0 in the 100 Å range, it appears that the approximate results (4.22) are in practice always valid under the conditions of ordinary solidification experiments, in which pulling velocities do not exceed, at most, a few hundred microns/sec. In this regime, the bifurcation diagram in the (G, V)-plane is therefore linear (Fig. 15).

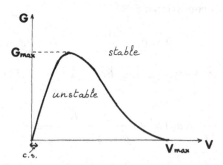

Figure 15: Bifurcation diagram in the (G, V)-plane.

It is physically clear (and can be proven formally by showing that, in this velocity range, $\max(\Omega(k))$ increases with V) that the planar front is stable for $V < V_c$: for a given V, an increase of the thermal gradient stabilizes the planar front. In this regime, the critical wavelength is obtained from Equation (4.22a). It varies, along the $G(V)$ curve, as $V_c^{2/3}$. For example, with $d_0 \sim 100$ Å, for a system with a critical velocity of a few microns/sec, λ_c lies in the 100 microns range.

Expression (4.22c) has been known in metallurgy for a long time as the *constitutional supercooling criterion* [58], and was first introduced on the basis of the following argument. Taking advantage of the fact that $dT_L^{(0)}/dz = G = C^{st}$, let us draw on the (T, C) diagram of Fig. 13 the concentration profile in the liquid (4.5b) associated with the planar solution. The corresponding curve is easily checked to penetrate into the $(L + S)$ coexistence region of the equilibrium phase diagram as soon as:

$$\left| \frac{dC_L^{(0)}}{dT_L^{(0)}} \right|_{z=0} = \frac{1}{G} \left| \frac{dC_L^{(0)}}{dz} \right|_{z=0} = \frac{|\Delta C_0|}{\ell G} > \frac{1}{|m_L|}, \tag{4.23}$$

which is identical to Eq. (4.22c) when $n = 1$.

This type of argument, based only upon equilibrium considerations, even though it happens here to yield the same result as the full theory, must be used with circumspection, since it does not correctly describe the kinetic effects which take place in the out-of-equilibrium system. Note moreover that it does not predict the right value for V_c when the two phases have different thermal conductivities ($n \neq 1$), nor when diffusion in the solid phase is non-negligible [59, 60, 61]. As will now appear, this criterion does not describe correctly, either, the bifurcation diagram in the large velocity regime.

Coming back to Eqs. (4.21) which define the position of the cellular bifurcation, one immediately checks that the r.h.s., $F(k)$, of Eq. (4.21b) is a decreasing function of k^2. So, this equation ceases to have a solution when:

$$\frac{d_0}{\ell} > F(k = 0) = \frac{1}{K}. \tag{4.24a}$$

So, when

$$V > V_{\max} = \frac{D}{d_0 K}, \tag{4.24b}$$

the bifurcation conditions can no longer be satisfied, and the planar front is stable whatever the value of the thermal gradient. The fact that $V_{\max} \propto d_0^{-1}$ shows that this high-velocity restabilization is essentially due to capillarity.

$V_{\max}$ is usually called the *limit of absolute stability* [55]. It is seen from Eq. (4.21b) that, when $V \to V_{\max}$, $k_c \to 0$. Then, from Eq. (4.21a), $\ell_T^{-1} \to 0$, so the point $V = V_{\max}$ on the bifurcation diagram has $G = 0$ (Fig. 15). A detailed study of Eqs. (4.21) shows that the shape of the $G(V)$ curve is that represented in Fig. 15 [55, 57, 61]. The variation of the critical wavevector k_c (given in the standard low velocity regime by Eq. (4.22a)) when moving along the bifurcation curve is sketched in Fig. 16.

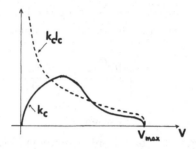

Figure 16: Variation of the critical wavevector k_c (full line) and of its reduced value $k_c\ell_c$ (dashed line) along the bifurcation diagram.

The value $G_{\max}$ of the thermal gradient beyond which the planar front is stable at all values of V is, in order of magnitude [61]

$$G_{\max} \sim \frac{m_L \Delta C_0}{d_0}. \qquad (4.25)$$

For mixtures with C_∞ of the order of a fraction of atomic per cent, $d_0 \sim 100$ Å, it is seen that $V_{\max}$ is at least in the meter/second range, while $G_{\max}$ is typically of order 10^5 K/cm. That is, these regions of the bifurcation diagram are inaccessible to ordinary directional solidification experiments.

4.4 Experimental studies of the bifurcation diagram

In view of these remarks, one can hope to observe *high-velocity restabilization* in two experimental situations

(i) *Rapid solidification*, where the surface of an alloy is swept with an electron or laser beam. Boettinger *et al.* [7, 62] have observed, under such conditions, the disappearance of the periodic impurity segregations in the resolidified alloy which are characteristic of cellular growth fronts. However, such experiments are difficult to analyze quantitatively in detail, due, in particular, to the complex (velocity-dependent) thermal configuration of the molten zone.

We should also insist that, at such high velocities ($V \sim 1$ m/sec), the minimal model becomes inadequate since, as discussed in § 2.2.D, the assumption of local equilibrium on the interface loses validity in this regime even for rough interfaces. It is easily checked that adding to the Gibbs-Thomson equation a kinetic correction (see Eq. (2.75)) results in an increase of $V_{\max}$. However, it is up to now impossible to predict its value since there is, to our knowledge, no available experimental information about the kinetic law at such high growth rates.

(ii) *The isotropic liquid/liquid crystal transition*

From Eq. (4.24b), V_{max} is proportional to the solute diffusion coefficient in the liquid, D_L. Now, due to molecular dimensions and shapes, chemical diffusion coefficients in the isotropic liquid phase of materials which form liquid crystal phases are in general noticeably smaller than in usual liquids. For these materials, typically, $D_L \sim 10^{-7}$ cm^2/sec. On the other hand, diffusion coefficients in the cold phase — the liquid crystal — D_{LC} are comparable with D_L, and one must extend the linear stability calculation of § 4.2 to take this into account [61]. The bifurcation curve is found to retain the same qualitative shape as in the *one-sided model* used above. In the linear, low velocity region, (4.22c) becomes

$$V_c = \frac{2D_L}{(n+1)m_L\Delta C_0}G\left[1 + \frac{D_{LC}}{D_L}K\right], \qquad (4.26)$$

and V_{max} is exactly given by expression (4.24b). For impurity concentrations, again, in the 10^{-2} range, restabilization should therefore occur at velocities of the order of, roughly, 100μm/sec. This is precisely what was observed by Bechhoefer *et al.* [44, 63] on the liquid to nematic transition of 8CB.

Trivedi *et al.* have also observed recently [64], for $V \simeq 650$ μm/sec, restabilization on distillated CBr$_4$ containing residual impurities with a small diffusion coefficient ($D_L \sim 2.10^{-6}$cm^2/sec) and $K \sim 1$.

It seems *a priori* much more easy to test experimentally the prediction (4.22c) of the position of the cellular bifurcation in the small velocity regime. In fact, precise measurements are difficult and have been performed only recently. Indeed, one should notice that, when V approaches V_c, the maximum of the $\Omega(k)$ curve approaches $\Omega = 0$. So, in the subcritical regime, possible front deformation transients generated when starting to pull the sample relax more and more slowly as V comes closer to V_c (this is the analogue of critical slowing down in the vicinity of a second order phase transition). Careful measurements of V_c therefore demand continuous observation of the front morphology, which is impossible in ordinary metallurgical experiments.

The recent quantitative studies have for this reason been performed on thin samples of transparent materials. It then appeared in the experiments of de Cheveigné *et al.* [40] on CBr$_4$ that, for a given value

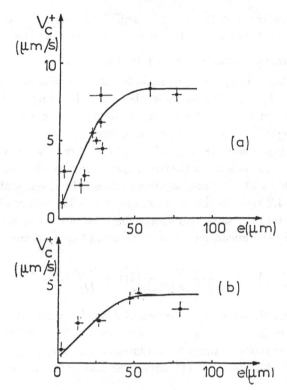

Figure 17a: Critical velocity versus sample thickness at two of the thermal gradients:
(a) $G = 120$ K/cm, (b) $G = 70$ K/cm (cf. [40]).

of the thermal gradient, the cellular threshold velocity V_c *depends on sample thickness* (Fig. 17a).

This phenomenon can be qualitatively interpreted as follows: the solidifying material is contained between transparent (glass) plates. The glass-liquid and glass-solid interface tensions, being different, impose, at the L-S-glass triple point, a contact angle which is in general $\neq \frac{1}{2}\pi$. So, the L-S interface deforms, along the thin direction of the sample, into a *meniscus* (Fig. 17b).

The shape of this meniscus depends on the sample thickness e but, also, for fixed e, on the pulling velocity. The mechanical equilibrium condition which fixes the contact angle acts on the system as an external static force driving front deformations of wavevectors $k_0 = 2\pi/e$ and its harmonics pk_0 (p =integer). When V approaches V_c, the critical modes

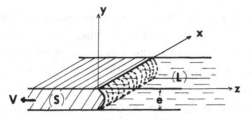

Figure 17b: Meniscus-shaped front in a thin sample.

with wavevector k_c calculated above (§ 4.3 b) become quasi-neutral thus giving rise to a resonant response,[†] the amplitude of which diverges at $V = V_c$ in the linear approximation, when $pk_0 = k_c$, i.e. when the sample thickness is a multiple of $\lambda_c = 2\pi/k_c$. Of course, non-linear effects limit these resonance effects, which we have mentioned here mainly in order to illustrate the influence of the pulling velocity on the shape of the solid-liquid meniscus. The cellular instability threshold measured by observing the front profile along the large dimension Ox (Fig. 17) of the sample is therefore, for a thin film, the instability threshold of a basic non-planar front. This explains qualitatively its thickness dependence. The only existing (analytic) calculation of this effect [67] was performed in the limit where the contact angle is very close to $\frac{1}{2}\pi$, its results are thus only indicative and do not allow quantitative predictions. The experimental results on CBr$_4$ [40] show that the thickness dependence of V_c saturates to the value predicted by the infinite system model (§ 4.3 b)) only when e becomes larger than, typically, 50 microns.

Besides indicating that much care must be taken when interpretating critical velocity measurements, the above discussion points to another important aspect of experiments on thin samples. As will become clear later, and as will also appear in the Chapter 4, on dendritic growth, predicting front shapes in strongly non-planar regimes resorts to very difficult non-linear problems. For this reason, most recent theoretical and numerical work has been performed on two-dimensional models. In order to test them, one is naturally tempted to perform experiments on very thin samples. The above remarks show that this is not necessarily

---

[†] The same type of resonant excitation of front deformations is respon-sible [65, 66] for the fact that, when V approaches V_c from below, the non-planarity of the front around the emergence lines of grain bound-aries is observed to increase considerably, so that these lines behave as 'nucleation centres' for the cellular deformation.

the best choice: as far as front shapes are concerned, a material does not systematically become 'more 2D' when used in very thin films — when e becomes very small, the transverse front curvature increases as $1/e$ and may eventually become the dominant — though unobserved — profile feature.

It may be useful in this respect to mention here two types of systems which have recently started to attract interest:

▷ The isotropic liquid-discotic phase transition: this has been studied by Oswald [68]. In the discotic phase the 'plate-shaped' molecules form a hexagonal lattice of 'piles of plates'. One can reasonably guess that these piles in general prefer to be perpendicular to the confining plates when these are properly prepared, so that the meniscus effect, and therefore the effect of sample thickness, should be weak. This has not been studied in detail yet. Let us simply mention that one of the main problems with these materials is the difficulty of their chemical synthesis.

▷ Amphiphilic monolayers deposited on water. These exhibit phase transitions, some of which at least are first order, and during which growth instabilities are observed [69, 70]. As far as their thickness is concerned, they are ideal 2D systems. However, some still unexplored questions should be raised about them. For example, is it 2D solute (impurity) diffusion in the monolayer which is the mechanism responsible for instabilities or the fact that, most probably, the amphiphilic molecular heads themselves move diffusively in the underlying water? Are these motions strongly coupled to the hydrodynamics of water? These are open questions to be elucidated by future experiments which are of interest from still another point of view: we may hope that they will open the possibility of using the recent progress about growth instabilities and morphologies to get some information about the material under study itself.

5

Directional solidification of mixtures: small amplitude cells

5.1 Nature of the bifurcation

In the preceding chapter we have shown that, beyond a threshold velocity V_c, the planar front becomes unstable. Infinitesimal front deformations are amplified if their wavevector k belongs to the band (k_1, k_2) of *linearly unstable modes* around the critical wavevector k_c (see Fig. 14). The wavevectors k_1, k_2 of the marginal (neutral) modes are defined by $\Omega(k_1) = \Omega(k_2) = 0$. The curve representing the variations of $k_{1,2}$ with the control parameter (we choose here the pulling velocity at G, C_∞ fixed) is called the *marginal curve*. It is sketched in Fig. 18. Linearly unstable modes correspond to the region inside this curve.

In the linearized approximation of Section 4, the time evolution of the amplitude, A_k, of a mode $z_{Ik}(x,t) = A_k \cos kx$ is:

$$\frac{\mathrm{d}A_k}{\mathrm{d}t} = \Omega(k)A_k, \qquad (5.1)$$

where $\Omega(k)$ is given by (4.17).

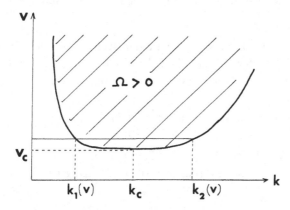

Figure 18: Sketch of the neutral Mullins-Sekerka curve.

For unstable modes, A_k grows exponentially, so the linear approximation only describes the initial evolution of the deformation. As A_k increases, non-linearities become more and more important, and the question arises of the value of the saturation amplitude, and of how the deformation amplitude evolves towards this value. This naturally leads to extending the linear perturbation calculation of Section 4 to higher orders in the deformation amplitude. One then expects such an amplitude expansion to generalize equation (5.1) into an equation of the Landau-Hopf type, namely:

$$\frac{\mathrm{d}A_k}{\mathrm{d}t} = \Omega(k)A_k - \alpha_1 A_k^3 + O(A_k^5). \tag{5.2}$$

Here we have taken advantage of the translational invariance of the basic planar front along Ox to anticipate the fact that (5.2) does not contain any term proportional to A_k^2. Indeed, due to the cosine form of $z_{Ik}(x, t)$, changing A_k into $(-A_k)$ simply amounts to a translation of the x-origin, along the front, by π/k. This must of course leave the amplitude evolution equation invariant, which entails that it only contains odd powers of A_k.

Such a perturbation expansion can only be valid in the vicinity of the bifurcation $((V - V_c)/V_c \ll 1)$, where the linearly unstable modes of interest here have $|k - k_c|/k_c \ll 1$. Since $\Omega(k)$ vanishes, on the marginal curve, it is important to retain its exact k-dependence in Eq. (5.2). On the other hand, the coefficient α_1 of the cubic term is in general finite at the bifurcation (see §§ 5.2, 5.3), so we may, as a first approximation, neglect its variations and take, in (5.2), $\alpha_1 \equiv \alpha_1(V_c, k_c)$.

Consider the case where $k = k_c$. Equation (5.2) has the stationary solutions:

(i) $A_{k_c} = 0$. This is simply the planar front solution. It is stable if $A_{k_c}^{-1}\mathrm{d}A_{k_c}/\mathrm{d}t < 0$ for A_{k_c} small (if the 'force' drives the 'particle' back towards $A = 0$), i.e. if Ω is negative, as seen in Section 4.

(ii) $$A_{k_c} = A_0 = \left[\frac{\Omega(k_c)}{\alpha_1}\right]^{1/2}, \tag{5.3}$$

which exists only if $\Omega(k_c)/\alpha_1 > 0$. This deformed solution is stable if, for small $\varepsilon = A_{k_c} - A_0$, $\varepsilon^{-1}(\mathrm{d}\varepsilon/\mathrm{d}t) < 0$. Linearizing Eq. (5.2), one gets

$$\frac{\mathrm{d}\varepsilon}{\mathrm{d}t} = (\Omega - 3\alpha_1 A_0^2)\varepsilon = -2\Omega\varepsilon, \tag{5.4}$$

and the A_0 solution is stable only if $\Omega > 0$.

So, we must distinguish between two cases, depending on the sign of α_1:

(a) $\alpha_1 > 0$

For $V < V_c$, $\Omega(k_c) < 0$, the only stationary solution is the planar front one.

For $V > V_c$, $\Omega(k_c) > 0$; the planar front is unstable. It is the deformed solution $A_0 = [\Omega(k_c)/\alpha_1]^{1/2}$ which is stable. From Eqs. (4.17), (4.21), one can easily calculate the value of $\Omega(k_c)$ close to the linear part of the bifurcation diagram, where Eqs. (4.22) hold:

$$\Omega(k_c) = k_c(V - V_c) + O[(V - V_c)^2]. \qquad (5.5)$$

The amplitude of the $k = k_c$ deformation mode thus evolves with the pulling velocity as shown in Fig. 19: it increases continuously, as

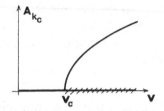

Figure 19: Velocity dependence of the deformation amplitude for a normal bifurcation. The cross-hatched line corresponds to unstable states.

$(V - V_c)^{1/2}$, beyond the bifurcation threshold. (This variation is the analogue of that of the order parameter close to a second order phase transition.) The bifurcation is said to be *normal* or *supercritical*.

(b) $\alpha_1 < 0$

The planar front is linearly unstable (resp. stable) for $V > V_c$ (resp. $V < V_c$). The deformed solution (5.3) exists for $V < V_c$ ($\Omega < 0$), but it is unstable. We therefore get the diagram of Fig. 20.

Since the planar front is unstable for $V > V_c$, it deforms into a structure which cannot be described by the perturbation expansion (5.2), i.e. which has a finite amplitude for $V \simeq V_c$. It corresponds to the dashed branch of the $A(V)$ curve in Fig. 20.

Such a bifurcation is called *inverted* or *subcritical*. It is the analogue for non-equilibrium systems of a first order phase transition. One then

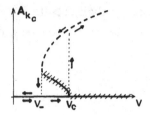

Figure 20: Velocity dependence of the deformation amplitude for an inverted bifurcation. The cross-hatched line corresponds to unstable states. The finite amplitude states corresponding to the dashed line cannot be described by the amplitude expansion.

expects a *hysteretic behavior* of the cellular front structure: if the pulling velocity is increased slowly, the planar front should remain stable up to $V = V_c$. As soon as $V > V_c$, the front should develop a finite deformation. This structure should persist, when V is slowly decreased, down to $V = V_-$ (Fig. 20) where the front becomes planar again. The stationary solution A_0, being unstable, plays the role of a deformation amplitude threshold to be 'jumped over', when $V_- < V < V_c$, in order for the front structure to change from planar to cellular (for example with the help of a large enough velocity jump).

5.2 Principle of the calculation of the coefficients α_1

In order to compute the cubic coefficient α_1 in Eq. (5.2), one must perform a perturbation expansion of Eqs. (4.2, 4.3) up to third order in the front deformation amplitude. This is a systematic, straightforward, but algebraically heavy calculation.[†] We will simply sketch out here its principle; the interested reader will find details in [57, 61].

The front deformation is expanded as:

$$z_I(x, t) = \sum_{n=1}^{\infty} \epsilon^n \zeta_n(x, t), \qquad (5.6)$$

[†] In mathematical language, this is called the reduction of a bifurcation to its normal form [71], which is complicated algebraically in the present case by the fact that we are dealing with a free-boundary problem, where non-linearities appear via boundary conditions on the front whose shape is to be determined and not, as is more usual, directly in the PDE describing the volume behavior.

where ε is a formal infinitesimal parameter which is used to separate orders in the mathematical expansion. Physically, ε measures the deformation amplitude ($\zeta_1 = O(1)$). Since we are interested in a mode of fixed wavevector k, we set:

$$\zeta_1(x,t) = A_k(t) \cos kx. \tag{5.7}$$

Similarly, all fields coming into play in the problem are expanded as:

$$\mathbf{v}(x,z,t) = \sum_{n=0}^{\infty} \varepsilon^n \mathbf{v}_n(x,z,t), \tag{5.8}$$

where $\mathbf{v} \equiv (C, T_L, T_S)$.

Expressions (5.7, 5.8) are then plugged into equations (4.2, 4.3). Separating terms of different orders in ε, one obtains the perturbation expansion of the growth problem at successive orders. Order zero corresponds to the planar stationary solution. Order ε is, of course, nothing but the linear calculation performed in detail in § 4.2 above, so that, to this order, we obtain:

$$\mathbf{v}_1(x,z,t) = A_k(t) \mathbf{v}_1(z) \cos kx, \tag{5.9a}$$

where

$$A_k(t) = \exp\left[\Omega(k)t\right], \tag{5.9b}$$

where $\Omega(k)$ is given by (4.17).

As soon as one gets to order ε^2, the superposition principle, which allows to separate terms corresponding to different wavevectors k, no longer applies since interface boundary conditions are non-linear. For example, the liquid concentration on the *deformed interface* which appears in the Gibbs-Thomson equation reads:

$$C_{\text{int}} = \left[C^{(0)} + \varepsilon C_1 + \varepsilon^2 C_2\right]_{z=\varepsilon\zeta_1+\varepsilon^2\zeta_2} \tag{5.10}$$

$$= C^{(0)}(0) + \varepsilon\left[C_1 + \zeta_1 \frac{\mathrm{d}C^{(0)}}{\mathrm{d}z}\right]_{z=0}$$

$$+ \varepsilon^2\left[C_2 + \zeta_1 \frac{\mathrm{d}C_1}{\mathrm{d}z} + \zeta_2 \frac{\mathrm{d}C^{(0)}}{\mathrm{d}z} + \frac{\zeta_1^2}{2} \frac{\mathrm{d}^2 C^{(0)}}{\mathrm{d}z^2}\right]_{z=0} + O(\varepsilon^3).$$

That is, the order two problem, from which $\zeta_2, C_2, T_{\alpha 2}$ must be determined, is not, as the linear one, homogeneous. It involves driving terms, such as $\zeta_1 \mathrm{d}C_1/\mathrm{d}z$ and $\frac{1}{2}\zeta_1^2 \mathrm{d}^2 C^{(0)}/\mathrm{d}z^2$ in Eq. (4.10), proportional

B. Caroli, C. Caroli, B. Roulet

to $\cos^2 kx$, i.e. with wavevectors 0, $2k$. These terms account for the generation, by the non-linearities of the system, of second order harmonics ($k \pm k$). Similarly, at order ε^3, the driving terms involve harmonics ($k \pm k \pm k$), i.e. $3k$ and k itself, thus reacting on the fundamental mode itself, and giving rise to the cubic correction (see Eq. (5.2)) to the evolution equation of its amplitude. One is thus led to setting:

$$\mathbf{v}_2(x, z, t) = A_k^2(t)\left[\mathbf{v}_{20}(z) + \mathbf{v}_{22}(z)\cos(2kx)\right], \qquad (5.11a)$$

$$\mathbf{v}_3(x, z, t) = A_k^3(t)\left[\mathbf{v}_{31}(z)\cos kx + \mathbf{v}_{33}(z)\cos(3kx)\right], \qquad (5.11b)$$

and analogous expressions for ζ_2, ζ_3, e.g.:

$$\zeta_2(x, t) = A_k^2(t)\left[\zeta_{20} + \zeta_{22}\cos(2kx)\right]. \qquad (5.11c)$$

It then appears that, in order to ensure the coherence of the perturbation expansion [57], one must set:

$$\frac{\mathrm{d}A_k}{\mathrm{d}t} = \Omega(k)A_k - \varepsilon^2\alpha_1 A_k^3(t), \qquad (5.11d)$$

i.e. precisely, as expected, equation (5.2) (with $\varepsilon A \mapsto A$).[†]

One then solves the two second order problems, i.e. calculates $(\mathbf{v}_{20}, \zeta_{20})$ and $(\mathbf{v}_{22}, \zeta_{22})$. Once the linear volume equations (4.1) have been solved, for each of the two problems the interface boundary conditions take the form of an inhomogeneous linear system, as discussed above.

The calculation of $(\mathbf{v}_{31}, \zeta_{31})$ has exactly the same structure, but a major difference appears: the characteristic determinant of the resulting inhomogeneous linear system vanishes for $\Omega = 0$, i.e. at the bifurcation, where we want to compute α_1. One must therefore impose on the r.h.s. part of the inhomogeneous system a *compatibility condition* ensuring that the system has a solution. It is this condition which determines the, up to now unknown, coefficient α_1.

This is the 'pedestrian method' for writing, for the third order problem, the Fredholm condition which systematically appears when building amplitude expansions close to a bifurcation. The above calculation

---

[†] This expression simply expresses the general phenomenon of amplitude-dependent frequency renormalization familiar in the theory of non-linear oscillators [72].

can also be performed with the help of formally different, but equivalent methods: amplitude expansion of the integral equation for the front shape [12, 59, 60, 73] to which the minimal model can be reduced in the one-sided and symmetric limits (see Appendix B); asymptotic expansion of Eqs. (4.1, 4.2) with a multiple-scale method [74].

It is important to realize what is the limit of validity of the amplitude expansion (5.11d). The calculation up to order ε^2 describes how the unstable k mode drives the uniform mode ((20) terms) and the second harmonic ((22) terms). It implies that these modes are 'slaved' by the slow unstable mode, i.e. that their amplitude adapts adiabatically to that of the driving term.[†] This is true only if these modes regress rapidly enough, i.e. if $|\Omega(0)|$, $|\Omega(2k)| \gg \Omega$. This condition is realized, for the spectrum given by (4.17) and with $V \simeq V_c$, except in two limit cases:

(i) $K \to 0$ (the solute is quasi-insoluble in the solid);

(ii) $V \to V_{\max}$ (absolute stability limit).

In these cases, the critical wavevector $k_c \to 0$, so that the harmonics $(0, 2k_c, 3k_c, \ldots)$ are themselves quasi-marginal, and the above expansion no longer converges, even for small amplitudes. Sivashinsky has shown that an asymptotic expansion in the new small parameter (K or $\delta V/V_{\max}$) can then be built. This results in a non-linear partial differential equation for the amplitude [75, 76] which generates very singular front shapes. This Sivashinsky regime, which has also been predicted to exist in other dissipative dynamical systems, has not been observed yet. Directional solidification of a mixture with a very small segregation coefficient seems to offer a possibility in this direction. However, this is not completely clear, in the absence of any explicit criterion about how small K should be in practice for the predicted singular effects to be observable.

5.3 Theoretical predictions and experimental results on the nature of the bifurcation

We will not give here the exact expression of α_1 valid at any point along the bifurcation curve [57, 61], which involves both the control and the

---

[†] Technically, this appears in the method sketched above as a condition on the order-2 determinants Δ, which must be large enough for the condition $|\Delta(\Omega) - \Delta(\Omega = 0)/\Delta(\Omega = 0)| \ll 1$ to be satisfied, so that $\Delta(\Omega) \simeq \Delta(0)$.

chemical parameters and is extremely heavy. It simplifies considerably in the linear region of the (G, V) bifurcation curve, where it may be shown to reduce to[†]:

$$\alpha_1 \cong \frac{D}{d_0 \ell_c^3} \left[-\frac{(n-1)}{8(n+1)} + \frac{K^2 + 4K - 2}{16K} \left(\frac{K^2 d_0}{4\ell_c} \right)^{1/3} \right]. \tag{5.12}$$

$(n - 1)$ is in general positive. If n is not too close to 1, the first term on the r.h.s. of Eq. (5.12) is dominant, since $d_0/\ell_c \ll 1$, and $\alpha_1 < 0$. Numerical calculations using the exact expression of α_1 show that it becomes positive at large values of the critical velocity [57] which are, for ordinary alloys, outside of the standard experimental regime. One therefore expects that inverted bifurcations should be commonly observed in directional solidification of bulk samples.

In the case of thin samples, a difference should be noticed: heat conduction usually takes place essentially through the (thicker) confining plates which therefore impose the thermal profile. The thermal gradient is the same in the two phases, which is equivalent to imposing $n = 1$ in the above calculations. Equation (5.12) then shows that α_1 should be positive (resp. negative) for $K > 0.45$ (resp. < 0.45).

Alexander *et al.* [77] have extended the calculation of α_1 to include the effect of latent heat production. As may be expected on the basis of the discussion of § 3.5, the parameter which controls the importance of the resulting shift of the bifurcation is $\xi = \dfrac{L}{C_p m_L \Delta C_0} \dfrac{D_L}{D_{\text{th}}}$, which is in general $\ll 1$. However, the influence of thermal effects on α_1 is not necessarily negligible [78] as the $-(n-1)/8(n+1)$ term in α_1 (Eq. (5.12)) is corrected into

$$-\frac{(n-1) - \xi(n+1)}{(1+\xi)(n+1)},$$

and this correction becomes important when $n \cong 1$. It can then shift the range where the bifurcation is supercritical towards lower values of the critical velocity [78].

Let us also mention that quasi-azeotropic alloys should offer an interesting opportunity from this point of view. Misbah [79] has shown that the scale of their (G, V) bifurcation diagram is strongly reduced

---

[†] Note that a factor $\frac{1}{16}$ is missing in the second term of the r.h.s. of Eq. (50) of [61]. We are grateful to P. Nozières for pointing this out to us.

as compared to that for dilute alloys ($V_{\max}$ and $G_{\max}$ are proportional respectively to δC_∞^2 and δC_∞^4, where δC_∞ is the departure of the mixture concentration from the azeotropic value). Moreover, the part of this diagram where the bifurcation should be supercritical is much wider.

Finally, the expression of α_1 is modified when diffusion in the cold phase is non-negligible [61]. Langer and Turski [59] have shown, for example, that when $D_s = D_L$ and $K = 1$, the bifurcation is always supercritical ($\alpha_1 > 0$).

Only recently, thanks to the development of experiments on thin samples of transparent materials, has it been possible to study the nature of the bifurcation. The first such study was that of de Cheveigné *et al.* [40], who found, in agreement with the theoretical prediction (5.12), a hysteresis of the cellular bifurcation in CBr_4 containing impurities with $K = 0.12$. Subcritical bifurcations have since been observed by Trivedi *et al.* [41, 80] on succinonitrile and on pivalic acid. On the other hand, the experiments of Bechhofer *et al.* [44, 63] on the liquid crystal 8CB ($K = 0.85$) have proved the bifurcation to be supercritical in this system. In all these cases, experimental observations agree with (5.12).

Metallurgical experiments do not permit us to conclude directly about the nature of the bifurcation, since direct observation is impossible. However, de Cheveigné *et al.* [81] have been able to analyze them by taking advantage of the following remark.

If the bifurcation is continuous (supercritical), when V is close above V_c, the wavevector of the cellular structure is necessarily close to the critical MS one, k_c. On the other hand, if it is discontinous (subcritical) there is no reason for the wavevector of the *finite amplitude* front structure which sets in at V_c to be close to k_c. (In this case the k_c deformation mode is simply the first unstable one.) Data about the cellular wavelength close above threshold indeed separate into two groups: one for which $\lambda \simeq \lambda_c = 2\pi/k_c$, another one for which, typically, $\lambda_c/\lambda \sim 2$. De Cheveigné *et al.* have shown that these two groups precisely correspond to cases where theory predicts that, α_1 should be respectively positive and negative.

5.4 The wavevector selection problem. The amplitude equation

The amplitude expansion discussed in §5.3 above unambiguously answers the question of the nature of the bifurcation. One would of

course also like to be able to predict front shapes beyond the bifurcation. Clearly, if this is subcritical, a perturbation expansion is of no
help since, precisely, the small amplitude stationary solution of (5.2) is
unstable, and one must tackle directly the problem of deep cell structures
(see Section 6 below).

When the bifurcation is supercritical, on the other hand, Eq. (5.2)
should permit us to study shallow cells close above threshold. The
central question of *wavevector selection* then comes to the fore: as soon
as $V > V_c$, there appears (cf. §5.1 and Fig. 18) a finite band (k_1, k_2) of
linearly unstable modes $(\Omega(k) > 0)$. Following (5.2), an infinite number
of periodic front structures

$$\zeta(x) = \sqrt{\frac{\Omega(k)}{\alpha_1}} \cos(kx + \varphi) \qquad (5.13)$$

of finite amplitude are therefore *a priori* possible. Are all of them
realizable, depending on initial conditions? Or, on the contrary, does the
system forget initial conditions to 'select' one only of these possibilities
— as is the case for systems at equilibrium which always evolve towards
the configuration with lowest free energy? This is the fundamental
wavevector selection problem.

Exhaustive studies of transient dynamics are of course impossible.
One may however, conversely, start from a more limited question,
namely: given a periodic structure of the form (5.13), is it linearly stable
(stable against small deformations)? In order to explore this question,
an equation describing the time evolution of slowly varying modulations[†]
of such a periodic basic structure is needed. This *amplitude equation* is
easily obtained by reinterpreting expansion (5.2). It is the analogue for
dissipative systems of Landau-Ginzburg equations for equilibrium ones.

Consider first the sinusoidal front profile:

$$\zeta_{\mathbf{q}}(\boldsymbol{\rho}, t) = a_{\mathbf{q}}(t) \exp\left[i(k_c\hat{\mathbf{x}} + \mathbf{q}) \cdot \boldsymbol{\rho}\right] + \text{c.c.}, \qquad (5.14)$$

where we have introduced the *complex amplitude* $a_{\mathbf{q}}$. Its modulus is the
deformation amplitude and its phase fixes the origin of x.

[†] They can be predicted to play an important part: for $\delta V/V_c \ll 1$, the
band of allowed states $(\Omega(k) > 0)$ is narrow (its width is proportional
to $(\delta V)^{1/2}$). A shape built by superposition of these — *a priori* most
dangerous — modes is therefore a slowly modulated sinusoidal one.

It is seen by comparison with (5.7) that, using (5.2):

$$\frac{da_{\mathbf{q}}}{dt} = \alpha_0(\mathbf{q}, \varepsilon) a_{\mathbf{q}} - \alpha_1 |a_{\mathbf{q}}|^2 a_{\mathbf{q}}, \tag{5.15}$$

where

$$\varepsilon = \frac{\delta V}{V_c} = \frac{(V - V_c)}{V_c}, \tag{5.16a}$$

$$\alpha_0(\mathbf{q}, \varepsilon) \equiv \Omega(\mathbf{q} + k_c \hat{\mathbf{x}}; \varepsilon). \tag{5.16b}$$

Equation (5.15) is valid only when $a_{\mathbf{q}}$ is small, i.e. close to the bifurcation ($q \ll k_c$, $\varepsilon \ll 1$).

Consider now a more general front profile representing a slow modulation of a sinusoidal shape with wavevector of order k_c. It can be written:

$$\zeta(\boldsymbol{\rho}, t) = \sum_{\mathbf{q}} \zeta_{\mathbf{q}}(\boldsymbol{\rho}, t), \tag{5.17}$$

where the $\zeta_{\mathbf{q}}$ are of the form (5.14), where $a_{\mathbf{q}}$ is non-negligible only for $q \ll k_c$.

One can then repeat for this basic shape a third order perturbation calculation formally identical with that of § 5.2. The analogue of the 'resonant term' $\alpha_1 A_k^3$ of equation (5.2) (or (5.11d)) originating from the third order driving term proportional to $\cos(kx)$ now has the form [82]:

$$\alpha_1 \sum_{\mathbf{q}_1, \mathbf{q}_2} a_{\mathbf{q}_1}(t) a_{\mathbf{q}_2}^*(t) a_{\mathbf{q} - \mathbf{q}_1 + \mathbf{q}_2}(t). \tag{5.18}$$

We now define the complex amplitude in real space:

$$A(\boldsymbol{\rho}, t) = \sum_{\mathbf{q}} a_{\mathbf{q}}(t) \exp(i\mathbf{q} \cdot \boldsymbol{\rho}), \tag{5.19}$$

which, due to the limitation $q \ll k_c$, varies in space on a scale $\xi \ll k_c^{-1}$. Performing the space Fourier transform $\mathcal{F}$ of Eq. (5.15), we obtain:

$$\frac{\partial A(\boldsymbol{\rho}, t)}{\partial t} = \mathcal{F}\Big[\alpha_0(\mathbf{q}, \varepsilon) a_{\mathbf{q}}(t)\Big] - \alpha_1 |A(\boldsymbol{\rho}, t)|^2 A(\boldsymbol{\rho}, t). \tag{5.20}$$

Since $q \ll k_c, \varepsilon \ll 1$, we may expand $\alpha_0(\mathbf{q}, \varepsilon)$ in the vicinity of $\mathbf{q} = 0$, $\varepsilon = 0$, i.e. of the bifurcation, where $\Omega(k)$ behaves as shown in Fig. 14.

To lowest order in q_x and in q_y

$$\big[\mathbf{q} + k_c \hat{\mathbf{x}}\big]^2 - k_c^2 = 2 k_c q_x + q_y^2. \tag{5.21}$$

Since Ω depends on k only through k^2, with Ω $(k_c,\ \varepsilon = 0) = 0$:

$$\alpha_0(\mathbf{q}, \varepsilon) \cong \varepsilon b - \beta \left(q_y^2 + 2k_c q_x\right)^2,\qquad (5.22)$$

where

$$b = \left.\frac{\partial\Omega}{\partial\varepsilon}\right|_{k_c,\varepsilon=0}, \qquad \beta = -\tfrac{1}{2}\left.\frac{\partial^2\Omega}{\partial(k^2)^2}\right|_{k_c,\varepsilon=0}. \qquad (5.23)$$

In the small velocity regime where (4.22) are valid, one gets:

$$b = k_c V_c, \qquad \beta = \frac{3Dd_0}{4k_c}. \qquad (5.24)$$

Finally, the amplitude equation which describes the evolution of front deformations of the form:

$$\zeta(\rho, t) = \mathrm{Re}\left[A(\rho, t)e^{ik_c x}\right] \qquad (5.25)$$

reads:

$$\frac{\partial A(\rho, t)}{\partial t} = \varepsilon b A(\rho, t) + \beta \left[2k_c\frac{\partial}{\partial x} - i\frac{\partial^2}{\partial y^2}\right]^2 A(\rho, t) \qquad (5.26)$$

$$- \alpha_1 \left|A(\rho, t)\right|^2 A(\rho, t).$$

The asymmetry between the x- and y-space derivative terms results from the asymmetry of the basic structure of which $A(\rho, t)$ is the slow envelope. Indeed this structure is (see (5.14)) a one-dimensional front deformation, and we have chosen the Ox-direction along its wavevector $(k_c \hat{x})$. A modulation with $\mathbf{q} = q_y \hat{y}$ therefore describes a 'zig-zag' deformation of the 1D cellular structure (Fig. 21), while a modulation with

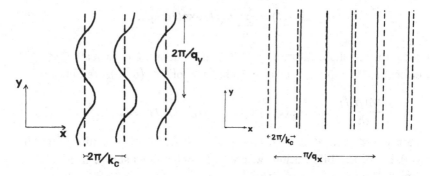

Figure 21: Zig-zag deformation (full lines) of the 1D cellular structure (dashed lines).

Figure 22: Compressional wave (full lines) in the 1D cellular structure (dashed lines).

$\mathbf{q} = q_x \hat{x}$ is a 'compressional wave' in the periodic structure (Fig. 22).

Such a deformation of the 1D cell pattern is called a *phase diffusion mode*. Indeed, it can be described by $\zeta \sim \mathrm{Re}\left[\exp(ik_c x + i\varphi(x, t))\right]$, where φ oscillates in space with wavevector $q_x \ll k_c$.

It is clear that these two types of modes are physically non-equivalent. So, it is seen that the functional form of the amplitude equation (5.26) only depends on the symmetries of the basic (non-planar) front pattern. It is thus a *generic equation*, which describes, close to threshold, slow amplitude and phase modulations of any 1D periodic structure appearing via a continuous bifurcation — e.g. the 1D roll structure in Rayleigh-Bénard convection, for which it was first established and studied in detail [82, 83]. The physical nature of the underlying physical problem only appears in this equation through the *values* of coefficients $\varepsilon b, \beta, \alpha_1$. Let us finally mention that it may also be obtained with the help of the multiple scale method, which operates directly in real space [84]. Due to the free boundary structure of the problem, it is rather heavy to implement it for solidification [74].

5.5 Phase diffusion. The Eckhaus instability

One immediately recovers, from Eqs. (5.25–5.26) the periodic stationary solutions described by Eq. (5.2). They correspond to:

$$A(\rho, t) = A_k e^{i(k-k_c)x}, \qquad (5.27)$$

with

$$A_k^2 = \frac{\varepsilon b - 4\beta k_c^2 (k - k_c)^2}{\alpha_1}, \qquad (5.28)$$

and exist only for modes within the unstable band (Fig. 18) (k_1, k_2) defined, close to threshold, by:

$$k_{\frac{1}{2}} = k_c \pm \sqrt{\frac{\varepsilon b}{4\beta k_c^2}} \equiv k_c \pm Q_m. \qquad (5.29)$$

We want to study here the linear stability of any one of these solutions against phase modulations. We thus set:

$$\zeta(x, t) = A_k e^{ikx}\left[1 + \xi(x, t)\right] + \text{c.c.} \qquad (5.30)$$

At the first order in which we are interested, $\xi_1 = \mathrm{Re}\xi$ is a modulation of the amplitude of the structure, $\xi_2 = \mathrm{Im}\xi$ a modulation of its phase with wavevector $\partial \xi_2 / \partial x$. (This can be checked from the linear expansion of $(A + r) \exp[i(k - k_c)x + i\varphi]$.) Once linearized in ξ, the amplitude equation (5.26) yields:

$$\frac{\partial \xi}{\partial t} = -\alpha_1 A_k^2 (\xi + \xi^*) + 4\beta k_c^2 \left[\frac{\partial^2 \xi}{\partial x^2} + 2i(k - k_c) - \frac{\partial \xi}{\partial x}\right]. \qquad (5.31)$$

The translational invariance of (5.31), together with the presence of ξ^* in its r.h.s., immediately indicate that we must look for eigenmodes of the form:

$$\xi(x, t) = e^{\omega t} \left[\xi_q e^{iqx} + \xi_{-q} e^{-iqx}\right], \qquad (5.32)$$

where ξ_q and ξ_{-q} are independent variables since ξ is *a priori* a complex quantity. Plugging expression (5.32) into (5.31), we find that the growth rates of the q-eigenmodes are the solutions of:

$$\omega^2 + 8\beta k_c^2 \left[Q_m^2 - Q^2 + q^2\right]\omega \qquad (5.33)$$
$$+ 2\left(4\beta k_c^2\right)^2 q^2 \left[Q_m^2 - 3Q^2 + \tfrac{1}{2}q^2\right] = 0,$$

where we have set

$$Q = k - k_c. \qquad (5.34)$$

We are interested in long wavelength modes ($q \ll Q_m$). Let us first consider the $q = 0$ limit. We immediately find, from (5.33), that the two homogeneous modes have growth rates:

$$\omega_a(0) = -8\beta k_c^2 \left[Q_m^2 - Q^2\right], \qquad (5.35a)$$

$$\omega_\varphi(0) = 0. \qquad (5.35b)$$

Mode (a) corresponds to $\mathrm{Im}\xi_0 = 0$: it is an *amplitude mode*. It is stable everywhere in the band $(Q^2 < Q_m^2)$ where the basic periodic stationary solution exists. Equation (5.35a) simply reproduces the result already obtained in § 5.1.

Mode (φ) corresponds to $\mathrm{Re}\xi_0 = 0$: it is the homogeneous *phase mode*, i.e. an infinitesimal translation of the basic structure along the planar front. It is therefore necessarily neutral.

For finite q, the two modes mix up amplitude and phase deformations. However, when q is small ($q \ll Q_m$), by continuity, the mode which evolves from the homogeneous amplitude one contains a small phase part and has $\omega \simeq \omega_a(0)$, i.e. is relaxing.

To order q^2, the amplification rate of the phase mode is given by:

$$\omega_\varphi(q) = -\mathcal{D}q^2, \qquad (5.36a)$$

with

$$\mathcal{D} = 4\beta k_c^2 \frac{Q_m^2 - 3Q^2}{Q_m^2 - Q^2}, \qquad (5.36b)$$

i.e. it is a diffusive mode — henceforth the term 'phase diffusion'. The *phase diffusion coefficient* $\mathcal{D}$ vanishes when:

$$Q^2 = Q_E^2 = \tfrac{1}{3}Q_m^2. \qquad (5.37)$$

That is, the band of allowed stationary 1D front patterns splits into two regions. The central subband $(-Q_E \leq k - k_c \leq Q_E)$ corresponds to locally stable structures $(\mathcal{D} > 0)$. Periodic front structures with k outside this subband (shaded region of Fig. 23) are unstable as regards phase diffusion: this is the *Eckhaus instability* [85] characteristic of 1D periodic structures in dissipative systems.

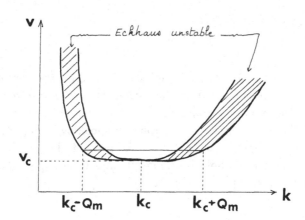

Figure 23: Neutral-Mullins-Sekerka (outside) curve and Eckhaus stability limit (inside) curve.

This instability is not limited to the immediate vicinity of the bifurcation examined here. One may also study phase diffusion in a strongly non-linear structure [86], but $\mathcal{D}$ must in general then be calculated numerically.

5.6 The zig-zag instability

We have looked in § 5.5 for longitudinal modes ($\mathbf{q} \parallel \mathbf{k}$). Let us now look for the transverse ones ($\mathbf{q} \perp \mathbf{k}$; $\xi \equiv \xi(y, t)$). Setting:

$$\xi(y, t) = e^{\omega t} \left[\xi_q \, e^{iqy} + \xi_{-q} \, e^{-iqy} \right], \tag{5.38}$$

we find, following the same steps which led to Eqs. (5.31–5.33) above, for small q :

▷ a stable amplitude mode with, clearly: $\tilde{\omega}_a(q) \cong \omega_a(0)$;

▷ a phase mode with amplification rate:

$$\omega_z(q) = -\mathcal{D}_z q^2, \tag{5.39a}$$

$$\mathcal{D}_z = 4\beta k_c Q. \tag{5.39b}$$

This phase mode, which is a zig-zag deformation (Fig. 21) is also diffusive. $\mathcal{D}_z$ is negative for $Q < 0$.

So, in a bulk sample, the $k < k_c$ half of the allowed band is unstable against zig-zags, and the only locally stable periodic 1D patterns are those with

$$0 < k - k_c < \frac{Q_m}{\sqrt{3}}. \tag{5.40}$$

On the other hand, in thin samples ($0 < y < e$), the contacts with the confining plates impose boundary conditions at $y = 0$, $y = e$. This amounts to the qualitative fact that modulations with $q_y < 2\pi/e$ are locked. In these experiments e is, roughly speaking, comparable with the critical wavelength $\lambda_c = 2\pi/k_c$. So, the plates forbid small q zig-zag deformations and the whole Eckhaus-subband ($|k - k_c| < Q_m/\sqrt{3}$) is stable with respect to phase diffusion.

5.7 Hexagonal cells

Up to now we have restricted our attention to 1D small amplitude front morphologies. These are only a subset among the periodic structures which are possible close to a normal bifurcation. Exactly at threshold, all the modes with wavevectors $\mathbf{k} = k_c \hat{u}$, where $\hat{u}$ is any unit vector of the (xy)-plane, are neutral.

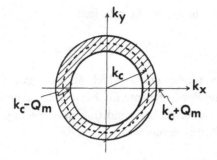

Figure 24: The shaded region corresponds to linearly unstable modes which may contribute to build up a 2D front deformation.

Similarly, beyond the bifurcation, the linearly unstable modes are those with $\mathbf{k}$ inside a circular shell (C) of the (k_x, k_y) space (Fig. 24), and we should look for stationary solutions of the general form:

$$\zeta(\rho) = \sum_{\mathbf{k} \in C} a_{\mathbf{k}} \exp(i\mathbf{k} \cdot \rho). \qquad (5.41)$$

In very thin samples $(\Delta y = e < \lambda_c)$, the only possible periodic solutions are those with $\mathbf{k} \;/\!/\; Ox$.

In metallurgic experiments in bulk samples, two types of cellular front structures appear [3]:

▷ quasi-periodic 1D patterns, called band-like cells, or bands, corresponding to the solutions studied in § 5.4–5.6;

▷ quasi-periodic hexagonal cells.

The nature (bands or hexagons) of these patterns depends on the distance to threshold and on the chemical nature of the alloy.

It is easily checked that a periodic hexagonal front structure is described by:

$$\zeta_k(\rho) = \sum_{i=1}^{3} \left(A_k \exp(ik\hat{u}_i \cdot \rho) + \text{c.c.} \right), \qquad (5.42)$$

where the $\hat{u}_i$ are three unit vectors of plane (xy) at angles $\frac{2}{3}\pi$ and $|k - k_c| \leq Q_m$. One must then generalize to such a form of ζ, the amplitude expansion. The technique of the calculation is the same as that sketched out in § 5.2. The following straightforward remark can be made: the driving terms in the order 2 problem, which are proportional to ζ_k^2, contain contributions such as:

$$A_k^{*2} \exp(ik(-\hat{u}_2 - \hat{u}_3) \cdot \rho) = A_k^{*2} \exp(ik\hat{u}_1 \cdot \rho), \qquad (5.43)$$

where we have made use of the relation:

$$\hat{u}_1 + \hat{u}_2 + \hat{u}_3 = 0. \tag{5.44}$$

The resonance phenomenon, which appears at order A^3 only for 1D structures, is therefore present, for hexagonal structures, at order A^2. This entails that a solvability condition must be imposed at this order, so that the amplitude expansion now reads:

$$\frac{\mathrm{d}A_k}{\mathrm{d}t} = \Omega(k)\,A_k + b_2\,A_k^{*2} - b_3|A_k|^2A_k, \tag{5.45}$$

i.e. it contains a quadratic term. b_2 is in general finite at the bifurcation so that the amplitude of hexagonal stationary solutions varies with V as indicated by Fig. 25: the bifurcation is asymmetric. The turning point of $A_{k_c}(V)$ corresponds to $\Omega(k_c) \cong k_c\delta V = -b_2^2/(4b_3)$, i.e. to a velocity gap $\delta V = \widetilde{V} - V_c$ which is in general not small.

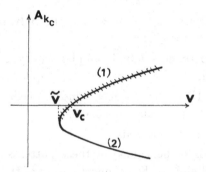

Figure 25: Asymmetric bifurcation of stationary hexagonal cellular fronts. The cross-hatched line corresponds to unstable states.

In order to examine the local stability of these hexagonal solutions, one must extend expansion (5.45) into an amplitude equation, describing slowly modulated structures:

$$\zeta(\rho, t) = \sum_{i=1}^{3}\Big[A_i(\rho, t)\exp(ik_c\hat{u}_i \cdot \rho) + \mathrm{c.c.}\Big], \tag{5.46}$$

with

$$A_i(\rho, t) = \sum_{\mathbf{q}} a_{i\mathbf{q}}(t)\exp(i\mathbf{q} \cdot \rho). \tag{5.47}$$

The evolution equations for the complex amplitudes A_i which generalize[†] (5.26) read [87]:

$$\frac{\partial A_i(\rho, t)}{\partial t} = \varepsilon b A_i + \beta \left[2k_c \frac{\partial}{\partial x_i} - i \frac{\partial^2}{\partial y_i^2} \right]^2 A_i \qquad (5.48)$$
$$- b_2 A_{i+1}^* A_{i+2}^* - b_3 |A_i|^2 A_i - b_3' \left(|A_{i+1}|^2 + |A_{i+2}|^2 \right) A_i,$$

with $i + 3 \equiv i$, and $x_i = \rho \cdot \hat{\mathbf{u}}_i$, $y_i = \rho \cdot \hat{\mathbf{v}}_i$, where $\hat{\mathbf{v}}_i$ is the unit vector perpendicular to $\hat{\mathbf{u}}_i$ in the (xy)-plane.

One can then perform, with the help of a method similar to that of §§ 5.4–5.6, a linear stability analysis for small amplitude periodic hexagonal fronts. Wollkind *et al.* [88] have shown that the solutions corresponding to branch (1) in Fig. 25 — which are the only solutions with arbitrarily small amplitude, for $V \simeq V_c$ — are unstable against amplitude fluctuations. The solutions corresponding to branch (2) have, in general, finite amplitudes. Only when the coefficient b_2 of the quadratic term in (5.48) becomes *accidentally* small is $|\tilde{V} - V_c|$ small enough for the perturbation expansion (5.45) to describe them correctly. Linear stability analysis then predicts that, for $V \simeq V_c$, locally stable hexagonal fronts are possible [88], although the band of allowed wavevectors is reduced by 2D phase instabilities of the Eckhaus type [87]. These results have been confirmed and extended by the numerical work of Mc-Fadden *et al.* [89]. They have in particular been able to extend the local stability analysis by studying, for example, the time evolution of a perturbation of hexagonal symmetry superimposed on a 1D band-like pattern. They show that, in Al-Cr alloys,[‡] banded cellular fronts are always unstable against hexagons and that, at a finite distance above threshold, rectangular cells should also exist. Such morphologies might have been indirectly observed [90].

All these results have primarily an indicative value, since they are only concerned with supercritical (bands) and weakly asymmetric bifurcations (hexagons), and the immediate vicinity of V_c. Moreover, stability results depend on material parameters. However, they are sufficient to get a flavor of the complexity of the problem of pattern selection. On

---

[†] Note that the amplitude equation (5.26) for 1D structures is obtained from (5.48) by imposing that one only of the three amplitudes A_i, e.g. A_1, be non-vanishing.

[‡] These alloys have a segregation coefficient $K > 1$. The bifurcation of 1D cells is supercritical, that for hexagons is rather weakly asymmetric.

the other hand, for a given class of structures (e.g. bands), a continuum (an infinite number) of locally stable periodic solutions with different wavevectors are possible. On the other hand, several types of patterns with different symmetries may coexist in some regions of parameter space. Local stability considerations reduce the infinite space of possible solutions, but they do not permit to decide about the relative[†] stability of various structures, i.e. to predict which pattern(s) is (are), in the presence of imperfections and noise, effectively realized. We will come back to this question in the next section.

6

Directional solidification of mixtures: deep cells

As discussed above, amplitude expansions only give information about weakly non-linear structures, and are therefore irrelevant when describing existing experiments on transparent materials, with the exception of the liquid/liquid crystal transition. However, the perturbation calculation sketched in § 5.2 provides qualitative information about the shape of a cellular front when non-linearities begin to be felt: it yields the amplitude of the first correction, $A_{22} \cos 2kx$, to the basic sinusoidal shape $\xi_1 = A_k \cos kx$. In the linear region of the (G, V)-bifurcation diagram, where $k_c \ell \gg 1$, one finds:

$$A_{22} = -\frac{k_c}{4K} A_k^2. \tag{6.1}$$

That is, non-linearities flatten the 'liquid side' of the front (Fig. 26) and deepen its troughs into grooves pointing into the solid. This is indeed what is observed in liquid crystal experiments [44, 63].

In systems with a subcritical cellular bifurcation, where the deformation amplitude just above threshold is finite, the cells have the same qualitative profile, but the groove depth increases rapidly with the distance to threshold $V = (V - V_c)/V_c$ (Fig. 27). While the grooves deepen,

---

[†] This is true even for periodic structures belonging to the same class and with nearly equal wavevectors $k' = k + \delta k$. Indeed, for two such structures, there always are regions of space (distant from $\sim (\delta k)^{-1}$) where the relative amplitude difference is of order 1.

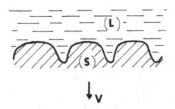

Figure 26: Schematic shape of a cellular front profile with finite deformation amplitude.

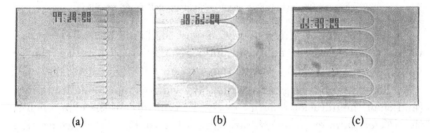

(a) (b) (c)

Figure 27: Front profile of CBr$_4$ in directional solidification. a) $V/V_c \simeq 1,8$; b) $V/V_c \simeq 3$; c) $V/V_c \simeq 6$. (Courtesy of S. de Cheveigné, G. Faivre, C. Guthmann, and P. Kurowski.)

their width decreases so much that, as soon as, typically, $\nu \gtrsim 2$, it becomes impossible to resolve optically their profile.

The only method which can then be used to make quantitative predictions about these large amplitude front structures is numerical study.

However, it is possible, at least in some limits, to build analytic approximations which shed some light on the problems of groove profile and of wavelength selection.

6.1 Analytic studies of deep cells

These all deal with 1D cell profiles — i.e. with solidification in an effectively 2D system — and with the case where the phases have identical thermal properties, i.e. where the temperature profile is linear: $T(z) = T_0 + Gz$.

It is clear that, in order to describe a front profile such as that of Fig. 28, one must distinguish between three regions:

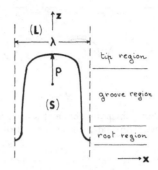

Figure 28: Schematic shape of a deep cell front profile.

▷ The cell tip or 'external' region, and the liquid beyond it. In this region, the front is reasonably flat and its characteristic length — the tip radius, ρ — is, roughly, at least of the order of the cellular space period λ.

▷ The 'internal' or groove region. Its width is much smaller than λ (and than the diffusion length $\ell = D_L/V$; typically, in the experiments, the Péclet number $p = \lambda/\ell$ varies from 0.2 to 1). So, concentration variations across the groove (in the x-direction) are necessarily very small.

▷ The 'root', i.e. the region where the groove closes up. Its dimensions along both x and z are very small and, roughly, comparable. Numerical simulations [91–94] of stationary shapes predict, and experiments [95] suggest[†] that the groove profile exhibits a bulge in that region (Fig. 29).

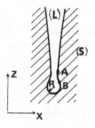

Figure 29: Shape of the groove in the root region.

[†] Available experimental results do not permit us to be completely affirmative about this. It seems that bulges at groove roots are unambiguously observed only when the root is unstable and pinches off to emit liquid droplets which get trapped into the solid before solidifying (see below).

Different approximation schemes must be developed to describe these different regions. A first approach, concerning the tip region, was proposed by Pelcé and Pumir [96]. They neglected capillarity and predicted that, when the Péclet number is small, and at least for small enough thermal gradients, the tip profile should essentially be that of a viscous Saffman-Taylor finger at zero surface tension, namely [97]:

$$\zeta(x) - \zeta_{\text{tip}} = \frac{\lambda}{2\pi}(1 - \alpha)\log\left[\tfrac{1}{2}\left(1 + \cos(2\pi x/\lambda\alpha)\right)\right], \qquad (6.2)$$

where α $(0 < \alpha < 1)$ is an adjustable (as long as the external profile is not matched to an internal one) parameter, to be interpreted as an effective groove width. Attempts to fit experimental profiles in CBr_4 [98] (see Fig. 30) as well as numerically simulated stationary shapes [99] are disappointing and indicate that, at least in the parameter range realized experimentally, this approximation is inadequate. It may be, in particular, that experimental Péclet numbers, which are not smaller than ~ 0.2, are too large for the corrections to terms of zeroth-order in p — which are the only ones retained in this calculation — to be negligible.

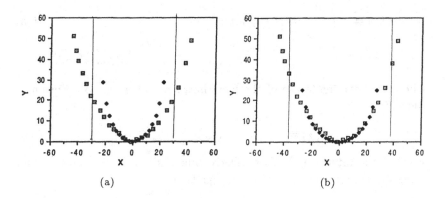

(a)   (b)

Figure 30: Attempt to fit experimental profile in CBr_4 with a Saffman-Taylor finger.
□ Empty position of the experimental front;
◇ Saffman-Taylor finger for (a) $\alpha = 0,5$ and (b) $\alpha = 0,6$ (cf. [98]).

This approach has been elaborated upon further by Dombre and Hakim [100], who consider the case of a mixture with constant concentration gap ΔC_0 ($K = 1$ limit), then by Weeks and van Saarloos [101] in the limit of small segregation coefficients ($K \to 0$). We follow here the presentation of [101].

(i) *Tip region:*

The front profile is a solution of the stationary version ($\partial C/\partial t = 0$, $\dot{z}_I = 0$) of Eqs. (4.1, 4.2).

Let $\phi(z)$ be the total concentration flux towards the solid across a horizontal line $0 < x < \lambda$. For a stationary state, ϕ is a conserved quantity, since no flux comes out of the side lines ($x = 0$, $x = \lambda$): the periodicity of the front pattern entails that $\partial C/\partial x(0, z) = \partial C/\partial x(\lambda, z) = 0$. In the liquid ($z > \zeta_{\text{tip}}$), since the profile is symmetric with respect to $x = \frac{1}{2}\lambda$:

$$\phi(z) = 2 \int_0^{\lambda/2} dx \left[V C_L(x, z) + D_L \frac{\partial C_L(x, z)}{\partial z} \right]. \qquad (6.3)$$

Using the lateral boundary conditions and the diffusion equation (4.1b), it is immediately checked that $\partial\phi/\partial z = 0$. So

$$\phi(z) = \phi(z \to \infty) = \lambda V C_\infty. \qquad (6.4)$$

For $z < \zeta_{\text{tip}}$:

$$\phi(z) = 2 \int_0^{x_I(z)} dx \left[V C_L + D_L \frac{\partial C_L}{\partial z} \right] \qquad (6.5)$$
$$+ 2KV \int_{x_I(z)}^{\lambda/2} dx\, C_L\big(x, \zeta(x)\big),$$

where we have made use of the fact that, since there is no diffusion in the solid:

$$C_S(x, z) \equiv C_S(x) = C_S(x)\big|_{\text{int}} = K C_L(x)\big|_{\text{int}}. \qquad (6.6)$$

$x_I(z)$ is the interface abscissa. Derivating (6.6) with respect to z and using (4.1b), one obtains the *exact* equation:

$$(1 - K)\frac{1}{\ell} C_L\big|_{\text{int}} + \frac{\partial C_L}{\partial z}\bigg|_{\text{int}} = \frac{d\zeta}{dx} \frac{\partial C_L}{\partial x}\bigg|_{\text{int}}. \qquad (6.7)$$

At the cell tip, due to the symmetry, $\dfrac{\partial \zeta}{\partial x} = \dfrac{\partial C_L}{\partial x}\big|_{\text{int}} = 0$. So the r.h.s. of Eq. (6.7) is small in the tip region. This remark suggests the assumption, in this external region, that $C_L(x, z) \cong C_L(z)$. Equation (6.7) can then be solved straightforwardly, yielding:

$$C_L\big|_{\text{int}} \equiv C_L(\zeta) = B \exp\left[-\frac{(1 - K)\zeta}{\ell} \right], \qquad (6.8)$$

where B is for the moment an arbitrary constant.

Plugging Eq. (6.8) into the Gibbs-Thomson equation (4.2d), one obtains the following non-linear differential equation for the front shape $\zeta(x)$ (measured from the planar front)

$$G\zeta(x) - m_L \frac{C_\infty}{K} = m_L B \exp\left[-\frac{(1-K)\zeta(x)}{\ell}\right] \tag{6.9}$$

$$+ m_L \Delta C_0 d_0 \frac{\zeta''}{(1+\xi'^2)^{3/2}},$$

which can be integrated numerically.

It may be shown [101] that this approximation is justified in two limits:

(1) $K \ll 1$: The concentration field (6.8) is an exact solution of (4.1b), (4.2c) for $K \to 0$. Equation (6.9) is the lowest order of an expansion in K. It is worth noticing that, as expected, small amplitude solutions of (6.9) are very similar to those obtained numerically from Sivashinsky's asymptotic expansion (see § 5.2), which is in principle valid in the same limit.

(2) $K \to 1$, $p \ll 1$: $\lambda \ll \ell$ [100]. In this limit, since

$$\frac{V \partial C_L / \partial z}{D_L \nabla^2 C_L} \sim p,$$

Equation (4.1b) reduces to a Laplace equation. Similarly, in (6.8), $\zeta \sim \zeta_{\text{tip}} \sim \rho$ and $C_L(z) \simeq B'z$, which is also an exact solution of (4.2c) for $K = 1$. In this case, (6.9) is the lower order of a p-expansion.

Note that, here again, the role of capillarity is essential. Although $d_0 \ll \ell$, λ, it is responsible for the presence of the derivative term in Eq. (6.9) — i.e., once more, capillarity appears as a *singular perturbation* which prohibits deformations with vanishingly small wavelengths by introducing a small length cut-off of order $\sqrt{d_0\,\ell}$ (see Section 3).

(ii) *Groove profile:*

Clearly, the above approximation cannot describe the groove region, where $d\zeta/dx$ becomes very large, and the r.h.s. of Eq. (6.7) is non-negligible. However, one can take advantage of the fact that, along the groove, the front curvature is small, so that, to lowest order, from the Gibbs-Thomson equation (4.2d), the interface concentration on the liquid side is given by:

$$u_{\text{int}} \equiv u\big(z, x_I(z)\big) \cong \frac{z}{\ell_T}, \tag{6.10}$$

where we have set:

$$u(x,z) = \frac{1}{\Delta C_0}\left[C_L(x,z) - \frac{C_\infty}{K}\right] \qquad (6.11)$$

and measured z from the planar front position.

In the liquid $(x_I(z) < x < \lambda - x_I(z))$ u obeys the diffusion equation (4.1b). On the other hand, since the groove is very narrow $(x_I(z) \ll \lambda, \ell)$, the variations of u across it are small, and u may be approximated by the first terms of its Taylor expansion:

$$u(x,z) \cong u(x_I(z),z) + \mu(x^2 - x_I^2(z)). \qquad (6.12)$$

From Eq. (4.1b), taking into account (6.10), we get:

$$\mu = -\frac{1}{2\ell\ell_T}, \qquad (6.13)$$

where we have neglected the terms proportional to dx_I/dz which would contribute to higher orders. Plugging expressions (6.12, 6.13) into the solute conservation equation (4.2c), we obtain, to lowest order in (dx_I/dz):

$$x_I\left[\frac{dx_I}{dz}\right]^{-1} = \ell_T - \ell - (1 - K)z, \qquad (6.14)$$

where ℓ_T is defined by Eq. (4.19).

The groove profile is given by[†]:

$$\zeta(x) = \frac{\ell_T - \ell}{1 - K} - Ax^{K-1} \qquad (K \neq 1), \qquad (6.15)$$

which reduces, for $K = 1$, to:

$$\zeta(x) = A + (\ell_T - \ell)\log x. \qquad (6.16)$$

A is an arbitrary constant which fixes the position of the groove along the z-axis and, thus, its width at a given z. This result was first obtained by Scheil and Hunt [102, 103].

(iii) *Matching between tip and groove:*

Dombre and Hakim [100] have shown that it is possible, with the help of a lubrication approximation applied to the condition of conservation of the flux ϕ (Eq. (6.3)) to extend systematically Scheil's approximation so

---

[†] Note that, since $V > V_C$, $\ell < \ell_T$ (see equation (4.22c)).

as to include the first corrections due to the curvature of the front profile. This introduces in the front equation (6.14) differential terms of higher orders proportional to d_0. For given values of the external constraints (given control parameters), a stationary periodic cellular profile exists if it is possible to perform, in the region where (i) and (ii) overlap, an asymptotic matching between a solution of the groove equation and a tip solution (Eq. (6.9)) with zero slope at $z = \frac{1}{2}\lambda$. Dombre and Hakim proved that, for $K = 1$, $p \ll 1$, this is always possible. Weeks and van Saarloos have extended this result to the case $K \ll 1$. So these authors show that, even in the strongly non-linear regime, there exists for given V, G, C_∞, an infinite family of cellular stationary solutions which differ by their wavevector. That is, as for small amplitude solutions, the question now arises of whether (and if so, how) the dynamics of the system leads to a selection among these front patterns.

(iv) *The root region*:

Far enough down the groove, Scheil's approximation ceases to be valid: the solute concentration in the solid, which in our one-sided approximation is the frozen trace of that on the solid side of the front at the same x, varies very rapidly with x for $x \simeq x_I(z)$ ($\partial u_s(x,z)/\partial x \sim K \partial u_L/\partial z|_{\text{int}} \cdot \mathrm{d}\zeta/\mathrm{d}x$, and $|\mathrm{d}\zeta/\mathrm{d}x| \to \infty$). Although the diffusion coefficient in the solid D_S is very small, the contribution to ϕ of the corresponding diffusion current is no longer negligible. Weeks and van Saarloos [101] show that for $K \ll 1$, Scheil's approximation should thus break down when

$$|\zeta| \sim \frac{D_s}{D} K \frac{K}{1-K} \ell \left(\frac{\mathrm{d}\zeta}{\mathrm{d}x}\right)^2. \tag{6.17}$$

In this groove-closure region, capillarity is, again, very important: the profile radius R is much smaller than ℓ, the concentration variation between points such as A and B (Fig. 29) is negligible. So it is capillary pressure variations which must compensate temperature variations, i.e.

$$\frac{d_0}{R} \sim \frac{R}{\ell_T}, \tag{6.18}$$

from which we get a rough estimate of root dimensions:

$$R \sim \sqrt{d_0 \ell_T}. \tag{6.19}$$

With $d_0 \sim 100$ Å, and ℓ_T in the 10^{-1}–10^{-2} cm range, this means that R should be of the order of a micron.

No more detailed analytic descriptions of this region have been possible yet. However, Weeks and van Saarloos give qualitative arguments supporting the fact that matching between this and the groove region should always be possible and that, consequently, groove closure should not introduce any further restriction on the number of stationary solutions.

6.2 Numerical studies

These are concerned with the shapes of *stationary* deep cells. The first ones were performed by Brown *et al.* [91, 104]. They have been confirmed and complemented, since, by the work of Kessler and Levine [93]. Both supercritical [93, 104] and subcritical [91, 92, 104] bifurcations have been studied, with the following results:

▷ For any set of control parameters V, G, C_∞ for which a deformed front exists ($V > V_c$ in the direct case, V larger than a marginal value $V_- < V_c$ in the inverse case), there is a continuum (an infinite number) of periodic cellular solutions which differ by their wavelength λ. This has also been confirmed by Ben Amar and Moussalam [105] in the case $K = 1, p \ll 1$.

Let us call $\lambda_{\min}$, $\lambda_{\max}$ the edges of the band of allowed wavelengths. Kessler and Levine [93] found that λ_m is proportional, for fixed G, C_∞ to $V^{-1/2}$. For a normal bifurcation, $\lambda_{\min}$ is quasi-equal to the $\lambda < \lambda_{\text{MS}}$ marginal wavelength [99].

▷ The cellular shapes obtained in the one-sided model with an inverted bifurcation up to values of $V/V_c \sim 3$ are in qualitative agreement with experimental observations [98, 99]. In particular, it is found that these profiles develop relatively deep grooves close above the Mullins-Sekerka threshold V_c.

In the symmetric model ($D_S = D_L$), the grooves are much less developed [93], in agreement with observations on liquid crystals.

As already discussed, grooves should bulge in the root. There, it is essential to take into account diffusion in the solid (see § 6.1), and all the more since, at a point like B (Fig. 29), the solid is melting locally. Kessler and Levine found that root dimensions are quite stable with respect to variations of D_S/D_L, in the region $0.2 \lesssim D_S/D_L < 1$, i.e. in a quasi-symmetric situation. Ramprasad *et al.* [104] found similar root shapes with $D_S/D_L = 10^{-3}$. It is impossible, for reasons of numerical precision, to study numerically cases where $D_S/D_L \lesssim 10^{-7}$, which would be more realistic for solidification.

6.3 Experimental studies of cell shapes and selection

While existing theoretical and numerical studies predict the existence of a continuum of periodic stationary cell patterns, they do not, at least up to now, provide information about their local nor relative stability. In order to study numerically structure selection, dynamical simulations following the evolution of an initial condition are needed. Such a program is presently being developed by Misbah *et al.* [99].

On the other hand, studying selection experimentally means trying to answer the following questions:

▷ For fixed external control parameters, does the system reach a truly periodic front structure?

▷ If so, is this structure independent of the history of the system (of the history of control parameter variations)? Is it unique for a given history?

▷ If the pattern is not completely periodic, how is it possible to characterize the cell-width dispersion and its dependence on system history?

Only recently have experimental studies been undertaken in this perspective. They reveal a wealth of results, which may be, at least tentatively, classified as follows:

(i) *Close above threshold:*

This corresponds, for thin samples of transparent materials like CBr$_4$ [40, 105] and succinonitrile [37, 41], in which the cellular bifurcation is subcritical, to $\delta V/V_c$ up to values, typically, of the order of 8.

In this region, the *average* space period of the cellular pattern, $\bar{\lambda}$, is *reproducible* and proportional, to a good accuracy, to $V^{-1/2}$. The wavelength (cell width) dispersion, $\Delta\lambda/\bar{\lambda}$ is, typically, $\lesssim$ 15–20% (see Fig. 31a).

Due to the presence of imperfections, such as grain boundaries in the solid, fluctuations of sample thickness, of the S-L plates contact angle and of external control parameters (in particular the pulling velocity), it is in practice impossible to decide whether this dispersion is, completely or partly, intrinsic, i.e. of dynamical origin.

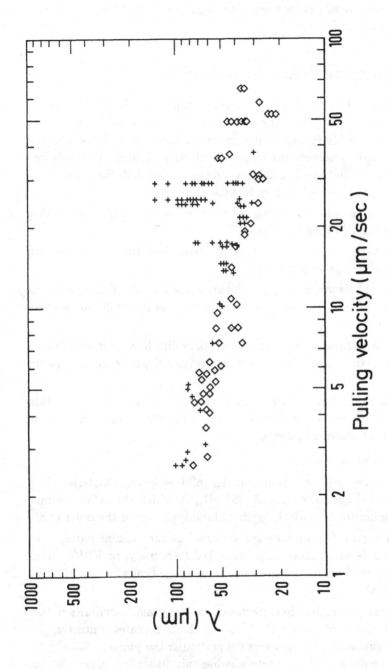

Figure 31a: Wavelength versus pulling velocity in CBr$_4$ for $G = 120$ K/cm ($\diamond$) and $G = 70$ K/cm ($+$). Note the rapid increase in cell width dispersion at $V/V_c \simeq 8$ (cf. [106]).

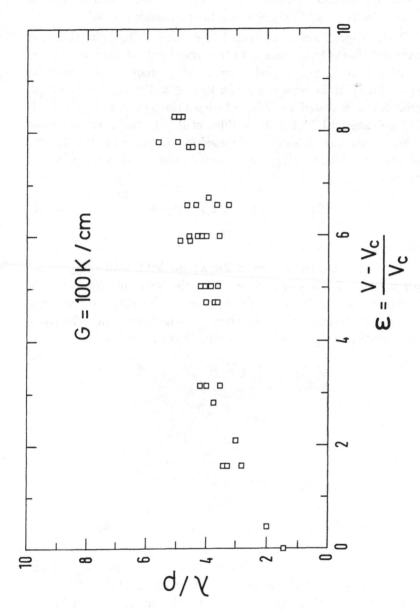

Figure 31b: Wavelength to tip radius ratio versus reduced velocity in CBr$_4$. (cf. [106]).

Liquid crystals, in which the cellular bifurcation is direct, exhibit in this low velocity range ($V/V_c \lesssim 3$) a similar behavior [44, 63].

Cellular shapes have been studied in detail in CBr$_4$ [106] (inverted bifurcation). In the immediate vicinity of the threshold, cells are shallow (Fig. 27a). As V is increased, they develop grooves which may be considered deep in the sense of § 6.2 for $V/V_c \gtrsim 3$. Cell tips become more pointed as V is increased, i.e. $\overline{(\lambda/\rho)}$ (where ρ is the tip radius and λ the cell width) increases with V (Fig. 31b). Billia *et al.* [107, 108] have performed a numerical analysis of the data obtained for succinonitrile [37, 41, 80]. They find that, in this region, $\bar{\lambda}$ is reasonably well fitted by the following expression:

$$\frac{\bar{\lambda}}{\ell}\left(\frac{\nu - 1}{\nu}\frac{K}{A}\right)^{1/2} = 8A^{-0.04}\nu^{1.09}, \tag{6.20}$$

where $\nu = \ell_T/\ell$; $A = Kd_0/\ell$.

In this regime, cellular shapes exhibit a remarkable, still unexplained, property [106]: for a given value of V, the value of λ/ρ, for cells with different widths is practically constant. This similarity between dispersed cells is even more strong: they are homothetic — i.e. cell shapes become identical when x and z are scaled on the local width λ.

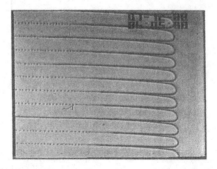

Figure 32: Groove instability in CBr$_4$. $V/V_c \simeq 6$ (cf. [95]).

For $V/V_c \gtrsim 5$, where the groove depth has become larger than typically 10λ, an *instability of groove roots* is observed [95]: the roots pinch off, thus emitting liquid droplets which remain trapped in the solid (Fig. 32) and then, later, solidify. This emission is periodic. Its frequency, ω_{drop}, is of the order of a fraction of a hertz, and it increases with groove length. Drop radii are in the micron range (which is in agreement with the estimate of root size (see § 6.2 above); it decreases when the groove depth increases. It is possible to trigger this instability by modulating the pulling velocity at frequencies of the order of 10^{-2} Hz.

Brattkus [109] has proposed an interpretation of this phenomenon in terms of an instability of the liquid jet type. Such instabilities do not exist in two-dimensional systems. So, this implies that, even in such thin samples, the roots of deep grooves are three-dimensional — which is also suggested by contrast analysis of optical observations. This interpretation is supported by the fact that it predicts a linear relation between drop radius and emission frequency, in agreement with experimental data.

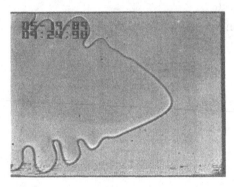

Figure 33: Dendritic cell in CBr_4 (courtesy of P. Kurowski).

(ii) *Dendritic region:*

At large pulling velocities ($V/V_c \gtrsim 12$), each cell has developed a true dendritic structure [37, 38, 41, 80] (Fig. 33) characterized by the quasi-periodic emission of side branches.

Branches do not, in general, start to appear at a unique and reproducible value of V/V_c. Although the branching threshold is, roughly, of order $10V_c$ [37, 38, 41, 106], its precise value depends on the cellular spacing [42, 106]. So cellular branching is not a global bifurcation of the whole front structure but, most probably, a rather local phenomenon, as suggested by the transient amplification scenario which is presently considered as explaining most data about free dendrite branching [110–114]. Eshelman *et al.* [80] have observed a hysteresis of this 'transition' in succinonitrile. We will not elaborate further here on the analysis of the dendritic regime, which is treated in Chapter 4 by Pomeau and Ben Amar.

(iii) *Predendritic region:*

In a rather narrow velocity range before branches start to appear, a quite spectacular change of cellular pattern is observed [106]. Experiments on CBr_4 show that, for $V/V_c \gtrsim 9$, cell-spacing dispersion increases abrupty (Fig. 31a). $\Delta\lambda/\bar\lambda$ reaches values of the order of 100%. The lower limit, λ_{min}, for the cellular spacing roughly continues the $\lambda \propto V^{-1/2}$ curve of regime (i). A similar phenomenon also seems to occur in succinonitrile (see Fig. 5 of [115]). In pivalic acid, which has a much more anisotropic interface tension, Bechhoefer *et al.* [42] observed a 'hysteresis' of cell-spacing which, in this regime, remains constant when a velocity jump is performed.

In the predendritic region, cells become much sharper, with much flatter sides (Fig. 34).

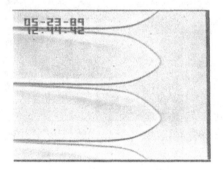

Figure 34: Predendritic cell in CBr_4 (cf. [106]).

These data raise the question of the nature of the mechanisms which are responsible for front pattern evolution. Observations on transparent materials point to the central role of several such mechanisms:

Figure 35a: Cell "birth" by tip splitting in CBr_4 ($V/V_c \simeq 2$).

▷ *Cell 'birth'* (Figs. 35a–35b) which reduces cell spacing. It occurs via tip-splitting of a preexisting cell. This mechanism is important not

Figure 35b: Time evolution of a "newborn" cell.

too far from threshold, for not too developed cells. It becomes less and less operative as V is increased and cells become more pointed.

▷ *Cell 'death'* (Fig. 35c). These events occur in the whole cellular velocity range for CBr_4, where they seem to be responsible for the large dispersion observed in the predendritic regime as, in this region, the wider cells resulting from a death event do not evolve on the time scale of the experiments [106].

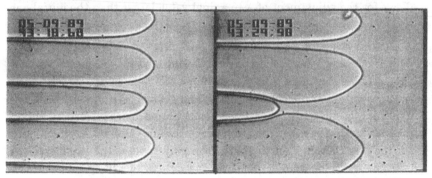

Figure 35c: " Cell death" in CBr_4. (Courtesy of S. de Cheveigné, G. Faivre, C. Guthmann and P. Kurowski.)

Phase diffusion: i.e. continuous evolution (compression or dilatation) of the cellular spacing. It is important in regime (i) and seems to be quasi-locked in the predendritic regime.

Many questions still remain unanswered, in particular as regards the predendritic region, for example:

▷ What, if any, is the role of capillary anisotropy in the locking of the cell-tip splitting mechanism far from threshold?

▷ Do interface kinetics and its anisotropy come into play in this velocity range?

▷ How does the phase diffusion coefficient (i.e. the time scale of continuous cell-spacing variations) evolve when the groove depth increases?
This question is now under study [116].

Finally, other mechanisms of cell-spacing adjustment have recently
been identified by Simon *et al.* [44] in their liquid crystal experiments.
In a small velocity range, they observe what they call a 'solitary wave'.
This is a small region where the cell-spacing is larger than that of the
environment along the front. This region drifts along the front with
constant velocity. After the front has been swept by such 'waves', the
dispersion of cell-spacings is noticeably reduced. Let us mention that
moving defects of the same type have been observed [117] on the lamellar
eutectic $CBr_4-C_2Cl_6$.

Simon *et al.* also observed, following a sudden velocity decrease, a
collective oscillation of the front deformation amplitude, with death of
several cells taking place when the amplitude is minimum. This process
allows for the adjustment of λ to a final value larger than the initial one.

All these observations point to the crucial role of dynamical defects
in pattern selection — a fact which was first brought to light by the
studies of Rayleigh-Bénard convection and of Taylor-Couette vortices.
This problem remains up to now quite an open one. In particular,
the observation of defects of the 'solitary wave' type, or of collective
amplitude oscillation modes acting as selection mechanisms, raises the
question of how 'generic' selection processes are — i.e. up to which point
can they be classified on the basis of purely topological considerations,
and how much do they depend on the details of the specific physical
problem under consideration. This, our understanding, is the central
problem to be elucidated by future research on patterns in dissipative
systems — among which exist solidification front patterns.

7

Coupling between solutal convection
and morphological instability

As discussed in Section 2, the minimal model of solidification which
we have used up to now neglects a physical effect which is often of

importance in practical solidification setups, namely convection in the liquid.

Fluid flows couple to front deformation. Assume that a solid mixture is grown by vertical directional solidification. Heat as well as an excess (or defect) of solute are produced on the moving interface which, as we have seen, gives rise, in the liquid ahead of the front, to concentration and temperature gradients. They superimpose on the — here vertical — imposed external thermal gradient. The density ρ_L of the liquid mixture depends on T and C, so the gradients induce buoyancy forces. If the lower liquid layers are the lighter ones, these forces can induce a Rayleigh-Bénard instability [35] corresponding to the appearance of a convective flow organized in periodic rolls. This flow induces horizontal temperature and concentration gradients which deform the, initially horizontal and planar, growth front.

Conversely, when the front undergoes the cellular instability, its deformation induces horizontal gradients, ∇T, ∇C, i.e. a horizontal density gradient. This always gives rises to convection in the liquid.

Finally, the density difference between the two phases induces advection currents, the flow lines of which are normal to the front. Front deformations distort these lines, this modifies thermal and chemical transport, thus reacting on the deformability of the front.

A description of solidification taking into account convection involves these coupled fields, namely liquid velocity, temperature and concentration. The problem of determining front morphology is then, obviously, highly complex [118] and can only be solved in general (especially when the liquid is being stirred) with the help of heavy numerical simulations. We will only examine here how the coupling between convection and the front deformations shifts the instabilities of the planar quiescent growth regime. Moreover, we only take into account solutal convection, and neglect thermal effects. This approximation notably simplifies theoretical formulation, thus permitting to identify and analyze more clearly the relevant physical mechanisms. Coriell *et al.* [119] have shown that it is justified (except at extremely small growth velocities) in the experimentally realizable case where the thermal gradient is stabilizing while the concentration one is destabilizing.

We assume — which does not modify the results qualitatively — that the liquid and the solid have identical thermal properties. Since $D_{\text{th}} \gg D$, the thermal field is linear (see § 2.2.F): $T(z) = T_0 + Gz$. Finally we neglect, as in Section 4, chemical diffusion in the solid ($D_S = 0$). We assume that the system is pulled vertically downwards (direction $-\hat{z}$) at velocity V, and that the interface is microscopically rough (local equilibrium on the front).

The equations describing this system read (see § 2.1, 2.2, where, in the momentum balance equation (2.16), the external force is the buoyancy force $-\rho_L \mathbf{g}$)

a) *In the liquid* ($z > \zeta(x,t)$):

▷ Solute diffusion:

$$\frac{\partial C}{\partial t} = D\nabla^2 C + (V\hat{z} - \mathbf{u}) \cdot \nabla C. \qquad (7.1)$$

$\mathbf{u}$ is the velocity of the liquid.

▷ Mass conservation:

$$\nabla \cdot \mathbf{u} = 0. \qquad (7.2)$$

▷ Navier-Stokes equation:

$$\rho_L \left\{ \frac{\partial}{\partial t} + (\mathbf{u} - V\hat{z}) \cdot \nabla \right\} \mathbf{u} = -\nabla p_L - \rho_L \mathbf{g} + \eta \nabla^2 \mathbf{u}. \qquad (7.3)$$

ρ_L, p_L, η are the density, pressure and dynamic viscosity of the liquid. As already mentioned, we neglect thermal convection, i.e. the dependence of ρ_L on temperature. That is, we take $\rho_L \equiv \rho_L(C)$.

b) *On the interface* ($z = \zeta(x,t)$):

▷ no-slip condition:

$$\mathbf{u} \wedge \hat{n} = 0. \qquad (7.4)$$

▷ mass conservation:

$$\rho_L(\mathbf{u} - \mathbf{v}_I) \cdot \hat{n} \equiv -\rho_S \mathbf{v}_I \cdot \hat{n} \equiv \mathbf{J} \cdot \hat{n}. \qquad (7.5)$$

$\mathbf{v}_I = (V + \dot{\zeta})\hat{n}$ is the local front velocity measured in the frame of the sample.

▷ solute conservation:

$$D\nabla C \cdot \hat{n} = (K - 1)C\mathbf{v}_I \cdot \hat{n}. \qquad (7.6)$$

▷ Gibbs-Thomson equation:

$$T_0 + Gz = T_M + m_L C - \frac{\gamma T_M}{\rho_S L}\mathcal{K}. \qquad (7.7)$$

As discussed in § 2.2.F, in Eq. (7.7) we have neglected kinetic energy and viscous stress corrections to the chemical potential, which are negligible for materials whith standard viscosities.

One immediately checks that Eqs. (7.1–7.7) have a stationary quiescent ($\mathbf{u} = 0$) solution whith a planar front. The corresponding concentration profile in the liquid is the same as in the standard MS problem (see § 4.1 and Eq. (4.5b)):

$$C^{(0)}(z) = C_\infty - \Delta C_0 \exp\left(-\frac{z}{\ell}\right), \qquad (7.8)$$

where the notations are those of Section 4.

One can then perform a linear stability analysis of this solution with the help of the usual first order perturbation method. The corresponding calculations are reported in detail in [120, 121]. Taking into account the fact that $D \ll \nu = \eta/\rho$,[†] one finds that the amplification rates Ω of fluctuations associated with a front deformation of horizontal wave vector $\mathbf{k}$ are the solutions of the equation:

$$B(R_S, k\ell, \Omega\tau) \qquad (7.9)$$
$$+ \left[K - 1 - \frac{K + \Omega\tau}{1 - \ell/\ell_T - d_0\ell k^2}\right] A(R_S, k\ell, \Omega\tau) = 0,$$

where the functions A and B are series of powers of the dimensionless parameter:

$$R_S = \frac{g\,\alpha_c\,(1 - K)\,D^2\,C_\infty}{K\,\nu\,V^3}. \qquad (7.10)$$

$\alpha_c = (1/\rho_L) \cdot (\partial\rho_L/\partial C)$ is the solutal expansion coefficient of the liquid, ν its kinematic viscosity.

R_S is the solutal Rayleigh number of our problem: the 'convection box' ahead of the front (the region where the concentration gradient is non-negligible) has a thickness $e \sim \ell = D/V$, the concentration gradient there $G_c \sim \Delta C_0/\ell$ so the Rayleigh number which is given by [34], $R_S = g\alpha_c G_c e^4/(\nu D)$ has the form (7.10).

One can immediatly draw from this the (*a priori* counter-intuitive), qualitative conclusion that, the smaller the growth velocity, the more important convection is: the increase of the thickness of the diffusive layer overcompensates the decrease of the gradient.

[†] This means that momentum diffusion in the liquid, which is governed by the kinematic viscosity η, is much faster than solute diffusion.

The spectrum resulting from (7.9) was studied numerically by Corriell *et al.* [119] and by Hurle *et al.* [122]. It can also be analyzed analytically [120, 121] in a weak coupling approximation which permits a simple physical interpretation, and which we will now briefly describe.

7.1 The uncoupled bifurcations

It is useful to consider first two simple limits:

a) $g = 0$:

This is the 'pure deformation' limit, since buoyancy forces are absent. In this case, where $R_S = 0$, it is found that $B/A = q\ell$, where $q\ell$ is given by Eq. (4.12). The spectrum (7.9) then reduces, as expected, to that of the standard MS problem (Eq. (4.17)). The only instability is the cellular one. For fixed thermal gradient G, the bifurcation curve in the (C_∞, V) plane has the shape represented in Fig. 36.

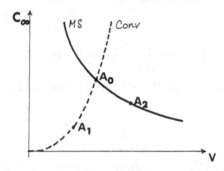

Figure 36: Instability diagram in the C_∞–V plane at a given thermal gradient for the uncoupled Mullins-Sekerka (full line) and convective (dashed line) bifurcations.

b) *Non-deformable front* $(d_0 \to \infty)$:

This is the 'pure convection' limit. Equation (7.9) then reduces to:

$$B(R_S, k\ell, \Omega\tau) + (K - 1)A(R_S, k\ell, \Omega\tau) = 0. \qquad (7.11)$$

In order to know whether a convective instability is possible, one must investigate the existence of marginal modes ($\Omega=0$ solutions of (7.11)). This can only be done numerically. It is found that a convective instability appears for $R_S \geq R_S^*(K)$, with a wave number at threshold

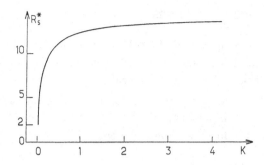

Figure 37: Critical Rayleigh number versus segregation coefficient for the decoupled convective instability.

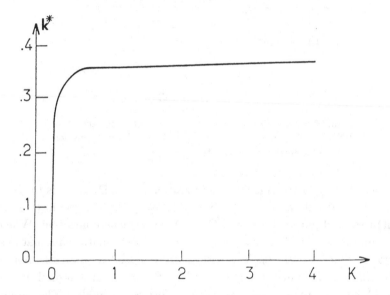

Figure 38: Critical convective wavenumber versus segregation coefficient.

$k^*(K)$, which gives the scale of the convective roll pattern. $R_S^*(K)$ and $k^*(K)$ are represented on Figs. 37 and 38.

It is seen that $k^*\ell \lesssim 0.35$, i.e. the spatial scale of the flow is larger than ℓ. The corresponding convective bifurcation curve $R_S = R_S^*(K)$ is sketched, in the (C_∞, V) plane, in Fig. 36 (dashed curve).

7.2 The coupled bifurcations

In the real system ($g \neq 0$, d_0 finite), the two instabilities are coupled. However it is possible to show, with the help of the following qualitative analysis, that this coupling is in general weak.

Consider the fluctuation spectrum of a system with control parameters corresponding to point A_1 (Fig. 36). In the absence of coupling, the spectrum takes the form as shown in Figure 39.

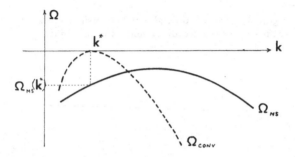

Figure 39: Amplification rate versus wavevector for uncoupled MS (full line) and convective (dashed line) bifurcations for control parameters to point A_1 of Fig. 36.

Since A_1 lies on the convective bifurcation curve, $\Omega_{\text{conv}}(k)$ is tangent to the $\Omega = 0$ axis at $k = k^*$. Since A_1 is in the MS stable (planar front) region of parameter space, $\Omega_{\text{MS}}(k)$ is everywhere negative. When the coupling is turned on, convective and front deformation fluctuations interact. Since we are dealing with a *linear* stability analysis, modes with different k values do not couple. So, a quasi-marginal mode $k \simeq k^*$ couples to a rapidly relaxing deformation mode. The larger $|\Omega_{\text{MS}}(k^*)|$, i.e. the farther A_1 is from the MS bifurcation curve, the more adiabatically the MS mode follows the slow convective one. That is, the less it shifts $\Omega_{\text{conv}}(k)$ and the smaller the effective coupling.

The situation is similar for a system with representative point A_2 (Fig. 36). In this case, it is the deformation mode $k = k_{\text{MS}}$ (Fig. 40) which is neutral and slaves the rapidly relaxing convective mode of growth rate $\Omega_{\text{conv}}(k_{\text{MS}})$.

When the representative point of the system moves along the bifurcation curves of Figure 36, the effective coupling increases as A_1 (resp. A_2) comes closer to the MS (resp. convective) bifurcation curve. It is maximum at the crossing point A_0 (Fig. 36). At this point, the zero-coupling

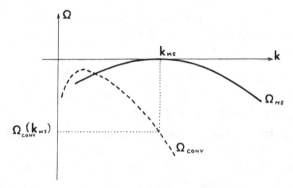

Figure 40: Same as Fig. 39 for point A_2 of Fig. 36.

spectrum contains *two* neutral modes (see Fig. 41). A_0 is a *codimension-2 bifurcation point.*

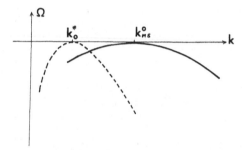

Figure 41: Same as Fig. 39 for point A_0 of Fig. 36.

One easily calculates, for fixed G, the coordinates V_0, $C_{\infty 0}$ of the crossing point A_0, from which one obtains, with the help of (4.22a), the value of the wavevector k_{MS}^0 of the corresponding neutral deformation mode. It is found [120, 121] that $k_{MS}^0 \gg 1/\ell_0$ (where $\ell_0 = D/V_0$) except in two limits, namely:

▷ very small thermal gradients (typically, $G \lesssim 1\text{K/cm}$);

▷ very small segregation coefficients (K $\ll 1$).

On the other hand, we have seen that $k_0^* \sim 1/\ell_0$. So in general, even at the point of maximum coupling,

$$k_0 \ll k_{MS}^0, \tag{7.12}$$

the two branches of the bare (uncoupled) spectrum are widely spaced, $|\Omega_{\text{conv}}(k_{MS}^0)|$ and $|\Omega_{MS}(k_0^*)|$ are large and *the effective coupling is weak.*

Moreover, it must be pointed out that the $\Omega_{MS}(k)$ curve is much flatter than $\Omega_{conv}(k)$: $d^2(\Omega_{MS}\tau)/d(k\ell)^2\big|_{k=k_{MS}}$ is of order $-(d_0/\ell)^{2/3}$, while $d^2(\Omega_{conv}\tau)/d(k\ell)^2\big|_{k=k^*} \sim 1$.

So at the point of codimension 2, A_0, $|\Omega_{conv}(k^0_{MS})| \gg |\Omega_{MS}(k^*_0)|$ and it can be predicted that the MS bifurcation should be much less shifted by the coupling than the convective one.

Since we are in a weak coupling situation, the spectrum (7.9), and therefore the shifts of the bifurcations, can be calculated analytically in a first order perturbation approximation. The results are displayed in Fig. 42 for the case of the Pb–Sn alloy growing in a thermal gradient $G = 200$ K/cm.

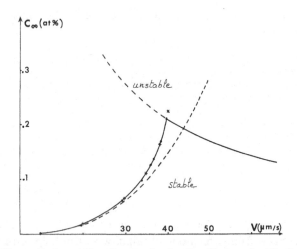

Figure 42: Instability diagram in the C_∞–V plane for Pb–Sn alloy at $G = 200$ K/cm, with (full lines) and without (dashed lines) coupling. × correspond to the perturbative analytic calculations, ● to numerical results (cf. [120]).

It is seen that the shifts are indeed smaller: at the crossing point A_0, the convective threshold is shifted by $(\delta V/V)_{conv} \simeq 9\%$. As expected, the displacement of the cellular threshold is considerably smaller. At A_0, $(\delta C_\infty/C_\infty)_{MS} \simeq 10^{-4}$. These shifts decrease when the system representative point moves away from A_0. For both bifurcations, the coupling is stabilizing (it reduces the unstable region). These results are in very good agreement with numerical studies [119].

The only possibility for the effective coupling to become large is that the two critical wavevectors k^*_0 and k^0_{MS} (Fig. 41) come close to each other, which is realized when $K \ll 1$ or for very small G. In this case, there is a strong mode-mixing in the k-region where the two branches of the uncoupled spectrum cross. The coupled spectrum may then exhibit a shape as sketched in Fig. 43. In this strong coupling region, Ω may become complex.

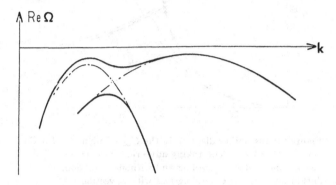

Figure 43: Possible shape of $\mathrm{Re}\,\Omega$ versus wavevector with (full lines) and without (dashed lines) strong coupling.

When $k_0^* \lesssim k_{\mathrm{MS}}^0$, the two branches of the spectrum cross close to the $\mathrm{Re}\,\Omega = 0$ axis, and it may happen that the first neutral mode is a mode belonging to the strong mixing region. If it has $\mathrm{Im}\,\Omega \neq 0$, the bifurcation becomes *oscillatory*. This is indeed what was found numerically by Schaefer *et al.* [123] for succinonitrile-ethanol mixtures $(K = 0.044)$ and for Pb-Sn alloys in the very small thermal gradient $G = 0.1$ K/cm.

A different analytical approach, based on Sivashinsky's asymptotic expansion, can be built in the small segregation coefficient limit, where both k_0^* and k_{MS}^0 become small. It has been developed by Young and Davis [124], who found that, in this limit also, convection stabilizes the planar front against deformations.

Finally, it is also possible to calculate the bifurcation shifts in the presence of the advective flow induced by the density difference $\delta\rho = \rho_S - \rho_L$. Since, typically, $\delta\rho/\rho_L \lesssim 10^{-1}$, this effect can also be treated perturbatively [125, 121].

Advection simply renormalizes the Rayleigh number

$$R_S \longmapsto R_S\left(1 + \frac{\delta\rho}{\rho_{\mathrm{L}}}\right)^{-3},$$

which induces an additional shift of the convective bifurcation $(\delta V/V)_{\mathrm{conv}} = -\delta\rho/\rho_L$. In the usual case where $\delta\rho > 0$, it stabilizes the system

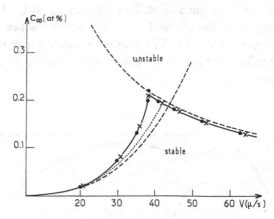

Figure 44: Instability diagram in the C_∞-V plane for Pb–Sn alloy at $G = 200$ K/cm, taking advection into account. dashed line: without coupling and without advection; dotted line: without coupling and with advection; full line: with coupling and with advection
X: perturbative analytic calculation,
●:   numerical calculation (cf. [125]).

against convection, and is comparable in magnitude to the shift due to the coupling to deformation modes (Fig. 44).

For usual values of the pulling velocity, advection shifts the cellular bifurcation by $(\delta C_\infty/C_\infty)_{\mathrm{MS}} \simeq -\delta\rho/\rho_L$, thus destabilizing the planar front. So this advective shift — which is present even in zero gravity — is much larger (typically by a factor $\sim 10^2$) than that due to convection.

All these results are only concerned with the position of instability thresholds. It would of course be desirable to get information about non-linear behavior beyond the bifurcations. Jenkins [126] has derived the coupled amplitude equations in the weak coupling limit. No detailed work about front shapes in the presence of convection has been performed yet. This demands sophisticated and heavy numerical studies, but would be of considerable practical interest.

8

Directional solidification of a faceted crystal

All our discussion of the cellular instability and front morphology has been concerned with materials with a rough solid-liquid interface — i.e. with quasi-instantaneous attachment kinetics and isotropic surface tension.

Extending this analysis of the cellular regime[†] to systems with a weak capillary anisotropy [89] and (isotropic or anisotropic) linear interface kinetics [127] does not bring in any new qualitative phenomenon.

Up to now, most theoretical works, as well as systematic experimental studies, of directional solidification have been dealing with solids which do not exhibit facets on their equilibrium shape. However, partly faceted cellular fronts have been observed in Bismuth alloys [128] and in semiconductors growing from the melt [129]. These partly faceted shapes, which have also been observed in free dendritic growth [130], obviously result from the competition between diffusion on the one hand, kinetics and capillarity on the other hand [131].

Let us consider the case where an impure solid exhibiting on its equilibrium shape, in the $\theta = 0$ direction, a facet matching tangentially with neighboring rough parts, is directionally solidified. The pulling velocity V is assumed to be perpendicular to the facet plane. We want to study the stability of this facet against front deformations and only consider, for the sake of simplicity, one-dimensional deformations.

The corresponding Wulff plot (Fig. 45) (see Chapter 1 by Nozières) therefore exhibits a cusp for $\theta = 0$, and the absence of missing orientations on the equilibrium shape entails that the *surface stiffness* $\Gamma(\theta) = [\gamma(\theta) + d^2\gamma(\theta)/d\theta^2]$ is positive for $\theta \neq 0$ ($\gamma(\theta)$ is the interface energy).

[†] The question of the importance of capillary anisotropy in the dendritic regime in directional solidification has not yet been elucidated.

268   B. Caroli, C. Caroli, B. Roulet

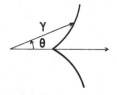

Fig. 45: Schematic shape of the Wulff
plot close to the facet orientation.

Assume for the moment that interface kinetics is everywhere instantaneous — this assumption is of course unrealistic for the facet and will be
relaxed later, but it does not modify the following qualitative argument.

In order to study the stability of the facet, it seems *a priori* natural
to extend to this anisotropic system the Mullins-Sekerka linear stability
analysis (Section 4). However, it must be noticed that, as mentioned
in § 2.2.B (Eq. (2.54)) since the capillary pressure is proportional to
the surface stiffness — and not to the surface energy — it is $\Gamma(\theta)$ which
now appears in the Gibbs-Thomson equation (4.2d). Now, a cusp
in $\gamma(\theta)$ gives rise to a singular contribution $\Delta\gamma'\delta(\theta)$ in $\Gamma(\theta)$ (where
$\Delta\gamma' = \left. d\gamma/d\theta\right|_{0+} - \left. d\gamma/d\theta\right|_{0-}$). Imagine for a moment that the cusp
of γ is smoothed on a width $\Delta\theta \sim \varepsilon$ (Fig. 46). $\Gamma(\theta)$ is now regular, with
a peak of width $\sim \varepsilon$; the equilibrium facet becomes a 'quasi-facet' of
angular width $\sim \varepsilon$.

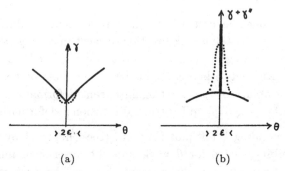

(a)   (b)

Figure 46: Schematic plot of the surface energy, (a), and surface
stiffness, (b), close to the facet orientation $\theta = 0$. The dotted
curves correspond to the smoothed cusp approximation.

The Gibbs-Thomson equation is now well-defined and a MS linear
analysis can be formally performed, yielding the standard result, where
the surface tension γ must be replaced by $\Gamma_0 = (\gamma + \gamma'')_{\theta=0}$. Note

that $\Gamma_0 \simeq \varepsilon^{-2}\gamma(0)$ is very large, so that the linearly unstable region of parameter space (cf. Eqs. (4.24b), (4.25)) becomes accordingly small.

The next question to be asked is that of the limits of validity of the linear approximation. Expanding the growth equations in powers of the deformation amplitude implies to Taylor-expand $\Gamma(\theta)$ in the Gibbs-Thomson equation. So, the linear approximation is valid only when $\theta^2(\mathrm{d}^2\Gamma/\mathrm{d}\theta^2)_0 \ll \Gamma_0$ (where $\theta = \tan^{-1}(\mathrm{d}\zeta/\mathrm{d}x)$, $\zeta(x)$ being the front profile). Since $(\mathrm{d}^2\Gamma/\mathrm{d}\theta^2)_0 \sim \varepsilon^{-2}\Gamma_0$, it is valid only when $\mathrm{d}\zeta/\mathrm{d}x \ll \varepsilon$, i.e. as long as the deformed front profile only contains orientations belonging to the quasi-facet. So, when $\gamma(\theta)$ exhibits a cusp ($\varepsilon \to 0$), the range of validity of the MS linear stability analysis vanishes.

This simply expresses the fact that any deformation of a facet implies the creation of steps, each of which corresponds to a finite energy cost $\gamma_0' a$ — where a is the atomic or molecular distance, and $\gamma_0' = (\mathrm{d}\gamma/\mathrm{d}\theta)_{0+}$. So, studying the stability of a facet would in principle necessitate to study the dynamical evolution from a variety of initial conditions. This is of course not realizable, and we limit ourselves here to a much more limited question [132, 133].

▷ Is it possible to find stationary periodic front shapes with a small deformation amplitude with respect to the infinite facet?

▷ If so, are these solutions locally stable as regards slow amplitude and phase modulations?

We assume, for simplicity, that the solid and liquid have the same thermal properties and — as usual — that thermal diffusion is quasi-instantaneous as compared with chemical diffusion (minimal model with $n = 1$). The thermal profile is linear, with $\mathrm{d}T/\mathrm{d}z = G$. Finally, we assume that $K = 1$ (constant concentration gap). The equations describing the solidifying systen then read:

(i) *In the liquid* ($z > \zeta(x,t)$):

$$\frac{\partial C}{\partial t} = D\nabla^2 C + V\frac{\partial C}{\partial z}. \tag{8.1}$$

(ii) *On the interface* ($z = \zeta(x,t)$):

▷ solute conservation:

$$D\hat{n} \cdot \nabla C = \Delta C_0\, n_z (V + \dot{\zeta}). \tag{8.2}$$

▷ generalized Gibbs-Thomson equation:

$$(\sigma + \sigma'')\mathcal{K} + \frac{\zeta}{\ell_T} + \mathcal{F}(V + \dot{\zeta}, \theta) = \frac{C}{\Delta C_0}, \tag{8.3}$$

where

$$\sigma(\theta) = \frac{\gamma(\theta) T_M}{\rho L m_L \Delta C_0}. \tag{8.4}$$

ζ is measured from the position of the $T = T_M$ isotherm, ℓ_T is defined by Eq. (4.19), and $\mathcal{K} = -\zeta''(1 + \zeta')^{-3/2}$; $\zeta' \equiv \partial\zeta/\partial x$; $\theta(x,t) = \tan^{-1}(\partial\zeta/\partial x)$. The $\mathcal{F}$ term in the generalized Gibbs-Thomson equation describes the undercooling associated with attachment kinetics. The kinetics is considerably slower on the facet than on the rough parts of the front where we assume it, for simplicity, to be instantaneous

$$\mathcal{F}(v,\theta) = \begin{cases} F(v) = \dfrac{\delta T_{\text{kin}}}{m_L \Delta C_0} & \theta = 0, \\ 0 & \theta \neq 0, \end{cases} \tag{8.5}$$

where δT_{kin} is the kinetic undercooling of the infinite facet growing at velocity v.

For orientations corresponding to a rough interface, $\theta \neq 0$, Eq. (8.3) is well defined and expresses *local* equilibrium on the front. For a facet, equilibrium can only be *global* (see Chapter 1 by Nozières). The shape of a facet is by definition locked (it is, in our 1D front geometry, a straight line joining the two endpoints of abscissa x_0, x_1), it can only climb globally. So, in order to give a physical meaning to Eq. (8.3), it is necessary to integrate it along the whole length ($x_0 < x < x_1$) of the facet. Since $\mathcal{K} = \mathrm{d}\theta/\mathrm{d}s$ (where s is the curvilinear abscissa along the front) and $\mathrm{d}s = \mathrm{d}x$ for our $\theta = 0$ facet, we thus obtain the equation expressing the global constraint on the growing facet, which was first written by Ben Amar and Pomeau [134] in their work on partly faceted needle-crystals:

$$\varepsilon_f \Delta\sigma' + \left[\frac{\zeta_f}{\ell_T} + F(V + \dot{\zeta}_f)\right](x_1 - x_0) = \int_{x_0}^{x_1} \mathrm{d}x \, \frac{C(x, \zeta_f, t)}{\Delta C_0}, \tag{8.6}$$

where $\Delta\sigma' = \sigma'_{0+} - \sigma'_{0-} > 0$.

ζ_f is the facet height, $(x_1 - x_0)$ its length. $\varepsilon_f = +1$ (resp. -1) for a 'quasi-convex' (resp. quasi-concave) solid facet. We call quasi-convex a facet such that $\theta(x_{0-}) = 0_+$, $\theta(x_{1+}) = 0_-$ (see Fig. 47).

For $F = 0$, Eq. (8.6) expresses the fact that the system free energy is stationary when the whole terrace climbs by one atomic distance a: the energy cost of the step thus created, $2|\sigma'_0|a = a\Delta\sigma'$, exactly compensates the bulk energy gained by transforming from liquid into solid volume $a(x_1 - x_0)$.

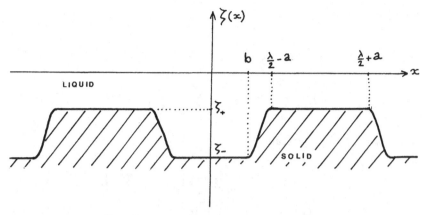

Figure 47: Periodic crenellated front profile.

Finally, since we have assumed that no orientation is missing on the equilibrium shape, mechanical equilibrium (which is instantaneous on the diffusive time scales of interest here) imposes that the curved parts of the front match tangentially with the facets.

The basic stationary solution corresponding to the infinite faceted front is immediately obtained by looking for a solution $C \equiv C^{(0)}(z)$ of (8.1, 8.2) and taking the $(x_1 - x_0) \to \infty$ limit of Eq. (3.6). It reads:

$$C^{(0)}(z) = C_\infty - \Delta C_0 \exp\left[-\frac{z - \zeta^{(0)}}{\ell}\right], \qquad (8.7)$$

with $\ell = D/V$. The facet height $\zeta^{(0)}$ is measured from the $T = T_M$ isotherm, and given by:

$$\zeta^{(0)} = \ell_T\left[\frac{C_\infty}{\Delta C_0} - 1 - F(V)\right]. \qquad (8.8)$$

We look for small amplitude non-planar solutions (Fig. 45). Since $\gamma(\theta)$ is even for small values of θ, we restrict ourselves to front shapes which are symmetric with respect to the center of a facet. Let $2a$, ζ_+ (resp. $2b$, ζ_-) be the length and height of a high (resp. low) facet, c the extension along Ox of each curved part or 'riser'. The period of such a 'crenellated' profile is $\lambda = 2(a + b + c)$.

8.1 Crenellated stationary front profiles [132]

Since we assume the deformation amplitude $(\zeta_+ - \zeta_-)$ to be small, we calculate these profiles by linearizing the diffusion problem (8.1, 8.2).

Setting:

$$\zeta^{(1)}(x) = \zeta(x) - \zeta^{(0)} = \sum_{n=-\infty}^{\infty} \zeta_n e^{inkx}, \tag{8.9a}$$

$$C(x,z) - C^{(0)}(z) = \sum_{n=-\infty}^{\infty} C_n^{(1)}(z) e^{inkx}, \tag{8.9b}$$

where $k = 2\pi/\lambda$, we obtain:

$$C_n^{(1)}(z) = -\Delta C_0 \frac{\zeta_n}{q_n \ell^2} \exp\left[-q_n(z - \zeta^{(0)})\right], \tag{8.10a}$$

$$q_n \ell = \tfrac{1}{2}\left[1 + \sqrt{1 + 4n^2 k^2 \ell^2}\right]. \tag{8.10b}$$

Inserting this result into the local equation (8.3) (with $\mathcal{K} \simeq -\zeta''$) and in the facet equations obtained from (8.6), we get, in the different parts of the profile:

a) *Low (quasi-concave) facet* ($-b < x < b$):

$$\Delta\sigma' + \frac{2b}{\ell}\left(1 - \frac{\ell}{\ell_T}\right)\zeta_-^{(1)} = \frac{2b}{\ell}\left\{\zeta_0^{(1)} + \sum_{n \geq 1} \frac{2\sin(nkb)}{nkbq_n\ell} \zeta_n^{(1)}\right\}. \tag{8.11}$$

b) *High (quasi-convex) facet* ($\tfrac{1}{2}\lambda - a \leq x \leq \tfrac{1}{2}\lambda + a$):

$$-\Delta\sigma' + \frac{2a}{\ell}\left(1 - \frac{\ell}{\ell_T}\right)\zeta_+^{(1)} = \frac{2a}{\ell}\left\{\zeta_0^{(1)} + \sum_{n \geq 1} \frac{2\sin(nka)}{nkaq_n\ell} \zeta_n^{(1)}\right\}. \tag{8.12}$$

c) *In each curved part* (e.g. $b < x < \tfrac{1}{2}\lambda - a$):

$$d_0\ell\frac{d^2\zeta^{(1)}}{dx^2} + \left(1 - \frac{\ell}{\ell_T}\right)\zeta^{(1)} + F(V) \tag{8.13}$$

$$= \zeta_0^{(1)} + \sum_{n \geq 1} \frac{2\cos(nkx)}{q_n\ell} \zeta_n^{(1)},$$

where $d_0 = (\sigma + d^2\sigma/d\theta^2)_{0+} = (\sigma + d^2\sigma/d\theta^2)_{0-}$ is the capillary length associated with the rough parts of the front.

The matching conditions read:

$$\zeta(b_+) = \zeta_-, \quad \zeta\left[(\tfrac{1}{2}\lambda - a)_-\right] = \zeta_+, \tag{8.14}$$

$$\zeta'(b_+) = \zeta'\left[(\tfrac{1}{2}\lambda - a)_-\right] = 0. \tag{8.15}$$

The set of Eqs.(8.11–8.13) is the equivalent, for the faceted case, of the linearized front integral equation (see Appendix B). We have been able to solve them analytically only in the *highly non-local limit* $\lambda \ll \ell$.

In this case, since $q_0\ell = 1$, $q_n\ell \simeq nk\ell \gg 1$ $(n \neq 0)$, one may neglect, up to errors of order $(\lambda/\ell)\log(\lambda/\ell)$, the $n \neq 0$ terms in Eqs. (8.11–8.13) and retain only the homogeneous term $\zeta_0^{(1)}$. This amounts to neglecting the distortion $C^{(1)}$ of the diffusion field induced by the deformation of the front and retaining, in $[C - C^{(0)}]_{\text{int}}$, only the term corresponding to exploration of the zeroth-order field. This would be exact if one could neglect the curvature of $C^{(0)}(z)$, i.e. if

$$C^{(0)}(z) \cong C^{(0)}\left(\zeta^{(0)}\right) + \Delta C_0 \frac{z - \zeta^{(0)}}{\ell}. \tag{8.16}$$

The next term in the expansion of $C^{(0)}$, $-\Delta C_0(z - \zeta^{(0)})^2/\ell^2$, gives rises to a defect of diffusion current at the interface of order $\Delta C_0 D\zeta^{(1)}/\ell^2$, which must be compensated by the current associated with the distortion $C^{(1)}$, of spatial scale λ, which is thus of order $DC^{(1)}/\lambda$. $C^{(1)}/\Delta C_0$ is therefore of order $\zeta^{(1)}\lambda/\ell^2 \ll \zeta^{(1)}/\ell$, as shown by Eq. (8.10a).[†]

Equation (8.13) for the curved parts of the profile then reduces to a differential equation, (8.11–8.12) to algebraic ones, and the problem can be solved straightforwardly. It is found that the curved parts are half arches of a sinusoid, the wavevector of which,

$$q = \left[\left(1 - \frac{\ell}{\ell_T}\right)\frac{1}{d_0\ell}\right]^{1/2}, \tag{8.17}$$

is that of the neutral mode of the standard cellular instability problem associated with rough off-facet orientations. Since we have assumed $\lambda \ll \ell$, it corresponds necessarily to the neutral mode with the larger wavevector $(q \gg q_{\text{MS}})$.

It immediately results from this that crenellated solutions only exist above the cellular threshold associated to rough orientations, i.e. for

$$V > V_{\text{MS}} = \frac{D}{\ell_T}. \tag{8.18}$$

[†] Note that this argument would not hold for the symmetric model $(D_S = D_L = D)$ where the dominant part of the current defect would be that in the solid, of order $(D/\lambda)\Delta C_0(\zeta^{(1)}/\ell)$, giving a correction to C of order $\zeta^{(1)}/\ell$ comparable with the main term in Eqs. (8.11–8.13).

The amplitude of the sinusoidal risers is imposed by matching them with the facet (Eqs. (8.14, 8.15)). It is found that, whatever λ, when condition (8.18) is fulfilled, there exists a stationary crenellated solution.

If interface dissipation is neglected, one finds that the high and low facets have the same length and the same deformation amplitude with respect to the planar front:

$$2a = 2b = \frac{\lambda}{2} - \frac{\pi}{q}, \tag{8.19a}$$

$$\zeta_+^{(1)} = \zeta_-^{(1)} = \ell\left[1 - \frac{\ell}{\ell_T}\right]^{-1} \frac{\Delta\sigma'}{2a}. \tag{8.19b}$$

In the presence of interface dissipation ($F \neq 0$), the hot (high) facets become longer than the cold ones, and the average front height is above that of the infinite facet: the rough parts of the front grow 'more easily', which, as expected, favors the hot quasi-convex facets. One finds that [132] $2a \sim [\zeta_+^{(1)} - \mathcal{L}]^{-1}$, $2b \sim [\zeta_+^{(1)} + \mathcal{L}]^{-1}$, where $\mathcal{L}$ is a length characteristic of interface dissipation:

$$\mathcal{L} = F\ell\left[1 - \frac{\ell}{\ell_T}\right]^{-1}. \tag{8.20}$$

Our method is justified only for $\lambda \ll \ell$, $q \gg q_{\text{MS}}$, with $\lambda \geq 2\pi/q$ ($a, b > 0$). This can be shown to entail that the above results are valid when:

$$\frac{V - V_{\text{MS}}}{V_{\text{MS}}} \gg \left[\frac{d_0}{\ell_{\text{MS}}}\right]^{1/3}. \tag{8.21}$$

So, we are unable to analyze the behavior of the facet in a vicinity of the rough cellular threshold $\delta V/V_{\text{MS}}$ of order, typically, 10^{-1}. For $V > V_{\text{MS}}$ outside of that zone, we can conclude that there exists, for a given set of values of the control parameters, a continuum of stationary crenellated solutions. For a given wavelength λ, the deformation amplitude decreases as V is increased. This model does not predict any upper velocity threshold for the existence of these solutions. However, it should be noticed that the model loses validity at very large velocities for two different reasons:

▷ the diffusion length D/V decreases, thus reducing the range of the highly non-local approximation;

▷ at high velocities the cusp in γ must be rounded-off by dynamical roughening effects (see Chapter 1 by Nozières), which must necessarily introduce a cross-over towards the standard MS regime. Of course, the larger the step energy (the deeper the cusp in $\gamma(\theta)$), the larger will be the cross-over velocity be.

8.2 Local stability of crenellated fronts [133]

We now want to study the spectrum of long wavelength $2\pi/p$ ($p\lambda \ll 1$) deformation modes of the stationary crenellated fronts. A linear analysis of the type that is used to study phase diffusion along rough cellular fronts is now possible, since we are only interested in the time evolution of front deformations which keep the number of facets constant. The shape of such a fluctuation is represented in Figure 48.

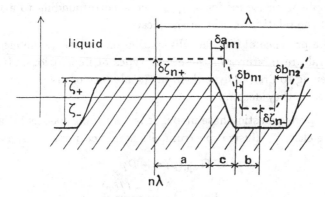

Figure 48: Full line: stationary periodic crenellated front. Dashed line: displaced profile.

When studying these fluctuation modes, it is no longer possible to simply approximate, as in the stationary calculation above, the concentration field ahead of the crenellated front by that, $C^{(0)}(z)$, of the average planar front. Indeed, if this is done, one is left with a single degree of freedom for the front, corresponding to global relaxation in the externally imposed thermal gradient.

For this reason, the spectrum can be calculated analytically only when the facets are much longer than the risers (rough parts). It is then legitimate to replace, in the r.h.s. of the generalized Gibbs-Thomson equation (8.6), the exact concentration field by that of a perfect crenel with vertical risers.

Considering the shape (Fig. 47) of the basic crenellated front, deformation modes can be separated qualitatively, for $p \to 0$, into four classes:

▷ Two amplitude modes involving primarily vertical displacements of the facets $\delta\zeta_{\pm n}$, namely:

- climbing modes for which $\delta\zeta_{n+} \simeq \delta\zeta_{n-}$;
- breathing modes in which hot and cold facets move out of phase $(\delta\zeta_{n+} \simeq -\delta\zeta_{n-})$;

▷ Two 'phase' modes involving primarily lateral displacements of the risers, δa_{ni}. They decompose, for $p \to 0$, into:

- 'acoustic' phase modes $\delta a_{n1} \simeq \delta a_{n2}$ (see Fig. 48). The $p = 0$ acoustical phase mode is a global translation of the front along Ox;
- 'optical' phase modes $\delta a_{n1} \simeq -\delta a_{n2}$ corresponding to a variation of the relative lengths of the facets.

In the presence of interface dissipation on the facet, the eigenmodes in general are mixtures of these four types of fluctuations. It is more enlightening to first analyze them in the absence of dissipation ($F = 0$). In this case, they decompose into:

(1) A pure acoustic phase mode. It is *stable* and has, as in the rough interface case, a diffusive spectrum. Its amplification rate is given by

$$\Omega_{\text{a.p.}}(p) = -\mathcal{D}p^2, \tag{8.22}$$

$$\mathcal{D} = \frac{\pi}{8} D \frac{\lambda}{\ell} \frac{(1 - \ell/\ell_T)}{\log(\lambda/c)}, \tag{8.23}$$

where $c = \pi/q$ is the x-extension of the riser ($c \ll \lambda$).

(2) A *stable* pure climbing mode

$$\Omega_{\text{cl}} \cong -\frac{D}{\ell\ell_T}. \tag{8.24}$$

(3) An *unstable* pure breathing mode

$$\Omega_{\text{br}} \cong \frac{\pi^2}{\phi_0} D \frac{d_0 q^2}{\lambda}. \tag{8.25}$$

where ϕ_0 is a dimensionless number $O(1)$.

(4) A *stable* optical phase mode coupled to breathing, so as to keep the average front position constant:

$$\Omega_{\text{o.p.}} \cong -2\pi D \frac{d_0 q^2}{\lambda \log(\lambda/c)}. \tag{8.26}$$

When interface dissipation on the facet is taken into account ($F \neq 0$), the acoustic phase mode remains pure and stable, the phase diffusion coefficient $\mathcal{D}$ remaining proportional to $[\log(\lambda/c)]^{-1}$.

The other three modes become admixtures of climbing, breathing and optical phase fluctuations. The admixture effect stems from the fact that, due to dissipation on the facet, a variation of crenel height imposes a change in the relative lengths of hot and cold facets.

It can be proved that, whatever the magnitude of F, there always is one and only one unstable mode. It always contains a breathing amplitude component, the magnitude of which depends on the value of F.

So, crenellated fronts are always unstable against a mode involving a fluctuation of their deformation amplitude. For $V > V_{MS}$, the planar front is unstable, and crenellated states must be considered as an amplitude threshold which must be jumped over in order for the system to reach a finite amplitude deformed front state. It may reasonably be guessed that these finite amplitude patterns are a Mullins-Sekerka structure of partly faceted cells.

Since crenellated states exist up to large growth velocities, one should observe a very large hysteresis of front morphologies, increasing with step energy.

Finally, it is of interest to comment on the form of the phase mode spectrum (8.22–8.23, 8.26):

▷ The phase diffusion coefficient $\mathcal{D}$ (Eq. (8.23)) is always positive: crenellated fronts are Eckhaus-stable. The 'local wavevector' q associated with the risers lies on the marginal MS curve, the corresponding structure would therefore be Eckhaus-unstable in the absence of facets. It is the facets which stabilize the front against phase diffusion.

▷ It is seen on Eqs. (8.23, 8.26) that any lateral motion of the risers induces a slowing down of the dynamics appearing via the factor $[\log(\lambda/c)]^{-1}$. This factor is the signature of the singularity of the diffusion field of a crenel in the non-local limit, where the diffusion equation reduces to Laplace's equation and where the jutting out corner produces a point effect. It is diffusion which is responsible for the low lateral riser mobility. We believe that this effect should appear in all situations involving motion of 'macrosteps' [131] along faceted fronts, e.g. directional solidification with a vicinal planar front.

There has been up to now, to our knowledge, no systematic experimental investigation of facet destabilization in directional solidification. It would be of much interest to try and observe the effects analyzed above, in particular the hysteresis of front morphology.

Acknowledgments

We have been introduced to this field of research by the works of J.S. Langer and R.F. Sekerka, and have benefited from their experience through many discussions. We are greatly indebted to C. Misbah with whom we have been collaborating for several years. The permanent interaction and cooperation which has been going on between our group and the experimental group formed by S. de Cheveigné, G. Faivre, C. Guthman, P. Kurowski and M.M. Lebrun has been invaluable. We are grateful to all of them.

References

[1] For a guide to the metallurgical literature on eutectic solidification, see Elliot, R., 1983, *Eutectic Solidification Processing*, Butterworth and Co.

[2] Hunt, J.D. and Jackson, K.A., 1966, *Trans. Met. Soc. AIME*, **236**, 843.

[3] Tiller, W.A., 1966, *The Art and Science of Growing Crystals*, J.J. Gilman Ed., J. Wiley and Sons, New York.

[4] Jackson, K.A., 1979, *Crystal Growth, A tutorial approach*, W. Bardsley, D.T.J. Hurle and J.B. Mullin Ed., North-Holland, Amsterdam.

[5] Sander, L.M., 1987, *The Physics of Structure Formation*, W. Güttinger and G. Dangelmayr Ed., Springer-Verlag, Berlin, and references therein.

[6] Schlitter, F.W., Eichkern G. and Fisher, H., 1968, *Electrochem. Acta*, **13**, 2063.

[7] Boettinger, W.J., Shechtman, D., Schaeffer, R.J. and Biancaniello, F.S., 1984, *Metall. Trans.*, **15A**, 55.

[8] Kurtz, W., Giovanola, B. and Trivedi, R., 1986, *Acta Metall.*, **84**, 823.

[9] Trivedi, R. and Kurz, W., 1986, *Acta Metall.*, **34**, 1663, and references therein.

[10] Jan-Houghton Brunn, 1950, *C.R. Acad. Sciences Paris*, **230**, 988.

[11] Haase, C.S., Chadam, J., Feinn D. and Orteleva, P., 1980, *Science*, **209**, 272.

[12] Langer, J.S., 1980, *Rev. Mod. Phys.*, **52**, 1.

[13] Langer, J.S. and Müller-Krumbhaar, H., *Acta Metall.*, 1978, **26**, 1681 and 1978, **26**, 1689.

[14] Seetharaman, V. and Trivedi, R., 1988, *Metall. Trans.*, **19A**, 2955.

[15] See for instance Brandle, C.D., 1979, *Crystal Growth, A tutorial approach*, W. Bardsley, D.T.J. Hurle and J.B. Mullin Ed., North-Holland, Amsterdam.

[16] Caroli, B., Caroli, C., Roulet, B. and Voorhees, P.W., 1989, *Acta Metall.*, **37**, 330.

[17] de Groot, S.R. and Mazur, P., 1969, *Non-equilibrium Thermodynamics*, North-Holland, Amsterdam.

[18] See for instance Landau, L.D. and Lifshitz, E.M., 1975, *Theory of Elasticity*, Pergamon Press, Oxford.

[19] Caroli, B., Caroli, C. and Roulet, B., *J. Crystal Growth*, 1984, **66**, 575 and 1985, **71**, 235.

[20] Caroli, B., 1987, *Cours de Thermodynamique Hors d'Équilibre*, Rapport G.P.S.-ENS n° 87-032.

[21] Bedeaux, D., Albano, A.M. and Mazur, P., 1976, *Physica*, **82A**, 438.

[22] Herring, C., 1951, *The Physics of Powder Metallurgy*, W.E. Kingston Ed., McGraw Hill, New York.

[23] Wollkind, D.J. and Maurer, R.N., 1977, *J. Crystal Growth*, **42**, 24.

[24] Castaing, B. and Nozières, P., 1980, *J. Phys.*, Paris, **41**, 701.

[25] de Gennes, P.G., 1971, *J. Chem. Phys.*, **55**, 572.

[26] Wolf, P.E., Edwards, D.O. and Balibar, S., 1983, *J. Low Temp. Phys.*, **51**, 489.

[27] Aziz, M.J., 1982, J. Appl. Phys., 53, 1158.

[28] Nozières, P., Preceding article and references therein.

[29] Caroli, B., Caroli, C. and Roulet, B., 1986, Acta Metall., 34, 1897, and references therein.

[30] See for instance Basset, D.C., 1981, Principles of Polymer Morphology, Cambridge University Press.

[31] Goldenfeld, N., 1987, J. Crystal Growth, 84, 601.

[32] Caroli, B., Caroli, C., Roulet, B. and Faivre, G., 1989, J. Crystal Growth, 94, 253.

[33] Mullins, W.W. and Sekerka, R.F., 1963, J. Appl. Phys., 33, 323.

[34] Landau, L.D. and Lifshitz, E.M., 1963, Fluid Mechanics, Pergamon Press, Oxford.

[35] Gershuni, G.Z. and Zhukhovitskii, E.M., 1976, Convective Stability of Incompressible Fluids, Keter Publ. House, Jerusalem.

[36] Huang, S.C. and Glicksman, M.E., 1981, Acta Metall., 29, 701.

[37] Somboonsuk, K., Mason, J.T. and Trivedi, R., 1984, Metall. Trans., 15A, 967.

[38] Esaka, H. and Kurz, W., 1985, J. Crystal Growth, 72, 578.

[39] Jackson, K.A. and Hunt, J.D., 1966, Trans. Met. Soc. AIME, 236, 1929.

[40] de Cheveigné, S., Guthmann, C. and Lebrun, M.M., 1986, J. Phys. Paris, 47, 2095.

[41] Eshelman, M.A., Seetharaman, V. and Trivedi, R., 1988, Acta Metall., 36, 1165, and Seetharaman, V., Eshelman, M.A. and Trivedi, R., 1988, Acta Metall., 36, 1175.

[42] Bechhoefer, J. and Libchaber, A., 1987, Phys. Rev., B35, 1393.

[43] Bilgram, J.H., Firmann, M. and Hürlimann, E., 1989, J. Crystal Growth, 96, 175.

[44] Simon, A.J., Bechhoefer, J. and Libchaber, A., 1988, Phys. Rev. Letters, 61, 2574.

[45] Honjo, H., Ohta, S. and Sawada, Y., 1985, Phys. Rev. Letters, 55, 841.

[46] Glicksman, M.E. and Schaeffer, R.J., 1967, J. Crystal Growth, 1, 297.

[47] See for instance Rubinstein, L.I., 1977, *The Stefan Problem*, American Mathematical Society.

[48] Coriell, S.R. and Parker, R.L., 1967, *Crystal Growth*, H.S. Peiser Ed., Pergamon, Oxford.

[49] Brush, L.N., Sekerka, R.F. and Mc Fadden, G.B., A numerical and analytical study of non-linear bifurcations associated with the morphological stability of two-dimensional single crystals, preprint, to appear in *J. Crystal Growth*.

[50] Goldbeck-Wood, G., 1986, Master thesis, Rheinisch-Westfälische Technische Hochschule, Aachen.

[51] Coriell, S.R., Hardy, S.C. and Sekerka, R.F., 1971, *J. Crystal Growth*, **11**, 53.

[52] Fujioka, T. and Sekerka, R.F., 1974, *J. Crystal Growth*, **24-25**, 84.

[53] Fujioka, T., 1978, Ph.D. Thesis, Carnegie-Mellon University, Pittsburg.

[54] Schaeffer, R.J. and Coriell, S.R., 1984, *Metall. Trans.*, **A15**, 2109.

[55] Mullins, W.W. and Sekerka, R.F., 1964, *J. Appl. Phys.*, **35**, 444.

[56] Rutter, J.W. and Chalmers, B., 1953, *Can. J. Phys.*, **31**, 15.

[57] Wollkind, D.J. and Segel, L.A., 1970, *Phil. Trans. Roy. Soc. London*, **268**, 351.

[58] See [56] and Ivantsov, G.P., 1951, *Dokl. Akad. Nauk*, **81**, 179.

[59] Langer, J.S. and Turski, L.A., 1977, *Acta Metall.*, **25**, 1113.

[60] Langer, J.S., 1977, *Acta Metall.*, **25**, 1121.

[61] Caroli, B., Caroli, C. and Roulet, B., 1982, *J. Phys. Paris*, **43**, 1767.

[62] Boettinger, W.J. and Coriell, S.R., 1985, *Nato Advanced Research Workshop on Rapid Solidification Technologies*, Theuern.

[63] Bechhoefer, J., 1988, Ph.D. Thesis, The University of Chicago.

[64] Trivedi, R., Sekhan, J.A. and Seetharaman, V., 1989, *Metall. Trans.*, **A20**, 769.

[65] Coriell, S.R. and Sekerka, R.F., *J. Crystal Growth*, 1973, **19**, 90 and 1973, **19**, 285.

[66] Ungar, L.H. and Brown, R.A., 1984, *Phys. Rev.*, **B30**, 3993.

[67] Caroli, B., Caroli, and Roulet, B., 1986, *J. Crystal Growth*, **76**, 31.

[68] Oswald, P., 1988, *J. Phys. Paris*, **49**, 1083.

[69] Suresh, K.A. and Rondelez, F., 1988, *Europhys. Letters*, **6**, 437.

[70] Suresh, K.A., Nittmann, J. and Rondelez, F., 1989, *Colloid and Polymer Science*, to appear.

[71] Guckenheimer, J. and Holmes, P., 1983, *Nonlinear Oscillations, Dynamical Systems and Bifurcations of Vector Fields*, Appl. Math. Sciences, Springer Verlag, **42**.

[72] See for instance Landau, L.D. and Lifshitz, E.M., 1969, *Mechanics*, Pergamon Press, Oxford.

[73] Dee, G. and Mathur, R., 1983, *Phys. Rev.*, **B27**, 7073.

[74] Misbah, C., 1989, *J. Phys. Paris*, **50**, 971.

[75] Sivashinsky, G.I., 1983, *Physica D*, **8**, 243.

[76] Brattkus, K. and Davis, S.H., 1988, *Phys. Rev.*, **B38**, 11452.

[77] Iwan, J., Alexander, D., Woolkind, D.J. and Sekerka, R.F., 1986, *J. Crystal Growth*, **79**, 849.

[78] Merchant, G.J. and Davis, S.H., 1989, Technical Report 8820, Northwestern University, to appear in *Phys. Rev. Letters*.

[79] Misbah, C., 1986, *J. Phys. Paris*, **47**, 1077.

[80] Eshelman, M.A. and Trivedi, R., 1987, *Acta Metall.*, **35**, 2443.

[81] de Cheveigné, S., Guthmann, C., Kurowski, P., Vicente, E. and Biloni, H., 1988, *J. Crystal Growth*, **92**, 616.

[82] Cross, M.C., 1980, *Phys. Fluids*, **23**, 1727.

[83] Newell, A.C. and Whitehead, J.A., 1969, *J. Fluid Mech.*, **38**, 279.

[84] See for example, Misbah, C., 1988, *Nonlinear Phenomena in Materials Science*, L. Kubin and G. Martin Ed., Trans. Tech. Publications.

[85] Eckhaus, W., 1965, *Studies in Nonlinear Stability Theory*, Springer Tracts in Natural Philosophy, **6**, Berlin.

[86] See for instance Joseph, D.D., 1976, *Stability of Fluid Motions II*, Springer Tracts in Natural Philosophy, **28**, Berlin.

[87] Caroli, B., Caroli, C. and Roulet, B., 1984, *J. Crystal Growth*, **68**, 677.

[88] Wollkind, D.J., Sriranganathan, R. and Oulton, D.B., 1984, *Physica*, D **12**, 215.

[89] Mc Fadden, G.B., Coriell, S.R. and Sekerka, R.F., 1988, *J. Crystal Growth*, **91**, 180.

[90] Segel, L.A., 1965, *J. Fluid Mech.*, **21**, 359.

[91] Ungar, L.H., Bennett, M.J. and Brown, R.A., 1985, *Phys. Rev.*, **B31**, 5923.

[92] Ungar, L.H. and Brown, R.A., 1985, *Phys. Rev.*, **B31**, 5931.

[93] Kessler, D.A. and Levine, H., 1989, *Phys. Rev.*, **A39**, 3208.

[94] Saïto, Y., Misbah, C. and Müller-Krumbhaar, H., 1988, *Nucl. Phys. B*, **5A**, 225.

[95] Kurowski, P., de Cheveigné, S., Faivre, G. and Guthmann, C., 1989, *J. de Phys. Paris*, **50**, 3007 and references therein.

[96] Pelcé, P. and Pumir, A., 1985, *J. Crystal Growth*, **73**, 337.

[97] Saffman, P.G. and Taylor, G., 1958, *Proc. Roy. Soc.*, **A245**, 312.

[98] de Cheveigné, S., Guthmann, C. and Kurowski, P., private communication.

[99] Misbah, C. and Müller-Krumbhaar, H., private communication.

[100] Dombre, T. and Hakim, V., 1987, *Phys. Rev.*, **A36**, 2811.

[101] Weeks, J.D. and Van Saarloos, W., 1989, *Phys. Rev.*, **A39**, 2772.

[102] Scheil, E., 1942, *Z. Metellkd.*, **34**, 70.

[103] Hunt, J.D., 1978, *Solidification and Casting of Metals*, Metals Society, London.

[104] Ramprasad, N., Bennett, M.J. and Brown, R.A., 1988, *Phys. Rev.*, **B38**, 583.

[105] Ben Amar, M. and Moussalam, B., 1988, *Phys. Rev. Letters*, **60**, 317.

[106] de Cheveigné, S., Guthmann, C. and Kurowski, P., to be published.

[107] Billia, B., Jamgotchian, H. and Capella, L., 1989, Unifying representations for cells and dendrites, preprint.

[108] Billia, B. and Trivedi, R., 1989, Cellular and dentritic regimes in directional solidification, preprint.

[109] Brattkus, K., 1989, *J. Phys. Paris*, **50**, 3020.

[110] Pelcé, P., 1986, Thèse d'Etat, Université de Provence, Marseille.

[111] Bensimon, D., Kadanoff, L.P., Liang, S., 1986, Schraiman, B.I. and Tang, C., *Rev. Mod. Phys.*, **58**, 1977.

[112] Caroli, B., Caroli, C. and Roulet, B., 1987, *J. Phys. Paris*, **48**, 1423.

[113] Barber, M., Barbieri, A. and Langer, J.S., 1987, *Phys. Rev.*, **A36**, 3340.

[114] Langer, J.S., 1987, *Phys. Rev.*, **A36**, 3350.

[115] Trivedi, R., 1984, *Metall. Trans.*, **15A**, 977.

[116] Brattkus, K. and Misbah, C., private communication.

[117] Faivre, G., de Cheveigné, S., Guthmann, C. and Kurowski, P., *Europhys. Letters*, to appear.

[118] For a survey see for instance Rosenberger, F., 1979, *Fundamentals of Crystal Growth I*, Springer Series in Solid-State Science, **5**, Berlin.

[119] Coriell, S.R., Cordes, M.R., Boettinger, W.J. and Sekerka, R.F., 1980, *J. Crystal Growth*, **49**, 13.

[120] Caroli, B., Caroli, C., Misbah, C. and Roulet, B., 1985, *J. Phys. Paris*, **46**, 401.

[121] Misbah, C., 1985, Thèse Doctorat de 3ème Cycle, Université Paris VII.

[122] Hurle, D.T., Jakeman, E. and Wheeler, A.A., 1983, *Phys. Fluids*, **26**, 624.

[123] Schaefer, R.J. and Coriell, S.R., *Materials Processing in the Reduced Gravity Environment of Space*, G.E. Rindone Ed., (Elsevier Science publishing Co).

[124] Young, G.W. and Davis, S.H., 1986, *Phys. Rev.*, **B34**, 3388.

[125] Caroli, B., Caroli, C., Misbah, C. and Roulet, B., 1985, *J. de Phys. Paris*, **46**, 1657.

[126] Jenkins, D.R., 1985, *J. Appl. Math.*, **35**, 145.

[127] Coriell, S.R. and Sekerka, R.F., 1976, *J. Crystal Growth*, **34**, 157.

[128] Jamgotchian, H., Billia, B. and Capella, L., 1983, *J. Crystal Growth*, **62**, 539.

[129] Shangguan, D.K. and Hunt, J.D., 1987, *Proceedings of the Conference on Solidification Processing*, Sheffield.

[130] Maurer, J., Bouissou, P., Perrin, B. and Tabeling, P., 1987, *Europhys. Letters*, **8**, 67.

[131] Chernov, A.A and Nishinaga, T., 1987, *Morphology of Crystals*, Part A, I. Sumagawa Ed., Terra Scient. Publ. Co, Tokyo.

[132] Bowley, A., Caroli, B., Caroli, C., Graner, F., Nozières, P. and Roulet, B., 1989, *J. Phys. Paris*, **50**, 1377.

[133] Caroli, B., Caroli, C. and Roulet, B., 1989, *J. Phys. Paris*, **50**, 3075.

[134] Ben Amar, M. and Pomeau, Y., 1988, *Europhys. Letters*, **6**, 609.

Appendix A:
Gibbs-Thomson equation for a binary alloy

Consider an alloy $1_{(1-c)}2_c$. Let its concentration at thermodynamic equilibrium (planar interface: $p_L = p_S$, temperature T_0) be $C_{\alpha 0}$ ($\alpha = L, S$) in phase α. The conditions for thermodynamic equilibrium in the reference state $P_0(p_L, T_0, C_{L_0}, C_{S_0})$ on the phase diagram read:

$$\mu_k^{(L)}(p_L, T_0, C_{L_0}) = \mu_k^{(S)}(p_L, T_0, C_{S_0}), \qquad (k = 1, 2). \qquad (A.1)$$

In the near equilibrium situation in which we are interested, the $\mu_k's$ can be developed to first order about the reference point P_0. We choose for the reference pressure the real pressure in the liquid, p_L. In the absence of Kapitza resistance effects, the interface temperature T_L is the same in both phases (see (2.65)), so that:

$$\Delta\mu_k = \mu_k^S(p_s, T_I, C_s) - \mu_k^L(p_L, T_I, C_L) \qquad (A.2)$$

$$= (T_0 - T_I)\Delta s_k - (p_L - p_S)V_k^S$$

$$+ (C_S - C_{S_0})r_k^S - (C_L - C_{L_0})r_k^L.$$

$V_k^S = (\rho_k^S)^{-1}$ is the specific volume of constituent k in the solid phase, and

$$\Delta s_k = s_{k_0}^S - s_{k_0}^L, \qquad r_k^\alpha = \left(\frac{\partial \mu_k^\alpha}{\partial C_\alpha}\right)_{p,T,C_{\alpha_0}}, \qquad (A.3)$$

are evaluated at the reference point P_0 on the phase diagram.

According to Eq. (2.59)

$$\Delta\mu_k = \Delta\tilde{\mu}_k - \left[\hat{n} \cdot \frac{\overline{\overline{\Pi}}}{\rho} \cdot \hat{n} + \tfrac{1}{2}(\mathbf{v} - \mathbf{v}_I)^2\right]_L^S \qquad (A.4)$$

where the $\Delta\tilde{\mu}_k$ are the kinetically induced jumps of effective potential at the interface (see Eqs. (2.60)).

In order to establish the relation between Δs_k and the $r_k^{\alpha\prime}s$, let us consider a point $P_0'(T_0', p_L, C_{L_0}', C_{S_0}')$ close to P_0 on the phase diagram.

Equation (A.1) holds for P_0' (with primed temperatures and concentrations), and Eq. (A.2) yields:

$$(T_0' - T_0)\Delta s_k + (C_{L_0}' - C_{L_0})r_k^L \tag{A.5}$$
$$- (C_{S_0}' - C_{S_0})r_k^S = 0, \qquad (k = 1, 2).$$

Since P_0' is close to P_0, the liquidus and solidus lines can be approximated in this neighborhood by their tangents, i.e.:

$$T_0' = T_0 + m_\alpha(C_{\alpha_0}' - C_{\alpha_0}), \qquad (\alpha = L, S), \tag{A.6}$$

with:

$$m_\alpha = \frac{dT_0}{dC_\alpha}\Big|_{C_{\alpha_0}}. \tag{A.7}$$

Moreover, the Gibbs-Duhem relations written at P_0 read:

$$(1 - C_{\alpha_0})r_1^\alpha + C_{\alpha_0} r_2^\alpha = 0, \qquad (\alpha = L, S), \tag{A.8}$$

so that, inserting Eqs. (A.6–A.8) into (A.5), one easily finds:

$$\Delta s_k = \frac{r_k^S}{m_S} - \frac{r_k^L}{m_L}, \qquad (k = 1, 2). \tag{A.9}$$

It is then easy to solve equations (A.2) for $T_I - T_0$ and $C_S - C_{S_0}$

$$C_S - C_{S_0} = \frac{m_L}{m_S}(C_L - C_{L_0}) \tag{A.10}$$
$$- \frac{C_{S_0}}{r_1^S \Delta s}\left\{\left[\Delta\mu_1 + \frac{p_L - p_S}{\rho_1^S}\right]\Delta s_2\right.$$
$$\left. - \left[\Delta\mu_2 + \frac{p_L - p_S}{\rho_2^S}\right]\Delta s_1\right\},$$

$$T_I - T_M = m_L(C_L - C_{L_0}) \tag{A.11}$$
$$- \frac{1}{\Delta s}\left\{(1 - C_{S_0})\Delta\mu_1 + C_{S_0}\Delta\mu_2 + \frac{p_L - p_S}{\rho_S}\right\},$$

where

$$\Delta s = (1 - C_{S_0})\Delta s_1 + C_{S_0}\Delta s_2 \tag{A.12}$$
$$= -\frac{L}{T_M} - (C_{L_0} - C_{S_0})\left(\frac{\partial s_L}{\partial C_L}\right)_{p_L, T_0, C_{L_0}}.$$

Here $L = T_M(s_{L_0} - s_{S_0})$ is the latent heat per unit mass and s_α the specific entropy of the alloy at thermodynamic equilibrium.

For a dilute alloy where $C_{\alpha_0} \ll 1$, Eqs. (A.10–A.12) reduce immediately to Eqs. (2.77, 2.78).

Appendix B:
The integro-differential front equation

Let us consider two semi-infinite media (1) and (2) separated by an interface Σ, and let D_i ($i = 1, 2$) be the (thermal or chemical) diffusion coefficient in medium (i). We will ultimately take medium 1 (resp. 2) to be the solid (resp. liquid) phase. In the frame of reference moving in the z-direction with constant velocity V, one can define in each medium reduced time and space variables by:

$$\mathbf{r}_i = \frac{\tilde{\mathbf{r}}}{\ell_i}, \quad t_i = \frac{\tilde{t}}{\tau_i}, \quad \ell_i = \frac{D_i}{V}, \quad \tau_i = \frac{D_i}{V^2}, \tag{B.1}$$

where $\tilde{\mathbf{r}}$ and $\tilde{t}$ are the physical variables. We introduce the following notations:

$$x_i = (\boldsymbol{\rho}_i, t) \; ; \; p_i = (\mathbf{r}_i, t), \quad p_{i\Sigma} = [x_i, \zeta_i(x_i)], \tag{B.2}$$

where $\boldsymbol{\rho} = x\hat{\imath} + y\hat{\jmath}$, $\mathbf{r} = \boldsymbol{\rho} + z\hat{k}$, and $\tilde{z} = \tilde{\zeta}(\tilde{x})$ is the equation of the interface Σ separating the two media.

Let $\hat{n}$ be the unit vector normal to the interface pointing from medium (1) ($\tilde{z} < \tilde{\zeta}(\tilde{x})$) into medium (2) ($\tilde{z} > \tilde{\zeta}(\tilde{x})$)

$$\hat{n} = \frac{\tilde{\nabla}\left(\tilde{z} - \tilde{\zeta}(\tilde{x})\right)}{\left|\tilde{\nabla}\left(\tilde{z} - \tilde{\zeta}(\tilde{x})\right)\right|}, \tag{B.3}$$

so that the outward normal to medium (i) is defined by the unit vector:

$$\hat{n}_i = (-1)^{i+1}\hat{n}, \quad (i = 1, 2). \tag{B.4}$$

1 General formalism

Let us define the diffusion field $u_i(p_i)$ in medium (i) in such a way that it satisfies the boundary conditions:

$$\lim_{z_i \to (-)^i \infty} u_i(p_i) = \lim_{z_i \to (-)^i \infty} \nabla u_i(p_i) = 0, \quad (i = 1, 2). \tag{B.5}$$

In the reference frame moving along Oz at constant velocity V, it evolves according to:

$$\left(\frac{\partial}{\partial t_i} - \frac{\partial}{\partial z_i} - \nabla_i^2\right) u_i(p_i) = 0. \tag{B.6}$$

The associated retarded Green's function $G_i(p_i, p_i')$ is the solution of:

$$\left(\nabla_i'^2 - \frac{\partial}{\partial z_i'} + \frac{\partial}{\partial t_i'}\right) G_i(p_i, p_i') \tag{B.7a}$$

$$= -\delta(\mathbf{r}_i - \mathbf{r}_i')\delta(t_i - t_i'),$$

$$G_i(p_i, p_i') \equiv 0, \quad \text{for} \quad t_i - t_i' < 0, \tag{B.7b}$$

so that

$$G_i(p_i, p_i') = \int_{-\infty}^{\infty} \frac{d\omega_i}{2\pi} \int \frac{d\mathbf{K}_i}{(2\pi)^3} \tag{B.8}$$

$$\times \frac{\exp\{i[\omega_i(t_i - t_i') + \mathbf{K}_i \cdot (\mathbf{r}_i - \mathbf{r}_i')]\}}{i\omega_i + \mathbf{K}_i^2 - iK_{iz}},$$

or

$$G(p_i, p_i') = \frac{\theta(t_i - t_i')}{[4\pi(t_i - t_i')]^{3/2}} \exp\left[-\frac{R_i^2}{4(t_i - t_i')}\right], \tag{B.9}$$

where $\theta(t)$ is the step function,

$$R_i^2 = (\rho_i - \rho_i')^2 + (z_i - z_i' + t_i - t_i')^2, \tag{B.10}$$

and

$$G(\mathbf{r}_i, t_i \; ; \; \mathbf{r}_i', t_{i-}) = \delta(\mathbf{r}_i - \mathbf{r}_i'). \tag{B.11}$$

Note that, setting in (B.8) $\mathbf{K}_i = \mathbf{k}_i + K_{zi}\hat{k}$, and performing the K_{zi}-integration, one finds:

$$G(p_i, p_i') = \int_{-\infty}^{\infty} \frac{d\omega_i}{2\pi} \exp[i\omega_i(t_i - t_i')] \tag{B.12}$$

$$\times \int \frac{d\mathbf{k}_i}{(2\pi)^2} \frac{\exp[i\mathbf{k}_i(\rho_i - \rho_i')]}{2q_i(k_i, \omega_i) - 1}$$

$$\times \exp\left[-\tfrac{1}{2}\left\{z_i - z_i' + |z_i - z_i'|(2q_i(k_i, \omega_i) - 1)\right\}\right],$$

where

$$q_i(k_i, \omega_i) = \tfrac{1}{2}\left[1 + \sqrt{1 + 4i\omega_i + 4k_i^2}\,\right]. \tag{B.13}$$

Integrating (B.7) over the volume $V_i'(t_i')$ of the (i) medium, and the time interval $-\infty \le t_i' \le t_{i-}$, one finds, using Green's theorem:

$$(-)^{i+1} \int_{-\infty}^{t_{i-}} dt_i' \int d\Sigma_i' \; \hat{n}' \cdot \boldsymbol{\nabla}_i' G_i(p_i, p_i') \Big|_{p_i' = p_{i\Sigma}'} \tag{B.14}$$

$$+ (-)^{i+1} \int_{-\infty}^{t_{i-}} dt_i' \int d\rho_i' \int_{\zeta_i(x_i')}^{(-)^i \infty} dz_i' \frac{\partial}{\partial z_i'} G_i(p_i, p_i')$$

$$+ \int_{-\infty}^{t_{i-}} dt_i' \left\{ \frac{d}{dt_i'} \int_{V_i'(t_i')} d\mathbf{r}_i' G_i(p_i, p_i') + (-)^i \int d\rho_i' \dot{\zeta}_i(x_i') G(p_i, p_{i\Sigma}') \right\} = 0,$$

So that, using Eq. (B.11):

$$(-)^{i+1} \int_{-\infty}^{t_{i-}} dt_i' \int d\Sigma_i' \; \hat{n}' \cdot \boldsymbol{\nabla}_i' G(p_i, p_i') \Big|_{p_i' = p_{i\Sigma}'} \tag{B.15}$$

$$+ (-)^i \int_{-\infty}^{t_{i-}} dt_i' \int d\rho_i \left[1 + \dot{\zeta}_i(x_i')\right] G_i(p_i, p_{i\Sigma}')$$

$$= -1 + (-)^i \int_{-\infty}^{t_{i-}} dt_i' \int d\rho_i \lim_{z_i' \to (-)^i \infty} G_i(p_i; x_i', z_i').$$

On the other hand, it is easy to show, using Eqs. (B.9), (B.10), that:

$$\lim_{z_i' \to (-)^i \infty} \int_{-\infty}^{t_{i-}} dt_i' \int d\rho_i' \; G(p_i \; ; \; x_i', z_i') \tag{B.16}$$

$$= \lim_{z_i' \to (-)^i \infty} \exp\left\{ -\tfrac{1}{2}\left(z_i - z_i' + |z_i - z_i'|\right) \right\}$$

$$= \delta_{i,2},$$

so that:

$$\int_{-\infty}^{t_{i-}} dt_i' \int d\Sigma_i' \; \hat{n}' \cdot \boldsymbol{\nabla}' G_i(p_i, p_i') \tag{B.17}$$

$$- \int_{-\infty}^{t_{i-}} dt_i' \int d\rho_i \left[1 + \dot{\zeta}_i(x_i')\right] G_i(p_i, p_{i\Sigma}') = (-)^i - \delta_{i,2}.$$

Returning now to Eqs. (B.6), (B.7), multiplying them respectively by $G_i(p_i, p'_i)$ and $u_i(p'_i)$, adding up and integrating over space and time, one finds with the help of Eq. (B.5):

$$u_i(p_i) = (-)^{i+1} \int_{-\infty}^{t_{i-}} dt'_i \int d\rho'_i [1 + \dot{\zeta}_i(x'_i)] u_i(p'_{i\Sigma}) G_i(p_i, p'_{i\Sigma}) \quad \text{(B.18)}$$

$$+ (-)^{i+1} \int_{-\infty}^{t_{i-}} dt'_i \int d\Sigma_i \, \hat{n}' \cdot \Big[G_i(p_i, p'_{i\Sigma}) \nabla'_i u_i(p'_i)$$

$$- u_i(p'_{i\Sigma}) \nabla'_i \, G_i(p_i, p'_i) \Big]_{p'_i = p'_{i\Sigma}} .$$

Applying to (B.17) and (B.18) the theorem on the discontinuity of the heat potential of a double layer [47], according to which:

$$\lim_{z_i \to \zeta_i(x_i) + (-)^i 0} \int_{-\infty}^{t_{i-}} dt'_i \int d\Sigma'_i \varphi_i(p'_{i\Sigma}) \quad \text{(B.19)}$$

$$\hat{n}' \cdot \nabla'_i \, G_i(p_i, p'_i)|_{p'_i = p'_{i\Sigma}}$$

$$= (-)^i \tfrac{1}{2} \varphi_i(p_{i\Sigma}) + \int_{-\infty}^{t_{i-}} dt'_i \int d\Sigma'_i \varphi_i(p'_{i\Sigma})$$

$$\hat{n} \cdot \nabla'_i \, G_i(p_{i\Sigma}, p'_i)_{P'_i = p'_{i\Sigma}}$$

one finds:

$$\int_{-\infty}^{t_{i-}} dt'_i \int d\Sigma'_i \, \hat{n} \cdot \nabla'_i \, G_i(p_{i\Sigma}, p'_i) \quad \text{(B.20)}$$

$$- \int_{-\infty}^{t_{i-}} dt'_i \int d^2\rho'_i \int_{\rho'_i}^{2} [1 + \dot{\zeta}_i(x'_i)] G_i(p_{i\Sigma}, p'_{i\Sigma}) = -\tfrac{1}{2}$$

and

$$\tfrac{1}{2} u_i(p_{i\Sigma}) = (-)^{i+1} \int_{-\infty}^{t_{i-}} dt'_i \int d\rho'_i [1 + \dot{\zeta}_i(x'_i)] \quad \text{(B.21)}$$

$$\times u_i(p'_{i\Sigma}) G_i(p_{i\Sigma}, p'_{i\Sigma})$$

$$+ (-)^{i+1} \int_{-\infty}^{t_{i-}} dt'_i \int d\Sigma'_i \, \hat{n}' \cdot \Big[G_i(p_{i\Sigma}, p'_{i\Sigma}) \nabla'_i \, u_i(p'_i)$$

$$- u_i(p'_{i\Sigma}) \nabla'_i \, G_i(p_{i\Sigma}, p'_i) \Big]_{p'_i = p'_{i\Sigma}} .$$

We now want to use Eqs. (B.20, B.21) to try to reformulate the equations describing the directional solidification of a mixture within the minimal model (Eqs. (2.86, 2.87)). The diffusion fields satisfying conditions (B.5) are defined, in terms of the concentration fields in each phase, as:

$$u_i(p) = \frac{C_i(p) - C_\infty}{\Delta C_0}, \qquad (B.22)$$

where the notations are those of Section 4.

In order for the thermal field to decouple completely from the growth problem (which is necessary for the reduction to a closed front equation to be possible), we assume that the thermal properties of the liquid and solid are identical ($n = 1$). Then:

$$T(\tilde{\mathbf{r}}) = T_0 + G\tilde{z}, \qquad (B.23)$$

where G is the applied thermal gradient

The interface boundary conditions then reduce to Eqs. (2.87c, d, e) which relate the interface values of the $u_i's$ to those of the $\nabla u_i's$, and the front curvature. It can be checked that, in the general case where the values of the diffusion coefficients D_S, D_L are unrelated, due to the form of the $\nabla' u_i$ term, Eq. (B.21) is not reducible to a closed equation for the front profile (Equation (2.87c) does not permit to eliminate both ∇u_1 and ∇u_2 from (B.21)). However, we will now show that such a reduction is possible in two particular cases, namely for the so-called one-sided and symmetric models.

2 Directional solidification of mixtures: the one-sided models

In this model, diffusion in one of the phases is assumed to be so slow that it can be neglected and, thus, this model is well suited to describe solidification of mixtures, for which the diffusing field is the concentration one.

Solidification is then described by Eqs. (4.1b, 4.2c, 4.2d, 4.3c) and (B.23). In order to ensure coherence between the directions of the outward normal in Eqs. (4.2) and (B.21), we have to choose medium (2) (resp. (1)) to be the liquid (resp. solid) phase.

According to Eq. (B.21), the (dimensionless) concentration field u_L (Eq. (B.22)) at the interface satisfies the equation:

$$\tfrac{1}{2} u_L(p_\Sigma) = - \int_{-\infty}^{t-} \mathrm{d}t' \int \mathrm{d}\rho' \left(1 + \dot{\zeta}(x')\right) u_L(p'_\Sigma) G(p_\Sigma, p'_\Sigma) \qquad (B.24)$$

$$- \int_{-\infty}^{t-} \mathrm{d}t' \int \mathrm{d}\Sigma' \, \hat{n}' \cdot \Big[G(p_\Sigma, p'_\Sigma) \nabla' u_L(p') - u_L(p'_\Sigma) \nabla' G(p_\Sigma, p') \Big]_{p'=p'_\Sigma}.$$

In our reduced units, the concentration balance equation (4.2c) and the Gibbs-Thomson equation (4.2d) become, with the help of (see Eq. (4.7)):

$$T_\Sigma(\zeta) \equiv T_S(\zeta) = T_L(\zeta) = T_M + m_L \frac{C_\infty}{K} + G\ell\zeta(x), \qquad (B.25)$$

and of Eq. (B.22):

$$\hat{n} \cdot \nabla u_L(p)_{p=p_\Sigma} = n_z(1 + \dot{\zeta}(x))\left[(K-1)u_L(p_\Sigma) + K\right], \qquad (B.26a)$$

$$u_L(p_\Sigma) = -1 + \frac{d_0}{\ell}\mathcal{K}(x) + \frac{\ell}{\ell_T}\zeta(x), \qquad (B.26b)$$

where d_0 is the solutal capillary length defined by Eq. (4.15) and ℓ_T the thermal length, defined by Eq. (4.19) with $n = 1$.

Adding Eqs. (B.20) and (B.24), and using Eqs. (B.26) in the resulting equation leads to the front integro-differential equation for the one-sided model:

$$\frac{1}{2}\left[2 - \frac{d_0}{\ell}\mathcal{K}(x) - \frac{\ell}{\ell_T}\zeta(x)\right] \qquad (B.27)$$

$$= \int_{-\infty}^{t^-} dt' \int d\rho'(1 + \dot{\zeta}(x'))$$

$$\times \left\{1 + K\left[\frac{d_0}{\ell}\mathcal{K}(x') + \frac{\ell}{\ell_T}\zeta(x')\right]\right\} \cdot G(p_\Sigma, p'_\Sigma)$$

$$- \int_{-\infty}^{t^-} dt' \int d\Sigma'\left[\frac{d_0}{\ell}\mathcal{K}(x') + \frac{\ell}{\ell_T}\zeta(x')\right]\hat{n}' \cdot \nabla' G(p_\Sigma, p')\big|_{p'=p'_\Sigma},$$

where we recall that x stands for (p, t), and that lengths and times are measured in units of $\ell = D/V$ and $\tau = D/V^2$.

Obviously, the planar front $\zeta(x) = \zeta = 0$ is a solution of (B.27). In order to give an example of how the front equation may be exploited, we will now rederive directly from Eq. (B.27) the MS linear spectrum (Eq. (4.7)) obtained in Section 4.

For this purpose, we linearize Eq. (B.27) about the planar front solution. Using:

$$\mathcal{K}(x) = -\nabla^2\zeta(x) + O(\zeta^3), \qquad (B.28a)$$

$$G(p_\Sigma, p'_\Sigma) = \left\{1 - \tfrac{1}{2}\left[\zeta(x) - \zeta(x')\right]\right\}G(x, x') + O(\zeta^2), \qquad (B.28b)$$

$$\hat{n}' \cdot \nabla' G(p_\Sigma, p')\big|_{p'=p'_\Sigma} = n'_z\left[\tfrac{1}{2} + O(\zeta)\right]G(x, x'), \qquad (B.28c)$$

where $G(x, x') = G(p, p')|_{z=z'=0}$, one obtains the linearized version of (B.27):

$$\left(1 - \frac{\ell}{\ell_T}\right)\zeta(x) + \frac{d_0}{\ell}\nabla^2\zeta(x) \tag{B.29}$$

$$= \int_{-\infty}^{t^-} dt' \int d\rho' \left\{ 2\dot{\zeta}(x') + \zeta(x') \right.$$

$$\left. + (2K - 1)\left[\frac{\ell}{\ell_T}\zeta(x') - \frac{d_0}{\ell}\nabla^2\zeta(x')\right] \right\} G(x, x').$$

Taking advantage of the translation invariance of the basic solution along the ρ-plane, we look for solutions of (B.29) of the form:

$$\zeta(x) = \zeta_k \, \exp\left[i(\mathbf{k} \cdot \boldsymbol{\rho} + \omega t)\right]. \tag{B.30}$$

We then get from Eq. (B.29) the dispersion relation:

$$\left(1 - \frac{\ell}{\ell_T}\right) - \frac{d_0}{\ell}k^2 \tag{B.31}$$

$$= \left[2i\omega + 1 + (2K - 1)\left(\frac{\ell}{\ell_T} + \frac{d_0}{\ell}k^2\right)\right]$$

$$\times \int_{-\infty}^{t^-} dt' \int d\rho' \, G(x, x') \exp(-i\left[\mathbf{k} \cdot (\boldsymbol{\rho} - \boldsymbol{\rho}') + \omega(t - t')\right]).$$

Since $G(p, p') = 0$ for $t' > t$ we can extend the t'-integration up to $+\infty$. Then, using for $G(x, x')$ equation (B.12) with $z = z' = 0$ one immediately finds (see Eq. (B.13)):

$$\left[\left(1 - \frac{\ell}{\ell_T}\right) - \frac{d_0}{\ell}k^2\right][2q(k, \omega) - 1] \tag{B.32}$$

$$= 2i\omega + 1 + (2K - 1)\left(\frac{\ell}{\ell_T} + \frac{d_0}{\ell}k^2\right),$$

which can be written as:

$$i\omega + K = \left(1 - \frac{\ell}{\ell_T} - \frac{d_0}{\ell}k^2\right)[q(k, \omega) + K - 1], \tag{B.33}$$

i.e. precisely the dimensionless version ($i\omega \mapsto -\Omega\tau$, $k \mapsto k\ell$, $q \mapsto q\ell$) of equation (4.17).

3 The symmetric model

In this model one assumes that the solute diffusion coefficients D_S and D_L are equal and that the concentration gap is constant ($K = 1$ limit). That is we now have:

$$D_1 = D_2 = D, \quad \ell_1 = \ell_2 = \ell = \frac{D}{V}, \quad \tau_1 = \tau_2 = \tau = \frac{D}{V^2}, \qquad (\text{B.34})$$

so that the reduced variables $\mathbf{r}_i$ and t_i, as well as the Green's functions $Gi(p_{i\Sigma}, p'_{i\Sigma})$ are now independent of i.

Summing equation (B.21) over the i index leads (with $d\rho = n_z\, d\Sigma$) to:

$$\frac{1}{2}\left[u_1(p_\Sigma) + u_2(p_\Sigma)\right] = \int_{-\infty}^{t_-} dt' \int d\Sigma' \qquad (\text{B.35})$$

$$\hat{n}' \cdot \left[\nabla' u_1(p'_\Sigma) - \nabla' u_2(p'_\Sigma)\right] G(p_\Sigma, p'_\Sigma)$$

$$- \int_{-\infty}^{t_-} dt' \int d\Sigma' \left[u_1(p'_\Sigma) - u_2(p'_\Sigma)\right]$$

$$\left\{\hat{n}' \cdot \nabla' G(p_\Sigma, p') - n'_z\left(1 + \dot{\zeta}(x')\right) G(p_\Sigma, p')\right\}_{p'=p'_\Sigma}.$$

The solidifying system, pulled at velocity V in the thermal gradient G, is now described by Eqs. (2.86b, 2.87c) with $D_S/D_L = 1$. Since we assume that the concentration gap on the phase diagram ΔC_0 is constant, Eq. (2.87d) becomes:

$$C_1(p_\Sigma) - C_2(p_\Sigma) = \Delta C_0, \qquad (\text{B.36})$$

and we must refer, in the Gibbs-Thomson equation, to the corresponding liquidus equation, so that Eq. (2.87e) now reads in reduced units

$$T_\Sigma = T_0 + m_L(C_2 - C_{20})_\Sigma - \frac{\gamma T_M}{\rho L \ell} \mathcal{K}, \qquad (\text{B.37})$$

where C_{20}, C_{10} are the concentrations of the liquid and solid at thermo-dynamic equilibrium (with a planar interface) at the reference temper-ature T_0. We choose, for convenience, T_0 to be such that $C_{10} = C_\infty$, $C_{20} = C_\infty - \Delta C_0$, and take the zero of ζ on the $T = T_0$ isotherm, so that

$$T_\Sigma = T_0 + G\ell\zeta(x). \qquad (\text{B.38})$$

Using definition (B.22), the interface boundary conditions become:

$$u_1(p_\Sigma) - u_2(p_\Sigma) = 1, \tag{B.39a}$$

$$u_2(p_\Sigma) = -1 + \frac{\ell}{\ell_T}\zeta(x) + \frac{d_0}{\ell}\mathcal{K}(x), \tag{B.39b}$$

$$\hat{n} \cdot \left[\nabla u_1 - \nabla u_2\right]_{p=p_\Sigma} = -n_z\left(1 + \dot{\zeta}(x)\right). \tag{B.39c}$$

Inserting Eqs. (B.39) into (B.35) yields, with the help of (B.20), the front equation for the symmetric model:

$$1 - \frac{\ell}{\ell_T}\zeta(x) - \frac{d_0}{\ell}\mathcal{K}(x) = \int_{-\infty}^{t^-}\!\!\mathrm{d}t' \int \mathrm{d}\rho'\left[1 + \dot{\zeta}(x')\right]G(p_\Sigma, p'_\Sigma). \tag{B.40}$$

The symmetric model, although leading to a more simple front equation than the one-sided one, does not provide a good quantitative description of solidification of an alloy. It is more adapted to the cases of liquid/liquid crystal and solid/solid[†] transitions, where diffusion coefficients in the mother and daugther phases are comparable.

Another case where it may provide a good approximation is that of free (thermally limited) growth of a pure material. In this case the diffusing field is the temperature one (see Section 4.1). In our reduced units, the energy balance and Gibbs-Thomson equations (4.2, 4.3) read:

$$\hat{n} \cdot \left[\nabla T_S - \nabla T_L\right]_\Sigma = n_z\frac{L}{C_p}\left[1 + \dot{\zeta}(x)\right], \tag{B.41a}$$

$$T_S(p_\Sigma) = T_L(p_\Sigma) = T_M - \frac{\gamma T_M}{\rho L\ell}\mathcal{K}. \tag{B.41b}$$

Calling T_∞ the temperature far from the front in the liquid and assuming, as usual, that $T = T_M$ far inside the solid, we define:

$$u_1(p_\Sigma) = \frac{C_p}{L}\left[T_S(p_\Sigma) - T_M\right], \quad u_2(p_\Sigma) = \frac{C_p}{L}\left[T_L(p_\Sigma) - T_\infty\right]. \tag{B.42}$$

Using Eqs. (B.41, B.42) together with (B.20) and (B.35), one immediately finds:

$$\Delta - \frac{d_0}{\ell}\mathcal{K}(x) = \int_{-\infty}^{t^-}\!\!\mathrm{d}t' \int \mathrm{d}\rho'\left[1 + \dot{\zeta}(x')\right]G(p_\Sigma, p'_\Sigma), \tag{B.43}$$

[†] It should, however, be kept in mind that a realistic description of a solid-solid transition must take into account stress effects (e.g. coherence stresses) which may be very important quantitatively.

where the reduced undercooling is

$$\Delta = \frac{C_p}{L}(T_M - T_\infty) \tag{B.44}$$

and d_0 is the thermal capillary length as given by Eq. (4.26).

It is Eq. (B.43) which has been the basis of most recent studies of dendritic growth (see Chapter 4 by Pomeau and Ben Amar).

CHAPTER 3

AN INTRODUCTION TO THE KINETICS OF FIRST-ORDER PHASE TRANSITIONS

J.S. Langer

Preface

In this chapter, I propose to discuss some of the most fundamental and well known topics in the theory of first-order phase transitions. My emphasis will be on the kinetics of these transitions, that is, on the way systems behave while out of equilibrium as phase transitions are taking place. In other words, the emphasis will be on processes rather than on steady-states.

To a large extent, I shall be reviewing old work, but with an eye toward discussing some important problems that I think have not yet been adequately solved or even adequately appreciated. Accordingly, the underlying theme for the chapter is the theory of fluctuations in complex systems. The fluctuations that I have in mind generally are large, slow events, especially the appearance and competitive growth of precipitates in binary solutions. The systems are 'complex' in the sense that there are many different, nearly stationary configurations in which they can remain for long periods of time.

The specific phenomena to be discussed are nucleation (here, only from the most old fashioned point of view), spinodal decomposition, and grain growth and coarsening. The logical progression is toward the unsolved problem of 'stretched exponential' decays of correlations in systems with spontaneously broken symmetries and, beyond that, toward a

theory of fluctuations in systems such as sandpiles or earthquake faults
which operate persistently at thresholds of instability. The latter topics
were presented in the style of seminars at Beg Rohu, but will not appear
in these notes.

I am indebted to B. Fourcade for preparing a first version of these
notes, to J.P. and E. Bouchaud for suggesting changes in that first
version, and to J. Donley for help in preparing the present version. I also
would like to express my gratitude to Claude Godrèche for his patient
and unrelenting encouragement in the completion of this project.

The preparation of this chapter was aided by support from U.S. Department of Energy Grant N° DE–FG03–84ER45108 and from National Science Foundation Grant PHY89–04035.

1
Qualitative Features of First-Order Phase Transitions

A brief review of some elementary models of first-order phase transitions
will serve to define notation and also to introduce some of the principal
theoretical questions. For general references to the topics discussed here,
see Ma [1981].

1.1 Ising ferromagnet

We start with the Ising ferromagnet. In fact, all of the models that
we shall consider belong to what is called the 'Ising universality class'.
This means that the local variables which describe their states of order
are simple one-component quantities, as opposed to vectors, tensors,
complex numbers, or the like. Moreover, we must be able to choose
these variables in such a way that their average values in any given
state of thermodynamic equilibrium can assume at most two values
corresponding to a maximum of two coexisting thermodynamic phases.
This is a very special class of models, but it will serve us nicely as a basis
for our study of the most fundamental nonequilibrium phenomena.

The Ising model consists of spins $S_i = \pm 1$ on a lattice of N sites labelled by the subscript i. The energy of this system is

$$E\{S\} = -\tfrac{1}{2} \sum_{i,j} J_{ij} S_i S_j - H \sum_i S_i, \qquad (1.1)$$

where H is the external field and the J_{ij} are interaction energies which fall off to zero as functions of the distance between the sites i and j. The free energy F is:

$$F(H,T) = -k_B T \ln\Big[\sum_{\{S\}} \exp(-E\{S\}/k_B T)\Big], \qquad (1.2)$$

where T is the temperature, k_B is Boltzmann's constant, and $\{S\}$ denotes configurations of the spins S_i. From F we can compute, for example, the magnetization M:

$$M = \sum_i S_i = -\frac{\partial F}{\partial H}. \qquad (1.3)$$

The equilibrium properties of this model are extremely well understood. The phase diagram in the H, T plane (Fig. 1.1) consists of a single line on the T axis between $T = 0$ and the critical point, $T = T_c$.

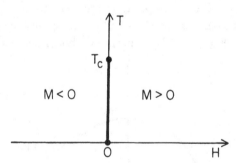

Figure 1.1: H, T phase diagram for the Ising model.

Points to the right of the T axis ($H > 0$) have positive M, those to the left ($H < 0$) have negative M. If we cross the transition line, say, by decreasing H at constant $T < T_c$ as in Fig. 1.2, we encounter a first-order phase transition in which M changes discontinuously from $+M_S$ to $-M_S$. Also shown by the dashed lines in Fig. 1.2 are the metastable continuations of the isotherms $M(H)$ for positive and negative H. We generally can reverse the field on a ferromagnet so that, for some time, it has positive M at negative H. Eventually, however, it will undergo the

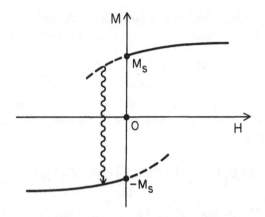

Figure 1.2: Schematic isotherm $M(H)$ for $T < T_c$. The dashed lines are metastable continuations, and the wiggly arrow indicates a transition from a metastable to a stable state.

transition to negative M shown by the wiggly arrow in the figure. The nature of this transition, and the rate at which it occurs as a function of the reversed field $-H$ and temperature T, is a central topic in this chapter.

A second form of the phase diagram that will be important in this chapter is the one in which we plot the spontaneous magnetization $M_S(T)$, as in Fig. 1.3. This function is shown here as a coexistence

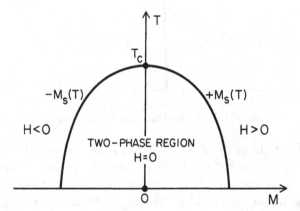

Figure 1.3: Schematic M, T phase diagram for the Ising model.

curve; at given T the magnetizations $\pm M_S(T)$ describe the two phases which coexist stably in contact with each other at $H = 0$. Points (T, M)

outside the coexistence curve identify equilibrium states of the system which can be achieved by applying nonzero fields H. Points inside the curve, in the so-called 'two-phase region', do not correspond to equilibrium states at all, and are strictly meaningless in a thermodynamic sense.

In principle, it is possible to quench a ferromagnet — by starting outside the two-phase region and suddenly decreasing T or changing H — so that it is instantaneously in a uniform state with M and T inside the coexistence curve. The subsequent behavior of the system is of some interest. We shall find such quenched states of greater interest, however, in cases where the analog of the magnetization is conserved as the system evolves, so that M rather than H is the quantity that can be controlled experimentally.

1.2 Lattice gas

A second common example of a first-order phase transition is the condensation of a vapor. The basic statement of equilibrium statistical mechanics for a system of classical molecules of mass m, whose positions are denoted $\mathbf{r}_1, \mathbf{r}_2, \ldots$, and which interact among themselves *via* a potential $U\{\mathbf{r}\}$, is

$$\frac{Vp(\mu)}{k_B T} = \ln\left\{ \sum_{\mathcal{N}=0}^{\infty} \frac{1}{\mathcal{N}!} \frac{e^{\mu \mathcal{N}/k_B T}}{\lambda^{d\mathcal{N}}} \int d\mathbf{r}_1 \cdots \int d\mathbf{r}_{\mathcal{N}} e^{-U/k_B T} \right\}, \qquad (1.4)$$

where p is the pressure, v is the volume, μ is the chemical potential, d is the dimensionality, and $\lambda = (h^2/2\pi m k_B T)^{1/2}$ is the thermal de Broglie wavelength. The factor $\lambda^{-d\mathcal{N}}$ is obtained by integrating out the kinetic degrees of freedom; thus we cannot really obtain any dynamic information from this formulation. However, we can relate (1.4) to the Ising ferromagnet via the conventional lattice-gas approximation, an exercise which is conceptually very useful.

The idea is simply to let $S_i = \pm 1$ denote, respectively, the presence or absence of a molecule at site i. That is, the occupation of site i is $n_i = \frac{1}{2}(1 + S_i) = 1$ or 0. The integration over configurations $\{\mathbf{r}\}$ in (1.4) is replaced by a sum over occupation states $\{n\}$, which is the same as the sum over spin states $\{S\}$:

$$\sum_{\mathcal{N}=0}^{\infty} \frac{1}{\mathcal{N}!} \int d\mathbf{r}_1 \cdots \int d\mathbf{r}_{\mathcal{N}} \cdots \longrightarrow \sum_{\{n\}} a^{d\mathcal{N}} \cdots \qquad (1.5)$$

where a is the lattice spacing. If U is a pairwise interaction of the form

$$U = -\tfrac{1}{2} \sum_{i,j} v_{ij} n_i n_j, \qquad (1.6)$$

then we have the following correspondences between the gas and the ferromagnet:

$$V p(\mu, T) \leftrightarrow -F(H,T) + NH + \tfrac{1}{2} \sum_{i,j} J_{ij},$$
$$v_{ij} \leftrightarrow 4 J_{ij}, \qquad (1.7)$$
$$\mu - d k_B T \ln(\lambda/a) + \tfrac{1}{2} \sum_j v_{ij} \equiv \mu - \mu_c \leftrightarrow 2H.$$

The important point is that the chemical potential μ, measured from its temperature-dependent critical value μ_c, plays exactly the same role as the external field H in the ferromagnet. The phase diagrams can therefore be transcribed directly from Figs. 1.1–1.3 by replacing H by $\tfrac{1}{2}(\mu - \mu_c)$ and M/N by $(2n-1)$, the quantity n/a^d now being understood as the number density.

1.3 Binary solution

The third version of this model that we shall need is the binary solid (or liquid) solution in which $S_i = -1$ denotes, say, an atom of species A at site i, and $S_i = +1$ denotes species B. If there are $\mathcal{N}_A$ A-atoms and $\mathcal{N}_B = N - \mathcal{N}_A$ B-atoms, then $\mathcal{N}_B - \mathcal{N}_A$ is the same as the magnetization M and, because

$$E - \mu_A \mathcal{N}_A - \mu_B \mathcal{N}_B = E - \tfrac{1}{2} M(\mu_B - \mu_A) - \tfrac{1}{2} N(\mu_A + \mu_B), \qquad (1.8)$$

the difference in chemical potentials $\tfrac{1}{2}(\mu_B - \mu_A)$ plays the same role as the field H. For a clustering or non-ordering alloy — the analog of a ferromagnet rather than an antiferromagnet — an appropriate order parameter is the concentration of B-atoms, which will be denoted by the symbol c, and which is the same as n_i/a^d for the lattice gas.

In what follows, we shall make frequent reference to the ultra-simplified T, c phase diagram in Fig. 1.4, the analog of the T, M diagram shown in Fig. 1.3. (To see what real alloy phase diagrams look

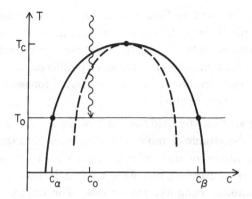

Figure 1.4: Schematic c, T phase diagram for a binary alloy. The solid line is the coexistence curve, and the dashed line is the spinodal. The wiggly arrow indicates a quench from some high temperature to $T = T_0$ for a system with concentration c_0 such that the point c_0, T_0 is in the two-phase region. As shown here, the point is in the unstable 'spinodal' region.

like, see Massalski [1986].) The basic process of interest is one in which we start at high temperature with a uniformly mixed solution of concentration c_0, and we quench to a lower temperature T_0 at which this solution is thermodynamically metastable or unstable against separating into coexisting regions of the stable concentrations c_α and c_β. The boundary between metastability and instability is called the 'spinodal' and is shown by the dashed curve in Fig. 1.4. If we have chosen, for example, c_0 to be only slightly larger than c_α, then we can visualize this process as one in which droplets of the β phase precipitate slowly out of the slightly supersaturated α phase. On the other hand, if c_0 is roughly equidistant from c_α and c_β, then the system undergoes a relatively fast 'spinodal decomposition' in which α-like and β-like regions appear in approximately equal proportions. In either case, it is interesting to ask about the rates of these processes and the morphologies of the various precipitation patterns which emerge.

1.4 Some remarks about realistic models

It must be stressed from the outset that all three of these elementary models are too simple to produce quantitatively accurate descriptions of most realistic, experimental situations. The underlying constituents of real magnets are quantum mechanical objects that couple in complex ways to each other and to their environments. While many ferromagnets, for example those with localized atomic moments in appropriately anisotropic environments, do belong to the Ising universality class, even these exhibit a wide variety of qualitatively different ways in which the magnetic degrees of freedom respond to external forces, deformations of the crystalline lattice, or the like.

Similar remarks can be made about real fluids and binary solutions. The details of the atomic or molecular structures, and the specific ways in which these atoms or molecules interact with each other or couple to other degrees of freedom can have profound effects on the kinetics of even the simplest Ising-like transitions. For example, the rates at which phase transformations occur in fluids are strongly dependent on the extent to which hydrodynamic motions compete with diffusion in transporting the chemical constituents or heat from one place to another within the system. In solids, lattice vibrations, spin waves, the diffusion of vacancies or impurities, all provide transport mechanisms which may be crucially important but are not contained in any of our overly simplified pictures. Moreover, the elastic forces in solids, because of their long-range nature, may qualitatively change the energetics of even the simplest phase transformations. These, too, will be omitted here.

The rationale for considering only the simplest models in this chapter is that these models are already highly non-trivial in their nonequilibrium behavior. Indeed, we shall see that, although much progress has been made in recent years, we have yet to achieve a complete understanding of phenomena such as grain growth and coarsening in even these conceptually simplest cases. Clearly, we need as complete as possible an understanding of the simple cases before we can have much confidence in dealing with more complex situations. This is not to say that those of us with more practical inclinations should be afraid of applying the conceptual insights gained so far to more realistic problems of technological importance. I think that some of the best current research is moving in just such directions, aided by recent theoretical progress plus, of course, the truly remarkable capabilities that have been provided for us by modern computers. My hope in presenting this chapter is that it will serve as a useful starting point for scientists preparing to work along these lines.

2

The droplet model of nucleation

2.1 Introduction

In this section we look at the liquid-vapor transition from the point of view of a model in which the transition from the vapor to the liquid is activated by the nucleation of a droplet of the condensed phase (Frenkel 1955, Zettlemoyer 1969). In the spin language of the first section, droplets are islands of down spins surrounded by up spins. In the following, however, we shall revert to the original picture of interacting molecules.

We most naturally think of condensing a vapor by either cooling or compressing it. That is, the temperature T and the pressure p are the natural control parameters. As seen in Section 1.2, it is most convenient theoretically to think about the phase as being controlled by the chemical potential μ. A discontinuous change from the vapor to the liquid state can be driven by changing the temperature or pressure in such a way that $\delta\mu(T,p) = \mu - \mu_c$ goes from $\delta\mu < 0$ in the vapor state to $\delta\mu > 0$. In the latter condition, the system may survive for some time as a metastable vapor; but eventually it will undergo a transition to the liquid phase.

We are interested in the neighborhood of the crossing point at $\delta\mu = 0$ with T below the critical temperature, so that the transition is first order. In the following sections, we first develop the droplet model in the framework of equilibrium statistical mechanics. For a vapor with $\delta\mu > 0$, we shall see that the equilibrium picture becomes internally inconsistent, and that non-equilibrium concepts are needed in order to understand the nature of the metastable state. We then adopt a nonequilibrium point of view for computing rates of nucleation.

2.2 The cluster approximation

Our thermodynamic variables have been chosen so that the vapor is in equilibrium near the crossing point when $\delta\mu < 0$. Well below the critical point, the vapor consists primarily of noninteracting single molecules. However, it can happen that two molecules are very close together so that they are nearly bound to one another. The same also holds for clusters of $\ell = 3, 4, \ldots$ molecules, but the probability of large clusters must be a rapidly decreasing function of the size ℓ. On the other hand, when we cross the line $\delta\mu = 0$, very large clusters become thermodynamically significant. We derive next the statistical thermodynamic properties of a gas of noninteracting clusters.

In the grand canonical ensemble, the partition function is

$$\Xi \equiv \sum_{\mathcal{N}=0}^{\infty} \frac{1}{\mathcal{N}!} \frac{e^{\mu\mathcal{N}/k_BT}}{\lambda^{3\mathcal{N}}} \int d\mathbf{r}_1 \cdots \int d\mathbf{r}_{\mathcal{N}} \, e^{-U/k_BT}, \qquad (2.1)$$

where λ is the thermal de Broglie wavelength, U is the potential energy of interaction between molecules located at $\mathbf{r}_1 \ldots \mathbf{r}_{\mathcal{N}}$, and we have specialized to the case of three dimensions. We now pretend that we have a non-interacting mixture of molecular clusters of sizes $\ell = 1, \ldots, \infty$, and that there are ν_ℓ clusters of size ℓ. That is, we think of each kind of cluster ℓ as a different species of molecule, with $\mathcal{N} = \sum_{\ell=1}^{\infty} \nu_\ell \ell$. Thus, we can write

$$\Xi \cong \sum_{\{\nu_\ell\}} \prod_{\ell \geq 1} \frac{q_\ell^{\nu_\ell}}{\nu_\ell!} e^{\mu\ell\nu_\ell/k_BT}, \qquad (2.2)$$

where q_ℓ is the canonical partition function for the species ℓ

$$q_\ell \simeq \frac{1}{\lambda^{3\ell}} \frac{1}{\ell!} \int d\mathbf{r}_1 \cdots \int d\mathbf{r}_\ell \, e^{-U/k_BT} \equiv \frac{V}{\lambda^3} e^{-\tilde{\epsilon}(\ell)/k_BT}. \qquad (2.3)$$

In Eq. (2.3), $\tilde{\epsilon}(\ell)$ is the free energy for species ℓ. It contains a bulk part of magnitude μ_c per molecule, and a surface term with a coefficient γ that is proportional to the surface energy:

$$\tilde{\epsilon}(\ell) \simeq \mu_c(\ell - 1) + \gamma(\ell - 1)^{2/3}. \qquad (2.4)$$

We have chosen the form of $\tilde{\epsilon}(\ell)$ in (2.4) so that both terms on the right-hand side vanish in the case of monomers ($\ell = 1$).

The sum appearing in Eq. (2.2) can now be evaluated:

$$\Xi = \exp\left\{ \sum_{\ell=1}^{\infty} q_\ell \, e^{\mu\ell/k_BT} \right\} \equiv \exp\left[\frac{pV}{k_BT} \right], \qquad (2.5)$$

where the pressure p appears in the right hand side as in (1.4).

In a more systematic derivation of (2.5), we can define a separate chemical potential for each species ℓ and replace the expression $\mu \mathcal{N}$ by $\sum_\ell \mu_\ell \nu_\ell$. Then it becomes clear that the average number $\langle \nu_\ell \rangle$ of clusters of size ℓ is

$$\langle \nu_\ell \rangle = k_B T \frac{\partial}{\partial \mu_\ell} \ln \Xi(\mu) \tag{2.6}$$

$$= \frac{V}{\lambda^3} \exp\left[-\frac{1}{k_B T}\left(\bar{\epsilon}(\ell) - \mu \ell\right)\right] \equiv \frac{V}{\lambda^3} \exp\left(-\frac{\epsilon(\ell)}{k_B T}\right)$$

with

$$\epsilon(\ell) \equiv (\ell - 1)(\mu_c - \mu) + \gamma(\ell - 1)^{2/3} \equiv -\delta\mu(\ell - 1) + \gamma(\ell - 1)^{2/3}. \tag{2.7}$$

The resulting pressure is

$$\frac{pV}{k_B T} = \sum_{\ell=1}^{\infty} \langle \nu_\ell \rangle. \tag{2.8}$$

$\langle \nu_\ell \rangle$ is plotted schematically as a function of ℓ in Fig. 2.1. For $\delta\mu < 0$, $\epsilon(\ell)$

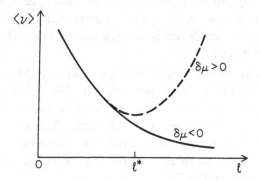

Figure 2.1: Average number of droplets as a function of size ℓ for $\delta\mu < 0$ and $\delta\mu > 0$.

is a monotonically increasing function of ℓ, and we recover the situation where the average number of clusters is a rapidly decreasing function of ℓ. However, for $\delta\mu > 0$, $\epsilon(\ell)$ has a maximum and $\langle \nu_\ell \rangle$ has a minimum when

$$(\ell - 1) > (\ell^* - 1) = \left(\frac{2\gamma}{3\delta\mu}\right)^3. \tag{2.9}$$

For $\ell > \ell^*$, $\langle \nu_\ell \rangle$ increases, and we face a totally illegal situation because the sum on the right-hand side of (2.8) diverges.

To cure the theory for $\delta\mu > 0$, we may cut off the sum in (2.8) at $\ell = \ell^*$. This leads to an isotherm for a metastable phase which is a smooth continuation of the isotherm for the stable phase (Langer 1967). The range of validity of this procedure is discussed below. Note in passing that a metastable state is stable against small perturbations; but large fluctuations such as clusters with $\ell > \ell^*$ may drive the system into a different macroscopic state. Metastable states therefore have finite lifetimes.

2.3 Nonequilibrium analysis (Becker-Döring theory)

Let us assume that we cool the system or compress it to a point such that $\delta\mu > 0$. How long do we have to wait before the liquid appears? This is the kind of question which must be answered in the framework of nonequilibrium statistical mechanics because the quantity of interest is the rate of nucleation.

To proceed, let us imagine an ideal process in which the metastable vapor is maintained perpetually in a steady state by a fictitious device which removes any liquid droplet as it forms, vaporizes it, and returns its constituent molecules to the system. The rate at which droplets are being processed by this device is the nucleation rate, which we shall denote by the symbol $\bar{I}$.

To get a rough idea of what $\bar{I}$ is, let us consider the size ℓ^* where the free energy $\epsilon(\ell)$ of a cluster is maximum. For $\ell > \ell^*$, the cluster lowers its energy by growing. On the contrary, a cluster of size $\ell < \ell^*$ decreases its energy by getting rid of material. Therefore, $\epsilon(\ell^*) \equiv \epsilon^*$ appears as an energy-barrier which gives the activation energy of the process. It is natural to think that the rate of nucleation has to be proportional to $\exp(-\epsilon^*/k_B T)$.

To obtain this result in more detail (Becker and Döring 1935; Frenkel 1955; Gunton et al. 1983), let us assume that the clusters grow or decay by the absorption or evaporation of only one molecule at a time. Let $I(\ell)$ be the rate at which clusters of size $(\ell - 1)$ are transformed into clusters of size ℓ by such processes. Then

$$\frac{d\nu(\ell)}{dt} = I(\ell) - I(\ell + 1). \tag{2.10}$$

Moreover, we can write $I(\ell)$ in the form

$$I(\ell) = a(\ell)\nu(\ell - 1) - b(\ell)\nu(\ell), \tag{2.11}$$

where $a(\ell)$ is the rate at which monomers are absorbed by a cluster of size $(\ell - 1)$, and $b(\ell)$ is the rate at which monomers are evaporated from a cluster of size ℓ. In the idealized steady-state achieved by our fictitious device, the rate $I(\ell)$ is a constant and is equal to the nucleation rate $\bar{I}$ that we wish to calculate.

The coefficients $a(\ell)$ and $b(\ell)$ in (2.11) can be inferred by invoking the principle of detailed balance; their ratio must be such as to drive the system toward thermal equilibrium. Thus,

$$\frac{a(\ell)}{b(\ell)} = \frac{\nu_\ell^{eq}}{\nu_{\ell-1}^{eq}} = \exp\left[-\frac{1}{k_B T}(\epsilon_\ell - \epsilon_{\ell-1})\right] \qquad (2.12)$$

where the superscript 'eq' has been employed to note that we use a property valid in principle only in an equilibrium state. If this is allowed, we find

$$I(\ell) = a(\ell)\left\{\nu_{\ell-1} - \exp\left(\frac{1}{k_B T}(\epsilon_\ell - \epsilon_{\ell-1})\right)\nu_\ell\right\}$$

$$\cong -a(\ell)\left\{\frac{1}{k_B T}\frac{d\epsilon}{d\ell}\nu_\ell + \frac{d\nu_\ell}{d\ell}\right\}, \qquad (2.13)$$

where we have used a continuum approximation valid for $\ell \gg 1$. Because $\bar{I}$ is independent of ℓ in steady state, (2.13) is an equation for the number of clusters of size ℓ. It consists of a 'diffusive' part proportional to the gradient in ℓ-space of the density ν_ℓ and a driven part proportional to the 'force' $d\epsilon/d\ell$. The original, fully time dependent equation becomes:

$$\frac{d\nu_\ell}{dt} = \frac{d}{d\ell}\left\{a(\ell)\left(\frac{1}{k_B T}\frac{d\epsilon}{d\ell}\nu_\ell + \frac{d\nu_\ell}{d\ell}\right)\right\}, \qquad (2.14)$$

which has the form of a Fokker-Planck equation. (We shall see this kind of equation again in Section 6.)

With $I(\ell) = \bar{I}$, Eq. (2.13) can be integrated to obtain:

$$\nu(\ell) = \bar{I}\int_\ell^\infty \frac{d\ell'}{a(\ell')}\exp\left(\frac{\epsilon(\ell') - \epsilon(\ell)}{k_B T}\right), \qquad (2.15)$$

which we now compare with the result of the equilibrium analysis, Eq. (2.6).

For large ℓ, the right hand side of (2.15) goes to zero exponentially fast and the divergent behavior of Eq. (2.6) is cured. To make a more precise statement, it suffices to consider the dominant part of the integrand.

For $\ell < \ell^*$, the range of integration in (2.15) includes the peak at $\ell = \ell^*$ where $\epsilon(\ell)$ is maximum, and the integral can safely be evaluated by a saddle-point approximation. This gives

$$\nu(\ell) \simeq \frac{\bar{I}}{a(\ell^*)} \left(\frac{2\pi k_B T}{\epsilon''(\ell^*)} \right)^{1/2} \exp\left(\frac{\epsilon(\ell^*) - \epsilon(\ell)}{k_B T} \right). \tag{2.16}$$

On the other hand, for $\ell \gg \ell^*$, the major contribution comes from the region beyond the peak, and a first-order expansion gives

$$\nu(\ell) \approx \frac{\bar{I}}{a(\ell)} \frac{k_B T}{|\mathrm{d}\epsilon/\mathrm{d}\ell|}, \quad \text{for } \ell \gg \ell^*. \tag{2.17}$$

If we require that Eq. (2.16) is equal to (2.6) for small ℓ, the nucleation rate can be written as,

$$\bar{I} \simeq \langle \nu_1 \rangle a(\ell^*) \left[\frac{\epsilon''(\ell^*)}{2\pi k_B T} \right]^{1/2} e^{-\epsilon(\ell^*)/k_B T}, \tag{2.18}$$

where

$$\epsilon(\ell^*) = \frac{4\gamma^3}{27(\delta\mu)^2}. \tag{2.19}$$

This is exactly the analytic form we have guessed before. The only factor which was not obvious from previous considerations is the quantity inside the square root, which is known as the Zel'dovich factor. $\bar{I}$ is proportional to the number of droplets which reach the activation barrier by thermal fluctuations, given by $\langle \nu_1 \rangle \exp(-\epsilon(\ell^*)/k_B T)$, times the rate at which they cross, as given by $a(\ell^*)$. The Zel'dovich factor accounts for the fact that not all droplets which reach size ℓ^* actually continue to grow.

Note that the saddle point approximation is justified by the fact that $|\delta\mu|$ has to be small in order for the droplet picture to make sense. To show this, it suffices to recognize that $\exp\left[(\epsilon(\ell^*) - \epsilon(\ell))/(k_B T) \right]$ is strongly peaked at $\epsilon(\ell^*)$ with a width of the peak equal to $\ell^{*2/3} = (2\gamma/(3|\delta\mu|))^2 \ll \ell^*$ for $\ell^* \to \infty$ (or $\delta\mu \to 0$).

To conclude this section, let us make some order-of-magnitude estimates for the various terms in (2.18). For the ordinary liquid-vapor transition, the rate $a(\ell^*)$ is approximately the rate at which molecules impinge on the surface of a sphere of radius R^*:

$$a(\ell^*) \approx (4\pi R^{*2}) \langle \nu(1) \rangle \left(\frac{k_B T}{m} \right)^{1/2}. \tag{2.20}$$

The thermal velocity $(k_B T/m)^{1/2}$ is about 10^5 cm sec^{-1}, and the vapor density $\langle \nu(1) \rangle$ is of order 10^{21} cm^{-3}. One can guess that $R^* \simeq 10^{-7}$ cm so that the critical cluster contains about 10^3 molecules. (We shall check this self-consistently.) To this rough approximation, the surface energy γ should be of order $k_B T$; thus the Zel'dovich factor is of order unity. Then

$$\bar{I} \simeq 10^{33} \, e^{-\epsilon^*/k_B T} \quad \text{sec}^{-1} \text{ cm}^{-3}. \tag{2.21}$$

To achieve an observable nucleation rate of, say, $\bar{I} = 1$ cm^{-3} sec^{-1}, we need $\epsilon^*/k_B T \simeq 70$, or $\ell^* \simeq 10^3$ as we guessed.

Equation (2.21) allows us to study the sensitivity of the rate of nucleation to variations of the chemical potential. Note that ϵ^* in (2.19) is inversely proportional to $(\delta\mu)^2$, and that $\bar{I}$ varies exponentially with ϵ^*. Thus, a one per cent change in $\delta\mu$ changes $\bar{I}$ by a factor of 5, and a ten per cent change induces a change in $\bar{I}$ of a factor 10^6 ! In this sense, the limit of observable supersaturation $\delta\mu$ is well defined; $\bar{I}$ passes from unobservably small values to unobservably large values within a very narrow range of values of $\delta\mu$.

3

Continuum models

3.1 Introduction

In modern statistical mechanics, thermodynamic properties are often described by a coarse-grained free energy which is a functional of an order parameter and whose minima are associated with the equilibrium states of the system. This approach allows us to use continuum models and thereby to bridge the gap between theories at the atomic level and macroscopic observations. It also allows us to take maximum advantage of the symmetries of the system, and to separate 'relevant' from 'irrelevant' aspects in the sense of renormalization group theory (Wilson and Kogut 1974, Ma 1976).

As has been shown in the preceding section, the study of the kinetics of phase transitions is necessarily a study of nonequilibrium phenomena, and therefore we shall have to extend the concept of a coarse-grained free energy to situations in which the system is not in thermal equilibrium. We review in this section the essential assumptions of this approach.

The systems of interest to us are not too far from equilibrium. That is, we are interested in processes which are slow on the scale of a local equilibration time, which we shall discuss below.

As usual, conservation laws play fundamental roles. Two general classes of systems must be distinguished from each other depending upon whether or not the order parameter of principal interest is a locally conserved quantity. Both cases will be studied here. To understand the nature of this dichotomy, note that a typical system of the first kind is an Ising ferromagnet with the magnetization playing the role of the nonconserved order parameter. An up-spin can be changed to a down-spin at some site, and there is no requirement that the magnetization that was lost in this process appear somewhere else in the system. On the other hand, a binary alloy made up of A and B atoms is a typical system in which the order parameter, say, the concentration of B atoms, is locally conserved. If we replace a B atom by an A atom at some site, the B must be put somewhere else, in most cases in the site vacated by the A. For simplicity, we shall consider only scalar order parameters, and we shall not consider situations in which more than one such order parameter is necessary for an adequate description of the system. We start with the nonconserved case.

3.2 Ising ferromagnet in the continuum limit

The relevant order parameter for a ferromagnet is the magnetization density, $m(\mathbf{r}, t)$, which is a function of position $\mathbf{r}$ and time t. The derivation of an equation of motion for $m(\mathbf{r}, t)$ starts with the definition of a coarse-grained free energy, which we now discuss. Two essential hypotheses for the validity of this approach must be emphasized. First, there is a spatial length scale ℓ, appreciably larger than the underlying lattice spacing a, within which $m(\mathbf{r}, t)$ varies smoothly. Second, equilibration of the system on length scales smaller than ℓ is much faster than the processes that we wish to describe (Langer 1974).

Let us consider the Ising model with a lattice spacing a. In principle, the partition function can be evaluated in two steps. We first divide the system into blocks of some mesoscopic size $\ell \gg a$, centered at positions $\mathbf{r}$. We restrict our attention to order parameters $m(\mathbf{r}, t)$ which vary slowly on the spatial scale ℓ. Our assumption of rapid equilibration on scales smaller than ℓ implies that it makes sense to compute the partial partition sum:

$$Z_\ell(\{m\}, H) = \sum_{\text{constrained } \{S\}} \exp(-E\{S\}/k_B T), \qquad (3.1)$$

where the sum over spin configurations $\{S\}$ is restricted by the constraint

$$\frac{1}{\ell^d} \sum_{\{S\} \text{ in block } \mathbf{r}} S_i = m(\mathbf{r}) \qquad (3.2)$$

for each of the coarse-grained positions $\mathbf{r}$. The resulting coarse-grained free energy, $F_\ell(\{m\}, H)$, may be expected to have the familiar Ginzburg-Landau form:

$$F_\ell(\{m\}, H) = -k_B T \ln Z_\ell(\{m\}, H)$$
$$\cong \int d\mathbf{r} [\tfrac{1}{2} K (\nabla m)^2 + f_\ell(m) - Hm]. \qquad (3.3)$$

The second step is to compute the thermodynamic free energy:

$$\bar{f}(m) = -\lim_{V \to \infty} \frac{k_B T}{V} \ln \int \mathcal{D}m \exp[-F_\ell\{m\}] \qquad (3.4)$$

where the functional integral denoted by $\int \mathcal{D}m$ is over the space of all remaining configurations $m(\mathbf{r})$ which are smooth on scale ℓ and which are consistent with a macroscopic magnetization $m = (1/V) \int m(\mathbf{r}) d\mathbf{r}$. Note that we have taken the limit of infinite volume in (3.4). The free energy $\bar{f}$ therefore has to be a convex function of the magnetization m. However, it is understood in Eq. (3.3) that the coefficients appearing in the expansion of the coarse-grained free-energy are computed at a mesoscopic scale ℓ. On this scale, $f_\ell(m)$ is a smooth function of the type shown in Fig. 3.1 and does not need to be convex.

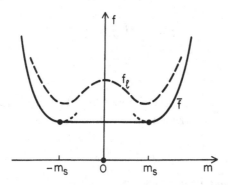

Figure 3.1: Schematic free energy $f(m)$ for the Ising model, showing both the coarse-gained free energy f_ℓ and the thermodynamic limit $\bar{f}$. The dotted lines are the metastable continuations of $\bar{f}$.

A physically motivated condition on ℓ is that it should be smaller than smallest length on which a phase can be defined. A natural candidate is therefore the correlation length ξ. If ℓ were greater than ξ, phase separation could occur within a block. Indeed, the greater ℓ becomes, the greater part of the partition sum is included in the first stage of the calculation, and $f_\ell(m)$ approaches the function $\bar{f}$ of Eq. (3.4). In a more field-theoretic language, ℓ is the ultraviolet cut-off which is implicit in any Ginzburg-Landau free energy.

3.3 Mean-field approximation for the coarse-grained free energy

A systematic, first-principles derivation of an explicit form for F_ℓ is extremely difficult; in general, it is preferable to determine F_ℓ by a combination of phenomenology and symmetry arguments. Some physical insight can be gained, however, from the naive mean-field approximation, which we summarize as follows.

Strictly speaking, the mean-field approximation makes sense only in the limit of a very long-ranged spin-spin interaction, and then only for magnetization fields $m(\mathbf{r})$ which are very slowly varying over this interaction range. We can then write the energy in the form:

$$E \cong -\tfrac{1}{2} \iint d\mathbf{r} \, d\mathbf{r}' J\big(|\mathbf{r} - \mathbf{r}'|\big) m(\mathbf{r}) m(\mathbf{r}')$$

$$\cong -\tfrac{1}{2} \bar{J} a^d \int d\mathbf{r} \, m^2(\mathbf{r}) + \tfrac{1}{2} K \int d\mathbf{r} (\nabla m)^2 + \cdots, \qquad (3.5)$$

where $J(r)$ is the obvious continuum limit of J_{ij} in (1.1),

$$\bar{J} = \frac{1}{a^d} \int J(r) \, dr, \qquad (3.6)$$

$$K = \frac{1}{2d} \int r^2 J(r) \, dr, \qquad (3.7)$$

and d is the dimensionality of the system. The gradient-energy coefficient K can further be written

$$K = \bar{J} a^d \xi_0^2, \qquad (3.8)$$

where ξ_0 is the range of interactions.

The assumption of large ξ_0 permits us to choose the coarse-graining length ℓ to be large, and then to estimate the total entropy as the sum of the entropies in blocks containing $\mathcal{N} \equiv (\ell/a)^d$ spins with magnetizations $\mathcal{M} \equiv m(\mathbf{r})\ell^d$. The entropy of one such block is

$$\mathcal{S}(\mathbf{r}) = k_B \ln\left[\frac{\mathcal{N}!}{(\frac{1}{2}(\mathcal{N} - \mathcal{M}))!\,(\frac{1}{2}(\mathcal{N} + \mathcal{M}))!}\right] \tag{3.9}$$

$$\cong k_B \left(\frac{\ell}{a}\right)^d \ln 2 - \tfrac{1}{2} k_B (\ell a)^d m^2(\mathbf{r}) + \tfrac{1}{12} k_B \ell^d a^{3d} m^4(\mathbf{r}) + \cdots.$$

In the approximate form of (3.9), we have used Stirling's formula and have assumed that $\mathcal{M}/\mathcal{N} = ma^d \ll 1$. The total entropy is then

$$S = \sum_{\text{blocks } (\mathbf{r})} \mathcal{S}(\mathbf{r}) \cong \int \frac{d\mathbf{r}}{\ell^d} \mathcal{S}(\mathbf{r}). \tag{3.10}$$

Combining terms in $F_\ell = E - TS$, and comparing to the standard Ginzburg-Landau form (3.3), we have K as given by (3.8), and

$$f_\ell(m) \cong \tfrac{1}{2}\tau m^2 + \tfrac{1}{4} u m^4 + \cdots, \tag{3.11}$$

with

$$\tau = a^d k_B (T - T_c), \quad u = \tfrac{1}{3} k_B T a^{3d}, \tag{3.12}$$

where $T_c \equiv \bar{J}/k_B$ is the mean-field critical temperature. Note that $f_\ell(m)$ has the characteristic shape shown in Fig. 3.1. Note also that the coarse-graining length ℓ does not appear in these results. We should not expect this to be the case for a more accurate representation. (From here on, however, we usually shall omit the subscript ℓ.)

3.4 Thermodynamic equation of motion

The simplest way of obtaining a nonequilibrium equation of motion for this system is to adopt a phenomenological, thermodynamic point of view. That is, we say that the local rate of displacement of the order parameter, $dm(\mathbf{r})/dt$, is linearly proportional to the local thermodynamic force, $\delta F/\delta m(\mathbf{r})$. The constant of proportionality, which we call Γ, is the response coefficient which defines a time scale for the system. (In principle, Γ may be a function of m; but we shall ignore that possibility for the present.) The general form of the equation of motion is therefore:

$$\frac{dm}{dt} = -\Gamma \frac{\delta F}{\delta m(\mathbf{r})}. \tag{3.13}$$

Note that (3.13) implies a purely dissipative system and not one, for example, with inertial Hamiltonian dynamics. Using the form (3.3) for F we find:

$$\frac{\partial m}{\partial t} = -\Gamma\left[-K\nabla^2 m + \frac{df}{dm} - H\right]. \qquad (3.14)$$

The total time derivative of the free energy is:

$$\frac{dF}{dt} = \int \frac{\delta F}{\delta m}\frac{dm}{dt}\,dr = \int \frac{\delta F}{\delta m}\left(-\Gamma\frac{\delta F}{\delta m}\right)dr$$

$$= -\Gamma \int \left(\frac{\delta F}{\delta m}\right)^2 dr \le 0, \qquad (3.15)$$

thus F is a strictly non-increasing function of time. This excludes any activated processes. To describe nucleation, a fluctuating term has to be added to the right-hand side of (3.14), as will be discussed in Section 6.

3.5 Flat interface

To illustrate the usefulness of Eq. (3.14), let us consider the case of a flat two-phase interface in zero magnetic field. In the one-dimensional situation depicted in Fig. 3.2, the order parameter varies between $m = -m_s$

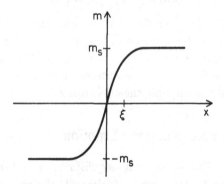

Figure 3.2: Ising-like two-phase interface.

for $x \to -\infty$ and $m = +m_s$ for $x \to +\infty$. One quantity of interest is the shape of the interface. In equilibrium, $\partial m/\partial t = 0$, so that (3.14) becomes

$$K\frac{d^2 m(x)}{dx^2} = \frac{df}{dm} = -\frac{dV_{\text{effective}}}{dm}. \qquad (3.16)$$

Here we have introduced the notation $f(m) = -V_{\text{effective}}(m)$ to emphasize that Eq. (3.16) is formally the same as an equation of motion for a point mass in a double-well potential with the following changes of variables:

$$\begin{cases} x \leftrightarrow \text{time}, & m \leftrightarrow \text{displacement}, \\ K \leftrightarrow \text{mass}, & V_{\text{effective}} \leftrightarrow -f. \end{cases}$$

This effective potential and the motion corresponding to $m(x)$ are shown in Fig. 3.3 (a). The motion being frictionless, the total 'energy' is a constant of the motion and is equal to $-f(m_s)$ because, for solutions of interest to us, the 'kinetic energy' vanishes at the 'tops of the hills' where $m = \pm m_s$. Therefore, $\frac{1}{2}K(\mathrm{d}m/\mathrm{d}x)^2 - f(m) = -f(m_s)$ is constant.

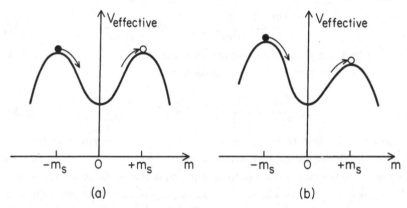

(a)   (b)

Figure 3.3: The mechanical analogs corresponding to: (a) a static interface in zero magnetic field and (b) a moving interface in nonzero field. The initial and final positions of the point mass are indicated by filled and open circles respectively.

The surface energy σ associated with the interface is computed by subtracting the bulk energy from the total energy:

$$\sigma = \int \left[\tfrac{1}{2}K \left(\frac{\mathrm{d}m}{\mathrm{d}x} \right)^2 + f(m) - f(m_s) \right] \mathrm{d}x = K \int \left(\frac{\mathrm{d}m}{\mathrm{d}x} \right)^2 \mathrm{d}x. \quad (3.17)$$

The width of the interface ξ (which is proportional to the correlation length in this theory) can be estimated by considering the asymptotic behavior of m near $\pm m_s$. Let $m = \pm m_s \mp u$. Then

$$K \frac{\mathrm{d}^2 u}{\mathrm{d}x^2} - \frac{\mathrm{d}^2 f}{\mathrm{d}m_s^2} u \cong 0$$

and $u \cong \exp[\mp x/\xi]$ with

$$\xi^2 = K \Big/ \left(\frac{\mathrm{d}^2 f}{\mathrm{d}m_s^2} \right) \sim (T_c - T). \quad (3.18)$$

Thus, the width of the interface exhibits the mean field divergence $(T_c - T)^{-1/2}$.

We are now able to take the first step toward non-equilibrium problems. If we apply a magnetic field to the system, the interface between the spin-up and spin-down phases must move. The question one must answer is how to compute the velocity v at which the interface is moving.

When a magnetic field is present, the effective potential of Eq. (3.16) becomes $V_{\text{effective}} = -f + mH$ [see Fig. (3.3b)] so that, in the mechanical analogy, the potential is asymmetric. Equation (3.14) can now be rewritten as

$$K\frac{\mathrm{d}^2 m}{\mathrm{d}x^2} = -\frac{\mathrm{d}}{\mathrm{d}m}(-f + mH) + \frac{1}{\Gamma}\frac{\mathrm{d}m}{\mathrm{d}t}, \qquad (3.19)$$

and we look for a solution moving at a constant velocity v : $m(x,t) = \widetilde{m}(x - vt)$. Equation (3.19) now becomes:

$$K\frac{\mathrm{d}^2 \widetilde{m}}{\mathrm{d}x^2} = -\frac{\mathrm{d}}{\mathrm{d}\widetilde{m}}(-f + \widetilde{m}H) - \frac{v}{\Gamma}\frac{\mathrm{d}\widetilde{m}}{\mathrm{d}x}. \qquad (3.20)$$

Note that the last term in Eq. (3.20) appears as an effective force of friction in the equation of motion. For $x \to \pm\infty$, one has $\widetilde{m} \cong \mp m_s$ as before. The friction term has to be such that the mass leaves the top of the highest hill and precisely stops on the top of the other one. There is just one value of the 'friction constant' v/Γ such that this solution exists, and this therefore determines the velocity v at which the interface is moving. Other solutions, such as the ones where the mass escapes from the inverse double well, have to be discarded. By integrating Eq. (3.20), we find:

$$\int \mathrm{d}x \frac{\mathrm{d}}{\mathrm{d}x}\left\{\tfrac{1}{2}K\left(\frac{\mathrm{d}\widetilde{m}}{\mathrm{d}x}\right)^2 - f + \widetilde{m}H\right\} = -\frac{v}{\Gamma}\int\left(\frac{\mathrm{d}\widetilde{m}}{\mathrm{d}x}\right)^2 \mathrm{d}x, \quad (3.21)$$

or
$$\Delta[f - \widetilde{m}H] \cong 2m_s H \cong \frac{v}{\Gamma}\frac{\sigma}{K}, \qquad (3.22)$$

where $\Delta[\ ]$ denotes the change in the quantity in square brackets when going from one side of the interface to the other. In the final forms of (3.22), we have made the additional assumptions that H is small and that σ is the equilibrium surface tension. From Eq. (3.22), we obtain the velocity of the moving interface:

$$v = \frac{2m_s K\Gamma}{\sigma}H. \qquad (3.23)$$

3.6 Spherical droplet

As a second application of Eq. (3.14), consider a droplet of up-spins immersed in a sea of down spins. A magnetic field which favors the up phase is applied to the system. The energy that we have to spend to form this droplet (for $d = 3$) consists of a bulk term and a surface term:

$$\Delta F = -2m_s H\left(\tfrac{4}{3}\pi R^3\right) + \sigma 4\pi R^2. \tag{3.24}$$

This energy is maximized for

$$R = R^* = \frac{\sigma}{m_s H},$$

or, more generally in d dimensions,

$$R^* = \tfrac{1}{2}(d-1)\frac{\sigma}{m_s H}. \tag{3.25}$$

Note that Hm_s plays a role analogous to a difference in chemical potential for a supersaturated system, and that (3.25) is a special form of the Gibbs-Thomson relation. Equation (3.24) is essentially identical to the droplet equation, (2.7), and has the same feature that R^* identifies an unstable maximum of ΔF.

The preceding example has rephrased the concept of an energy barrier for the nucleation process. Next, we want to know how the droplet actually moves under these conditions. To do that, we need only a slight modification of the preceding analysis for the planar interface.

If the problem is spherically symmetric, the kinetic equation (Eq. (3.14)) can be written as

$$\frac{\partial m}{\partial t} = \Gamma\left[K\frac{\partial^2 m}{\partial r^2} + K\frac{(d-1)}{r}\frac{\partial m}{\partial r} - \frac{df}{dm} + H\right], \tag{3.26}$$

where the magnetic field favors the $m = +m_s$ phase. If we look for a solution of the type $m = \tilde{m}(r - R(t))$ we are led to

$$-\frac{1}{\Gamma}\frac{dR}{dt}\frac{d\tilde{m}}{dr} = K\frac{d^2\tilde{m}}{dr^2} + K\frac{(d-1)}{r}\frac{d\tilde{m}}{dr} - \frac{df}{d\tilde{m}} + H. \tag{3.27}$$

Note that our solution describes an imperfectly sharp interface of width ξ in the neighborhood of $r \cong R$. If ξ is sufficiently small, i.e. $\xi \ll R$, we can approximate $r \simeq R$ because the variations of $\tilde{m}$ are only relevant for $R - \xi \leq r \leq R + \xi$. With these approximations, Eq. (3.27) can be solved in the spirit of the preceding section. To make this connection more explicit, we can rewrite Eq. (3.27) as:

$$K\frac{\mathrm{d}^2\tilde{m}}{\mathrm{d}r^2} + \left(\frac{K(d-1)}{R} + \frac{v}{\Gamma}\right)\frac{\mathrm{d}\tilde{m}}{\mathrm{d}r} - \frac{\mathrm{d}f}{\mathrm{d}m} + H = 0, \qquad (3.28)$$

where $v = \mathrm{d}R/\mathrm{d}t$. The preceding analysis applies if we replace v by $v + \Gamma K(d-1)/R$. From Eq. (3.27), we deduce that the velocity v at which the interface is moving is:

$$v = \frac{\mathrm{d}R}{\mathrm{d}t} = \frac{2\Gamma m_s H}{\sigma} - \frac{\Gamma K(d-1)}{R^2(t)} \equiv \Gamma K(d-1)\left(\frac{1}{R^*} - \frac{1}{R(t)}\right). \qquad (3.29)$$

As a consequence, we retrieve the results of Section 2: the droplet collapses for $R < R^*$ and grows for $R > R^*$. The unstable stationary situation, $v = 0$, occurs when $R = R^*$ as it should according to the Gibbs-Thomson relationship. To conclude, note that for $H = 0$, $R^* = \frac{1}{2}(d-1)\sigma/(m_s H)$ becomes infinite, so that all droplets of finite size collapse.

The preceding discussion can be generalized to apply to curved interfaces which are not simply spheres. The normal velocity at any point on such an interface will be given by (3.29) with $(d-1)/R$ replaced by the local curvature $\mathcal{K}$. The only restriction remains that $\mathcal{K}\xi \ll 1$.

3.7 Domain coarsening

A frequently encountered physical situation is one in which an Ising-like system is quenched at zero external field to some temperature below its critical point. Both coexisting equilibrium phases, spin-up and spin-down or $m = \pm m_s$, are equally likely to appear under these conditions. Ordinarily, after the quench has been completed, the system will exhibit a pattern of interpenetrating regions or 'domains' of the two phases. The characteristic size of these domains may at first be very small, but the pattern must coarsen as the system moves toward states of lower interfacial energy. We shall discuss the coarsening problem in some detail for systems with conserved order parameters later in the chapter. For the case of the nonconserved order parameter now under consideration, however, we shall look only at one simple but important result.

It is easiest to start by supposing that, at some time after the quench, the system consists of spherical domains of one phase embedded in a background of the second phase. More realistically, the domains are not spherical, and the two-phase topology is one in which there are domains inside of domains inside of domains, etc.such details are not really essential for the following argument, however. We simply need to think of the radius of a spherical domain as being in reality the length scale that characterizes the coarseness of the pattern, that is, a typical radius of curvature of the two-phase interfaces. And we can think of Eq. (3.29) as being valid not just for a truly spherical droplet but also, at least in a scaling sense, for determining the motion of an arbitrary piece of interface whose radius of curvature is R. Note that, for $H = (d - 1)\sigma/(2R^*) = 0$ in (3.29), dR/dt is always negative and all convex domains shrink.

Now ask the following question. At a time t after the quench, what is the minimum initial size R_0 of domains which have not yet shrunk to zero? Solving (3.29) for $H = 0$, we find

$$R^2(t) = R_0^2 - 2\Gamma K(d - 1)t, \qquad (3.30)$$

so that, when $R(t) = 0$, $R_0 = [2\Gamma K(d - 1)t]^{1/2}$. The largest domains shrink most slowly according to (3.29), whereas shrinkage accelerates as R decreases. Our result (3.30) therefore tells us that the only domains that remain in the system at time t are those whose sizes are of order $R_0 \sim \sqrt{t}$. This scaling law has been observed in numerical simulations (Safran *et al.* 1983) and also, for example, in the growth of antiphase domains in ordered FeAl alloys as analyzed by Allen and Cahn (1979).

3.8 Completion formula

Let us consider again the domain coarsening of the preceding section but with a positive magnetic field. In that case, the spin-up phase eventually will fill the sample. An important quantity to compute is the fraction of the volume transformed at time t after the quench (Kolmogorov 1937, Avrami 1939).

The simplest case to consider is that in which the up-phase is nucleated at constant rate $\bar{I}$ and spreads with constant speed $v \cong 2\Gamma m_s H/\sigma$ for $R \gg R^*$. The problem is most easily solved by asking an inverted question: Take a point O at random. What is the probability P that O has *not* been transformed at time t? Because the up-phase spreads at speed v, O can only be reached by droplets which are nucleated at distances $r < vt'$ from O, with $0 \leq t' \leq t - r/v$. Thus, for $d = 3$,

$$P = \prod_{\substack{r<vt \\ 0<t'<t-r/v}} (1 - \bar{I}\, dr\, dt') \qquad (3.31)$$

$$= \exp\left[-\bar{I} \int_0^{vt} 4\pi r^2\, dr \int_0^{t-r/v} dt'\right] = \exp\left[-\tfrac{1}{3} v^3 \bar{I}\pi t^4\right].$$

The completion curve, $1 - P(t)$, has a sigmoidal shape which is very familiar, for example, in the metallurgical literature.

It must be emphasized that (3.31) is valid only under very special circumstances. It is obviously incorrect for a conserved order parameter, in which case both $\bar{I}$ and v decrease with time because of depletion effects. Such situations will be described in later sections. There are other circumstances, more nearly within the framework of the present calculation, in which $\bar{I}$ may depend on the state of transformation. For example, Chandra [1989] recently has analyzed a ferroelectric transition in which the relaxation of elastic stress as the transformation proceeds causes $\bar{I}$ to increase with time. The result is a completion curve that is even sharper than that given by (3.31) — a sudden, 'explosive' transition.

4
Continuum model with an Ising-like conserved order parameter

4.1 Introduction

In this section, we consider the kinetics of growth phenomena in a situation where the order parameter is conserved. An example of such a system is a binary solid solution made up of atoms of types, say, A and B. A convenient choice for the order parameter, analogous to the magnetization $m(\mathbf{r}, t)$ in Section 3, is the number of B atoms per unit volume, $c(\mathbf{r}, t)$. As before, this concentration $c(\mathbf{r}, t)$ must be understood to be averaged over a coarse-graining volume ℓ^d. The coarse-grained free energy is again written as

$$F_\ell\{c\} = -k_B T \ln\left[\sum_{\substack{\text{constrained} \\ \text{configurations}}} \exp(-E/k_B T)\right] \qquad (4.1)$$

$$\cong \int d\mathbf{r}\left[\tfrac{1}{2} K\left(\nabla c(\mathbf{r})\right)^2 + f_\ell(c)\right],$$

where the sum over constrained configurations includes only those microscopic states of the system which are consistent with $c(\mathbf{r})$. The coarse-graining length scale ℓ again must be smaller than relevant correlation lengths. In Eq. (4.1), f_ℓ is the free energy at scale ℓ whose typical variation as a function of c for a temperature $T < T_c$ is drawn in Fig. 4.1.

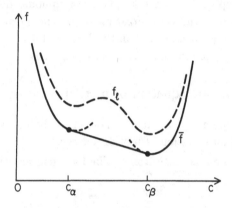

Figure 4.1: Schematic free energy for a binary alloy, showing both the coarse-gained free energy f_ℓ and the thermodynamic limit $\bar{f}$. The dotted lines are the metastable continuations of $\bar{f}$.

Because the thermodynamic limit $\ell \to \infty$ has not been taken in (4.1), f_ℓ is again a non-convex function. Note that there is no *a priori* symmetry of f_ℓ for this case. As in the preceding section, the thermodynamic free energy $\bar{f}(c)$ is

$$\bar{f}(c) \equiv -k_B T \lim_{V\to\infty} \frac{1}{V} \ln\left[\int \mathcal{D}c\exp(-F\{c\}/k_B T)\right], \qquad (4.2)$$

where the functional integration is over smooth configurations such that $c = (1/V)\int c(\mathbf{r})\, d\mathbf{r}$. Let us remark that the chemical potential does not appear explicitly in Eq. (4.1); it will appear as a derived quantity in the next stage of our analysis.

The two coexisting phases, denoted by c_α and c_β in Fig. 4.1, are obtained by a common-tangent construction whose significance we shall discuss shortly. Because the thermodynamic limit has been taken in (4.2), $\bar{f}(c)$ has a linear section between c_α and c_β. The metastable free energies, obtained in principle by analytic continuation through c_α and c_β, are indicated by dotted lines in the figure.

The physical situation to be considered is analogous to the one we have studied previously. The system initially is uniformly mixed at some high temperature and is quenched to a lower temperature such that phase separation occurs. This may be achieved if the concentration of B atoms in the initially homogeneous phase is chosen such that the system is quenched from outside to inside the coexistence region, as shown in Fig. 1.4.

This phase separation can take place via spinodal decomposition or nucleation. Because the concentrations of A and B atoms are locally conserved, the simplest and most natural model for us to develop is one in which the transport mechanism is purely diffusive.

4.2 Thermodynamic equation of motion

As in Section 3, we adopt a phenomenological, thermodynamic point of view in deriving an equation of motion for $c(\mathbf{r}, t)$.

Let $\mathbf{j}$ denote the flux of B atoms. The local conservation condition is

$$\frac{\partial c(\mathbf{r}, t)}{\partial t} = -\nabla \cdot \mathbf{j}.$$

The thermodynamic force which drives the flux $\mathbf{j}$ is the gradient of the chemical potential μ (more accurately, $\mu_B - \mu_A$), which is obtained from F via the relation:

$$\mu(\mathbf{r}) = \frac{\delta F}{\delta c(\mathbf{r})}. \tag{4.3}$$

The associated linear response coefficient is a mobility, which we denote by the symbol $M(c)$ — here emphasizing the possibility that M may be concentration dependent. Thus

$$\mathbf{j}(\mathbf{r}) = -M(c)\nabla\mu, \tag{4.4}$$

and
$$\frac{\partial c(\mathbf{r}, t)}{\partial t} = \nabla \cdot \left[M(c) \nabla \frac{\delta F}{\delta c(\mathbf{r})} \right] \qquad (4.5)$$

$$= \nabla \cdot \left[M(c) \nabla \left(-K \nabla^2 c + \frac{\partial f}{\partial c} \right) \right].$$

Equation (4.5) is well known as the 'Cahn-Hilliard' equation (Cahn 1968, Gunton *et al.* 1983).

The gradient energy, and thus the term $-K\nabla^2 c$ in (4.5), is relevant only to behavior on the spatial scale of the correlation length. To begin to understand the physical significance of Eq. (4.5), we may momentarily neglect this term and linearize the equation about some uniform concentration c_0. With the further assumption that $M(c)$ is a constant, (4.5) becomes a simple diffusion equation with a diffusion constant $D(c_0) = M(c_0) \partial^2 f / \partial c_0^2$.

As before, one cannot use the thermodynamic equation of motion (4.5) to describe activated processes. To see this, it suffices to integrate the Cahn-Hilliard equation by parts to show that

$$\frac{dF}{dt} = - \int M \left(\nabla \frac{\delta F}{\delta c} \right)^2 d\mathbf{r} \leq 0. \qquad (4.6)$$

In other words, there is no way to overcome an energy barrier because the total derivative of the free energy with respect to the time is never positive.

4.3 Planar interface

In analogy to our development for the case of a non-conserved order parameter, we next consider the stationary solutions of (4.5) (Cahn and Hilliard 1958). If $\delta F / \delta c \equiv \mu$ is constant, then $\partial c / \partial t = 0$, and, moreover, no flux enters or leaves the system. Our basic equation becomes

$$\frac{\delta F}{\delta c} - \mu = -K\nabla^2 c(\mathbf{r}) + \frac{\partial f}{\partial c} - \mu = 0, \qquad (4.7)$$

or equivalently,

$$\frac{\delta}{\delta c} \left[F - \mu \int c(\mathbf{r}) \, d\mathbf{r} \right] = 0. \qquad (4.8)$$

Thus, the equation which determines stationary states for this system is the same as for the case of the non-conserved order parameter, with the chemical potential μ playing the role of the magnetic field H.

Let us look at a situation where the phases α and β are in equilibrium with each other at a flat interface. Because there is no flux across the interface, μ must be the same constant, say μ_s, in both phases. In addition, the variational form (4.8) tells us that $f(c) - \mu_s c$ must have the same value on both sides of the interface. Examination of Fig. 4.1 should make it apparent that the preceding two conditions are equivalent to the common-tangent construction which fixes c_α and c_β.

The concentration profile in the x-direction, normal to the flat interface, is determined by the equation

$$-K\frac{d^2c}{dx^2} + \frac{\partial}{\partial c}(f - \mu_s c) = 0, \qquad (4.9)$$

which is essentially the same as (3.16). Thus, in analogy to (3.17), the surface tension σ is

$$\sigma = K \int \left(\frac{dc}{dx}\right)^2 dx. \qquad (4.10)$$

4.4 Spherical droplet

Next consider a spherical droplet of the β phase immersed in supersaturated α phase. The corresponding concentration profile is shown in Fig. 4.2. So long as we are looking only at stationary states, our anal-

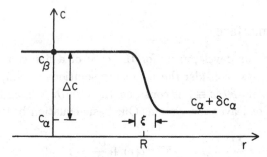

Figure 4.2: Concentration profile for a stationary droplet of β phase immersed in a supersaturated α phase. Because $\xi \ll R$, the point $r = 0$ is too far to the left to be shown in this figure.

ysis must remain identical to that in Section 3. In particular, a curved surface can be stationary only if the chemical potential differs from μ_s. Inside the droplet, the β phase has a concentration of B atoms approximately equal to c_β. Outside the droplet, the B atoms in the α phase are in excess and their concentration is $c_\alpha + \delta c_\alpha$. It follows that the chemical potential μ outside the droplet is no longer equal to its equilibrium value μ_s but is increased by a factor proportional to δc_α:

$$\mu - \mu_s \cong \frac{\partial \mu}{\partial c} \delta c_\alpha \equiv \chi_\alpha^{-1} \delta c_\alpha, \tag{4.11}$$

where we have defined a susceptibility χ_α which relates the thermodynamic driving force per unit volume and the variation of the order parameter. The droplet being stationary, no flux comes into the system and Eq. (4.7) applies.

In our spherical geometry, we have:

$$K \frac{\mathrm{d}^2 c}{\mathrm{d}r^2} + \frac{K(d-1)}{r} \frac{\mathrm{d}c}{\mathrm{d}r} - \frac{\mathrm{d}f}{\mathrm{d}c} = -\mu. \tag{4.12}$$

To obtain an analog of (3.25), it is convenient in this case to integrate both sides of (4.12) across a region of width $2\epsilon \gg \xi$ which includes the position of the interface at $r = R$:

$$\int_{R-\epsilon}^{R+\epsilon} \mathrm{d}r \left[K \frac{\mathrm{d}^2 c}{\mathrm{d}r^2} + \frac{K(d-1)}{r} \frac{\mathrm{d}c}{\mathrm{d}r} - \frac{\mathrm{d}f}{\mathrm{d}c} \right] \frac{\mathrm{d}c}{\mathrm{d}r} = - \int \mathrm{d}r \, \mu \frac{\mathrm{d}c}{\mathrm{d}r}, \tag{4.13}$$

from which it follows that

$$\int_{R-\epsilon}^{R+\epsilon} \mathrm{d}r \frac{\mathrm{d}}{\mathrm{d}r} \left[\tfrac{1}{2} K \left(\frac{\mathrm{d}c}{\mathrm{d}r} \right)^2 - f + \mu c \right] = \frac{K(d-1)}{R} \int \left(\frac{\mathrm{d}c}{\mathrm{d}r} \right)^2 \mathrm{d}r. \tag{4.14}$$

If $\mathrm{d}c/\mathrm{d}r = 0$ far from the interface, the left-hand side of Eq. (4.14) is simply the jump in the quantity $f - \mu c$ across the interface. The condition for a stationary droplet is thus:

$$\delta(f - \mu c) \cong \delta\mu \Delta c = \frac{(d-1)}{R} \sigma, \tag{4.15}$$

where $\Delta c \equiv c_\beta - c_\alpha$ and $\delta\mu = \mu - \mu_c$. Equation (4.15) is another form of the Gibbs-Thomson relation. Using (4.11), we can rewrite it in the more convenient form:

$$\delta c_\alpha = \frac{(d-1)}{R\Delta c} \sigma \chi_\alpha. \tag{4.16}$$

Consider now the case of a growing droplet. Growth implies a flux of B atoms and, thus, a gradient of concentration in the region outside the interface, $r \gtrsim R$. Within the interface region, that is, for $r - R$ small of order ξ, we need the full Cahn-Hilliard equation in order to

328   *J.S. Langer*

describe the system. Far outside the interface, however, where $r-R \gg \xi$, the diffusion approximation is valid. Thus there are two distinct length scales that are important in this situation: the microscopic scale ξ which describes correlations and interfacial structure, and the much larger scale — usually of order R — which describes the diffusion field. The two scales are shown schematically in Fig. 4.3. As shown in that figure, the concentration at R is fixed at $c_\alpha + \delta c_\alpha$ where $\delta c_\alpha(R)$ is given by (4.16). Far away from the growing interface, $c(r)$ approaches $c_\infty \equiv c_\alpha + \delta c_\infty$.

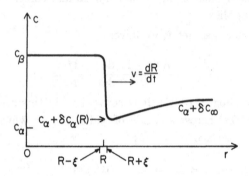

Figure 4.3: Concentration profile for a growing droplet. The length scale along the r-axis is chosen so as to show the droplet radius R and the range of the diffusion field, in comparison to which ξ is very small.

This situation is called 'diffusion controlled'. In order for the droplet to grow, material has to be transported by diffusion from the far field where $c \cong c_\infty$, across the depletion zone, to the surface of the droplet. Other situations are possible. For example, in some of the cases considered in the lectures by Nozières and Caroli, the process of attachment of atoms at the interface is much slower than diffusion in the bulk, and we have what is known as 'interface control'. In general, the slowest process controls the motion of the interface.

Near the moving interface, we can write the concentration of B atoms as $c(r,t) \cong c(r - R(t))$ so that

$$\frac{\partial c}{\partial t} \cong -\dot{R}\frac{\partial c}{\partial r}. \qquad (4.17)$$

The chemical potential must be continuous at the interface; a discontinuity in μ would imply an infinite flux. Conservation of the flux of B atoms at the interface implies:

$$-\int_{R-\epsilon}^{R+\epsilon} \frac{\partial c}{\partial t}\,dr = \Delta c\dot{R} = D\frac{\partial c}{\partial r}\Big|_{R+\epsilon}. \qquad (4.18)$$

In the quasi-stationary approximation, the concentration of B atoms outside the droplet is a slowly varying function of time, and one can assume that

$$\frac{\partial c}{\partial t} = D\nabla^2 c = 0; \quad \text{for } r > R, \qquad (4.19)$$

the hypothesis being that the natural diffusion time scale defined by R^2/D is much smaller than the growth time scale $R/\dot{R}$. For $d = 3$, Eq. (4.19) implies that

$$C(r) = c_\infty - \frac{\text{Const}}{r} \quad \text{for } r > R. \qquad (4.20)$$

The constant in (4.20) can be evaluated by requiring that $c(R) = c_\alpha + \delta c_\alpha$:

$$c(r) = c_\infty - \frac{R}{r}(\delta c_\infty - \delta c_\alpha). \qquad (4.21)$$

We then find that

$$\left.\frac{dc}{dr}\right|_{R+\epsilon} = \frac{\delta c_\infty - \delta c_\alpha}{R}. \qquad (4.22)$$

Finally,

$$\dot{R} = \frac{D}{R}\left(\Delta - \frac{2d_0}{R}\right), \qquad (4.23)$$

where

$$\Delta = \frac{\delta c_\infty}{\Delta c} \qquad (4.24)$$

is the dimensionless supersaturation and

$$d_0 = \frac{\sigma \chi_\alpha}{(\Delta c)^2} \qquad (4.25)$$

is a capillary length. Note that (4.23) implies that the quasistationary approximation, $R^2/D \ll R/\dot{R}$, is equivalent to $\Delta - 2d_0/R \ll 1$. That is, the supersaturation must be small ($\Delta \ll 1$).

Remember that we have enforced spherical symmetry in this discussion. In fact, sufficiently large droplets, growing under the conditions described here, are morphologically unstable. For more information, see the discussion of the Mullins-Sekerka instability in the chapter by Caroli, Caroli and Roulet.

Returning to Eq. (4.23), note that $R^* = 2d_0/\Delta$ sets a length scale; it is, of course, the familiar critical radius for nucleation. For $R < R^*$ the droplet shrinks; for $R > R^*$, it grows.

In the early stages of phase separation after a shallow quench, the system consists of many droplets growing independently from each other with $R \gg R^*$. In this case, the growth law is:

$$\dot{R} \cong \frac{\Delta D}{R}, \qquad (4.26)$$

so that $R \sim t^{1/2}$.

In the later stages of growth, the situation becomes more complicated. Eventually, a significant fraction of the B atoms disappear from the supersaturated α phase, the supersaturation Δ decreases, R^* increases, and one reaches the regime of competitive growth. The transition between the two regimes occurs when some droplets find themselves with $R \lesssim R^*$. Such droplets are decreasing in size; B atoms are evaporating from them and diffusing toward droplets with $R > R^*$. The assumption is that the average size of the droplets in this regime is of order R^*, and that R^* is the only relevant length scale in the problem. Thus, both terms on the right-hand side of (4.23) are of the same order of magnitude, $\Delta \sim d_0/R$, and the equation may be written in the form:

$$R^2 \dot{R} \cong \text{Const} \times d_0 D. \qquad (4.27)$$

It follows that the growth law at late stages of coarsening is the Lifshitz-Slyozov-Wagner law, $R \sim t^{1/3}$ (Lifshitz and Slyozov 1961, Wagner 1961). We shall discuss this regime in greater detail in Section 8.

5

Spinodal decomposition: basic concepts

5.1 Introduction

Spinodal decomposition is the mechanism by which phase separation occurs in a mixture that is quenched into a thermodynamically unstable state. In this section, we consider spinodal decomposition from the point of view of the Cahn-Hilliard theory. As before, the homogeneous binary mixture with a concentration c_0 of B atoms, initially in equilibrium at some temperature above the coexistence curve, is quenched to a temperature T_0 inside the two-phase region (see Fig. 1.4). Consequently, the alloy decomposes into two phases with concentrations c_α and c_β, respectively. Because the order parameter is conserved, the decomposition is controlled by relatively slow diffusive transport.

The first object of this section is to study the problem at initial times in the framework of a linear stability analysis. In Section 5.2, we shall use the linear theory to compute the structure factor, i.e. the X ray or neutron scattering intensity, for this system shortly after the quench. We shall see that phase separation manifests itself in the structure factor as a sharp peak at small momentum transfers. This peak has a useful scaling form. Although the linear theory is valid only at very short times after the quench, it provides a useful intuitive basis for understanding subsequent developments. Some general observations concerning the physics of domain coarsening are presented in Section 5.4.

5.2 Linear instabilities

Because we are interested in the fluctuations of the concentration field, it is useful to define a new variable $u(\mathbf{r}, t)$:

$$c(\mathbf{r}, t) = c_0 + u(\mathbf{r}, t), \qquad (5.1)$$

where $u(\mathbf{r}, t)$ can be expanded as a sum of Fourier modes

$$u(\mathbf{r}, t) = \frac{1}{V} \sum_{\mathbf{k}} \hat{u}(\mathbf{k}, t) \exp(i\mathbf{k} \cdot \mathbf{r}). \qquad (5.2)$$

In (5.2), the sum is over all wave vectors $\mathbf{k}$ in the first Brillouin zone with the convention that $\hat{u}(\mathbf{k} = 0) = 0$. In this Fourier representation, the linearized Cahn-Hilliard equation becomes:

$$\frac{d}{dt}\hat{u}(\mathbf{k}, t) = \omega(k)\hat{u}(\mathbf{k}, t), \qquad (5.3)$$

where

$$\omega(k) = -Mk^2 \left[\frac{Kk^2 + \partial^2 f}{\partial c_0^2}\right]. \qquad (5.4)$$

If the initial state has a concentration c_0 such that $\partial^2 f/\partial c_0^2 > 0$, the amplitude of any fluctuation decreases in time ($\omega(k) < 0$), and the system is stable. On the other hand, if $\partial^2 f/\partial c_0^2 < 0$, fluctuations are unstable [$\omega(k) > 0$] for wave vectors in the range $0 < k < k_0 = [-(\partial^2 f/\partial c_0^2)/K]^{1/2}$. These two characteristic behaviors of $\omega(k)$ are illustrated in Fig. 5.1. The spinodal line, defined in the (c, T)-

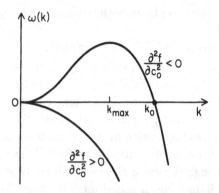

Figure 5.1: Amplification factor ω as a function of wavenumber k for
an unstable situation (upper curve) and a stable one (lower curve).

plane by the condition $\partial^2 f / \partial c_0^2 = 0$ and shown schematically in Fig. 1.4,
separates the region of metastable states from the region of unstable
states where growing modes drive phase separation.

Shortly after the system has developed these instabilities, however,
the linear approximation breaks down. The spinodal line is therefore
not a physically well-defined construct. At best, it gives us a rough idea
of where a smooth transition takes place between thermally activated
processes, i.e. nucleation, and spinodal processes for which there is no
activation barrier. To understand the smoothness of this transition,
note that, on the metastable side of the spinodal line, the activation
barriers may be much smaller than $k_B T$, and thus cannot be thought of
as barriers at all. Conversely, on the unstable side of the line, activated
processes with barriers of order $k_B T$ or less are likely to be more effective
in initiating phase separation than the very slow modes that first become
nominally unstable at long wavelengths. It follows, therefore, that the
linear approximation is particularly inaccurate in the neighborhood of
the spinodal line and, accordingly, that the spinodal line itself is not well
defined.

5.3 The structure factor

The quantity that is measured in elastic scattering experiments is:

$$\hat{S}(\mathbf{k},t) \equiv \left\langle |\hat{u}(\mathbf{k},t)|^2 \right\rangle = \int \left\langle u(\mathbf{r},t) u(\mathbf{r}',t) \right\rangle e^{i\mathbf{k}\cdot(\mathbf{r}-\mathbf{r}')} \, d\mathbf{r} \, d\mathbf{r}', \qquad (5.5)$$

where the angular brackets denote a statistical average about which
we shall have much more to say in the following. $\hat{S}(\mathbf{k},t)$ is known as

the 'structure factor'. It is proportional to the scattering intensity at angle θ, where $k = (4\pi/\lambda)\sin(\frac{1}{2}\theta)$ and λ is the wavelength of the incident radiation.

According to the linearized Cahn-Hilliard equation (5.3),

$$\hat{S}(\mathbf{k}, t) \equiv \langle |\hat{u}(\mathbf{k}, t = 0)|^2 \rangle \exp[2\omega(\mathbf{k})t]. \tag{5.6}$$

The resulting prediction for $\hat{S}(\mathbf{k}, t)$ is shown schematically in Fig. 5.2.

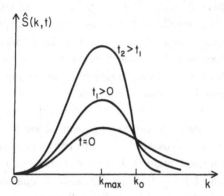

Figure 5.2: Structure factor according to the linearized Cahn-Hilliard theory for three successive times t.

According to Eq. (5.6), all modes with $k < k_0$, and especially those in the neighborhood of $k = k_{\max} \equiv k_0/\sqrt{2}$ where $\omega(k)$ is a maximum, make exponentially growing contributions to $\hat{S}(\mathbf{k}, t)$. On the other hand, small-wavelength fluctuations with $k > k_0$ are exponentially damped. Note that the structure factor must vanish at all times for $k = 0$ because concentration is conserved. (This last condition can be easily derived by setting $\mathbf{k} = 0$ in Eq. (5.5) and remembering that $\int u(\mathbf{r})\,d\mathbf{r} = 0$ by definition.)

The key prediction of the linearized Cahn-Hilliard theory is thus a peak located at $k_{\max}$ in the structure factor, whose height increases exponentially with time, and whose location is stationary. This prediction can be valid, at best, only for the very early stages of the spinodal process. In fact, we shall see that it usually fails even then. To construct a better theory, however, we shall need to include nonlinear effects and shall also need to account explicitly for thermal fluctuations. Before plunging into the mathematical development that will be necessary, it will be useful to consider the later stages of spinodal decomposition from a more general point of view.

5.4 Some general observations about the late stages of phase separation

At times sufficiently long after a quench, our spinodally decomposing system will exhibit a pattern of interspersed regions of coexisting α and β phases. For systems of the kind we have been discussing, this precipitation pattern is characterized by only two length scales. First, there is the range of correlations in the two phases, $\xi_s (s = \alpha, \beta)$, where $\xi_s^2 = K/(\partial^2 f/\partial c_s^2)$. These are time-independent, microscopic lengths, usually of the same order of magnitude, which — just as in Eq. (3.18) — also characterize the thickness of the interfaces which separate the α and β regions. Second, there is the time-dependent length $L(t)$ which characterizes the size of these regions. For symmetric quenches, with roughly equal volumes of α and β, $L(t)$ is the scale size of an interpenetrating network of the two phases. For asymmetric quenches, $L(t)$ might be the average size of isolated regions of the minority phase, which necessarily scales in the same way as the average separation between these regions. In either case, we expect $L(t)$ to be an increasing function of time; that is, the pattern 'coarsens'. (Note that the asymmetric situation could equally well emerge from nucleation as from spinodal decomposition. It is generally impossible to distinguish between the two mechanisms simply by looking at a late-stage precipitation pattern.)

Because each of these two coexisting phases is — very nearly — internally in thermodynamic equilibrium, and because we assume that we have quenched far enough that we are not probing critical fluctuations, we may use an Ornstein-Zernicke approximation to compute the structure factor for scattering from either one of these phases. That is, the linearized Cahn-Hilliard free energy has the form $\frac{1}{2} \sum_\mathbf{k} \epsilon_s(k) |\hat{u}_s(\mathbf{k})|^2$ where $\epsilon_s(k) = Kk^2 + \partial^2 f/\partial c_s^2$, $(s = \alpha, \beta)$; thus, by equipartition,

$$\hat{S}_s(k) \equiv \langle |\hat{u}_s(\mathbf{k})|^2 \rangle \cong \frac{k_B T}{\epsilon_s(k)} = \frac{k_B T}{K(k^2 + 1/\xi_s^2)}. \qquad (5.7)$$

In the fully phase-separated state, the scattering intensity will be the incoherent sum of the intensities of scattering from each of the phases, weighted by their respective volume fractions p_s:

$$\hat{S}(\mathbf{k}, t) \approx p_\alpha \hat{S}_\alpha(\mathbf{k}) + p_\beta \hat{S}_\beta(\mathbf{k}), \quad (kL \gg 1), \qquad (5.8)$$

where, because $p_\alpha c_\alpha + p_\beta c_\beta = c_0$,

$$p_\alpha = \frac{c_\beta - c_0}{c_\beta - c_\alpha} \; ; \quad p_\beta = 1 - p_\alpha = \frac{c_0 - c_\alpha}{c_\beta - c_\alpha}. \qquad (5.9)$$

This approximation makes sense, however, only for large wavevectors such that $kL(t) \gg 1$; the scattering at smaller wavevectors is sensitive to inhomogeneities on the scale of the precipitation pattern, $L(t)$.

If the scattering at small wavevectors is, in fact, dominated by only the single length $L(t)$, then we can write this part of the structure factor in a useful scaling form (Binder and Stauffer 1974, Marro *et al.* 1979, Gunton *et al.* 1983). To do this, we use the sum rule:

$$\int d k \, \hat{S}(k,t) = \langle u^2(\mathbf{r}) \rangle = \langle [c(\mathbf{r}) - c_0]^2 \rangle \tag{5.10}$$

$$\cong p_\alpha (c_\alpha - c_0)^2 + p_\beta (c_\beta - c_0)^2 = (c_\beta - c_0)(c_0 - c_\alpha).$$

So long as we are not too close to the critical point, the contribution to the integral over $\mathbf{k}$ in (5.10) from the large-k region in (5.8) will be small; the short-wavelength equilibrium fluctuations are relatively weak. (Remember that the maximum value of k in the integral in (5.10) is the inverse coarse-graining length or, at the most, the inverse lattice spacing.) Therefore, we may expect that the dominant contribution to (5.10) will come from a peak in the neighborhood of $k = L^{-1}(t)$. That is,

$$\hat{S}(k,t) \approx L^d(t) \mathcal{F}[kL(t)] \times \text{Const}, \quad kL(t) \lesssim 1, \tag{5.11}$$

where the constant is of order $(c_\beta - c_0)(c_0 - c_\alpha)$ and $\mathcal{F}$ is some time-independent function of its argument. The overall structure factor predicted by (5.8) and (5.11) is shown schematically in Fig. 5.3. The general features illustrated here, as well as the scaling law (5.11), have been observed in both numerical simulations (Amar *et al.* 1988) and laboratory experiments (Gaulin *et al.* 1987).

6

Fluctuation theory

6.1 Introduction

We have seen in the preceding section that we must take into account thermal fluctuations in order to characterize spinodal decomposition in a satisfactory way. Clearly we cannot do this within the framework of equilibrium statistical physics; the processes that we want to describe take

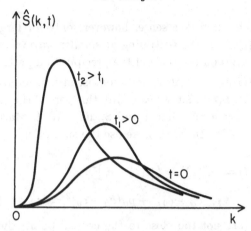

Figure 5.3: Schematic structure factor, at three successive times,
as deduced from general considerations discussed in Section 5.4.
Note that the growing peak shifts toward small wavenumbers k
and becomes narrower at later times. Note also that $\widehat{S}$ approaches
a nonzero limiting form for large k.

place well away from thermodynamic equilibrium. Our approach will
continue to be more phenomenological than rigorous; in fact, nonequilibrium statistical mechanics is generally less well developed mathematically than the equilibrium theory. Accordingly, this section provides a
nonrigorous conceptual introduction to the Langevin and related Fokker-
Planck approaches to the theory of nonequilibrium phenomena. (For a
general reference, see van Kampen 1984.)

6.2 The Langevin approach: basic concepts

Rather than working directly in the continuum representation, it will be
simplest for present purposes to assume that our system can be described
by a denumerable set of variables. Let the symbol ϕ_α denote a fluctuating
variable that describes the state of the system at a coarse-grained site α.

(In this section, we shall use lower case Greek subscripts to identify
state variables. We do this in order to emphasize that these indices may
be interpreted more generally than the lattice site indices i, j, that we
used in Section 1. The reader should not confuse these subscripts with
those used in Sections 4 and 5 to denote thermodynamic phases.) For
example, ϕ_α could be the magnetization per site in an Ising ferromagnet
or the local concentration in a binary alloy. The ϕ_α could also be

components of vectors; in more fully dynamical situations, some of them could be coordinates and others momenta.

The Langevin form of the equation of motion for ϕ_α is

$$\frac{\mathrm{d}\phi_\alpha}{\mathrm{d}t} = U_\alpha\{\phi\} + \eta_\alpha(t), \qquad (6.1)$$

where U_α denotes the deterministic part of the motion and η_α is the noise, that is, the stochastic 'Langevin force.' For systems of the kind we have been discussing, U_α might have the form

$$U_\alpha = -\sum_\beta \Gamma_{\alpha\beta} \frac{\partial F}{\partial \phi_\beta}, \qquad (6.2)$$

where $F\{\phi\}$ is a coarse-grained free energy and $\Gamma_{\alpha\beta}$ is a matrix of transport coefficients. In a nonconserving situation, the choice $\Gamma_{\alpha\beta} = \Gamma\delta_{\alpha\beta}$ produces the discrete version of (3.13). For the conserving case, $\Gamma_{\alpha\beta} = MD_{\alpha\beta}$, where $D_{\alpha\beta}$ is the discrete representation of $-\nabla^2$, produces (4.5). However, Eq. (6.2) is not necessary in general. One does not always need to assume that the flow described by U_α is derivable from a potential, although that will be the case for the thermodynamic systems of interest to us here.

The noise $\eta_\alpha(t)$, in our context, is supposed to mimic the thermally fluctuating environment in which the degrees of freedom of primary interest — the magnetic moments, the local concentrations, etc. — are immersed. Its physical origin might be the lattice vibrations in a solid or the analogous thermal fluctuations of the molecules in a liquid — whatever it is that constitutes a 'heat bath' and determines the temperature of the system as a whole. Accordingly, there are two conditions that we impose on the $\eta_\alpha(t)$. First, their characteristic time scale must be very fast compared to the motions of the $\phi_\alpha(t)$. When observed over times in which the $\phi_\alpha(t)$ are changing appreciably, the $\eta_\alpha(t)$ must look like white noise. Second, the $\eta_\alpha(t)$ must be such that they drive the system toward thermodynamic equilibrium.

Neither of these conditions is absolutely required by the Langevin equation itself. The noise could be correlated over long time scales and could be characteristic, say, of loud music being played outside the laboratory instead of intrinsic thermal fluctuations in the experimental sample. For systems undergoing first-order phase transformations, however, it is intuitively appealing — if not yet mathematically rigorous —

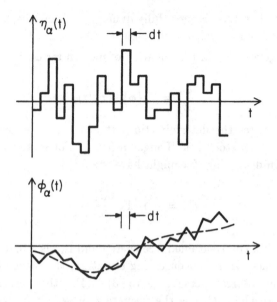

Figure 6.1: A characteristic noise function $\eta_\alpha(t)$ (upper graph) and a corresponding behavior of the variable $\phi_\alpha(t)$ (lower graph). The dashed curve in the lower graph represents the slow, deterministic part of $\phi_\alpha(t)$.

to think of the systems as being separated distinctly into slowly moving modes that describe the properties of direct interest, and rapidly fluctuating noise.

The situation described by (6.1) is shown schematically in Fig. 6.1. The function $\eta_\alpha(t)$ is represented by a sequence of uncorrelated square pulses of magnitudes $\eta_{\alpha,n}$ at times $t_n = n\,dt$, where dt is the vanishingly small correlation time for the noise. The function $\phi_\alpha(t)$ exhibits noisy fluctuations on the scale dt superimposed upon its slow deterministic motion. Consider the noise-induced change in ϕ_α in a time $t = N\,dt$, where dt is small and $N = t/dt$ is large:

$$\delta\phi_\alpha(t) \equiv \phi_\alpha(t) - \phi_\alpha(0) = \sum_{n=1}^{N} \eta_{\alpha,n}\,dt. \qquad (6.3)$$

Because the η's are uncorrelated, $\delta\phi_\alpha$ performs a random walk, with

$$\left\langle \delta\phi_\alpha^2(t) \right\rangle_W = \left\langle \eta_\alpha^2 \right\rangle_W N\,dt^2. \qquad (6.4)$$

Here $\langle\ \rangle_W$ denotes an average over the ensemble of functions $\eta_\alpha(t)$. The subscript W is the statistical weight, which we shall define more precisely later, for functions in this ensemble. In order for $\langle \delta\phi_\alpha^2 \rangle_W$ to be finite and nonzero in the limit $dt \to 0$, we must have $\langle \eta_\alpha^2 \rangle_W = \gamma/dt$, where γ is some constant.

More generally, in this limit, we can write:

$$\langle \eta_\alpha(t) \rangle_W = 0 \ ; \quad \langle \eta_\alpha(t)\eta_\beta(t') \rangle_W = \gamma_{\alpha\beta}\delta(t - t'), \qquad (6.5)$$

where $\gamma_{\alpha\beta}$ is the equal-time correlation function for the η's, and the delta function is the limiting form of the square pulse of magnitude dt^{-1}. Equation (6.5) completes as much of a definition as we shall need for the terms in the Langevin equation (6.1). It is important to recognize, however, that (6.5) defines a very special kind of noise.

6.3 The Langevin approach: an illustrative example

To see how the Langevin equation works in practice, consider a one-dimensional Brownian particle of mass m, moving at speed $v(t)$, in a stationary fluid of light molecules. Collisions with the molecules of the fluid produce both resistance to the motion of the particle and fluctuations in its trajectory. On average, the particle experiences a decelerating force $-\alpha v$, where α is a friction coefficient. The Langevin equation is therefore

$$m\frac{dv}{dt} = -\alpha v + \eta(t), \qquad (6.6)$$

where

$$\langle \eta(t)\eta(t') \rangle_W = \gamma\delta(t - t'). \qquad (6.7)$$

The second condition mentioned above, that $\eta(t)$ must be chosen so that the system approaches thermodynamic equilibrium, can now be used to determine a relationship between γ, α and the temperature T.

The solution of (6.6), if we assume that any initial transient has died out, is

$$v(t) = \frac{1}{m}\int_{-\infty}^{t} \exp\left[-\frac{\alpha}{m}(t - t')\right]\eta(t')\,dt'. \qquad (6.8)$$

Thus

$$\langle v^2(t) \rangle_W = \frac{1}{m^2}\int_{-\infty}^{t} dt' \int_{-\infty}^{t} dt'' \exp\left[-\frac{\alpha}{m}(2t - t' - t'')\right] \langle \eta(t')\eta(t'') \rangle_W$$

$$= \frac{\gamma}{2m\alpha}. \qquad (6.9)$$

In thermodynamic equilibrium, we must have $\langle v^2 \rangle = k_B T/m$, therefore:

$$\gamma = 2\alpha k_B T. \tag{6.10}$$

This is the simplest form of a fluctuation-dissipation theorem; it relates the fluctuation strength γ to a dissipation coefficient α. We shall see a more general form of this relationship in the following paragraphs.

6.4 The Fokker-Planck equation

Our next step is to start from the Langevin equation (6.1) and derive an equation of motion for the probability distribution over the space of configurations $\{\phi\}$. We denote this distribution by $\rho(\{\phi\}, t)$, and let $T_{dt}(\{\phi\} \leftarrow \{\phi'\})$ be the probability for going from a configuration $\{\phi'\}$ at time t to a configuration $\{\phi\}$ at later time $t + dt$. The time evolution of ρ is therefore determined by an equation of the form:

$$\rho(\{\phi\}, t + dt) = \int \mathcal{D}\{\phi'\} T_{dt}(\{\phi\} \leftarrow \{\phi'\}) \rho(\{\phi'\}, t'), \tag{6.11}$$

where

$$\int \mathcal{D}\{\phi'\} \cdots \equiv \prod_\alpha \int d\phi'_\alpha \cdots \tag{6.12}$$

denotes the integration over configuration space.

For any given set of noise functions $\eta_\alpha(t)$, the motion is purely deterministic. Thus, from (6.1), the probability for going from $\{\phi\}$ at time t to $\{\phi'\}$ at $t' = t + dt$ is

$$\prod_\alpha \delta[\phi_\alpha - \phi'_\alpha - U_\alpha\{\phi'\} dt - \eta_\alpha(t') dt].$$

We now use in an essential way our assumption that the $\eta_\alpha(t)$ are uncorrelated over time intervals of order dt. With this assumption, we can write:

$$T_{dt}(\{\phi\} \leftarrow \{\phi'\}) = \left\langle \prod_\alpha \delta\left[\phi_\alpha - \phi'_\alpha - U_\alpha\{\phi'\} dt - \eta_\alpha(t') dt\right] \right\rangle_W. \tag{6.13}$$

It is now useful — although still not absolutely necessary — to be specific about our choice of the weight function $W\{\eta(t)\}$. To the order of accuracy that is needed here, we lose no generality by choosing W to be a Gaussian distribution. At the nth time interval, $t = n\,dt$, we can write

$$W\{\eta\} = \overline{W} \exp\left[-\tfrac{1}{2}\,dt \sum_{\alpha,\beta} \gamma_{\alpha\beta}^{-1} \eta_\alpha \eta_\beta\right], \qquad (6.14)$$

where $\gamma_{\alpha\beta}^{-1}$ is the inverse of the correlation matrix $\gamma_{\alpha\beta}$ defined in (6.5) and $\overline{W}$ is a normalization factor that is easily found to be

$$\overline{W} = \left[\det\left(\frac{dt}{2\pi}\gamma^{-1}\right)\right]^{1/2}. \qquad (6.15)$$

The entire distribution for $N = t/dt$ time intervals is, in the limit $dt \to 0$,

$$W^{(t)}\{\eta\} = \overline{W}^N \exp\left[-\tfrac{1}{2} \sum_{\alpha\beta} \int_0^t dt'\, \gamma_{\alpha\beta}^{-1} \eta_\alpha(t')\eta_\beta(t')\right]. \qquad (6.16)$$

Note that, in accord with our expectation, (6.14) implies $\langle \eta^2 \rangle \sim \gamma/dt$; the fluctuations become larger for smaller dt. Also note that the normalization of $W\{\eta\}$ assures the required normalization of T_{dt}:

$$\int \mathcal{D}\{\phi\} T_{dt}(\{\phi\} \leftarrow \{\phi'\}) = 1. \qquad (6.17)$$

The equation of motion for $\rho(\{\phi\}, t)$, Eq. (6.11), is now

$$\rho(\{\phi\}, t + dt) \qquad (6.18)$$

$$= \int \mathcal{D}\{\phi'\} \left\langle \prod_\alpha \delta(\phi_\alpha - \phi'_\alpha - U_\alpha\{\phi'\}\,dt - \eta_\alpha\,dt) \right\rangle_W \rho(\{\phi'\}, t).$$

We want to turn this into a differential equation for ρ. Formally, because η_α is of order $dt^{-1/2}$, the change in ϕ_α, $\delta\phi_\alpha \equiv \phi'_\alpha - \phi_\alpha$, is of order $dt^{1/2}$. If we want to keep all terms of order dt on the right-hand side of (6.18), we must keep terms in $\delta\phi_\alpha$ of orders up to and including the second. Accordingly, we write

$$\rho(\{\phi'\}, t) = \rho(\{\phi\}, t) + \sum_\beta \frac{\partial \rho}{\partial \phi_\beta} \delta\phi_\beta \qquad (6.19)$$

$$+ \tfrac{1}{2} \sum_{\beta\beta'} \frac{\partial^2 \rho}{\partial \phi_\beta \partial \phi_{\beta'}} \delta\phi_\beta \delta\phi_{\beta'} + \cdots.$$

At zeroth order in (6.19), the contribution to the right-hand side of (6.18) is

$$\int \mathcal{D}\{\phi'\}\Big\langle \prod_\alpha \delta(\phi_\alpha - \phi'_\alpha - U_\alpha\{\phi'\}\,dt - \eta_\alpha\,dt)\Big\rangle_W \rho(\{\phi\}, t) \qquad (6.20)$$

$$\cong \prod_\alpha \Big(\frac{1}{1 + \partial U_\alpha/\partial\phi_\alpha\,dt}\Big)\rho(\{\phi\}, t) \cong \Big[1 - \sum_\alpha \frac{dU_\alpha}{d\phi_\alpha}\,dt\Big]\rho(\{\phi\}, t).$$

The first-order contribution is:

$$\int \mathcal{D}\{\phi'\}\Big\langle \prod_\alpha \delta(\phi_\alpha - \phi'_\alpha - U_\alpha\{\phi'\}\,dt - \eta_\alpha\,dt)\Big\rangle_W \sum_\beta \frac{\partial\rho}{\partial\phi_\beta}\delta\phi_\beta$$

$$\cong -\sum_\beta \frac{\partial\rho}{\partial\phi_\beta}(U_\beta + \langle\eta_\beta\rangle w)\,dt = -\sum_\beta \frac{\partial\rho}{\partial\phi_\beta}U_\beta\,dt; \qquad (6.21)$$

and, at second order, we find:

$$\tfrac{1}{2}\int \mathcal{D}\{\phi'\}\Big\langle \prod_\alpha \delta(\phi_\alpha - \phi'_\alpha - U_\alpha\{\phi'\}\,dt - \eta_\alpha\,dt)\Big\rangle \sum_{\beta,\beta'} \frac{\partial^2\rho}{\partial\phi_\beta\partial\phi_{\beta'}}\delta\phi_\beta\delta\phi_{\beta'}$$

$$\cong \tfrac{1}{2}\sum_{\beta,\beta'} \frac{\partial^2\rho}{\partial\phi_\beta\partial\phi_{\beta'}}\langle\eta_\beta\eta_{\beta'}\rangle w\,dt^2 = \tfrac{1}{2}\sum_{\beta\beta'}\gamma_{\beta\beta'}\frac{\partial^2\rho}{\partial\phi_\beta\partial\phi_{\beta'}}\,dt. \qquad (6.22)$$

Finally, expanding the left-hand side of (6.18) in dt and collecting terms of first order on both sides, we obtain the desired differential equation:

$$\frac{\partial\rho}{\partial t} = \sum_\alpha \frac{\partial}{\partial\phi_\alpha}\Big(-U_\alpha\rho + \tfrac{1}{2}\sum_\beta \gamma_{\alpha\beta}\frac{\partial\rho}{\partial\phi_\beta}\Big). \qquad (6.23)$$

This is the general form of the Fokker-Planck equation. Note immediately that (6.23) can be written as a continuity equation in ϕ-space:

$$\frac{\partial\rho}{\partial t} = -\sum_\alpha \frac{\partial J_\alpha}{\partial\phi_\alpha}, \qquad (6.24)$$

where

$$J_\alpha = U_\alpha\rho - \tfrac{1}{2}\sum_\beta \gamma_{\alpha\beta}\frac{\partial\rho}{\partial\phi_\beta} \qquad (6.25)$$

is a conserved probability flux. The ϕ-space vector U_α is the drift velocity, and $\tfrac{1}{2}\gamma_{\alpha\beta}$ is a matrix of ϕ-space diffusion coefficients.

Just as in the case of the Brownian particle described in Section 6.3, the matrix $\gamma_{\alpha\beta}$ may be determined by insisting that the stationary solution of (6.24) be the thermodynamic equilibrium distribution — assuming, of course, that the system being described does indeed move toward thermal equilibrium and that the $\eta_\alpha(t)$ are intended to represent only thermal noise. The simplest situation is that in which U_α is given by (6.2) and the function $F\{\phi\}$ shown there is the same free energy that appears in the equilibrium distribution:

$$\rho_{\text{eq}}\{\phi\} = \frac{1}{Z}\exp\left[-F\{\phi\}/k_B T\right]. \tag{6.26}$$

Then the choice

$$\gamma_{\alpha\beta} = 2k_B T \Gamma_{\alpha\beta} \tag{6.27}$$

is sufficient to ensure that $J = 0$ for $\rho = \rho_{\text{eq}}$.

Equation (6.27) is a fluctuation-dissipation theorem directly analogous to (6.10); it relates the fluctuation strength $\gamma_{\alpha\beta}$ to the transport coefficients $\Gamma_{\alpha\beta}$ and the temperature T. Note that (6.27) is not a necessary condition.

Two final formal observations are useful. First, use (6.24) and (6.25) to write an equation of motion for $\langle\phi_\alpha\rangle_\rho$, where $\langle\ \rangle_\rho$ denotes an average with respect to the time-dependent distribution function ρ. We find:

$$\frac{\partial}{\partial t}\langle\phi_\alpha\rangle_\rho = \int \mathcal{D}\{\phi\}\phi_\alpha\frac{\partial\rho}{\partial t} = -\int \mathcal{D}\{\phi\}\phi_\alpha\sum_{\alpha'}\frac{\partial J_{\alpha'}}{\partial\phi_{\alpha'}}. \tag{6.28}$$

Assuming that the integrand is localized in ϕ-space, i.e. that there are no sources or sinks of probability flux at infinity, we can integrate by parts to obtain

$$\frac{\partial}{\partial t}\langle\phi_\alpha\rangle_\rho = \int \mathcal{D}\{\phi\}J_\alpha = -\sum_\beta\left\langle\Gamma_{\alpha\beta}\frac{\partial F}{\partial\phi_\beta}\right\rangle_\rho. \tag{6.29}$$

So far no approximation has been made. We have recovered a statistical average of the deterministic equation (6.1) with no explicit Langevin noise — not a directly useful result because we do not know ρ. The result is reassuring, however, because we expect the approximation

$$\left\langle\Gamma_{\alpha\beta}\frac{\partial F}{\partial\phi_\beta}\right\rangle_\rho \approx \Gamma_{\alpha\beta}\frac{\partial F\{\langle\phi\rangle\}}{\partial\langle\phi_\beta\rangle} \tag{6.30}$$

to be valid when the fluctuations are very small; and this is indeed just the deterministic equation of motion.

A more practically useful result along these same lines is obtained by using (6.24) to derive an equation of motion for the equal-time correlation function $\langle \phi_\alpha \phi_\beta \rangle_\rho$:

$$\frac{\partial}{\partial t}\langle \phi_\alpha \phi_\beta \rangle_\rho = \int \mathcal{D}\{\phi\}\,\phi_\alpha \phi_\beta \frac{\partial \rho}{\partial t} = -\int \mathcal{D}\{\phi\}\,\phi_\alpha \phi_\beta \sum_\gamma \frac{\partial J_\gamma}{\partial \phi_\gamma}. \qquad (6.31)$$

Integrating by parts again, we find

$$\frac{\partial}{\partial t}\langle \phi_\alpha \phi_\beta \rangle_\rho = \int \mathcal{D}\{\phi\}(\phi_\alpha J_\beta + \phi_\beta J_\alpha) \qquad (6.32)$$

$$= -\sum_\gamma \left\langle (\phi_\alpha \Gamma_{\beta\gamma} + \phi_\beta \Gamma_{\alpha\gamma})\frac{\partial F}{\partial \phi_\gamma} \right\rangle_\rho + 2k_B T \Gamma_{\alpha\beta}.$$

In the last part of (6.32), we assume that $\Gamma_{\alpha\beta}$ is a symmetric matrix and that it is independent of the ϕ's. If F is a quadratic form in the ϕ's, then (6.32) is a closed equation for $\langle \phi_\alpha \phi_\beta \rangle$ — only quadratic terms appear on the right-hand side. Otherwise, (6.32) can be used as the beginning of a hierarchy of coupled equations for higher-order correlation functions. We shall return to this equation in the following sections.

7

Spinodal decomposition: fluctuations and nonlinear effects

7.1 Introduction

Equipped now with a mathematical method for dealing with thermal fluctuations in systems that are not in thermodynamic equilibrium, we return to Section 5 and take up again the discussion of spinodal decomposition. We start by using (6.32), a formally exact equation for the equal-time correlation function, to obtain an equation of motion for the structure factor $\hat{S}(k, t)$ defined in Eq. (5.5). We then look at a linear approximation for this equation, check its limits of validity — thereby obtaining a 'Ginzburg criterion' for spinodal decomposition — and, finally, look at some not altogether successful attempts at computing nonlinear corrections.

7.2 Equation of motion for the structure factor

To make contact between the formalism used in Section 6 and the continuum representation of Section 5, we identify the fluctuating variables φ_α with the deviations from the average concentration c_0,

$$\varphi_\alpha(t) \leftrightarrow u(\mathbf{r}, t) = c(\mathbf{r}, t) - c_0. \tag{7.1}$$

The subscript α denotes the coarse-grained position $\mathbf{r}$. The rest of the table of correspondences is as follows:

$$F\{\phi\} \leftrightarrow F\{u\} = \int d\mathbf{r}\left[\tfrac{1}{2}K(\nabla u)^2 + f(c_0 + u)\right]; \tag{7.2}$$

$$\Gamma_{\alpha\beta} \leftrightarrow -M\nabla^2; \tag{7.3}$$

$$J_\alpha \leftrightarrow J(\mathbf{r}, t) = -M\nabla^2\left[\frac{\delta F}{\delta u(\mathbf{r})}\rho + k_B T\frac{\delta\rho}{\delta u(\mathbf{r})}\right]. \tag{7.4}$$

In the continuum representation, the correlation function for the noise (Eqs. (6.5) and (6.27)) becomes

$$\begin{aligned}
\langle\eta_\alpha(t)\eta_\beta(t')\rangle_W &\leftrightarrow \langle\eta(\mathbf{r}, t)\eta(\mathbf{r}', t')\rangle_W \\
&= -2k_B T M\nabla^2\delta(\mathbf{r} - \mathbf{r}')\delta(t - t').
\end{aligned} \tag{7.5}$$

In order to evaluate the probability flux in (7.4), we need

$$\frac{\delta F}{\delta u(\mathbf{r})} = -K\nabla^2 u + \frac{\partial f(c_0 + u)}{\partial u} \tag{7.6}$$

$$= -K\nabla^2 u + \frac{\partial f}{\partial c_0} + \frac{\partial^2 f}{\partial c_0^2}u + \cdots + \frac{1}{(n-1)!}\frac{\partial^n f}{\partial c_0^n}u^{n-1} + \cdots.$$

The structure factor is the spatial Fourier transform of the two-point, equal-time correlation function:

$$\langle\varphi_\alpha(t)\varphi_\beta(t)\rangle_\rho \leftrightarrow \langle u(\mathbf{r}, t)u(\mathbf{r}', t)\rangle_\rho \equiv S(|\mathbf{r} - \mathbf{r}'|, t); \tag{7.7}$$

and

$$\hat{S}(\mathbf{k}, t) = \int d\mathbf{r}\, S(r, t)e^{-i\mathbf{k}\cdot\mathbf{r}} = \langle\hat{u}(-\mathbf{k}, t)\hat{u}(\mathbf{k}, t)\rangle_\rho, \tag{7.8}$$

where the Fourier amplitudes $\hat{u}(k, t)$ were defined in (5.2). The equation of motion for $\hat{S}(\mathbf{k}, t)$ is therefore obtained from (6.32), which we can write in the form

$$\frac{\partial\hat{S}(\mathbf{k}, t)}{\partial t} = -2Mk^2\left\langle\hat{u}(-\mathbf{k}, t)\frac{\delta F}{\delta\hat{u}(-\mathbf{k}, t)}\right\rangle_\rho + 2k_B T M k^2. \tag{7.9}$$

So far, no approximation has been made.

To make further progress, the first thing to do is to truncate the
expansion for $\delta F/\delta u$ in (7.6) at the term linear in u. We obtain:

$$\frac{\partial \hat{S}(\mathbf{k},t)}{\partial t} \cong 2\omega(k)\hat{S}(\mathbf{k},t) + 2k_B T M k^2, \qquad (7.10)$$

where

$$\omega(k) = -Mk^2\left(Kk^2 + \frac{\partial^2 f}{\partial c_0^2}\right), \qquad (7.11)$$

as in (5.4). This is the situation anticipated at the end of the preceding section.

The linear approximation for $\delta F/\delta u$, that is the quadratic approximation for F, produces a closed equation of motion for the two-point
correlation function rather than an infinite hierarchy.

The solution of (7.10) is:

$$\hat{S}(\mathbf{k},t) \cong \hat{S}(\mathbf{k},0)\exp[2\omega(k)t] - \frac{Mk_B T k^2}{\omega(k)}[1 - \exp(2\omega(k)t)]. \quad (7.12)$$

This result was first obtained by H. Cook (1970). If the system is outside
the spinodal region, $\partial^2 f/\partial c_0^2 > 0$, (7.12) converges to:

$$\hat{S}(\mathbf{k},t) \underset{t\to\infty}{\longrightarrow} \frac{k_B T \chi_0}{1 + k^2\xi^2}, \qquad (7.13)$$

where the susceptibility χ_0 is defined by

$$\chi_0^{-1} = \frac{\partial^2 f}{\partial c_0^2} = \frac{\partial \mu}{\partial c_0}, \qquad (7.14)$$

and ξ is the correlation length; $\xi^2 = K\chi_0$. On the other hand, if
the system is inside the spinodal, $\partial^2 f/\partial c_0^2 < 0$, $\omega(k)$ is positive for
$k < k_0 = \left[-(\partial^2 f/\partial c_0^2)/K\right]^{1/2}$, and $\hat{S}$ increases exponentially without
bound for $k < k_0$, much as described in Section 5, with a peak remaining
at $k = k_{\max} = k_0/\sqrt{2}$. At $k = k_0$, where $\omega(k) = 0$, $\hat{S}$ increases linearly
in t. For $k > k_0$, where $\omega(k) < 0$,

$$\hat{S}(\mathbf{k},t) \underset{t\to\infty}{\longrightarrow} \frac{k_B T}{K(k^2 - k_0^2)}. \qquad (7.15)$$

While these results are marginally better than (5.6), which predicts
that $\hat{S}$ vanishes at late times for $k > k_0$, they remain qualitatively
inconsistent with our physical expectations developed in Section 5.4.
The approximate structure factor in (7.12) does not exhibit coarsening,
nor does it approach the expected equilibrium Ornstein-Zernicke function (5.7) at large wavenumbers. To improve this situation, we shall have
to include the nonlinear terms so far neglected in the expansion (7.6).

7.3 The onset of nonlinear effects

If we maintain our apparently reasonable assumption that the coarse-grained free energy $f(c)$ is a smoothly varying function of c as shown in Fig. 4.1, with minima near c_α and c_β and not much structure otherwise, then the only important scale of concentration is $\Delta c \equiv c_\beta - c_\alpha$. The criterion for linearity of the equation of motion or, equivalently, validity of the quadratic approximation for f, must therefore have the form $\langle u^2 \rangle \ll (\Delta c)^2$ where $\langle u^2 \rangle$ is the mean-square fluctuation of the field u. More precisely, in order to be consistent with our correspondence (7.1) between $u(\mathbf{r}, t)$ and the fluctuating variable $\varphi_\alpha(t)$, we must understand $u(\mathbf{r}, t)$ to be the average value of the field in a coarse-graining volume $\ell^d \sim k_{\max}^{-d}$. Thus,

$$\langle u^2(\mathbf{r}, t) \rangle \cong \int d\mathbf{k} \langle \hat{u}(\mathbf{k}, t) \hat{u}(-\mathbf{k}, t) \rangle e^{i\mathbf{k} \cdot \mathbf{r}} \sim k_{\max}^d \hat{S}(k_{\max}, t). \quad (7.16)$$

Note that $\langle u^2 \rangle$ is necessarily sensitive to the presumably self-consistent choice of the coarse-graining length ℓ or, equivalently, to the upper cutoff in the integration over $\mathbf{k}$ in (7.16). It follows that a first condition for validity of the linear theory described in the previous paragraphs is:

$$k_{\max}^d \hat{S}(k_{\max}, t) \ll (\Delta c)^2. \quad (7.17)$$

To estimate $\hat{S}(k_{\max}, t)$ in (7.17), note that the interesting behavior occurs when the first term on the right-hand side of (7.10) is much greater than the second, that is, when

$$\hat{S}(k_{\max}, t) \gg \frac{k_B T M k_{\max}^2}{|\omega(k_{\max})|}, \quad (7.18)$$

so that the intrinsic dynamics of the system is more important than the extrinsic thermal fluctuations. The inequality (7.18) is the condition for validity of the exponential growth of correlations as in (5.6) or, equivalently, the dominance of the first term on the right-hand side of (7.12).

Combining (7.17) and (7.18), and using (7.11) to evaluate $\omega(k_{\max})$, we find

$$\frac{k_B T k_{\max}^{d-2}}{K(\Delta c)^2} = \frac{2 k_B T k_{\max}^d}{|\partial^2 f/\partial c_0^2|(\Delta c)^2} \ll 1, \quad (7.19)$$

a condition that was first discussed by Binder (1983, 1984, 1990).

It is interesting to examine (7.19) in two limiting cases. First, suppose that we are near a critical point, $T \cong T_c$. In this case, the only relevant length scale is the correlation length $\xi \sim k_{max}^{-1}$. The quantity $K(\Delta c)^2/\xi$ is essentially (in a scaling sense) the surface tension σ, as can be seen by applying the scaling argument to the expression for σ in (4.10). Thus (7.19) becomes

$$\frac{k_B T_c}{\sigma \xi^{d-1}} \ll 1. \tag{7.20}$$

But the usual critical-point scaling arguments (as well as experimental data, see Moldover 1985) tell us that $\sigma \xi^{d-1}/k_B T_c$ should be a universal constant of order unity, at least for $d < 4$. Thus (7.20) can never be satisfied for near-critical quenches; the nonlinear behavior is always relevant. The thermal fluctuations immediately drive the system out of the region where the linearization is valid, so that the exponential growth of correlations predicted by the linearized Cahn-Hilliard theory cannot be observed.

The second interesting limit occurs when the system is quenched to a state so far from the critical point that the mean-field estimates derived in Section 3 are accurate. From (3.8) and (3.12), we have:

$$K \cong k_B T_c a^d \xi_0^2 ; \tag{7.21}$$

$$\left| \frac{\partial^2 f}{\partial c_0^2} \right| \cong a^d k_B (T_c - T) ; \tag{7.22}$$

and
$$(\Delta c)^2 \cong \frac{3}{a^{2d}} \left(\frac{T_c}{T} - 1 \right). \tag{7.23}$$

Here, a is the microscopic lattice spacing and ξ_0 is the range of interactions. Now (7.18) becomes

$$\left(\frac{a}{\xi_0} \right)^d \left(\frac{T}{T_c} \right)^2 \left(1 - \frac{T}{T_c} \right)^{d/2-2} \ll 1. \tag{7.24}$$

For $d < 4$, this inequality can be satisfied either for sufficiently deep quenches, $T \ll T_c$, or for sufficiently long-ranged interactions, $\xi_0 \gg a$. Equation (7.24) is the same as the Ginzburg criterion for the validity of the mean-field approximation in equilibrium statistical mechanics (Ma, 1976). Note that, for dimensions $d > 4$, the mean-field theory remains valid all the way in to the critical point, $T \to T_c$.

It is important to recognize that (7.24) is a necessary but not a sufficient criterion for validity of the linearized theory. If (7.24) is satisfied, then there may exist some finite time interval after a quench during which exponential growth of the kind predicted by Cahn and Hilliard can be observed. That time interval can be estimated by using the exponential approximation (5.6) or (7.12) in the left-hand side of (7.17). See Binder (1983, 1984, 1990). However, unbounded growth of correlations obviously is inconsistent with the ultimate formation of coexisting stable phases; thus, the inequality (7.17) must eventually be violated, and the nonlinear behavior described qualitatively in Section 5.4 must become dominant.

7.4 Nonlinear approximations

The full equation of motion for the structure factor, (7.9), can be written in the following form with use of the expansion (7.6):

$$\frac{\partial \hat{S}(\mathbf{k}, t)}{\partial t} = -2Mk^2 \left[\left(Kk^2 + \frac{\partial^2 f}{\partial c_0^2} \right) \hat{S}(\mathbf{k}, t) \right. \tag{7.25}$$

$$\left. + \sum_{n=3}^{\infty} \frac{1}{(n-1)!} \frac{\partial^n f}{\partial c_0^n} \hat{S}_n(\mathbf{k}, t) \right] + 2Mk^2 k_B T,$$

where $\hat{S}_n(k, t)$ is the Fourier transform of the n-th order, two-point correlation function:

$$S_n(\mathbf{r} - \mathbf{r}', t) \equiv \left\langle u^{n-1}(\mathbf{r}, t) u(\mathbf{r}', t) \right\rangle_\rho. \tag{7.26}$$

A complete solution of the problem would require writing and solving equations of motion for each of the $\hat{S}_n$; and these equations, in turn, would involve higher-order, multi-point correlation functions of the form $\left\langle u^n(\mathbf{r}, t) u^m(\mathbf{r}', t) u^\ell(\mathbf{r}'', t) \right\rangle_\rho$, etc., where n, m, ℓ are integers. Obviously, some truncation is necessary.

The simplest such truncation is equivalent to a random-phase approximation (RPA). For simplicity, let us specialize to the symmetric case where only terms with even values of n survive in (7.25). Then the first nonlinear term is $\hat{S}_4$, which we write in the form

$$S_4 \left(|\mathbf{r} - \mathbf{r}'|, t \right) = \left\langle u^3(\mathbf{r}, t) u(\mathbf{r}', t) \right\rangle_\rho \tag{7.27}$$

$$\cong 3 \left\langle u^2(\mathbf{r}, t) \right\rangle_\rho \left\langle u(\mathbf{r}, t) u(\mathbf{r}', t) \right\rangle_\rho$$

or, equivalently,

$$\hat{S}_4(\mathbf{k}, t) \cong 3 \left\langle u^2(t) \right\rangle_\rho \hat{S}(\mathbf{k}, t), \tag{7.28}$$

with

$$\left\langle u^2(t) \right\rangle_\rho = \frac{1}{(2\pi)^d} \int d\mathbf{k} \, \hat{S}(\mathbf{k}, t). \tag{7.29}$$

(As before, $\langle u^2 \rangle$ must be self-consistently cutoff-dependent.) The RPA decoupling in (7.27) is equivalent to an assumption that $\rho\{u\}$ is a simple Gaussian distribution in the u's, centered at $u = 0$. This is not a very plausible assumption for a situation in which the characteristic values of u are expected to be moving away from zero toward $c_\alpha - c_0$ and $c_\beta - c_0$, but it is a useful starting point.

With these approximations, (7.25) becomes:

$$\frac{\partial \hat{S}(\mathbf{k}, t)}{\partial t} \cong -2Mk^2 \left[Kk^2 + \frac{\partial^2 f}{\partial c_0^2} + \frac{1}{2} \frac{\partial^4 f}{\partial c_0^4} \langle u^2(t) \rangle_\rho \right] \hat{S}(\mathbf{k}, t)$$
$$+ 2Mk_B T k^2, \tag{7.30}$$

which we can rewrite in the form:

$$\frac{\partial \hat{S}(\mathbf{k}, t)}{\partial t} \cong -2Mk^2 K \left[k^2 - k_0^2(t) \right] \hat{S}(\mathbf{k}, t) + 2Mk_B T k^2, \tag{7.31}$$

where
$$k_0^2(t) = \frac{1}{K} \left[\left| \frac{\partial^2 f}{\partial c_0^2} \right| - \frac{1}{2} \frac{\partial^4 f}{\partial c_0^4} \langle u^2(t) \rangle_\rho \right]. \tag{7.32}$$

With (7.29), Eqs. (7.31) and (7.32) constitute a closed set of equations for $\hat{S}(\mathbf{k}, t)$ which can be solved easily by numerical methods. The qualitative features of this solution, however, are easy to deduce directly from the structure of the equations. As the correlations grow, $\langle u^2(t) \rangle_\rho$ must increase, which means that $k_0^2(t)$ must decrease as a function of time. In this sense, we are seeing coarsening; the maximum amplification of $\hat{S}(\mathbf{k}, t)$ occurs at $k_{max}(t) \cong k_0(t)/\sqrt{2}$, thus the peak must move to smaller k as k_0 decreases. The time dependence of k_0 also implies that $\hat{S}$ will grow less rapidly than exponentially at fixed values of k. In fact, closer inspection reveals that k_0 approaches zero at large t and that, as it does so, $\hat{S}$ must approach $k_B T / Kk^2$ for all $k > k_0$. A sharp peak does emerge near $k = 0$, but the k^{-2} behavior is not a correct description of the correlations in the equilibrating stable phases as discussed in Section 5.4. Moreover, even the very early-stage coarsening, that is, the position of the peak in $\hat{S}(k, t)$ and the value of $\hat{S}$ at its maximum, are poorly predicted by the RPA approximation in comparisons with experimental data or numerical simulations.

A modest but useful improvement of this situation has been achieved by what has come to be known as the LBM method (Langer, Baron, and Miller 1975). The idea here, as in the RPA, is to truncate the hierarchy of correlation functions at S_4, but to use a decoupling approximation that allows the distribution function $\rho\{u\}$ to have more structure than a simple Gaussian, in particular, to be peaked at nonzero values of u. To do this, we introduce the two-point distribution function

$$\rho_2\left[u(\mathbf{r}), u'(\mathbf{r}')\right] \equiv \int \mathcal{D}\{u\}\rho\{u\}\delta\left[u(\mathbf{r}) - u\right]\delta\left[u(\mathbf{r}') - u'\right], \qquad (7.33)$$

and the associated one-point function:

$$\rho_1(u) = \int \rho_2\left[u(\mathbf{r}), u'(\mathbf{r}')\right]\mathrm{d}u' \qquad (7.34)$$

($\rho\{u\}$, ρ_1, and ρ_2 all are functions of time t. For a translationally invariant system, ρ_1 is independent of position $\mathbf{r}$). In terms of ρ_1 and ρ_2, we have

$$S_n(\mathbf{r} - \mathbf{r}', t) = \iint \mathrm{d}u\,\mathrm{d}u'\,u^{n-1}u'\rho_2\left[u(\mathbf{r}), u'(\mathbf{r}')\right], \qquad (7.35)$$

and

$$\left\langle u^n(t)\right\rangle_\rho = \int \mathrm{d}u\,u^n\rho_1(u). \qquad (7.36)$$

To account for phase separation, we expect that $\rho_1(u)$ will develop peaks away from $u = 0$. To close the hierarchy at this level, we need to make an extra assumption that will allow us to express ρ_2 in terms of ρ_1.

The LBM ansatz that serves this purpose has the form

$$\rho_2\left[u(\mathbf{r}), u'(\mathbf{r}')\right] \cong \rho_1(u)\rho_1(u')\left[1 + g\left(|\mathbf{r} - \mathbf{r}'|, t\right)uu'\right]. \qquad (7.37)$$

The overall factor, $\rho_1\rho_2$, would be correct without modification if fluctuations at the points $\mathbf{r}$ and $\mathbf{r}'$ were completely uncorrelated; the second term in brackets corrects at least for weak correlations. Specifically,

$$S_n\left(|\mathbf{r} - \mathbf{r}'|, t\right) = \left\langle u^n\right\rangle_\rho\left\langle u^2\right\rangle_\rho g\left(|\mathbf{r} - \mathbf{r}'|, t\right) \qquad (7.38)$$

$$= \frac{\left\langle u^n(t)\right\rangle_\rho}{\left\langle u^2(t)\right\rangle_\rho}S\left(|\mathbf{r} - \mathbf{r}'|, t\right).$$

The resulting equation (for the symmetric case again) is formally identical to (7.31) except that, now,

$$k_0^2(t) = \frac{1}{K}\left[\left|\frac{\partial^2 f}{\partial c_0^2}\right| - \frac{1}{6}\frac{\partial^4 f}{\partial c_0^4}\frac{\left\langle u^4(t)\right\rangle_\rho}{\left\langle u^2(t)\right\rangle_\rho}\right]. \qquad (7.39)$$

The algebraically difficult part of the LBM method is a self-consistent calculation of $\rho_1(u)$, which is necessary in order to be able to compute the moments $\langle u^2(t) \rangle_\rho$ and $\langle u^4(t) \rangle_\rho$ appearing in (7.39). This calculation is performed by going back to the Fokker-Planck equation for the full $\rho\{u\}$, integrating out all but one of the variables $u(\mathbf{r})$ to obtain the first in another hierarchy of equations for the ρ_n, and then using the ansatz (7.37) once again to obtain a closed equation involving only $\rho_1(u)$ and integrals over $\hat{S}(\mathbf{k}, t)$. For details, the reader should consult the original paper or subsequent reviews. The resulting $\rho_1(u)$ does indeed develop a doubly-peaked structure as expected.

It should be obvious from the above discussion that the LBM method does no better than the RPA in describing the correct short-range correlations in the emerging stable phases. The function $k_0^2(t)$ still approaches zero at large times, thus the structure function still approaches $k_B T / K k^2$ for $k > k_0(t)$ instead of assuming the correct Ornstein-Zernicke form. For short times, however, LBM is appreciably more accurate in its agreement with numerical simulations than is RPA. The operative definition of the term 'short times', in this context, is that the size $L(t)$ of the emerging single-phase regions has grown to no more than a few correlation lengths ξ. Beyond that early stage of phase separation, as $L(t)$ and ξ become increasingly disparate length scales, the coarsening mechanism becomes very different from the mechanism by which the short-range correlations within the coexisting stable phases are coming into equilibrium. So far as I know, we have as yet no satisfactory way of computing the structure function at times intermediate between the early-stage decomposition and the coarsening regimes. Nor, for that matter, do we have a satisfactory first-principles method for computing the late-stage scaling function $\mathcal{F}$ that appears in Eq. (5.11).

8

Late stages of phase separation

8.1 Introduction

At various points throughout this chapter we have looked briefly at what happens during the late stages of phase separation — the so-called "coarsening regime". For the case of the non-conserved order parameter, we obtained the $t^{1/2}$ law and the Kolmogorov-Avrami formula in Sections 3.7 and 3.8. We anticipated the Lifshitz-Slyozov-Wagner $t^{1/3}$ law for the case of the conserved order parameter in Section 4.4, and introduced the idea of a scaling description of late-stage precipitation patterns in Section 5.4. The present section is devoted entirely to the important problem of late-stage coarsening in systems with conserved order parameters.

From a practical point of view, the late-stage behavior is usually the most interesting part of the phase transformation. With the exception of some very slowly diffusing systems such as polymer blends, phase separating materials seldom remain long enough in their unstable early-stage configurations to be useful — or even easily observable — in such states. Moreover, late-stage configurations may have little or no dependence on the early-stage mechanisms by which they are formed. That is, the distinction between nucleation and spinodal decomposition at onset may have little relevance to late-stage domain-growth and coarsening mechanisms.

What does seem to matter in the late-stage processes are the relative volume fractions of the emerging coexisting phases. A precipitation pattern that looks like a dilute gas of small droplets, possibly — but not necessarily — formed by nucleation after a shallow, off-critical quench, will coarsen somewhat differently from one in which both emerging phases occupy comparable volumes and are interspersed among each other in complex, multiply connected geometries. The first case, that is, the limit of small volume fraction, is the subject of the Lifshitz-Slyozov-Wagner (LSW) theory (Lifshitz and Slyozov 1961, Wagner 1961). This is theoretically the simplest case but, as we shall see, even this limit is non-trivial.

8.2 The Lifshitz-Slyozov-Wagner theory

Our starting point is an A-B mixture of the kind described in Section 4, in which a dilute dispersion of spherical β-phase droplets has emerged in a background of slightly supersaturated α phase. The behavior of a completely isolated droplet was discussed in Section 4.4, and we shall make extensive use of formulas derived there. Specifically, the growth rate of a β-droplet of radius R into an α-phase of supersaturation δc_∞ is

$$\dot{R} = \frac{D}{R}\left(\Delta - \frac{2d_0}{R}\right), \tag{8.1}$$

where
$$\Delta = \frac{\delta c_\infty}{\Delta c} = \frac{2d_0}{R^*}, \tag{8.2}$$

R^* being the critical radius for nucleation.

The LSW theory uses Eqs. (8.1) and (8.2) as the basis for a conceptually simple approximation for the evolution of a coupled system of many such droplets. The idea is to let the supersaturation $\delta c_\infty(t)$ or, equivalently, the critical radius $R^*(t)$ serve as a time-dependent mean field determined self-consistently by global conservation of the numbers of A and B atoms. This concept captures much of the physics of the situation. According to (8.1) and (8.2), droplets with R greater than R^* are growing; B atoms are diffusing onto their surfaces. Conversely, droplets with R less than R^* are becoming smaller; B atoms are diffusing away from them. The net effect is a competition in which larger droplets are growing at the expense of smaller ones, and this diffusion of material from small droplets to big ones is accounted for at least approximately by the self-consistent R^*. As the system moves toward equilibrium, the supersaturation δc_∞ decreases; thus R^* increases, and the characteristic size of the droplets must increase accordingly. What is not clear from these considerations is why the mean-field approximation should be valid in the limit of small volume fraction of the β phase. We shall return to that question in the next section.

Let $\nu(R, t)$ be the size distribution of the droplets; that is, $\nu\,dR$ is the number of droplets per unit volume with radii between R and $R + dR$. If nucleation has ceased, and if the system is evolving according to the purely deterministic rules outlined in the last two paragraphs, then the equation of motion for ν is

$$\frac{\partial \nu}{\partial t} = -\frac{\partial}{\partial R}\left[\dot{R}(R)\nu\right], \tag{8.3}$$

and the conservation condition is

$$\Delta_0 = \Delta(t) + \tfrac{4}{3}\pi \int_0^\infty R^3 \nu(R, t)\,dR, \tag{8.4}$$

where $\Delta_0 = \Delta(0) = \delta c_\infty(t = 0)/(\Delta c)$ is the initial dimensionless supersaturation. Equation (8.4) says that all of this initial supersaturation Δ_0

must be contained either in the current supersaturation $\Delta(t)$ or in the droplets.

The LSW solution of this set of equations is a scaling form that can be shown to be asymptotically valid in the limit of long times. That is, the solution is stable and essentially all initial distributions of droplets will approach this solution if one waits long enough. As we have argued previously, once any transients associated with the initial distribution have died out, the only relevant length scale in the system should be $R^*(t)$. Thus we should write

$$\nu(R, t) \approx \frac{1}{[R^*(t)]^4} \Phi\left[\frac{R}{R^*(t)}\right],$$ (8.5)

where Φ is some function to be determined. The fourth power of R^* in the prefactor is fixed by the normalization of ν as follows. We know that $\Delta(t)$ in (8.4) must vanish as t becomes large; therefore, with (8.5),

$$\int_0^\infty R^3 \nu(R, t)\, \mathrm{d}R \approx \int_0^\infty x^3 \Phi(x)\, \mathrm{d}x$$

approaches a constant in this limit. The scaling function Φ can then be obtained by solving a first-order ordinary differential equation, essentially (8.3). The detailed analysis is elegant but nontrivial, and is somewhat more elaborate than is appropriate for these notes. (I encourage the interested reader to go directly to the original paper by Lifshitz and Slyozov.) The result is that Φ has a pronounced maximum in the neighborhood of $R/R^* \cong 1$ and vanishes for $R/R^* > \frac{3}{2}$. A specially famous and important result is that

$$R^*(t) \approx \left(\tfrac{8}{9} D d_0 t\right)^{1/3},$$ (8.6)

and that

$$\bar{R}(t) = \int_0^\infty \nu(R, t) R\, \mathrm{d}R \approx R^*(t).$$ (8.7)

This is the anticipated $t^{1/3}$ law with a specific and presumably testable prediction for the constant that we defined in Eq. (4.27).

The power $\frac{1}{3}$ in the coarsening law (8.6) is almost universally confirmed by experiments. As we have seen, it arises from very general arguments that are based on little more than dimensional analysis. The conditions for its validity are:

(1) Coarsening must be driven by surface tension. That is, the system must be evolving in such a way as to minimize its surface area.

(2) Transport must be by diffusion through the bulk. This is ordinarily the case for relatively simple systems in which the effective order parameter is a locally conserved quantity. In more complex systems such as fluids or multi-crystalline solids, the argument can be invalidated by mechanisms such as convection (Siggia 1979) or rapid diffusion along the interphase boundaries (Huse 1986). Even in the latter situations, however, the $t^{1/3}$ law may often be observed during some stages of the phase transformation.

(3) The length scale that describes the coarsening process — e.g. the average droplet radius or, equivalently, the average spacing between droplets — must be the only relevant length scale in the system. In the simplest cases, this means that we must wait long enough for this scale to be much larger than microscopic correlation lengths; but other length scales, such as those that govern convection, or those associated with the strains in elastic solids, may intervene in more complex situations.

Nowhere in the above criteria have we mentioned the volume fraction. Indeed, $t^{1/3}$ behavior has been observed in situations that look not at all like the dilute dispersion of small droplets envisioned by LSW. For example, Voorhees and Glicksman (1985) report measuring $t^{1/3}$ quite accurately for the coarsening of dendritic sidebranches in solidification of a pure crystal from its melt. Here, the controlling transport mechanism is the diffusion of latent heat, and the volume fractions of liquid and solid are roughly equal to one another.

On the other hand, the actual LSW distribution is seldom observed, even for situations where the volume fraction is quite small (Marder 1987). Moreover, the rate constant — the analog of the factor $\frac{8}{9}$ in (8.6) — seems to depend in an important way on the volume fraction. Accordingly, we look next at one of the central LSW assumptions — that there are no significant correlations between the positions and sizes of the droplets in the coarsening distribution — and shall see that the volume fraction is playing a sensitive role.

8.3 Correlations and screening

In the LSW picture, before making the mean-field approximation, we are looking at a dilute distribution of, say, N droplets of radii R_i, centered at positions r_i, $i = 1, \ldots, N$. If the i-th droplet is growing (or shrinking) at rate $\dot{R}_i$, then it is a sink (or source) of B atoms of strength $-4\pi(\Delta c)R_i^2\dot{R}_i$. With slightly less confidence than in Section 4.4, we can make a quasistationary approximation in solving the diffusion equation, and write

$$\delta c(\mathbf{r}) \cong \delta c_\infty - \frac{(\Delta c)}{D}\sum_{i=1}^{N}\frac{R_i^2\dot{R}_i}{|\mathbf{r}-\mathbf{r}_i|}, \qquad (8.8)$$

where $\delta c(\mathbf{r})$, as previously, is the excess concentration of B atoms in the α phase at position $\mathbf{r}$. Note that we are still assuming that the droplets are point-like objects so far from one another that the diffusion field is not affected by any distortions in their shapes.

The uncertainty regarding the quasistationary approximation, which allows us to write the diffusion kernel in (8.8) in the time-independent form $(4\pi D|\mathbf{r}-\mathbf{r}_i|)^{-1}$, arises from the fact that we are proposing to use (8.8) to compute interactions between well separated droplets. Because diffusion fields transmit signals very slowly across large distances, one might worry about neglecting retardation. As we shall see, the approximation is saved by the screening effect.

Given (8.8), we can write a modified version of the rate equation, (8.1), in which we recognize that the i-th droplet actually sees the field $\delta c(\mathbf{r}_i)$ as opposed to simply δc_∞. In this way, we obtain a set of coupled, first-order differential equations for the $R_i(t)$:

$$\begin{aligned}
\dot{R}_i &\cong \frac{D}{R_i}\left[\frac{\delta c(\mathbf{r}_i)}{\Delta c} - \frac{2d_0}{R_i}\right] \\
&\cong \frac{D}{R_i}\left(\Delta - \frac{2d_0}{R_i}\right) - \frac{1}{R_i}\sum_{j\neq i}\frac{R_j^2\dot{R}_j}{|\mathbf{r}_i-\mathbf{r}_j|}.
\end{aligned} \qquad (8.9)$$

This equation was first obtained by Weins and Cahn (1973). See also Kawasaki and Ohta (1983). In principle, one can choose a set of positions $\mathbf{r}_i$ and initial radii R_i, and then simply solve (8.9) to find the subsequent evolution of the system. There are problems associated with the sum because of the infinitely long-ranged diffusion kernel and, at this level, there is some ambiguity concerning the precise interpretation of Δ. For more information, see Marder (1987). But these are not our principal concerns at the moment. The essential point is that, in order for (8.9) to be equivalent to (8.1) as used in the LSW theory, the sum in (8.9) must vanish. That is, the sources and sinks in the sum over distant droplets in (8.8) or (8.9) must very nearly cancel each other, and the residual contribution, if any, must be independent of the position r_i of the droplet at which the sum is evaluated. This is most likely to happen if there are many terms in the sum so that fluctuations are not important, which is precisely the criterion for validity of the mean-field approximation.

Suppose that the sum in (8.8) does indeed contain many terms so that a continuum approximation is valid. It then turns out to be easiest to analyze the situation by going back to the original, quasistationary diffusion equation and making a crude but sensible analysis. (One can obtain the same results more systematically and take the calculation much further by working directly with (8.9). See Marder (1987).) Let the total number of droplets per unit volume be $n(\mathbf{r})$, and let the average radius of these droplets be R. We are interested in correlations and screening; therefore the appropriate strategy is to perturb the system by inserting a source of B atoms of strength unity at $\mathbf{r} = 0$ and looking at the response $U(\mathbf{r})$; $\delta c(\mathbf{r}) \cong \delta c_\infty + U(\mathbf{r})$. The diffusion equation for U is:

$$D\nabla^2 U \cong \delta(\mathbf{r}) + 4\pi(\Delta c)\,n\,R^2\delta\dot{R}, \qquad (8.10)$$

where the last term on the right-hand side is our rough approximation for the distributed source associated with the growing droplets. More precisely, this term is the incremental change in the source that is induced by the perturbation U, and our assumption is that this linear change in the local concentration makes a corresponding linear change in the growth rate as in (8.1) or the first line of (8.9):

$$\delta\dot{R} \cong \frac{DU}{(\Delta c)R}. \qquad (8.11)$$

Inserting (8.11) into (8.10), we find an equation of the form

$$\nabla^2 U \cong \frac{1}{\xi_s^2}U + \frac{1}{D}\delta(\mathbf{r}) \qquad (8.12)$$

where ξ_s is a screening length:

$$\xi_s = \left(\frac{1}{4\pi nR}\right)^{1/2}. \qquad (8.13)$$

The solution of (8.12) is the screened, quasistationary diffusion kernel:

$$U(r) = \frac{1}{4\pi Dr}\,e^{-r/\xi_s}. \qquad (8.14)$$

Now note the following. The volume fraction of precipitate is

$$\phi = \tfrac{4}{3}\pi R^3 n; \tag{8.15}$$

thus the ratio of the droplet radius to the screening length is

$$\frac{R}{\xi_s} = (3\phi)^{1/2}. \tag{8.16}$$

Equivalently, the number of droplets within a sphere whose radius is ξ_s is

$$\tfrac{4}{3}\pi\xi_s^3 n = \frac{1}{(3\phi)^{1/2}}. \tag{8.17}$$

As anticipated, the screening length diverges in the limit of vanishing ϕ.

Because (8.17) gives us an estimate of the number of terms in the sum in (8.9), the divergence justifies our assertion that the LSW mean-field approximation is valid in this limit.

It is important to recognize the physical meaning of the correlations implied by the exponentially screened kernel (8.14). A larger than average droplet grows at the expense of its neighbors; thus, the droplets nearby tend to be somewhat smaller. Observed from a distance, the growing larger droplet is a sink of B atoms that is screened by the sources — the smaller droplets — induced in its neighborhood. It is this correlation between positions and sizes of droplets that broadens the LSW distribution and modifies the rate constant at nonzero volume fractions.

Unfortunately, we have as yet no entirely satisfactory analytic approach to the theory of intermediate and late-stage coarsening at arbitrary volume fractions.

At least two ambitious attempts have been made to develop expansions at small ϕ based on Eq. (8.9) (Enomoto *et al.* 1986, Marder 1987). Both take advantage of the insight gained from considerations of screening and correlations. The idea is that in equations of motion for $\nu(R)$ and its generalizations to higher-order distribution functions, the diffusion kernel must always appear in its screened form (8.14).

This automatically solves the divergence problems and also constrains our use of the quasistationary approximation. For the latter approximation to be valid, the droplet growth rate, $\dot{R}/R$, must be much smaller than D/ξ_s^2, the relaxation rate for the diffusion field over the largest relevant distance ξ_s. From (8.1), we know that $\dot{R}/R \sim Dd_0/R^3$; thus, using (8.16), we find the quasistationary condition to be

$$d_0 \ll \phi R. \tag{8.18}$$

Because d_0 is ordinarily a microscopic length scale, this condition can be violated only at very early times when R may also be microscopic — i.e. only during the transient initial stage of phase separation. This stage lasts a long time for small ϕ, however; widely spaced droplets grow independently for quite a while before they start interacting with each other via their diffusion fields. Such non-trivial considerations must be kept carefully in mind in trying to develop generalizations of the LSW theory.

The expansions at small ϕ are technically very complex and will not be described in any detail here. It is also not clear — to me — how systematic they are. The apparently appropriate expansion parameter is $\phi^{1/2}$, but it is not easy to understand just which functions can be expanded in powers of this parameter. To illustrate the difficulty, I mention one intriguing aspect of Marder's results. He derives a generalization of the LSW equation of motion (8.3) for the distribution function $\nu(R, t)$ in the form

$$\frac{\partial \nu}{\partial t} = -\frac{\partial}{\partial R} \left[\dot{R}(R)\nu - \mathcal{D}\frac{\partial \nu}{\partial R} \right], \qquad (8.19)$$

where the 'diffusion constant' $\mathcal{D}$ is defined self-consistently in terms of computed correlation functions and turns out to be of order $\phi^{1/2}$. The extra diffusion in R space broadens the LSW distribution function in such a way that Marder is able to explain some results of numerical simulations and also some metallurgical observations. The most important point, however, may be that $\phi^{1/2}$ appears in (8.19) as the coefficient of the highest derivative in the equation. If this is true, and not some artifact of the approximations leading to (8.19), then the volume fraction is technically a singular perturbation in this problem. The behavior of the system at any nonzero ϕ may be qualitatively different than it is at precisely $\phi = 0$.

To conclude this discussion, I would like to pose a few questions regarding directions for future research in this area. We clearly do not yet have a deep theoretical understanding of what happens during the intermediate and late stages of phase separation. Is such an understanding possible? Might we be facing a 'no-theory' situation in which the best we can do in principle is to simulate increasingly complex phenomena on a computer? Do we really need more theory? Or might the computer simulations be sufficient?

My sense is that we ought to be able to make more progress in solving several basic problems, and that those solutions will be essential for further work. For example, it ought to be possible to compute a late-stage scaling distribution of droplet sizes, analogous to (8.5), for small but nonzero volume fraction. It ought also to be possible to generalize the definition of this distribution function so that it makes sense when the droplet picture is no longer valid, for example, when the precipitates form plates or rods or, perhaps, even dendrites. (Glicksman has suggested defining a distribution of surface area as a function of local curvature.) In the same vein, it ought to be possible to compute the late-stage scaling form for the structure factor, as in (5.11), for arbitrary volume fraction. (See Mazenko 1990, Ohta and Nozaki 1989.) One potentially very useful result that should emerge from such analyses would be a better understanding of the screening length. Might the fact that droplets can shift their positions and deform their shapes in response to changes in the diffusion field enhance the screening effect?

If questions like these could be answered, we might have a great deal more confidence as we move away from the specially simplified models discussed in this chapter and begin to tackle the problems posed by realistic complications such as fluid flow or the coupling between composition variations and elastic strains in a crystalline solid. (See recent work by Onuki for treatments of both of the latter effects.) Ultimately, I am sure that we shall have to rely on the computer for quantitative, predictive analyses of realistically complex phenomena; but I suspect that there is still more fundamental progress to be made.

References

Allen, S.M. and Cahn, J.W. (1979), Acta Metall. **27**, 1085.

Amar, J., Sullivan, F. and Mountain, R. (1988), Phys. Rev. **B37**, 196.

Avrami, M. (1939), J. Chem. Phys. **7**, 1103.

Becker, E. and Döring, W. (1935), Ann. Physik **24**, 719.

Binder, K. (1983), J. Chem. Phys. **79**, 6387.

Binder, K., (1984), Phys. Rev. **A29**, 341.

Binder, K. (1990), in 'Materials Science and Technology', Vol. 5 of *Phase Transitions in Materials*, edited by P. Haasen (V.C.H. Verlagsgesellschaft, Weinheim, Germany, in press).

Binder, K. and Stauffer, D. (1974), Phys. Rev. Lett. **33**, 1006.

Cahn, J.W. (1968), Trans. Metall. Soc. AIME **242**, 166.

Cahn, J.W. and Hilliard, J.E. (1958), J. Chem. Phys. **28**, 258.

Chandra, P. (1989), Phys. Rev. **A39**, 3672.

Cook, H.E. (1970), Acta Metall. **18**, 297.

Enomoto, Y., Tokuyama, M. and Kawasaki, K. (1986), Acta Metall. **34**, 2119.

Frenkel, J. (1955), *Kinetic Theory of Liquids* (Dover, New York).

Gaulin, B.D., Spooner, S. and Morii, Y. (1987), Phys. Rev. Lett. **59**, 668.

Gunton, J.D., San Miguel, M. and Sahni, P.S. (1983), in *Phase Transitions and Critical Phenomena*, edited by C. Domb and J. Lebowitz (Academic Press, New York).

Huse, D.A. (1986), Phys. Rev. **B34**, 7845.

Kampen, N.G. van (1984), *Stochastic Processes in Physics and Chemistry* (North Holland, Amsterdam).

Kawasaki, K. and Ohta, T. (1983), Physica **118A**, 175.

Kolmogorov, A.N. (1937), Bull. Acad. Sci. U.S.S.R., Phys. Ser. **3**, 355.

Langer, J.S. (1967), Ann. Phys. (N.Y.) **41**, 108.

Langer, J.S. (1974), Physica **73**, 61.

Langer, J.S., Baron, M. and Miller, H.D. (1975), Phys. Rev. **A11**, 1417.

Lifshitz, I.M. and Slyozov, V.V. (1961), J. Phys. Chem. Solids **19**, 35.

Ma, S.K. (1976), *Modern Theory of Critical Phenomena* (W.A. Benjamin, Inc., London)

Ma, S.K. (1981), *Statistical Mechanics* (World Scientific, Singapore).

Marder, M. (1987), Phys. Rev. **A36**, 858.

Marro, J., Lebowitz, J.L. and Kalos, M.H. (1979), Phys. Rev. Lett. **43**, 282.

Massalski, T. (1986), Ed., *Binary Alloy Phase Diagrams* (American Society for Metals).

Mazenko, G. (1990), University of Chicago preprint.

Moldover, M. (1985), Phys. Rev. **A31**, 1022.

Ohta, T. and Nozaki, H. (1989), in *Space-Time Organization in Macromolecular Systems*, Springer Series in Chemical Physics, Vol. 51, edited by F. Tanaka, M. Doi, and T. Ohta (Springer-Verlag, Berlin).

Safran, S., Sahni, P. and Grest, G. (1983), Phys. Rev. **B**, 2705.

Siggia, E.D. (1979), Phys. Rev. **A20**, 595.

Voorhees, P.W. and Glicksman, M.E. (1985), J. Cryst. Growth **72**, 599.

Wagner, C. (1961), Z. Elektrochem. **65**, 581.

Weins, J.J. and Cahn, J.W. (1973), in *Sintering and Related Phenomena*, edited by G.C. Kuczynski (Plenum, London).

Wilson, K. and Kogut, J. (1974), Physics Reports **C12**, 77.

Zettlemoyer, A.C., Ed. (1969), *Nucleation* (Dekker, New York).

CHAPTER 4

DENDRITIC GROWTH AND RELATED TOPICS

Y. Pomeau and M. Ben Amar

1

Introduction

In this chapter we answer various questions related to the growth of crystals from their melt. We shall mostly take the point of view that these growth phenomena are fully described by macroscopic equations, to be written below, so avoiding reference to the details of the microscopic molecular behavior. This point of view is perhaps not the most common in the literature on the subject, and the real shape of growing crystals — particularly with facets — is often interpreted by means of microscopic theories, although macroscopic phenomena due to thermal convection, for instance, play an important role very often and are difficult to model because they depend on parameters such as the shape of the container.

Besides exotic experimental situations (to be considered at the end of section 7), the growth of crystals is slow enough to permit a local equilibration, even in the vicinity of the growing interface. To be more specific one might consider the ratio of the velocity of growth of crystals, very often less than centimeters per second to typical molecular speeds, to be of the order of kilometers per second at room temperature.

In the case of the very fast cooling rates needed to reach noticeable kinetic phenomena, it is also very likely that the solids themselves will no longer be perfect crystals and could even become glassy, making the whole picture completely different, and certainly very interesting (for instance we do not know of any systematic study of the interface between a glass and its melt). Even within this restricted subject (that is when growth occurs in a quasiequilibrium way), a tremendous variety of results and problems have appeared recently. It is the aim of this chapter to present this information to as wide a community as possible.

We shall try to do this by keeping the mathematical level as low as possible, even though this task is not always very easy. Unfortunately in this field a host of important results can be explained only by using rather sophisticated maths. We have restricted references to the literature mostly to that which is directly relevant to our developments. The interested reader will find rather extensive bibliographies of this subject in Langer [1987], Pelcé [1988] and Kessler *et al.* [1988].

The organization of this chapter reflects more or less that of the lectures given at the Beg-Rohu Summer School and is as follows:

In section 2 the Stefan-Lamé equations and boundary conditions (b.c.) for describing the growth limited by diffusion are derived. As we shall see this is just a way of writing conservation relations in a convenient form. The problem common to many parts of this subject is that even though the equations look fine, it is not obvious that they have solutions. This is fundamentally because they are what mathematicians call free boundary problems and it is not obvious at all that there are enough equations and boundary conditions to solve an evolution problem, given the initial data. Thus we will show that this set of equations and b.c. has solutions in — at least — one space dimension, as was found first by Lamé and Clapeyron [1831] more than a century ago. The extension to higher space dimensions is difficult, at least for arbitrary geometries.

A major result in this area is that after a finite time initially smooth initial data generate singularities. These singularities involve an infinite curvature of the liquid-solid interface. This situation is clearly not possible near equilibrium since the Laplace capillary pressure will diverge. We will show how those capillary effects enter into the game, through the so-called Gibbs-Thomson condition, as well as its extension to the anisotropic situation.

The Gibbs-Thomson condition is discussed in section 3. As crystals are anisotropic, we will discuss how to modify this condition to take into account possible anisotropies of surface tension: this will introduce what we call the Gibbs-Thomson-Herring (GTH) condition. We will show, as an example, how it helps to explain the remarkable phenomenon of spiral growth. Furthermore the possibility of faceting can be incorporated into a generalized GTH condition. Microscopic effects, such as the kinetics of attachment of particles from the melt to the crystal lattice, are often held responsible for everything in the growth of faceted crystals. This differs from the philosophy of the approach used here, which is basically macroscopic.

Section 4 is devoted to a qualitative analysis of the growth of needle crystals. Needle crystals are often observed in experiments and yield an interesting and reasonably clean object of study. There we shall present a scaling approach, aimed at obtaining relationships between various physical parameters by dimensional analysis of the equations. This allows in particular a rather complete picture of what happens when molecular diffusion combines with the effects of a flow parallel to the axis of the needle crystal. This qualitative approach will hopefully help in the understanding of the mathematical developments of section 5.

In section 5, we show how to reduce the full Nash-Glicksman equation [1974] (the integro-differential equation relevant for the free needle crystal problem) to a nonlinear eigenvalue problem in the (mostly realistic) limit of low dimensionless undercooling. This reduction leaves a still complicated problem. In principle, however, it gives all the physical parameters of a needle crystal once the proper physical quantities are known.

Before solving the Nash-Glicksman equation, we shall consider in section 6 a so-called geometrical model, as well as its mathematical analysis by Martin Kruskal and Harvey Segur. This analysis is a cornerstone of the subject and so we discuss it in depth, even though it might sometimes be difficult.

In section 7, we explain how to apply the Kruskal-Segur method to the nonlinear eigenvalue equation derived in section 6. We also explain what happens beyond this low undercooling limit, as this is at the basis of an important successful test of this approach, allowing us to describe quantitatively a recent experiment on the fast solidification of nickel.

2

The moving solidification front

From the point of view of thermodynamics an undercooled melt is in *metastable* equilibrium. This means that, if one waits long enough, in principle a seed of the most stable thermodynamic phase (the crystal) will appear spontaneously by thermal fluctuations and grow. This is *not* the kind of problem we shall consider; instead we shall discuss a crystal growing in an undercooled melt, the undercooling being small enough to make the spontaneous transition from liquid to crystal very unlikely.

To take an example, below the roughening transition it is often argued that flat facets grow by nucleation of terraces, a process depending on a Kramers-Zeldovich jump over a barrier by thermal activation. It is unclear — at least to us — that this is always compatible in real life experiments with the absence of nucleation in the bulk of the melt. As explained before, this chapter will describe crystal growth without any so-called 'kinetic phenomenon', except for a brief account in section 7.

Thus we shall assume that the growth is limited by diffusion, either of latent heat or of impurities usually not miscible in the solid phase. Before we jump into the rather intricate problem of needle crystal growth considered in the coming sections, we shall consider in this section some of the basic physics involved and how it translates into equations and b.c. As often done when studying macroscopic phenomena, we shall introduce first some dimensionless quantities brought about by combining various physical parameters. The first quantity of this sort is the dimensionless undercooling, hereafter called Δ. We shall mostly consider growth limited by latent heat. The situation of impurity-limited growth yields a lot more quantities, and that is one of the reasons why we shall not consider this case, even though it is perhaps more important for many experiments.

Let T_{eq} be the equilibrium temperature between the solid and its melt; there is undercooling when the melt temperature, say T_0, is uniform and less than T_{eq} very far from the growing solid. Then Δ is a dimensionless measure of the thermal desequilibrium and is defined as:

$$\Delta = \frac{c_p(T_{eq} - T_0)}{L},$$

where c_p is the heat capacity of the melt (for example) and L the latent heat — both given per unit volume. Neglecting the temperature dependence of material properties, one can say that $\Delta = 1$ ($-80°$ C for water) is the undercooling such that the latent heat generated by the phase transformation is just enough to heat the melt at the equilibrium temperature. As seen below the solutions of the Stefan-Lamé equations change qualitatively at $\Delta = 1$. Let us consider now the standard (Stefan) problem of propagation of a plane solidification front in an undercooled melt.

There is of course nothing new in this solution, but it helps in the understanding of the connection between the equations and the physics. The temperature field is $T(x,t)$, where the x position is perpendicular to the solid surface. This temperature field is the solution of the Fourier heat equation:

$$T_t = DT_{xx},$$

where D is the heat diffusivity coefficient, assumed to be equal in the melt and the solid (this is the assumption made in the so-called symmetrical models), and subscripts x and t are for derivatives. This temperature field satisfies two b.c.: $T \to T_0$ at $x \to +\infty$ (in the melt), and $T = T_{eq}$ on the solid/melt interface, that is located at $X(t)$, a function of time which must be found.

There is a supplementary condition, usually called Stefan's condition, here Stefan-Lamé, that expresses the conservation of the total energy when some matter is transformed from the liquid into the solid state. Let us write this condition in arbitrary geometries:

$$Lv_n = \mathbf{n} \cdot \left[(Dc_p \nabla T)_{solid} - (Dc_p \nabla T)_{liquid} \right],$$

where v_n is the normal velocity of the interface toward the liquid phase, and $\mathbf{n}$ is the normal to the same interface also directed toward the liquid. This condition expresses the energy conservation, because the left hand side is the energy released per unit time by the phase transformation, whereas the right hand side gives the difference between the molecular heat (or energy, as we neglect volume changes) flux across the interface.

In one dimension, and with our notations and assumptions, this becomes:

$$L\frac{dX}{dt} = -Dc_pT_x,$$ (2.1)

where T_x is taken on the liquid side, by assuming that the initial condition (and this remains true at any positive time) for the temperature on the solid side is $T = T_{eq} = $ constant and where $X(t)$ is the position of the interface at time t. The diffusion equation in the frame of reference of the solidification front reads:

$$T_t = \dot{X}T_{x'} + DT_{x'x'},$$ (2.2)

where $x' = x - X(t)$. From this equation it is natural to seek a solution depending on x' and t through the combination $z = x'/\sqrt{Dt}$ so that all terms in (2.2) have the same dimension. The same condition leads us to try $X(t) = \beta\sqrt{Dt}$, where the pure number β will be a function of the dimensionless undercooling Δ that we shall now find.

This transformation of the coordinates is due to Boltzmann. Later we shall briefly discuss the more general solution of the Stefan problem. The one we are currently investigating belongs to the class of 'similarity' or self-similar solutions; it is the asymptotes of any initial condition where the solid fills a half space, whatever the temperature distribution, given that it has the same asymptotes as the similarity solution (see Barenblatt [1979] for a general discussion of this kind of reduction). The equation for $T(z)$ reads:

$$\beta T_z + zT_z = -2T_{zz},$$ (2.3)

with the b.c. $T \to T_0$ (in the melt) at $z \to +\infty$ and $\frac{1}{2}L\beta = -c_pT_z$ at $z = 0$. This last condition is the transform of (2.1). This introduces into the equations the dimensionless undercooling, because it combines the temperature T and the latent heat L. Thus one puts:

$$\Theta(z) = \frac{c_p(T - T_{eq})}{L},$$

so that (2.3) becomes:

$$0 = -\tfrac{1}{2}\beta - \Theta(z) \quad \text{at } z = 0.$$ (2.4)

The rest of the calculation seeks to obtain a relation between β (unknown) and Δ (given). For that purpose one solves (2.3) for Θ, related linearly to T to obtain:

$$\Theta(z) = \tau_0 \int_0^z dz \exp\left(-\tfrac{1}{2}\beta z - \tfrac{1}{4}z^2\right),$$

τ_0 unknown (for the moment). The lower bound for the integral is imposed by the condition $T = T_{eq}$ or $\Theta = 0$ at $z = 0$. The other b.c. are $\Theta \to -\Delta$ at $z \to +\infty$ and $\Theta_z = -\frac{1}{2}\beta$ at $z = 0$. This gives the following implicit relation between β (the speed of growth) and Δ (the dimensionless undercooling):

$$\phi(\beta) = \Delta,$$

where $\phi(\beta)$ is the function defined as:

$$\phi(\beta) = \frac{1}{2}\beta \int_0^\infty \mathrm{d}z \exp\left(-\frac{1}{2}\beta z - \frac{1}{4}z^2\right).$$

Indeed the function $\phi(\cdot)$ can be expressed by means of known transcendental functions, but we shall content ourselves with two limits: β tending to $+\infty$ and to $-\infty$.

For β tending to $+\infty$ (very large speed of growth of the solid), the integral is dominated by the contribution coming from the term $\frac{1}{2}\beta$ in the argument of the exponential (that is for z small: in physical terms this is the contribution of a thin boundary layer near the interface). This gives $\phi(\beta)$ and $\Delta \to +1$ as β tends to $+\infty$.

This marginal situation, $\Delta = 1$, is the one already alluded to above: the undercooling is just large enough to heat the melt to the equilibrium temperature. Indeed the problem so formulated loses any physical meaning as kinetic effects must become important if such a large undercooling in a bulk phase can ever be reached.

In the absence of such kinetic (or microscopic) effects, it seems that the Stefan problem could still have a solution for initial conditions with a continuous distribution of temperature, but with an asymptotic behavior of $X(t)$ depending on the details of those initial conditions (as in the general class of KPP problems of propagation in unstable media, Pelcé [1988]), contrary to what happens for Δ less than 1.

In the opposite limit (β tending to $-\infty$, that is the solid melts very fast), the integral defining $\phi(\cdot)$ is dominated by the contribution of a saddle point at $z = \beta$, and has the asymptotic form:

$$\phi(\beta) \approx \beta\pi^{1/2}\exp(\tfrac{1}{4}\beta^2),$$

which implies a logarithmic growth of β as a function of Δ at very large overheating. This solution has a rather interesting physical structure: as β increases, and as Δ becomes very large and negative (that means an overheated melt), the temperature drop occurs at a large distance from the physical interface but its thickness is independent of β (this thickness is the range of variation of z near the saddle point which depends on the quadratic term in the exponential only).

To summarize, we have shown that a 1D solidification/melting front moves as $X(t) = \beta(\Delta)(Dt)^{1/2}$ for large t and that $\beta(\Delta)$ tends to $+\infty$ as $\Delta \to 1$ and to $-\infty$ as Δ tends to $-\infty$.

Let us comment further and extend this solution. First, as we have already said, this solution can be included in a much more general class, at least for this symmetrical model. The Fourier equation being linear, and assuming that the diffusion coefficients on both sides of the solidification front are the same, the temperature field at any time is the sum of the one given by the solution of the initial condition for the heat equation in homogeneous medium plus the one generated by the latent heat. This latter condition can be computed by considering a moving heat source of strength $L(\mathrm{d}X/\mathrm{d}t)$ located on the interface, that is at $x = X(t)$. The position of the interface is finally given by the condition that the temperature there is T_{eq}.

As we shall use again and again this idea that the b.c. can be computed by adding a moving heat source on the liquid/solid interface, we discuss it here. By considering a heat source at time t and strength $S(t)$, we can replace the Fourier equation by:

$$T_t - DT_{xx} = S(t)\delta[x - X(t)], \tag{2.5}$$

where $\delta(\cdot)$ is the Dirac delta function. Integrating (2.5) on space near $x = X(t)$, one gets a contribution of the l.h.s. from the highest space derivative, that gives a jump of the first derivative T_x. The r.h.s. gives a finite contribution proportional to $S(t)$, so that (2.5) is equivalent to the usual Fourier equation, i.e. without the r.h.s. plus the b.c.:

$$\left[-DT_x\right]_{x=X(t)_+} - \left[-DT_x\right]_{x=X(t)_-} = S(t) \quad \text{at} \quad x = X(t),$$

which is equivalent to the Stefan-Lamé b.c. in the so-called symmetric case, when the thermal properties of the solid and the melt are assumed to be the same.

The interesting part of this computational trick is that (2.5) is solvable by standard application of the Green's function method:

$$T(x,t) = \int_{-\infty}^{t} \frac{\mathrm{d}t'\, S(t')}{\left[4\pi D(t-t')\right]^{1/2}} \exp\left[\frac{-(x - X(t'))^2}{4D(t-t')}\right]. \tag{2.6}$$

Writing everything with the dimensionless quantities introduced before, the desired integral equation for $X(t)$ reads:

$$0 = \int_{-\infty}^{+\infty} \frac{dx'\,\Theta(x',0)}{[4\pi Dt]^{1/2}} \exp\left[\frac{-(X(t)-x')^2}{4Dt}\right] \tag{2.7}$$

$$- \int_0^t \frac{dt'}{[4\pi D(t-t')]^{1/2}} \frac{dX}{dt'} \exp\left[\frac{-(X(t)-x(t'))^2}{4D(t-t')}\right],$$

where the first term comes from the free evolution of the initial conditions [the temperature field $\Theta(x',0)$], and the second one from the latent heat released at the interface. The previous solution is recovered by taking $X(t) = \beta(Dt)^{1/2}$ and $\Theta(x,0)$ as a step function: $\Theta_{\text{step}} = 0$ for $x < 0$ and $\Theta_{\text{step}} = \Delta$ for $x > 0$. Indeed this equation yields that the temperature on the free surface, i.e. at $x = X(t)$, is the equilibrium temperature.

Following a suggestion by Bernard Derrida, it is also possible to get from this expression two corrections to the Stefan-Lamé law, that is to the asymptotes $X(t) = \beta(Dt)^{1/2}$ at $t \to \infty$. These corrections arise from distortions of the initial temperature field out of the step function. Let $\Theta_{\text{step}}(x)$ be the step function defined before and let us expand the first integral on the r.h.s. of (2.7) at large t, for x near the interface:

$$\int_{-\infty}^{+\infty} \frac{dx'\,\Theta(x',0)}{[4\pi Dt]^{1/2}} \exp\left[-\frac{(X(t)-x')^2}{4Dt}\right] \tag{2.8}$$

$$\approx \int_{-\infty}^{+\infty} \frac{dx'\,\Theta_{\text{step}}(x')}{[4\pi Dt]^{1/2}} \exp\left[-\frac{(X(t)-x')^2}{4Dt}\right]$$

$$- \frac{X(t)}{2Dt[4\pi Dt]^{1/2}} \exp\left[-\frac{X(t)^2}{4Dt}\right] \int_{-\infty}^{+\infty} x'dx'\,[\Theta(x',0) - \Theta_{\text{step}}(x')].$$

The corrective term, i.e. the last one in (2.8), is of order t^{-1} as compared to the first dominant one [$X(t)$ is of order $t^{1/2}$ although the denominator is like $t^{3/2}$]. This must be compensated for by a change in the last term on the r.h.s. of (2.8), that is by a change of dX/dt of order t^{-1}, or by changing $X(t)$ by a perturbation of order $\ln t$. To end this section let us notice that this similarity solution is stable for one dimensional perturbations, whatever it means for unsteady solutions. Furthermore, the linear stability analysis, again restricted to one dimensional perturbations, may be carried out completely by using as independent variables

$x'' = x(Dt)^{-1/2}$ and $\ln t$, so that the stability problem becomes formally autonomous in time and eigenperturbations depend on time and position as $x^{\mu}\phi_{\mu}(x'')$, μ being a linear eigenvalue.

3

Effects of surface tension

In the one dimensional case considered in section 2, surface tension is unimportant, as there is no creation of surface between the two thermodynamic phases during the growth. However, these surface tension effects become important as soon as the interface is no longer planar, for a number of different reasons. Surface tension is important because it introduces a length scale (the capillary length), of the order of a few angstroms, although the formulation of the Stefan problem outlined in section 2 omits any 'intrinsic' or molecular length. This length scale introduces microscopic phenomena into the problem, contrary to what was claimed in the introduction, but it also introduces a short scale (ultraviolet...) cut-off absent in the previous mathematical formulation of the free boundary problem.

The need for this short scale cut-off appears in the following simple reduction of the Stefan problem. In more than one dimension, and at low undercooling, one may replace the diffusion equation by Laplace's equation (i.e. neglecting the time derivative) because then the dynamics is slow. Then the Stefan problem may be cast into a simple form [that is exact for the Saffman-Taylor instability (Saffman and Taylor [1958]; Pelcé [1988]), because then Laplace's equation gives the pressure field and does not arise from a quasistatic limit of a diffusion equation].

Let $G(t)$ be a closed curve (or surface in 3D), bounding the solid at time t, and let $\phi(\mathbf{r})$ be the potential (the mathematical idealization of the temperature field) solution of Laplace's equation with the b.c. $\phi = 0$ on G and ϕ tending to the undercooling Δ at infinity (this has to be amended slightly in 2D because of log divergence at infinity). This is supplemented by the Stefan-Lamé condition, which states that the outward normal velocity to the surface is proportional to the local (normal) gradient of ϕ. In two dimensions of space, this problem can be explicitly solved (Shraiman and Bensimon [1984]), by noticing that it is equivalent to the dynamics of the Riemann function conformally mapping G on a unit circle, and that the equation of evolution of this analytical map is solvable explicitly in a large class of examples. In particular this shows that singularities occur after a finite break-up time, because the Riemann function becomes singular on G itself.

This kind of solution is very interesting: first it shows that the diffusion equation, in its quasistatic limit, plus the Stefan-Lamé b.c. are enough to specify the evolution at least during a finite period of time; it also shows that this is an ill-posed problem for a 'large' class of initial conditions; finally, by looking at the solution just before it breaks up, one sees that its radius of curvature tends to zero near a point. Coming back to what was said before, one expects then that neglected small scale phenomena, all described under the heading of 'surface tension effects', will take over before break-up occurs in the real world. This break-up of solutions after a finite time may be understood qualitatively by adapting an argument due to Birkhoff.

Let $f(s,t)$ be a function describing the evolution of a perturbation depending on space (variable s, the curvilinear coordinate on G normalized to have unit circumference) and time t. Suppose too that $f(s,t=0)$ is a meromorphic function of s in a band up to the (nonzero) distance s_0 from the real axis. Then the Fourier expansion of $f(s,t=0)$ reads:

$$f(s,t=0) = \sum_{0}^{\infty} f_n e^{2i\pi ns},$$

where the Fourier components f_n decay as $e^{-2\pi ns_0}$ at large n. Notice now that linear stability theory predicts that the Fourier component f_n is unstable with a rate of instability proportional to n, then $f_n(t) = f_n(0)e^{unt}$, where u has the dimension of a velocity. This is the dependence of the growth rate of the Mullins and Sekerka [1964] instability, which is relevant to the present problem. Plugging this time dependence into the Fourier expansion of $f(s,t)$, one obtains that in this linear approximation:

$$f(s,t) = \sum_{0}^{\infty} f_n e^{2i\pi ns} e^{unt}. \tag{3.1}$$

Recalling now that the Fourier coefficients f_n decay as $e^{-2\pi ns_0}$ at large n, one sees that the Fourier series on the r.h.s. of (3.1) diverges at $t > s_0/u$. This reasoning is quite approximate, because it tries to extend to finite amplitude perturbations the result of linear stability analysis. However it is basically correct (as far as the estimate of the break-up time is concerned for instance), and applies also to other kinds of instabilities where the rate of growth is proportional to the wave number.

There have been some attempts to make the formulation more rigorous without solving the full nonlinear problem, for instance for the Kelvin-Helmholtz instability on a vortex sheet, but this is a very difficult question. The present analysis shows that if one adds the effect of 'surface tension' the break-up disappears: as shown for instance by the linear stability analysis of Mullins and Sekerka [1964], the short scale (large n) fluctuations are damped instead of being unstable, and so the above reasoning does not predict further break-up after a finite time.

We know of no precise explanation of the effect of the capillary terms (not yet explicitly introduced) on the break-up of the curve $G(t)$, although it is likely that the limit of a small capillary effect could make sense, because again there is an exact solution in the absence of those capillary terms. We guess this is not a simple matter: in the very closely related case of the *stationary* Saffman-Taylor fingers, it took a great deal of effort to understand what happens under weak but non-zero capillary effects. (See below the analysis of this limit for the case of the free needle crystal, which is quite similar, and the paper by Combescot *et al.* [1988].)

Thus it is likely that the evolution problem is even more complicated. Furthermore, although it is most likely that the Stefan problem as such yields break-up after a finite time for initially regular conditions in more than one dimension of space, the problem in the quasistatic limit is more difficult. Just before the break-up, the interface experiences a very large velocity in this quasistatic limit and it is not obvious that this quasistatic (slow dynamics) limit still applies there, even if the undercooling at infinity stays small.

Let us introduce more specifically the capillary phenomena. As we know from thermodynamics, capillary phenomena change the b.c. and make the local equilibrium conditions dependent on the curvature: this is the so-called Gibbs-Thomson (GT) condition. There are various levels of sophistication in this condition. At the fundamental level, one replaces the equilibrium assumption $T = T_{eq}$ at the solid/liquid interface by:

$$\frac{T - T_{eq}}{T_{eq}} = -\frac{\gamma T_{eq}}{L}\left(\frac{1}{R_1} + \frac{1}{R_2}\right). \tag{3.2}$$

In this GT formula, T is the actual temperature at the interface, γ the surface tension, L the latent heat and R_1 and R_2 are the principal radii of curvature of the interface. As noticed by Herring [1951, a], this formula is not valid for crystals, since then the surface tension γ depends on the orientation of the surface with respect to the crystal axis. The complete formula by Herring for 3D anisotropic crystals is quite complicated: it is derived in the book by Pelcé [1988]. We shall give the 2-dimensional formula only. Then γ is a function of an angle θ measuring the orientation of the interface with respect to the crystalline axis, and (3.2) is to be replaced by:

$$T - T_{\text{eq}} = -\frac{T_{\text{eq}}}{LR}\left[\gamma(\theta) + \frac{\mathrm{d}^2\gamma}{\mathrm{d}\theta^2}\right]. \tag{3.3}$$

There is a 'third level' of complication [the introduction of the GT formula (3.2) being the first one, and the equation (3.3), that is the Gibbs-Thomson-Herring (GTH) formula, the second one]. It corresponds to the introduction of facetting. Then a simple mathematical trick allows us to derive from (3.3) the length of the facets near equilibrium. When crystals are faceted, and when their surface is smooth (that is when all possible orientations are represented continuously on the crystal surface), one may extend the GTH formula to facets as follows.

Suppose that facets occur with the orientation θ_0. From the classical theory of Wulff [L. Landau and I.M. Lifshitz [1967, b], this corresponds to a value of the argument of γ where this function is continuous, but has a discontinuity in its first derivative: $\mathrm{d}\gamma/\mathrm{d}\theta$ has different limits, whether θ tends to θ_0 by upper or lower values. In (3.3) the radius of curvature R can also be written as $\mathrm{d}s/\mathrm{d}\theta$, s being the curvilinear coordinate along the surface of the crystal. Furthermore, let $T(s)$ be the temperature field along the facet, then one can write (3.3) as:

$$\mathrm{d}s[T(s) - T_{\text{eq}}] = -\frac{T_{\text{eq}}}{L}\left[\gamma(\theta) + \frac{\mathrm{d}^2\gamma}{\mathrm{d}\theta^2}\right]\mathrm{d}\theta. \tag{3.4}$$

Integrating this along the facet, one gets on the l.h.s. a contribution equal to the average temperature shift from T_{eq}. The contribution to the r.h.s. is dominated by the second derivative, since it is singular at the orientation where the first derivative of γ has a finite jump. All this yields the following condition:

$$\int_{\text{facet}} \mathrm{d}s \frac{L[T(s) - T_{\text{eq}}]}{T_{\text{eq}}} = -\left[\left(\frac{\mathrm{d}\gamma}{\mathrm{d}\theta}\right)_+ - \left(\frac{\mathrm{d}\gamma}{\mathrm{d}\theta}\right)_-\right], \tag{3.5}$$

where the r.h.s. represents the jump of the derivative of $\gamma(\theta)$ at θ_0. Indeed this jump is related to the step energy in the molecular theory of surface tension. The extension of (3.5) to three dimensions has been carried out by Ben Amar and Pomeau [1988, b].

As shown by recent work by Bowley *et al.* [1989] on the instability of growth of faceted crystals, this condition (3.5) is sufficient to define completely the b.c. for a crystal with facets in an undercooled melt, because it provides just the right number of b.c. for the diffusion equation (this is still the Stefan-Lamé condition, left unchanged by this introduction of capillary phenomena), and for the shape of the crystal itself. Other theories pioneered by Burton *et al.* [1951] introduce a mobility of the interface, imposing a definite relationship between the velocity of growth and the undercooling on the interface.

Such theories are not compatible with the solution of the diffusion equation in the melt, if one imposes *everywhere* on the facet the mobility condition, as this gives far too many conditions for the diffusion equation: one is elliptic and of second order and is not compatible with Dirichlet-like (the Gibbs-Thomson condition) and Neuman-like (the Stefan-Lamé condition) b.c. at the same time. In the case of the smooth (non-faceted) crystals, there are two such b.c. but one has enough degrees of freedom because of the arbitrary shape of the interface, these degrees of freedom obviously missing in the case of facets. It is reassuring to see here the consistency of the laws of physics.

A last extension of the Gibbs-Thomson law, due to Herring [1951, b], concerns the case where not all orientations are represented on the crystal surface. This happens (but is not restricted to this case) when facets merge with either smooth parts of the surface or other facets at an angle. Then the Herring formula (using our notations and the 2D geometry) is:

$$\int_{\text{facet}} ds \, \frac{L[T(s) - T_{\text{eq}}]}{T_{\text{eq}}} = \frac{\gamma_1}{\sin \theta} - \frac{\gamma_0}{\tan \theta}, \qquad (3.6)$$

where γ_1 is the surface tension of the surface merging with the facet, γ_0 is the surface tension of the facet itself and where θ is the angular jump of orientation of the surface at the boundary of the facet. This angular jump exists as soon as $\gamma + \gamma''$ is negative for crystal orientations (called forbidden orientations) near the facet. From Herring's work we can deduce the angular jump θ by assuming local equilibrium in the curved parts of the crystal, near the facet. It is obtained from the Wulff plot:

$$\gamma'(\theta) + \frac{\gamma_0}{\sin \theta} - \frac{\gamma(\theta)}{\tan \theta} = 0.$$

This relation gives the θ value required in (3.6). In the limit of small angular jump, θ is the solution of:

$$\gamma(\theta) + \gamma''(\theta) = 0 \quad \text{when} \quad \gamma(0) + \gamma''(0) < 0.$$

In the last part of this section we will show using a rather spectacular example how surface tension does influence the shape of growing crystals. This comes from experiments by McConnell *et al.* [1986]; Weis and McConnell [1984]. They studied the growth of needle crystals on so-called Langmuir monolayers floating on water. By increasing the surface pressure these 2D monolayers crystallize, and in some cases the crystals grow in the form of spirals. This pattern is related to the optical activity of the molecules making up the layer: the sense of rotation of the spirals (Fig. 1) changes with the sign of the optical activity and

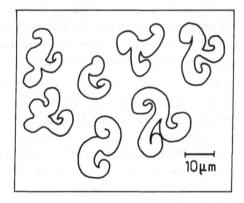

Figure 1: Spiral crystals, reproduced from a photograph of RDPPC at the air-water interface (McConnell *et al.*, [1986]).

there are no spirals without this activity. The fact that optical activity is linked to the external shape of crystals has been known since Pasteur separated left from right tartric acid using the fact that their crystals are not their own mirror image. As the molecules in Langmuir monolayers have a definite vertical orientation, the loss of mirror symmetry may be restricted to plane symmetries, i.e. to symmetry with respect to a line. But the occurrence of spiral crystals is not as directly related to the optical activity as it is in Pasteur's experiments.

The spirals are certainly in a very far from equilibrium configuration as the usual Wulff construction gives convex crystals only, and spirals are not convex. It has been suggested too that these spirals are due to long range forces that are sensitive to the optical activity, and not considered in the minimization of energy leading to the usual Wulff construction. This is quite unlikely, because the size of the spirals is of the order of micrometers, very much larger than the range of molecular forces, especially the ones sensitive to optical activity that involve high multipoles and so decay rather quickly with intermolecular distance.

The proposed explanation for spiral growth (Pomeau [1987]) makes use of the GTH condition for local curvature at the interface. This relates the local radius of curvature to the undercooling, as we have seen. However, in the experiments under consideration, undercooling plays a negligible role, and the growth is limited by the diffusion of impurities. But the GTH condition may be extended to show that the concentration of impurities on the interface also depends on the local curvature of this interface. This relation is changed in the presence of anisotropy, leading to something akin to (3.3). This implies that the concentration of impurities in the melt and the solid on both sides of the interface depends on the geometry of this interface, and of its orientation with respect to the crystal axis as in formula (3.4).

If the crystal has no mirror symmetry, as is the case for optically active materials, the two sides of a growing needle crystal will not be equivalent under a mirror symmetry. Then, from the GTH formula, more impurities will be incorporated into the crystal on one side than on the other. But, as shown in Fig. 2, this leads to a bending of the lattice, because of the gradient of foreign impurities in the lattice from one side to the other, and because in general these impurities locally distort the lattice when they try to fit into it. We leave to the interested reader to look at the original papers (Pomeau [1987] and refs. therein) for more information on this subject of spiral crystals.

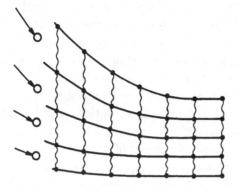

Figure 2: Spiral growth: bending of the lattice due to impurities.

4

Scaling laws for the needle crystal

In this section, we begin to study the problem of growth of needle crystals. This subject has a rather long and rich history, summarized in part in the *Review of Modern Physics* paper by Langer [1980]. More recent developments are to be found in the book by Pelcé [1988], in the course given by Langer at Les Houches [1987] and in the recent review article by Kessler *et al.* [1988]. In this section we shall follow a path slightly different from the usual one, and emphasize as much as possible simple approximations which allows us to understand where the problems are. The subject started with the exact solution by Ivantsov [1947] of the needle crystal problem (it seems to have been known earlier in some form but we have not been able to track down a relevant reference for this).

To understand the physics of this solution, one should note that the 1D solution of the Stefan-Lamé problem presented in section 2 is not 'optimal' (in a rather vague sense) because under uniform external conditions (the undercooling of the melt at infinity) it gives a speed of the solidification front tending to zero as time goes on. This is because in 1D it becomes more and more difficult as time goes on to get rid of the latent heat by molecular diffusion. The situation is far better in any higher dimension of space, because a needle crystal may 'radiate' (by diffusion) this latent heat on its sides, and so grow at constant speed.

As we shall see, this can be explained rather simply in the low undercooling limit, without resorting to the complete (and nontrivial) Ivantsov solution. It is interesting to ask if other solutions exist, for instance for crystals growing faster than constant speed. We would be inclined to think so, although those solutions would probably not survive the perturbation by surface tension (GT condition), and so be relevant for transient behavior only in real life.

Below we shall derive the proper scaling laws for the needle crystal from simple boundary layer estimates. We will do so first in the usual case when molecular diffusion only limits the growth. Then we shall consider the effect of an imposed axial flow on the growth, again in the framework of scaling arguments and boundary layer analysis. This will lead to the recognition of several regimes depending on the mutual values of the fluid velocity and of the undercooling, as they combine into a single dimensionless parameter denoted N.

As already mentioned, the foundations of this subject were laid many years ago by Ivantsov, when he derived his exact solution of the coupled diffusion and Stefan-Lamé equations for parabolic and paraboloidal shapes. Although this solution is correct, it does not say everything about the real physics, as a continuum of solutions comes out of this analysis simply because the physical data lack a length scale and so the solution is invariant under a continuous one parameter group (this incompleteness was noted by Ivantsov). It was noticed quite early too that by adding the GT condition, this spurious invariance is lost and a unique, or at least discretely many, solutions are selected. Even though this idea was true, a great deal of effort was needed to prove it. We try to review this work in sections 5 to 7.

Below (in section 4.1), we explain the simple scaling of the Ivantsov solution in the low undercooling limit and then how the GT relation leads to a selection of the needle crystal parameters, satisfying the famous relationship between the radius of curvature ρ at the tip and the growth velocity u_c:

$$\rho^2 u_c = \text{constant} \quad \text{(independent of the undercooling).}$$

However, it must be recalled that this scaling approach is not a complete answer as a difficult mathematical nonlinear eigenvalue problem for computing dimensionless constants still needs to be solved. It turns out precisely that this has no solution without anisotropy in surface tension between the solid and the melt (see sections 5 to 7).

The situation of a needle crystal growing in an axial flow already presents nontrivial questions of scalings and this too is an important problem for applications. The difference from the standard case (no flow) comes from the fact that the concentration field (or the excess of latent heat) is now both spread by molecular diffusion and advection. The balance between these two phenomena depends in principle on a dimensionless number, the so-called Péclet number: $\text{Pe} = \lambda u_\infty / D$ where u_∞ is the fluid velocity far from the crystal and λ a geometrical length scale. When Pe is large (and this can happen quite easily because the molecular diffusion coefficients in liquids — D — is usually very small), the convection takes over from the diffusion.

However things are more complicated than that because the length λ in the Péclet number depends itself on the efficiency of the convection-diffusion through the Stefan-Lamé and GT(H) relations. Thus we consider below how all these effects combine to produce various well defined regimes where different physical phenomena do not balance in the same way. This leads to three different regimes, delineated by the value of a single dimensionless quantity called N, which combines the imposed undercooling and flow velocity at infinity and various molecular parameters, such as the capillary length and molecular transport coefficients.

This question of growth in an axial flow was studied by Ben Amar *et al.* [1988] and by Ananth and Gill [1988, a, b], and a nice exact parabolic solution was found for an inviscid fluid flow. We plan here to consider an axial and steady flow of viscous fluid which creates a Blasius boundary layer along the crystal. In a first step (section 4.1), we rederive known scaling laws for both the growth rate and the tip radius of the free needle crystal as a function of the undercooling and in the absence of flow, then we extend the results to the case when a forced flow introduces an important perturbation in a sense which has to be defined. The calculations are presented for the 2D case; the results for the tridimensional case are indicated without any detailed derivation when they differ from the 2D case. Since our aim is to establish scaling laws, the equalities do not have to be understood rigorously; they indicate order of magnitude or equivalence only.

4.1 Scaling law without axial flow

Let us determine, as a first step, scaling laws for the boundary layer in the asymptotic part of the growing needle crystal. In the stationary frame of the crystal, the diffusion equation for the temperature field (or the impurity concentration field) reduces to:

$$u_c \frac{\partial \Theta}{\partial y} = D \frac{\partial^2 \Theta}{\partial x^2}, \tag{4.1}$$

where u_c is the growth velocity, D is the molecular diffusion coefficient (either thermal or solutal), and where x (resp. y) is the direction perpendicular (resp. parallel) to the crystal axis. Finally Θ is either the temperature or the concentration field. For small undercooling (this limit will be defined later on), the transverse width of the crystal is negligible compared to the thickness of the thermal boundary layer. Then, in the asymptotic part of the crystal, one can replace it by a straight line ($x = 0$, $y < 0$) to compute the diffusion field. This boundary layer corresponds to a similarity solution of (4.1) such that Θ is a function of the dimensionless quantity $Dy/u_c x^2$ only. For large arguments, or far away from the crystal, this function tends to Δ, the dimensionless undercooling previously defined as $(T_{eq} - T)c_p/L$.

The local thickness of the boundary layer is, as usual, the average distance over which a particle diffuses during the time y/u_c spent since it began to feel the needle crystal: $e_{th}(y) = (Dy/u_c)^{1/2}$. The crystal width $e_{cr}(y)$ at the distance y from the tip is found by considering the growth rate as given by the Stefan-Lamé law during this typical time y/u_c: as the crystal grows at the speed u_c, this is again the time elapsed since the crystal began to grow at a given location 'downstream'. Assuming the local temperature gradient along the x axis to be of order $\Delta/e_{th}(y)$, one gets:

$$e_{cr}(y) = \Delta e_{th}(y).$$

For small Δ, the crystal width is small compared to the thermal length. Near the tip, length scales along x or y become the same. The thickness of the diffusion boundary layer is fixed there by $e_{th}(y) = y = e_{th}$, so

$$e_{th} = \frac{D}{u_c}. \tag{4.2}$$

To obtain both the growth rate u_c and the tip radius, one has to consider the Stefan-Lamé and GT relations (Langer [1987], Kessler *et al.* [1988]). A first condition can be found by noticing that, near the tip, the convection term becomes negligible in the diffusion equation: the characteristic length scale is the tip radius, assumed to be small compared to e_{th}. As a consequence, the temperature field satisfies Laplace's equation with a Dirichlet condition on the crystal. For a half straight line, this field has a square-root singularity near the tip and behaves locally as

$$\Theta = \Delta\left(\frac{r}{e_{th}}\right)^{1/2} \quad \left[3D : \Theta = \frac{\Delta\ln(r/\rho)}{\ln(e_{th}/\rho)}\right],$$

with r the length scale near the tip. The denominator e_{th} allows us to match the diffusion field with the outer boundary layer. The gradient of Θ near the tip is of order:

$$\Delta\left(\frac{1}{\rho e_{th}}\right)^{1/2} \quad \left[3D : \frac{\Delta}{\rho\ln(e_{th}/\rho)}\right]. \tag{4.3}$$

Combining this with the Stefan-Lamé law, one finds the laws for the growth rate u_c:

$$u_c = \frac{D\Delta}{(\rho e_{th})^{1/2}} \quad \left[3D : u_c = \frac{D\Delta}{\rho\ln(e_{th}/\rho)}\right], \tag{4.4}$$

equivalent to the Ivantsov relation between the Péclet number $(Pe = u_c\rho/D)$ and D:

$$Pe = \Delta^2 \quad \left[3D : Pe = -\frac{\Delta}{\ln\Delta}\right].$$

It is perhaps of interest to notice that this approach is consistent with a non-constant speed of growth. Assuming $u_c(t) \approx t^\alpha$ at large t, $\alpha > 0$, one sees that the radius of curvature at the tip, ρ, decreases as $t^{-\alpha}$ at large t. This could even be consistent with a parabolic shape, at least in the limit $\Delta \to 0$. Indeed such exotic solutions should disappear under the effect of surface tension.

The Gibbs-Thomson law provides the second relation needed to fix u_c and ρ independently:

$$\theta = \frac{\Gamma}{\rho},$$

where Γ is the capillary length that is proportional to the surface energy (resp. tension). Then:

$$\Delta\left(\frac{\rho}{e_{\text{th}}}\right)^{1/2} = \frac{\Gamma}{\rho} \quad \left[3\text{D} : \frac{\Gamma}{\rho} = -\frac{\Delta}{\ln \text{Pe}}\right]. \qquad (4.5)$$

When combined with (4.4), Equation (4.5) gives, both in the 2D and 3D case, the famous:

$$\rho^2 u_c = D\Gamma. \qquad (4.6)$$

This eliminates the undercooling, often difficult to measure experimentally. It does not exhibit the thermal length e_{th} at the tip. To get (4.6), we used the hypothesis of small undercooling and assumed a parabolic tip. In particular (4.6) does not require the crystal to behave like a parabola in its asymptotic part. This is why it is observed to hold in experiments wherein the crystal sides exhibit secondary sidebranchings. Equation (4.6) is no longer valid as soon as the undercooling is not small, as shown by Ben Amar [1990].

Below we shall use the same method as above to establish the relevant scaling laws when a steady and axial flow is imposed in the melt (assumed to be a viscous fluid). Then it is necessary to distinguish several limits depending on the relative order of magnitude between the three important length scales: ρ, the tip radius of the crystal; e_{th} and e_{hy}: thermal and hydrodynamical lengths, estimated near the tip. Other quantities relevant for the asymptotic part of the crystal are the same in different cases, so they will be derived in section 4.3. We will use the same notation for the same physical quantities although the algebraic expressions of this quantity differ most of the time according to the considered approximation.

4.2 Scaling laws for the growth in a forced flow when $\rho < e_{\text{th}} < e_{\text{hy}}$

Here, we shall not consider the case where the flow is a small perturbation: then, obviously, the results of section 4.1 remain approximately valid and can be applied in a first approximation. This subsection is devoted to the case where the flow is an important perturbation in a sense to be defined later. The convection-diffusion equation (4.1) has now to be changed into:

$$u_c\frac{\partial\Theta}{\partial y} + \mathbf{u}_h \cdot \nabla\Theta = D\frac{\partial^2\Theta}{\partial x^2}, \qquad (4.7)$$

where u_h is the flow field in the vicinity of the needle crystal measured in its own moving frame. Contrary to recent work (Ben Amar *et al.* [1988]), we shall consider the case — which is *a priori* realistic if the value of u_h at infinity is large enough — where the first term on the left side of Eq. (4.7) is negligible compared to the second one and where the Blasius boundary layer is much thicker than the thermal one. This last condition physically means that the vorticity diffuses more quickly than heat (or the scalar field Θ). It then implies that the Prandtl number (or the Schmidt number for impurities) is large. The first condition depends on the dimensionless ratio: u_c/u_∞) where u_∞ is the flow velocity far away from the solid. Evaluating the viscous boundary layer thickness in the same way as done for the thermal layer, we get

$$e_{\text{hy}}(y) = \left(\frac{\nu y}{u_\infty}\right)^{1/2}, \tag{4.8}$$

where ν is the fluid kinematic viscosity. This estimate is consistent with the Stokes approximation, and also with the Blasius one where the Reynolds number is arbitrary. The following calculation based on this hypothesis would become inconsistent if, near the tip, the hydrodynamic boundary layer characterized by $e_{\text{hy}} = \nu/u_\infty$ is slimmer than the tip radius. We will come back later to this condition: $\nu/u_\infty \gg \rho$. In the asymptotic branches of the crystal, the axial component of the hydrodynamic speed is given by $u_y = u_\infty x/e_{\text{hy}}(y)$, as far as $x < e_{\text{hy}}(y)$. By substitution of this estimate into the convection-diffusion equation near the parabola, we derive:

$$u_y \frac{\partial \Theta}{\partial y} + u_x \frac{\partial \Theta}{\partial x} = D \frac{\partial^2 \Theta}{\partial x^2}, \tag{4.9}$$

where u_y and u_x are the Cartesian components of the flow velocity. By noticing that, due to the incompressibility of the liquid phase, both terms on the left side of (4.9) have the same order of magnitude, we get the estimate:

$$e_{\text{th}}(y) = (D^2\nu)^{1/6}\left(\frac{y}{u_\infty}\right)^{1/2} \tag{4.10}$$

for the thermal length along the x-axis. This is consistent with the assumptions if $e_{\text{th}}(y) \ll e_{\text{hy}}(y)$, or $u/D) \gg 1$ as mentioned above. We can neglect the first term in (4.7), as done before if

$$\frac{u_c}{u_\infty} \ll \left(\frac{D}{u}\right)^{1/3}. \tag{4.11}$$

Let us now study the vicinity of the tip. The thickness of the thermal boundary layer is then e_{th} such that $e_{\text{th}}(e_{\text{th}}) = e_{\text{th}}$ where we use the form of $e_{\text{th}}(y)$ given by (4.10). It reads:

$$e_{\text{th}} = \frac{(D^2\nu)^{1/3}}{u_\infty}$$

which defines for this case the typical length around the tip analogous to the length e_{th} already introduced for the case without flow. The following calculation for various estimates in the vicinity of the tip uses the same strategy as before, but in the absence of axial flow. On one hand it leads to the relation (4.6) between the tip radius and the growth rate and, on the other hand, it leads to the following dependence of these two quantities in the undercooling:

$$u_c = \frac{u_\infty^{2/3} D^{5/9} \Delta^{4/3}}{\Gamma^{1/3} \nu^{2/9}} \quad \left[3D : u_c = \frac{D\Delta^2}{\Gamma \ln[(D^2\nu)^{1/3}\Delta/u_\infty\Gamma]} \right]. \quad (4.12)$$

The condition (4.11) and the initial assumption ($\rho < e_{\text{th}}$) can be expressed in terms of the control parameters: the undercooling, and the hydrodynamical velocity. This is equivalent to the following inequalities:

$$1 < N < \frac{1}{\Delta^3} \text{ in 2D} \quad \text{and} \quad 1 < N < \frac{1}{\Delta} \text{ in 3D}$$

with N defined by $N = \Delta(D^2\nu)^{1/3}/u_\infty\Gamma$.

4.3 Scaling laws for the growth with an axial flow such that $e_{\text{th}} < \rho < e_{\text{hy}}$

When the velocity u_∞ is such that the three typical lengths of the problem, that is ρ, e_{th}, e_{hy}, are such that $e_{\text{th}} < \rho < e_{\text{hy}}$ (contrary to the former case where they are ordered as $\rho < e_{\text{th}} < e_{\text{hy}}$), it is necessary to modify the estimates which have just been presented. The inequality $e_{\text{th}} < \rho$ means that the thermal boundary layer is much thinner than the characteristic width of the crystal (measured for example by the radius of curvature at the tip). The length scale e_{th} cannot be found as previously, where we considered the needle crystal as a simple half line, at least for the determination of the thermal field in its vicinity. Our starting point is again the boundary layer approximation for the hydrodynamical velocity field which keeps the following dependence in x and y:

$$u = u_\infty F\left[\frac{xu_\infty^{1/2}}{(\nu y)^{1/2}}\right],$$

where the function F is deduced from the solution of the classical Blasius similarity equation. Due to the hydrodynamic b.c., this function must go to 1 for large arguments and vanish for small arguments. Near zero, its Taylor expansion begins with a linear term. Previously, this kind of argument led us to an estimate of the thermal boundary layer thickness since we always assumed that this one was nearly parabolic:

$$x = (e_{\text{th}}y)^{1/2}.$$

When this estimate is put in the above expression for the velocity u, we conclude that this parabolic dependence holds for small arguments of the function F and that:

$$u = u_\infty \left(\frac{e_{\text{th}}}{e_{\text{hy}}}\right)^{1/2}.$$

This estimate, when put into the convection-diffusion equation leads to the previous result for the thermal boundary layer. In the case under consideration, the thermal boundary layer is *a priori* thinner than the tip radius. The order of magnitude of the velocity field is always given by the boundary layer approximation but now it has to be evaluated near the parabola, which physically represents the crystal, given by:

$$x = (\rho y)^{1/2}.$$

So the new estimate for the magnitude of the velocity field is:

$$u = u_\infty \left(\frac{\rho}{e_{\text{hy}}}\right)^{1/2}.$$

In the thermal boundary layer and near the tip, we need an estimate for the gradient of the velocity field since the velocity must vanish on the crystal. Then, along the crystal surface and near the tip, the gradient of the velocity within the boundary layer approximation is given by:

$$\text{grad}\, u = \frac{u_\infty}{(\rho e_{\text{hy}})^{1/2}}.$$

From this, we deduce an estimate of u in the thermal boundary layer of width e_{th} (unknown for the moment):

$$u = \frac{u_\infty}{(\rho e_{hy})^{1/2}} e_{th}.$$

(4.13)

When this estimate for the velocity is introduced into the convection-diffusion equation, we get:

$$e_{th} = \frac{D^{1/2}\rho^{1/4}e_{hy}^{1/4}}{u_\infty^{1/2}}.$$

(4.14)

We must now establish the needle crystal parameters, by use of the Stefan-Lamé and Gibbs-Thomson relations and determine the range of validity of the present calculation:

$$u_c = \frac{D\Delta}{e_{th}} \quad \text{and} \quad \rho = \frac{\Gamma}{\Delta},$$

so
$$u_c = \Delta^{5/4}u_\infty \left(\frac{D}{\nu}\right)^{1/2}\left(\frac{\nu}{\Gamma u_\infty}\right)^{1/4}.$$

(4.15)

Equation (4.15) is included in order to specify dimensionless quantities. This second case is relevant if $(D/\nu)^{2/3} < N < 1$. The formula for the 3D case is identical if x is the positive distance from the axis.

4.4 Scaling laws for the growth with an axial flow such that $e_{th} < e_{hy} < \rho$

Let us now consider the last case, where the hydrodynamical velocity is large enough to make the thickness of the thermal and the hydrodynamic boundary layers much less than the tip radius as given by the GT condition:

$$\rho = \frac{\Gamma}{\Delta}.$$

Near the tip, the boundary layer does not obey the Stokes approximation, since the Reynolds number calculated with the tip radius is $\text{Re} = \rho/e_{hy}$, much larger than 1 by hypothesis. We are in the classical situation of a viscous boundary layer, assumed to be attached, which is certainly true in the vicinity of the stagnation point in front of the crystal. The external flow when calculated at distances of the order of the tip radius ρ is the stationary Euler flow around the tip, the no-slip condition being ensured by the viscous boundary layer. The classical calculation of hydrodynamical boundary layers gives for the vicinity of the stagnation point a typical length of order:

$$e_{\text{hy}} = \left(\frac{\rho^2 \nu}{u_\infty}\right)^{1/3}.$$

If the Péclet (or Schmidt) number which describes the diffusion of the scalar quantity Θ is small, one can neglect e_{hy} compared to the thermal length. In this case, it is necessary to apply the method of Ben Amar *et al.* [1988] to solve this problem but one needs to apply moreover the GT condition. We have throughout been interested in the other limit, where the diffusion of Θ is much slower than that of vorticity. Then, the thermal boundary layer is inside the hydrodynamic boundary layer where we can evaluate the velocity by $u_\infty(e_{\text{th}}/e_{\text{hy}})^2$. The square comes from the Taylor expansion of the flow field near a stagnation point that begins by second order terms. Inserting this approximation into the convection-diffusion equation one gets:

$$e_{\text{th}} = \frac{D^{1/3}\rho^{4/9}\nu^{2/9}}{u_\infty^{5/9}}, \tag{4.16}$$

which, together with the relation $u_c = D\Delta/e_{\text{th}}$, leads to the approximate value of the growth rate:

$$u_c = \frac{\Delta^{13/9}D^{2/3}u_\infty^{5/9}}{\nu^{2/9}\Gamma^{4/9}}. \tag{4.17}$$

Knowing the growth rate, this applies to flows such that

$$N < \left(\frac{D}{u}\right)^{2/3}.$$

4.5 Scaling laws for the asymptotic branches

As suggested by the geometry of the problem itself, it is unlikely that the asymptotic branches of the crystal depend greatly on the physical phenomenon fixing the curvature of the tip. As shown below, these parts are dependent on local parameters and also on the growth rate u_c, which will be left undefined and should be replaced by the convenient expressions derived above. The main result of this section will be that the asymptotic parts are always parabolic and have in Cartesian coordinates:

$$y = \frac{x^2}{E_c},$$

where E_c, the length scale, is to be found. If the needle crystal were defined by a unique length scale E_c, it should be of the order of magnitude of the tip radius. This is what happens in the first two cases considered above: growth in 2D with $\rho < e_{th} < e_{hy}$ or without flow. On the other hand, in the other cases, E_c does not have the same order of magnitude as ρ: if the inner (= near tip) problem has an asymptote for $x \approx \rho$, we expect that the matching is achieved along this parabola for $y \approx \rho^2/E_c$, which is consistent whatever the value of ρ and E_c. At the distance y from the tip, the width of the viscous boundary layer is:

$$e_{hy}(y) = \left(\frac{y\nu}{u_\infty}\right)^{1/2},$$

from which we deduce by the same argument as before that:

$$e_{th}(y) = (yE_{th})^{1/2},$$

with $E_{th} = (D\nu^{1/2})^{2/3}/u_\infty$. The crystal width $e_c(y)$, at distance y from the tip, is calculated by noticing that the growth rate in the x-direction is given by the Stefan-Lamé law, with a gradient of temperature of order $\Delta/e_{th}(y)$, that is:

$$u_c(y) = \frac{D\Delta}{e_{th}(y)}.$$

We get $e_c(y)$ now by integration of the growth rate during the growth time at the distance y, that is y/u_c. This yields:

$$e_c(y) = (yE_c)^{1/2} \quad \text{with} \quad E_c = \frac{D^2\Delta^2}{E_{th}u_c^2},$$

which is the most general expression for this length scale and applies also to the case without flow, in 2D. In the 3D case, because of curvature effects, it is only valid if $E_c > E_{th}$. In 2D and with an external flow, one has the following general approximation for E_c:

$$E_c = \frac{\Delta^2 D^{4/3}u_\infty}{\nu^{1/3}u_c^2}. \tag{4.18}$$

In this section 4, we have derived from simple scaling arguments the relationship between the observable parameters of a needle crystal: its tip radius and velocity of growth and the control parameters, that is the undercooling Δ and the flow velocity at infinity u_∞. There are three regimes of growth depending on the value of a single quantity N, a combination of control parameters and of molecular parameters such as the diffusion coefficient D, the viscosity ν and the capillary length Γ. Depending on the values of N, we have found different scaling laws for the growth rate which can in principle be checked experimentally (Glicksman *et al.* [1986], Bouissou *et al.* [1989]). Obviously, precise comparisons with experimental measurements require the determination of numerical constants, out of reach of this kind of dimensional analysis. To get them, it is necessary to solve free boundary problems of this kind, as studied in sections 5 to 7.

5

The integral equation of Nash and Glicksman and its low undercooling limit

In this section we present the integral equation that has to be solved in order to find the shape of a needle crystal. This equation, and also some of its variants, were written first by Nash and Glicksman [1974]. They have since been the object of rather intensive studies. Unfortunately, as mentioned in the introduction, many of their most important properties depend on non-trivial mathematics. To try to convey them, we have chosen to write the Nash-Glicksman equations first, then to reduce them in the low undercooling limit. Postponing the study of those equations by themselves to section 7, we shall expose in section 6 the basic idea of the relevant mathematical method, due to Kruskal and Segur [1985], which deals with a simplified 'geometrical' model.

The starting point of the theory by Nash and Glicksman is to notice that one can solve analytically the heat equation for arbitrarily distributed heat sources in a uniform medium, which is the situation for a symmetric model. This is done by using a Green's function method. This is interesting in that it transforms a problem with two unknown functions, the temperature field and the shape of the growing crystal, into one with a single unknown function, i.e. the shape of the crystal. This kind of reduction is often used when a simple solution of the field equation is known, for instance in radiation of antenna with a nonsimple shape. Here indeed we know how to solve the heat equation in a uniform medium by this Green's function technique.

The price to be paid is that the integral equation depends *a priori* on the whole history of the deformation of the growing crystal, since the temperature field depends on the release of latent heat from the very beginning. This is however not a problem if one assumes, as we shall do, that the shape of the needle crystal is invariant in a moving frame of reference.

Our starting point will be the equations of Nash and Glicksman that read in 2D:

$$\Delta + \frac{\xi_{xx}}{(1+\xi_x^2)^{3/2}} = \frac{1}{2\pi} \int_0^\infty \mathrm{d}\tau \int_{-\infty}^\infty \mathrm{d}x' \frac{\mathrm{d}\zeta(x', t - \tau)}{dtau} \tag{5.1}$$

$$\times \exp\left\{ -\frac{1}{2\tau} \left[\left(\xi(x,t) - \xi(x', t - \tau)\right)^2 + (x - x')^2 \right] \right\}.$$

In this equation, as often done in the field, the unit length has been taken as the capillary length, $\Gamma = \gamma/L$ (L being an energy per volume and γ an energy per area, Γ has the dimension of a length, and is typically of order of the angstroms). The temperature unit is c_p/L, the time scale is $\gamma^2/(L^2 D)$. The unknown quantity in the equation (5.1) is the history of the shape of the needle crystal defined by the curve of the Cartesian equation $y = \xi(x, t)$, x and y being the rectangular coordinates as in section 4, but in a fixed frame of reference here. The equation (5.1) expresses that the temperature on the surface of the growing solid satisfies the GT condition. The second term on the l.h.s. is the curvature written in Cartesian coordinates, the first term represents the contribution to the temperature field from the undercooling at infinity, and the r.h.s. the change of temperature due to the release of latent heat by solidification. The two contributions are added because of the linearity of the heat equation. The equation (5.1) is non-local in time, as it relates the present shape to the whole history of the growth (τ-integral).

We shall consider here a simple situation, that of a needle crystal growing at constant speed, where the function $\xi(x, t)$ depends on space and time as $\xi(x) - Ut$. Then one can perform the integral over time (τ) in equation (5.1) and gets:

$$\Delta + \frac{\xi_{xx}}{(1+\xi_x^2)^{3/2}} = \frac{U}{2\pi} \int_{-\infty}^\infty \mathrm{d}x' \exp\left[-\tfrac{1}{2}U\big(\xi(x) - \xi(x')\big)\right] \tag{5.2}$$

$$\times K_0\big(\tfrac{1}{2}U|R(x) - R(x')|\big).$$

The quantity $|R(x) - R(x')|$ is for the distance between two points of the interface: $[(x - x')^2 + (\xi(x) - \xi(x'))^2]^{1/2}$, and K_0 is the modified Bessel function of first order, the other notations and units being the same as in (5.1).

The Ivantsov solution to this equation is of the form:

$$\xi_{\mathrm{Iv}} = -ax^2.$$

It is a solution of (5.2), but without the curvature term (the second on the l.h.s.). Dropping this term is equivalent to the b.c. $T = T_{\mathrm{eq}}$ on the interface, which is also equivalent to neglecting the curvature in the GT relation. It is not obvious that such a parabolic form is a solution of (5.2), and this depends on non-trivial relations between special functions. It is slightly easier to show that this holds for (5.1), that is by plugging in $\xi(x,t) = -ax^2 - Ut$, and then making various changes of variable of integration to show that the final result is a constant (instead of a function of x). This is done in Pelcé and Pomeau [1986], and was done by Ivantsov by a direct solution of the Stefan problem .

The final result is a relation between the two dimensionless quantities U and Δ. This neglect of the curvature term in (5.1–5.2) is not an obviously consistent approximation, since it neglects a term formally of order 1. There is another related difficulty with this limit: (5.1–5.2) have been written by taking the capillary length as the unit length, and neglecting the curvature in those equations amounts precisely to discarding the physical effects where capillary phenomena enter. This means in particular that if one drops this GT term, one can rewrite the equation (5.1) with exactly the same form, but with a different length scale.

This is the invariance alluded to before: equation (5.2) keeps the same form if one multiplies the inverse velocity $1/U$, ξ and x by the same arbitrary constant. This means in particular that whatever the solution of the corresponding equation, it defines only a relationship between U/a (actually the Péclet number $\mathrm{Pe} = U/aD$) and the dimensionless undercooling Δ:

$$\Delta = \tfrac{1}{2}(\pi\,\mathrm{Pe})^{1/2} \exp\!\left(\tfrac{1}{4}\,\mathrm{Pe}\right) \mathrm{Erf}\!\left(\tfrac{1}{2}\,\mathrm{Pe}^{1/2}\right). \tag{5.3}$$

This last expression is valid for an arbitrary Péclet number, Erf being the error function. It has two interesting limits: the small Pe limit, which shows that $\mathrm{Pe} \approx \Delta^2$, a result already found in subsection 4.1 of section 4. The other limit is Pe tending to infinity and Δ tending to 1, a situation somewhat reminiscent of the singular behavior of the Stefan problem in 1D at $\Delta = 1$. As said before, in most experimental situations, the undercooling is quite small, so that the first limit (Δ and Pe small) is the most relevant one. Thus, it is tempting to try to find a similar limit (Δ and $U > 0$) for the 'true' equation (5.2) with the curvature term included. This needs a trick, because if one takes the limit $U \to 0$ in (5.2) as it is, one gets a very strong divergence of the integral term: the function K_0 diverges logarithmically for small arguments.

To take this limit (small undercooling, small Δ) without getting into such divergence problems, one notices that the capillary effects are negligible far away from the tip where the needle crystal should become approximately parabolic. On those asymptotic branches, the curvature term in (5.2) becomes negligible, and the integral equation becomes identical to the one for the Ivantsov parabola; whence the solution must tend to an Ivantsov parabola:

$$\xi(x) \propto \xi_{\text{Iv}}(x) \propto -ax^2 \quad \text{when} \quad x \to \pm\infty. \tag{5.4}$$

Recalling that ξ_{Iv} is a solution of (5.2) without the curvature term on the left hand side, one may replace the constant Δ by the integral on the r.h.s. to which it is equal, from Ivantsov's solution. Inserting the integral on the r.h.s. instead, we derive the following more suitable equation for analytical and numerical treatment — including at finite Pe, and completely equivalent to (5.2):

$$\frac{\xi_{xx}}{(1+\xi_x^2)^{3/2}} = \frac{C}{2\pi} \int_{-\infty}^{\infty} dx' \exp\left\{-\tfrac{1}{2}\text{Pe}\left[\xi(x) - \xi(x')\right]\right\} \tag{5.5}$$
$$\times K_0\left(\tfrac{1}{2}\text{Pe}\left|R(x) - R(x')\right|\right) - (\xi \to \xi_{\text{Iv}}),$$

where $(\xi \to \xi_{\text{Iv}})$ means the same expression as the former one, but with ξ_{Iv} replacing ξ. Furthermore we have introduced for convenience a new quantity C equal to U/a^2. The equation (5.5) has C as a nonlinear eigenvalue, depending *a priori* on Pe. The imposed undercooling Δ is hidden in the continuous parameter Pe: the subtracted Ivantsov [1947] solution has the same undercooling and thus the same Pe as the sought solution (Pe being defined by the value of a on the asymptotic branches). In order to calculate the growth rate, one has to solve (5.5) and find the Péclet dependence of the eigenvalue spectrum, since

$$v = \text{growth rate} = U\frac{D}{\Gamma} = \frac{\text{Pe}^2}{C(\text{Pe})}\frac{D}{\Gamma}. \tag{5.6}$$

Much progress has been made in the last three years (this being written in 1989) by looking at two different limiting values of the undercooling Δ : $\Delta \to 0$ (Ben Amar and Pomeau [1986, a, b], Ben Amar and Moussallam [1987], Kessler *et al.* [1986]) or $\Delta \to 1$ (Caroli *et al.* [1986], Barbieri [1987]). The first limit (vanishing Péclet number limit) seems relevant in many experiments of solidification induced by thermal diffusion, except the very recent experiment on nickel by Willnecker *et al.* [1989], whose results will be briefly discussed in section 7.

To come to the limit we are interested in, that is Pe small, the diffusion length as given by D/U is quasi-infinite, as compared to the tip radius. This would be equivalent to replacing the diffusion equation by Laplace's equation. But we know that in 2D this has solutions with a logarithmic growth at large distances. This is what is reflected here in the limiting behavior of K_0 at small arguments (large diffusion length):

$$K_0(x) \propto -\ln(\tfrac{1}{2}x) \quad \text{as} \quad x \to 0^+.$$

If one inserts this limit form into (5.2), one would get a logarithmically diverging integrand (at small U). But the subtraction leading from (5.2) to (5.5) now makes the integrand formally convergent, and with a well defined limit at Pe $\to 0$. The physical meaning of this substitution is that one computes the far temperature field by replacing the exact solution by the Ivantsov parabola with the same asymptotic branches. Similar methods may be used to analyze the diffusion limited growth (again at small Δ) of crystals with a compact form: one replaces the growing crystal by a circle with the same area as the real crystal to compute the far field and then makes the subtraction as done here to compute the near temperature field on the actual crystal by solving Laplace's equation. In the present problem, one is left with the following integral equation:

$$\frac{\xi_{xx}}{(1+\xi_x^2)^{3/2}} = -\frac{C}{2\pi} \int_{-\infty}^{+\infty} dx' \ln \frac{|R(x) - R(x')|}{|R_{\mathrm{Iv}}(x) - R_{\mathrm{Iv}}(x')|}. \tag{5.7}$$

The dependence on Pe has disappeared in (5.7), after the introduction of C. The rest of the notes will be mostly devoted to the analysis of this integral equation as well as to various extensions of it. First it is important to understand why (5.7) is truly a nonlinear eigenvalue problem. This means that if one imposes the condition that the solution tends to the Ivantsov parabola at large distances, and if one assumes too, as seems reasonable, that this solution is a symmetric needle crystal, i.e. that $\xi(x)$ is an even function of x, then there is one more condition that they are 'free parameters' and this supplementary condition can be accounted for only by specific choices of the eigenvalue C. It must be understood too that such a 'counting argument' does not imply that a solution exists or does not exist: if one thinks of the bound states of the Schrödinger equation in a 1D localized potential, it is one thing to show that the energy is an eigenvalue, it is another to prove that such a bound state exists or not. Such a relevant eigenvalue exists if the potential is attractive somewhere.

To show that (5.7) is an eigenvalue problem, one considers the expansion of the solution $\xi(x)$ at infinity. This expansion can be done order by order, but a supplementary condition must be added, playing the rôle of a 'quantization' condition. However, this leaves things in a quite unsatisfactory state, as all this does not say whether this eigenvalue problem has solutions or not. Numerical work, by Meiron [1986] and by Ben Amar and Moussallam [1987] indicated strongly that such a solution does not exist. This was confirmed later by a detailed analysis of this equation in the two limits C tending to infinity (Ben Amar and Pomeau [1986a]) and C tending to zero (Ben Amar and Pomeau [1986b]). This is the limit that we are going to consider from now on.

The difficulty (and interest) of this limit is that it depends on effects of asymptotes 'beyond all orders' in the terminology of Kruskal and Segur [1985]. A regular order by order expansion in $1/C$ does not show any indication of quantization of the solution: it depends on terms that are of transcendentally small order in $1/C$ at large C (they are actually of order $\exp(-KC^{1/2})$, K pure number). As the matter is quite complicated, we shall explain first the basic ideas in section 6. Then, in section 7, we shall show how to apply this to (5.7) and how solutions are recovered by adding a small anisotropy in surface tension.

6

Asymptotes beyond all orders in the geometrical model

In this section we present the Kruskal and Segur method for the same model as chosen by those two authors. This belongs to the class of so-called geometrical models of crystal growth. In the previous section we have shown that, when the growth is limited by diffusion, the diffusion field introduces a sort of nonlocal (in space and time) interaction between the various positions on the surface of the growing solid. Geometrical models try to avoid this kind of complication by reducing the interaction to a purely local one. This reduction becomes close to the physics when the diffusion field has fast and local dynamics, compared to that of the surface itself.

This means in particular that the thickness of the thermal boundary layer is much less than the radius of curvature of the interface. This happens when the dimensionless undercooling Δ is close to 1^-. It is even possible in this limit to derive this local dynamics from the Nash-Glicksman equations [1974] in a formal way. However this is not completely obvious and the final result remains quite difficult to handle. The simpler geometrical model we are going to investigate is derived by ad hoc assumptions, based in part on 'natural' symmetries that such a local equation should have.

In subsection 6.1 we shall present the model, which reduces to the solution of the third order nonlinear ordinary differential equation (6.3). Counting the number of free parameters, given the b.c., there are too many b.c., given the order of the equation, and the rest of this section is devoted to this point. The equation (6.3) depends formally on a dimensionless parameter called δ that can be seen as the capillary length. The solution of (6.3) is expanded formally in powers of this parameter. But it turns out that this formal expansion is diverging. This is shown by extension in the complex plane of the variable.

In particular the formal perturbation expansion breaks down near a singularity in the complex plane because the terms of increasing order are more and more singular near this point. The resummation of the most diverging terms is carried out by a boundary layer method and yields the 'inner equation' (6.13): this inner equation is to be satisfied by the solution of the original equation near the singularity.

The connection between this and the solution of (6.3) is made in subsection 6.3: when linearizing the original equation near its formal solution, one gets an homogeneous WKB-term, with an amplitude formally unrelated to the rest of the regular expansion at any order in powers of δ. But near the singularity in the complex plane this WKB-term and the regular part of the expansion become of the same order and so mix together. This allows to find this amplitude through the asymptotic behavior of the solution of the inner equation. Finally this WKB-term is responsible for a violation of one of the b.c. by a term of transcenden-

tally small order in the expansion parameter δ out of reach of ordinary
perturbation expansion.

6.1 Elementary properties

Using as far as possible the same notations as Kruskal and Segur, we
wish to introduce a phenomenological law for a curved solidification front
moving at constant velocity. Let this velocity be the unit velocity and θ
be the local orientation of the surface (we shall consider a 2D situation
so that this surface will be actually a line and θ an angle). Then the
condition that the moving front keeps the same shape as time goes on,
reads:

$$v_n = 1 \cdot \cos\theta, \qquad (6.1)$$

where v_n is the normal velocity at the orientation θ, which is the angle
of the normal to the front with the direction of growth, $\theta = 0$ being the
orientation of the tip and $\theta = \pm\frac{1}{2}\pi$ the orientation of the asymptotic
branches. This kinematic condition for the shape invariance is independent of the model under consideration. The estimated condition for the
dependence of v_n on the local parameters of the surface now follows.
This is to model physical processes that depend *a priori* on the intrinsic
properties of the 'surface' (actually a line), so that v_n has to depend on
the parameters as the curvature κ, and its derivatives with respect to
the curvilinear distance s. We shall write derivatives of any function of s
as $\dot{f} = \mathrm{d}f/\mathrm{d}s$. The choice of the phenomenological relation between v_n
and κ and its derivatives is:

$$v_n = \kappa + \delta^2\, \ddot{\kappa}\,. \qquad (6.2)$$

In equation (6.2), the coefficient of κ has been set to 1; this can always
be done by a proper rescaling of lengths for instance. Furthermore there
is no first derivative with respect to s, because it would lack invariance
under a change of orientation of s, and finally the coefficient in front of
the second derivative has the dimension of a length square, δ being a
length scale with a rôle similar to that of the capillary length. We shall
study the case of a small 'capillary length', that is the limit $\delta \to 0$. The
choice of sign for the coefficient of the second derivative on the r.h.s.
of (6.2) is dictated by the fact that this second order term stabilizes the
dynamics of small scale fluctuations, as it would follow from a standard
linear stability analysis of the solution $\theta = \frac{1}{2}\pi$, $v = \kappa = 0$. The

other sign choice for the second derivative in (6.2) would instead lead to catastrophic amplification of the short scale fluctuations. From (6.1) and (6.2) and from $\kappa = \dot{\theta}$, we get:

$$\delta^2 \, \dddot{\theta} + \dot{\theta} = \cos\theta. \tag{6.3}$$

The boundary conditions for this ordinary differential equation are as follows: the tip may be chosen (arbitrarily) as the point of curvilinear coordinate $s = 0$, and as already discussed, $\theta = 0$, at that point. Another condition is that θ is an odd function of s, in order to represent a symmetric needle crystal. As equation (6.3) is of third order, the free parameters at $s = 0$ may be chosen as $\theta \, (= 0)$, $\dot{\theta}$, and $\ddot{\theta}$. An odd function has vanishing second derivative at $s = 0$ (and any even derivative in general, but this holds formally true for solutions of (6.3) as soon as θ and its second derivative at $s = 0$ vanish, because (6.3) is compatible with an odd solution: odd derivatives on the l.h.s. are even functions of σ and the cosine of an odd function is even). Thus a condition on the relevant solution of (6.3) is $\ddot{\theta} = 0$ at $s = 0$. Finally the conditions on the asymptotic branches are $\theta \to \pm\frac{1}{2}\pi$ as $s \to \pm\infty$.

As the equation (6.3) is of third order with respect to the variable s, *a priori* no more than three independent conditions can be imposed on its solution. Let us count the number of free parameters we need to satisfy the conditions imposed on the solution to represent a symmetric needle crystal. The parity condition imposes one parameter, as it says that the second derivative is zero when the function itself vanishes. The counting of the asymptotic conditions is done as follows. Let us linearize the solution near its asymptotic behavior, say near $s \to +\infty$: $\theta = +\frac{1}{2}\pi + \theta'$, with θ' small. This perturbation is the solution of the third order linear equation:

$$\delta^2 \, \dddot{\theta}' + \dot{\theta}' + \theta' = 0. \tag{6.4}$$

The corresponding secular polynomial is:

$$\delta^2 \sigma^3 + \sigma + 1 = 0,$$

for perturbation $\theta' \sim e^{\sigma s}$. For small δ, the limit we are interested in, the roots of this polynomial are:

$$\sigma_0 = -1 + \delta^2 - 3\delta^4 + \cdots,$$
$$\sigma_\pm = \pm\frac{i}{\delta} + 1 + \cdots.$$

The amplitude of the σ_0-mode vanishes as s tends to $+\infty$ and so is arbitrary. On the contrary, the amplitudes of the two modes corresponding to the roots $\sigma_\pm$ have to vanish because otherwise they would increase exponentially as s tends to $+\infty$, so contradicting the b.c. that θ' tends to zero at $s \to +\infty$. By symmetry there is a mirror condition at s tending to minus infinity. This makes four conditions for the behavior at plus and minus infinity. This is indeed already too much as the equation (6.3) is of third order only. However things are not so simple, because, as we shall see a perturbative solution in powers of δ fits all b.c. order by order. We can already grasp the reason why this perturbative approach fails: the modes corresponding to the roots $\sigma_\pm$ of the secular equation have an amplitude $e^{\pm is/\delta}$, certainly out of reach of a perturbative approach in powers of δ. However there is a rather subtle link between the power expansion (in δ) and those contributions of a transcendental order.

6.2 Perturbative solution and boundary layer analysis of its singularity

Let us define first the regular expansion of the solution of (6.3) as:

$$\theta(s, \delta) = \theta_0(s) + \delta^2 \theta_1(s) + \delta^4 \theta_2(s) + \cdots. \qquad (6.5)$$

The dominant term, i.e. $\theta_0(s)$, is the solution of:

$$\dot\theta_0 = \cos\theta_0. \qquad (6.6)$$

This is solved by elementary means and yields, together with the relevant b.c.:

$$\sin\theta_0 = \frac{e^{2s} - 1}{e^{2s} + 1} = \frac{e^s - e^{-s}}{e^s + e^{-s}}. \qquad (6.7)$$

Writing the solution in this way makes obvious a number of things: first the solution satisfies the b.c., as $\theta_0(0) = 0$, where θ_0 is an odd function, and finally it tends to $\frac{1}{2}\pi$ as s tends to $\pm\infty$. Another property of this solution is also important due to the following: it becomes singular when the denominator of the r.h.s. vanishes, that is at the roots of $e^{2s} + 1 = 0$. This has two roots in the complex plane, at $s = \pm\frac{1}{2}i\pi$. One gets the behavior of $\theta_0(s)$ near any one of those two singularities ($s = \pm\frac{1}{2}i\pi$ for instance) by putting $s = \frac{1}{2}i\pi + \zeta'$, with ζ' small. This yields the following expansion for θ near the singularity:

$$\theta_0 \approx -i\ln\tfrac{1}{2}\zeta' - \tfrac{1}{2}\pi + \cdots. \qquad (6.8)$$

This behavior is consistent also with equation (6.6). The next order term in the expansion in powers of δ^2 is the solution of:

$$\dot{\theta}_1 + \theta_1 \sin \theta_0 = - \dddot{\theta}_0. \qquad (6.9)$$

The b.c. for θ_1 (and more generally for any θ_j at $j \neq 0$) at $\pm\infty$ are $\theta_1 = 0$, and this must be an odd function of s. The algebra for computing the successive terms in the expansion is greatly simplified by noticing that

$$\dot{f} + f\,\mathrm{th}(s) = \frac{1}{\mathrm{ch}(s)} \frac{\mathrm{d}}{\mathrm{d}s}\left[f(s)\mathrm{ch}(s) \right]$$

together with

$$\dddot{\theta}_0 = - \cos \theta_0 \cos 2\theta_0.$$

This yields

$$\theta_1 = \frac{2\mathrm{th}(s) - s}{\mathrm{ch}(s)}.$$

It is a matter of proper book-keeping to prove that at any order there is a θ_j with the correct b.c. and symmetry. Now comes a question that is not so often posed in physics; what is the nature of the convergence of the series so obtained? It turns out that this series is almost surely diverging: we have already seen that θ_0 has a logarithmic singularity at s close to $\pm\frac{1}{2}i\pi$. By looking at θ_1 one can see that this has an inverse power law singularity near the same value of the variable. The next order contributions would be more and more diverging at increasing order. Near the singularity one of the largest inhomogeneous terms in the equation for θ_j will come from the third derivative acting on θ_{j-1}, because the highest derivative yields the largest negative power. This leads one to try as an expression for the dominant order term in θ_j:

$$\theta_j \approx \frac{A_j}{\zeta^{2j}}.$$

Inserting this into

$$\dot{\theta}_{j+1} + \theta_{j+1} \sin \theta_0 = - \dddot{\theta}_j \qquad (6.10)$$

as this captures part of the dominant terms in θ_{j+1} near the singularity, we get a recursion relation for the coefficients A_j's:

$$A_{j+1} = -(2j)(2j+1)A_j,$$

which shows that the coefficients of the resulting series grow as a factorial of j at large order, so that the radius of convergence in z of the series $\sum A_j/\zeta^{2j}$ is zero.

Now two things can be done: one may either try to redo the above estimate more carefully, and then sum the diverging series by a Borel summation (Combescot *et al.* [1988]), or — as we shall do — use (6.4) instead of the original differential equation.

Coming back to the solution of the original equation, we see that the most diverging terms are of the form $A_j \delta^{2j}/\zeta'^{2j}$, where $\{Aj\}$ is a set of pure numbers (i.e. independent of the variable and of the parameter δ). The sum of this series, whatever this means, is a function of the so-called stretched variable $\zeta = \zeta'/\delta$. We shall try now to derive a differential equation for the sum of the most divergent terms near $s = \pm\frac{1}{2}i\pi$ by a convenient transformation of the original equation.

At finite ζ, and small δ, the dominant term in the series for θ is θ_0, as it behaves as $\ln \delta$ at small δ although all the other terms are formally finite. This leads one to try a form of θ near $\delta = 0$ and ζ finite that matches this logarithm as well as the full series:

$$\theta = -\tfrac{1}{2}\pi + i\ln\frac{2}{\delta} + i\phi(\zeta,\delta), \qquad (6.11)$$

where ϕ is of order 1 in the domain of the ζ and δ under consideration. From the original equation, ϕ is the solution of:

$$\frac{d\phi}{d\zeta} + \frac{d^3\phi}{d\zeta^3} = \left[e^\phi - \left(\tfrac{1}{2}\delta\right)^2 e^{-\phi}\right]. \qquad (6.12)$$

In the limit δ small, one can drop the last term in this expression, and one is left with the 'boundary layer' inner equation, which is parameterless:

$$\frac{d\phi}{d\zeta} + \frac{d^3\phi}{d\zeta^3} = e^\phi. \qquad (6.13)$$

Recall that this equation describes the behavior of θ near its singularities in the complex plane, and when δ is small, and once the proper rescalings and changes are made. It also remains to define the b.c. for the solution of (6.13). This is the heart of the problem.

6.3 The WKB contribution 'beyond all orders' and the absence of needle crystal solution of (6.3)

It is perhaps a slight overextension of terminology to call WKB the kind of problem we are going to consider here. However the case of quantum mechanics has been studied a lot in this framework and it will probably be a simple task for most readers to lay their hands on an exposition of this subject in textbooks of quantum mechanics, such as by Landau and Lifshitz [1967, a] for instance. The present problem definitely has a WKB taste because the small parameter is in front of the highest derivative. But contrary to what happens in quantum mechanics, this singular perturbation remains compatible with a formal expansion of the solution in powers of this small parameter.

In physical terms, this regular expansion assumes that the order of magnitude of a contribution to the perturbation expansion is independent of the order of derivation of this contribution, which is obviously wrong if this contribution depends on the smallness parameter (here δ) as a transcendental, usually either an exponential-function of $(s\delta^{-a})$ where a is either positive, or has a positive real part. On the other hand, the algebraic machinery of the order by order expansion in a small parameter never shows such terms with a nonalgebraic dependence in the smallness parameter.

As a consequence, the amplitude of those 'transcendentally small' WKB contributions is given by a rather nontrivial analysis. So before we expose this in some detail, we shall explain how these WKB terms do show up, but with an unspecified amplitude, and then we explain how this amplitude follows from the solution of the inner equation (6.13).

Before we start this calculation of the WKB perturbation, we have to specify the solution around which the perturbation is made. Actually, we saw in the previous section that there is a good chance that the equation (6.3) has no solution satisfying all b.c. In particular, the radius of convergence of the regular expansion (6.5) is probably zero, and this cannot be taken as a good definition of the basic solution around which we want to make a perturbation.

This 'good solution' is found by imposing the condition that at $s \to -\infty$, $\theta \to -\frac{1}{2}\pi$, which counts for two conditions, because the unstable manifold of this solution is two dimensional. This completes the number of possible conditions, since the equation is autonomous. The supplementary condition could be (for instance) that at $s = 0$, $\theta = 0$. This is OK for the order of the equation (6.3), but says nothing about the parity of this solution, as $\ddot{\theta}$, that should be zero to ensure an odd solution, being left unspecified by those conditions.

As noticed by Kruskal and Segur, this second derivative $\ddot{\theta}$ at $s = 0$ can be taken to be the relevant parameter that we wish to calculate as a function of δ: if this is exactly zero, then a solution exists with the right symmetry; if on the contrary it is non-zero, such a solution does not exist. But we noticed that the formal expansion in powers of δ^2 satisfies order by order every condition. In particular, the second derivative $\ddot{\theta}(s=0)$, is a function of δ that is not zero at $\delta = 0$, but is smaller than any power of δ. We are now going to compute this second derivative for the solution of the half line problem just defined.

Anticipating the result, we shall assume that, on the half line $-\infty < s < 0$, and for δ small, the solution is close to θ_0. Contrary to the case of the ordinary perturbation expansion, we assume that the order of magnitude of the WKB perturbation (with respect to δ) and of its derivative with respect to s are not the same. All this results in the introduction of a perturbation to θ_0, called θ_{WKB} in the following, that is the solution of:

$$\delta^2 \, \dddot{\theta}_{\mathrm{WKB}} + \dot{\theta}_{\mathrm{WKB}} = -\theta_{\mathrm{WKB}} \sin \theta_0. \tag{6.14}$$

This equation (6.14) is nothing else but (6.3) linearized around a 'solution' θ_0. Actually, one should linearize instead near the exact solution, but this one is close to θ_0 for δ small. By assumption, this WKB solution has a fast variation with respect to its argument s. More precisely, the two terms on the left hand side are of the same order of magnitude if θ_{WKB} is a function of s/δ. Then the l.h.s. is of order $1/\delta$, which dominates the r.h.s., and, at the dominant order at δ small, this WKB solution takes the form:

$$\theta_{\mathrm{WKB}} \approx \alpha'_{\pm}(s) \, \mathrm{e}^{\pm is/\delta}.$$

At the next order one gets the (sometimes called the van Vleck prefactor) functions $\alpha'_{\pm}(s)$, up to constants, denoted as $\alpha_{\pm}$:

$$\alpha'_{\pm}(s) = \alpha'_{\pm}(\mathrm{e}^s + \mathrm{e}^{-s})^{1/2}.$$

We recover the exponential growth already found for the possible asymptotic behavior of the perturbation to a solution. This calculation, if pursued at any order in the expansion parameter and as being typical of this kind of WKB theory leaves the plus and the minus modes uncoupled. It also leaves the wrong impression that it is enough to set to zero the two constants $\alpha_{\pm}$ to get rid of the exponential growth at plus and minus infinity. Actually those two constants are determined in a sense by the b.c. imposed on the solution and they depend on the domain in the complex s-plane considered. The boundaries between these domains are the so-called Stokes lines, on which a given exponential changes from increasing to decreasing. In the present case these lines have the equations $s'' = \pm\frac{1}{2}s'\delta$, where s' and s'' are the real and the imaginary parts of the variable $s = s' + is''$, considered as complex.

These Stokes lines are drawn in the part of the complex s-plane wherein the WKB part of the solution is manifestly different from the regular part. In the present case, these two parts differ from each other almost everywhere except in the near vicinity of the singularity at $s = \frac{1}{2}i\pi$: near this singularity the basic solution θ_0 diverges logarithmically and so it is no longer feasible to separate a slowly varying part from a WKB part. Also near this singularity the Van Vleck prefactor of the WKB solution diverges and so cannot be considered any longer as varying slowly with respect to the exponential. Thus the Stokes lines are defined away from this singularity, that is very far from the inner region near this singularity: in this inner region the WKB and the regular part of the solution mix together.

Furthermore, and we shall return to this point later, the equation we are considering are nonlinear, so that there is no simple relation between the coefficients of the exponentially growing and the coefficients of the exponentially decreasing WKB parts, as defined in various Stokes sectors: the exponentially increasing solution depends on the nonlinear part of the equation (this would not happen for a purely linear differential equation; for such nonlinear equations complications arise because of the nonuniformity of the exponents) and not the decreasing WKB part. In other words, the 'constants' $\alpha_\pm$ are not true constants and take different values in different Stokes sectors. (We mean by Stokes sectors the angular regions inside an angle drawn by two Stokes half lines.) In the present problem the geometry of the Stokes lines forbids in particular the use of the same coefficients $\alpha_\pm$ at plus and minus infinity on the real axis, because one has to cross Stokes lines when going from $-\infty$ to $+\infty$ on this real axis.

In the WKB limit of the one dimensional linear Schrödinger equation, the coefficients similar to $\alpha_\pm$ are related to each other in various Stokes sectors by a so-called linear monodromy. In the present case, the equation is nonlinear, and so the equivalent of this monodromy relation must be a nonlinear mathematical object. Here the rôle of the monodromy matrix will be an implicit nonlinear relationship following from the solution of the inner equation (6.13).

To show this, we notice first that, as the problem under consideration is nonlinear, it is possible to have $\alpha_\pm$ equal to zero in a Stokes sector and not zero in another one, contrary to what would happen for a linear monodromy. This is precisely the kind of solution we are looking for: it has no exponentially growing part in the sector of the negative real axis (in the complex s-plane), but it could eventually have a nonzero growing part on the positive real axis. The link between the various possible WKB behaviors is, again in a 1D Schrödinger equation, near a turning point, where the WKB part, as well as the terms coming from the regular expansion (6.5), become all of the same order. Near this turning point the solution of (6.3) is completely described in the small δ limit by the inner equation (6.13). This equation plays a rôle very similar to that of the Airy equation near classical turning points in the Schrödinger equation.

We look therefore for a solution of this inner equation with the relevant asymptotic properties. As it is of third order one may impose three conditions. Indeed one of these conditions is the absence of a WKB growing mode in the sector of the negative real axis. The other condition is to ensure that at $s = 0$, the solution is odd at any algebraic order in the expansion parameter δ. The Stokes lines merge at $s = \frac{1}{2}i\pi$, since outside of a neighborhood of size δ near this point a WKB solution will have reached its asymptotes. Thus $s = 0$ does not belong to the same Stokes sector as the negative axis at minus infinity, and the vanishing of the WKB part on the negative real axis does not imply that this WKB part is zero at $s = 0$. To be more precise, let us consider the behavior of the WKB part, supposed to be small on the imaginary axis, $s = is''$, with s'' real: from the general form of this WKB part, we have:

$$\theta_{\mathrm{WKB}} \approx \alpha_+ e^{-s''/\delta} + \alpha_- e^{s''/\delta}. \tag{6.15}$$

We have neglected in this last expression the subdominant contribution to the increments. The two coefficients $\alpha_\pm$ may now be considered as parameters to be specified by some b.c. But these parameters have to be consistent too with the matching with the inner equation. As this equation is parameter free, this matching requires that, near the singularity, θ_{WKB} depends on the variable and on δ in a way that is consistent with the so-called stretching transformations leading to (6.13). The stretched variable is $\zeta = \delta(s + \frac{1}{2}i\pi)$; this is consistent with the $1/\delta$ terms in the exponentials on the r.h.s. of (6.15).

Introducing the same variable ζ into (6.15), we get that α_+ (resp. α_-) must be equal to $\alpha_+'' e^{-\pi/(2\delta)}$ (resp. $\alpha_-'' e^{\pi/(2\delta)}$), $\alpha_\pm''$ being two numbers of order 1 as δ tends to zero, again because of the matching with (6.13) that is formally independent of δ. The contribution to θ_{WKB} proportional to α_-'' must vanish, because it is exponentially large as δ tends to zero at $s = 0$, although we are looking at a solution of (6.3) that vanishes there. Thus one has to impose the b.c. $\alpha_- = 0$ to the inner equation, which becomes a condition to be imposed on its asymptotic behavior in the Stokes sector including the origin.

Now we have used all the free conditions we had: the vanishing of the two WKB amplitudes in the sector of the infinite negative real axis and the coefficient α_-. These remains thus a non-zero WKB amplitude in the neighborhood of $s = 0$, the one proportional to $\alpha_+'' e^{-\pi/(2\delta)}$. Its contribution to θ can always be cancelled by a small real shift of the origin of s, but when doing so, one cannot also cancel the WKB contribution to $\dot{\theta}$ at the same place, and one is left with a WKB contribution to $\ddot{\theta}$ that is of order

$$\frac{\alpha_+''}{4\delta^{5/2}} \exp\left(-\frac{\pi}{2\delta}\right),$$

with α_+'' being a constant in the $\delta \to 0$ limit (the replacement of δ^2 by $\delta^{5/2}$ comes from a Van Vleck prefactor with a power law behavior near the singularities).

This is the sought after transcendentally small contribution to the second derivative at $s = 0$. Recall that this should precisely vanish for the solution to represent a symmetric needle crystal. The calculation of α_+'' is not a trivial problem, either by a purely numerical approach, as done by Kruskal and Segur, or by a resummation of the asymptotic series for the solution at large distance of the inner equation. Both methods however do indicate that this number is not zero, and thus that no solution to the initial problem exists with the relevant b.c. and symmetries.

In section 7, we shall explain in a less detailed fashion how the same method shows that the integral equation (5.7) has no solution when C is large. This, however, proves the existence of solutions of (5.7), but with a small anisotropy in surface tension added. This last result is quite important for this application, as it explains why crystals of anisotropic solids grow and why they do so along the direction of lowest surface tension, as observed. This is one indication that this quite refined mathematical theory is relevant for real life experiments. We shall comment at the end of section 7 upon other related questions.

7

Solution of the integral equations for the needle crystal

In section 6 we explained why the geometrical model given by equation 6.3) lacks any solution satisfying the proper boundary conditions. As we saw, this is the result of a rather lengthy and nontrivial analysis, since an ordinary perturbation expansion in powers of δ seems to yield such a solution order by order. Perhaps this should have been expected, as the small parameter is in front of the largest derivative. However things are not that simple, as by changing the sign in front of the third derivative in (6.3) and by redoing all the analysis one can see that there is a solution to the equation, in agreement with the b.c. and with the third derivative included.

As the model studied in section 6 is not so close to the integral equation we want to solve, it is not obvious that the analysis done for this geometrical model is of any help for the 'real' problem, that is for solving either the integral equation (5.2) at arbitrary Péclet number or its limit (5.7) at low Péclet and low undercooling. However it happens that the mathematical machinery of section 6 is precisely the same one that is needed to solve (5.2) and (5.7).

In the coming subsection 7.1, we shall consider the low undercooling limit, i.e. the solution of (5.2). Indeed as already noticed, this integral equation has no small or large parameter, and so it seems unlikely that a perturbative method as developed in section 6 may be of any help. Actually we shall consider the limit $C \to \infty$, where the formally highest derivative (coming from the GT term) is small. It happens, as for the geometrical model, that a solution to the integral equation (5.2) can be found formally order by order by an expansion in inverse powers of C. The beginning of this expansion is given by cancelling the integral term on the r.h.s. of (5.2), that is of course realised by setting to 1 the argument of the logarithm, or by assuming that $\xi(x)$ is precisely equal to $\xi_{Iv}(x)$.

The next order term in the expansion, as well as those following, is much less obvious, but it turns out that the corresponding linear integral equation can be explicitly solved. Were this series solution convergent, this would indicate that a continuum of solutions exists, at least for large C's, a rather surprising (and actually wrong) result in view of the fact that this theory was developed to show that capillary phenomena pick up special solutions in the Ivantsov continuum.

This 'explicit' solution plays for this problem the same rôle as the perturbative solution (6.5) for the geometric model. In particular the perturbation gets into trouble in the complex extension with respect to the x-dependence near the value of x making infinite the curvature of the Ivantsov parabola (at $x = \pm\frac{1}{2}i$, with our notations). The rest of the story is rather similar to the one of the geometrical model, but with some more technical complications. One shows that the perturbative expansion misses WKB terms, and that these terms may be computed by a method very similar to the one used for the geometric model of section 6.

Finally this gives, as in section 6 a transcendentally small (in the expansion parameter $1/C$) contribution to the slope of the needle crystal at $x = 0$, this being contrary to the existence of a symmetric needle crystal. This is not the end of the story, as this does not preclude that solutions exist for C finite (instead of small). In this domain the numerical works of Meiron [1986] and of Ben Amar and Moussallam [1987] indicate however that no such solutions exist in this range; finally a scaling argument (Ben Amar and Pomeau [1986, a, b]) also proves that no solutions exist at large C. For many people working in that field, this result was quite a surprise, and it is probably still not accepted by everybody. It should be said that this goes a bit against common sense: one has the impression that the ultimate nonlinear evolution of the Mullins-Sekerka instability should always lead to something like a needle crystal.

As we shall see, however, a small amount of crystal anisotropy in surface tension is enough to yield a nice solution, with C becoming bigger and bigger as the anisotropy gets smaller and smaller. Unfortunately this anisotropy of surface tension is difficult to measure accurately in real crystals and it is thus difficult to make precise comparisons between a theory where this anisotropy is crucial and real experiments, although much progress has been done recently in this direction (Bouissou [1989]). Indeed it is tempting to think of an experiment where this anisotropy would be absent, and with steady needle crystals as described here. One might think either of growing smectic liquid crystals perpendicular to the layers or of glassy solids. We do not know of experiments in needle crystal growth with any of those solids.

Finally in subsection 7.2, we consider the case of a finite undercooling Δ and show that this is not very different from the small undercooling limit. In particular, this yields quantitative predictions that have been checked against a recent experiment on fast solidification by Willnecker *et al.* [1989].

Our starting point will be a slight extension of the equation (5.5) that takes into account possible effects of a four-fold crystal anisotropy in surface tension, through what we called the Gibbs-Thomson-Herring term. From (3.3) this amounts to replacing γ in the usual GT formula by the combination $\gamma(\theta) + \mathrm{d}^2\gamma/\mathrm{d}\theta^2$, where $\gamma(\theta)$ is the orientation dependent surface tension, θ being the angle of the normal to the surface of the crystal with a fixed direction. This four-fold anisotropy changes (5.5) into:

$$\frac{\xi_{xx}}{(1+\xi_x^2)^{3/2}}\left[1 - \varepsilon\cos(4\theta)\right] = \tag{7.1}$$

$$\frac{C}{2\pi}\int_{-\infty}^{\infty} \mathrm{d}t \, \exp\left[-\tfrac{1}{2}\mathrm{Pe}\big(\xi(x) - \xi(t)\big)\right] K_0\big(\tfrac{1}{2}\mathrm{Pe}|R(x) - R(t)|\big)$$

$$-\frac{C}{2\pi}\int_{-\infty}^{\infty} \mathrm{d}t \, \exp\left[-\tfrac{1}{2}\mathrm{Pe}\big(\xi_{\mathrm{Iv}}(x) - \xi_{\mathrm{Iv}}(t)\big)\right] K_0\big(\tfrac{1}{2}\mathrm{Pe}|R_{\mathrm{Iv}}(x) - R_{\mathrm{Iv}}(t)|\big).$$

The notations are the same as in (5.5), except for the new quantity called ε in front of $\cos(4\theta)$ on the left hand side, this being proportional to the anisotropy of surface tension. Due to the form of the GTH condition, the function $\gamma(\theta)$ is then equal to a constant times $[1 + \frac{1}{15}\varepsilon/\cos(4\theta)]$. In order to calculate the growth rate, one has to solve (7.1). This is the problem we consider below, first in the low undercooling-low Péclet number limit, then for finite values of those quantities.

7.1 Solution of the integral equation for low undercooling

Using the method explained in section 5, one gets the limit form of (7.1) at small Pe and Δ:

$$\frac{\xi_{xx}}{(1+\xi_x^2)^{3/2}}(1-\varepsilon\cos(4\theta)) = \tag{7.2}$$

$$-\frac{C}{2\pi}\int_{-\infty}^{+\infty} dt \ln \frac{|R(x)-R(t)|}{|R_{\mathrm{Iv}}(x)-R_{\mathrm{Iv}}(t)|},$$

where the dependence on Pe has disappeared. As noticed before, this integral equation has a structure *a priori* different from the ordinary differential equation (6.3). In particular its form requires the unknown function $\xi(x)$ on the full real axis for x, forbidding *a priori* the imposition of boundary conditions on the half line only as we did for (6.3). This is not a big problem, however, because one can transform (7.2) into an integral equation for a function $\xi(x)$ defined on the half line (x positive for instance), then replace in the integrand the same function for negative arguments by $\xi(-x)$, as we are looking for even solutions, and finally check whether one has got a possible solution by computing the derivative $d\xi/dx$ at $x = 0$ [$= \xi_x(0)$] which should vanish for an even function.

This integral equation posed on the half positive line reads:

$$\frac{\xi_{xx}}{(1+\xi_x^2)^{3/2}}(1-\varepsilon\cos(4\theta)) = -\frac{C}{4\pi}K[\xi,x], \tag{7.3}$$

where, by definition:

$$K[\xi,x] = \int_0^\infty dt\Big\{L\big[\xi(t),t,x\big] + L\big[\xi(-t),-t,x\big]\Big\}, \tag{7.4}$$

with

$$L\big[\xi(t),t,x\big] = \ln\Big\{1 + \frac{[\xi(x)-\xi(t)]^2}{(x-t)^2}\Big\} - \ln\big[1+(x+t)^2\big].$$

One may check that by putting $\xi(x) = -x^2$ into L and finally into K, one gets zero. This means that at large C's the left hand side of the integral equation may be considered as a small perturbation for a solution given at the dominant order by the Ivantsov parabola $\xi(x) = -x^2$. This indicates that one should try to find a solution of (7.3) by expansion near the Ivantsov solution. There is still a free parameter, that is the direction of growth with respect to the crystal axis, or equivalently the choice of origin for the definition of the angle θ. We shall choose this direction such that $\theta = \tan^{-1}|d\xi/dx|$. With this choice, it is convenient to introduce as an expansion parameter σ such that $\sigma^2 = (1 - \varepsilon)/C$, so that the perturbative solution of (7.3) reads:

$$\xi(x) = -x^2 + \sigma^2 y_1(x) + \cdots + \sigma^{2n} y_n(x) + \cdots, \qquad (7.5)$$

n being a natural integer. At each order the unknown function $y_i(x)$ is given by the solution of a linear inhomogeneous equation of the Fredholm type. There are many sources of inhomogeneous terms in this expansion, but it turns out that the most important one for our purpose is simply obtained by inserting into the left hand side of (7.3) the previous term of the expansion (7.5). This gives (again this does not represent the complete equation for y_j):

$$\frac{d}{dx}\left[A(x)\frac{dy_{j-1}}{dx}\right] = \frac{1}{2\pi}\int_{-\infty}^{+\infty} dt \frac{[y_j(x) - y_j(t)]}{x - t} \frac{x + t}{1 + (x + t)^2}, \qquad (7.6)$$

where

$$A(x) = \left[1 + \frac{4\alpha x^2}{(1 + 4x^2)^2}\right](1 + 4x^2)^{-3/2}, \quad \alpha = \frac{8\varepsilon}{1 - \varepsilon}.$$

This equation is written for functions $y_j(x)$ defined on the full real line, because this does not introduce any trouble here. It so happens that the linear integral equation (7.6) can be solved explicitly in Fourier transform (Ben Amar and Moussallam [1987]). It can be shown in particular that the corresponding solution is a smooth and even function of x and that it satisfies the 'quantization' condition at any order. We saw that this condition is imposed by the large distance behavior of solutions of integral equations as (7.2) and its various transforms. Here this condition amounts to the simple condition:

$$\int_{-\infty}^{+\infty} dt\, y_j(t) = 0 \quad \text{at any } j.$$

Thus, as for the Kruskal-Segur model, one has the impression that the perturbation yields a good solution. However, for both models, the situation is far more complex. This is again basically because this perturbative solution becomes ill-defined somewhere in the complex extension in x, and because this perturbation neglects WKB terms that cannot be imposed to be zero uniformly on the real line.

Let us begin with the derivation of the inner equation, aiming at describing the behavior of the complex extension of the solution of (7.3) near the complex value of x such that the expansion (7.5) loses its meaning because the successive terms become larger and larger. In a terminology used sometimes in theoretical physics, searching for this inner equation is equivalent to making a resummation of the most diverging terms of an expansion. As is fairly obvious from (7.6), this expansion fails in the neighborhood of values of x such that the denominator $(1 + 4x^2)$ becomes large (i.e. near $x = \pm\frac{1}{2}i$) in the inhomogeneous part of the equation (that is the l.h.s. of (7.6)). As a first step we shall limit ourselves to the isotropic case, hopefully to make things simpler.

Let us denote by Φ the difference between the Ivantsov solution and the true solution:

$$\Phi(x) = \xi(x) + x^2.$$

We expect this quantity to become large near $x = \frac{1}{2}i$ (what happens near $x = -\frac{1}{2}i$ is completely symmetric for a real function on the real x-axis), when computed by a regular expansion in powers of $1/C$. As for the equation studied by Kruskal and Segur, we wish to get a parameterless inner equation for $\Phi(\cdot)$ by a stretching transformation. As usual, this transformation is not completely obvious from the form of the equation itself, and it is found partly by trial and error.

To explain this, we try first a stretching transformation such that the contribution of the basic solution (i.e. $-x^2$) and of Φ to the second derivative of ξ is of the same order of magnitude in the neighborhood of the singularity. This is achieved by defining a local variable u near $x = \frac{1}{2}i$ as $x = \frac{1}{2}i + C^\beta v$, with v of order 1 and $\beta < 0$ unknown, and a scaling of Φ as $\Phi = C^{2\beta}F(v)$, F of order 1. Thanks to the choice of the same β in the definitions of v and of $F(\cdot)$, the second derivative of Φ near $x = \frac{1}{2}i$ is of order 1.

The exponent β is found now by balancing the powers of C on the left hand side of (7.3) and on the right hand side, when everything is computed near $x = \frac{1}{2}i$. The integral on the r.h.s. is calculated as follows.

One notices first that, in this inner domain, the function $\xi(x)$ itself remains close to the Ivantsov solution $-x^2$, except indeed when computing derivatives, because of the stretching in the inner variable. Because of this, one computes the r.h.s. of (7.3) by retaining the term linear in Φ

in an expansion near this Ivantsov solution. This gives:

$$K[\xi, x] \sim \int_{-\infty}^{+\infty} dt \frac{[\Phi(x) - \Phi(t)](x+t)}{(x-t)[1+(x+t)^2]}. \tag{7.7}$$

In this inner region, in a sense, this integral is dominated by the local contribution. In other words, there is an obvious first contribution obtained by computing formally the integral as a principal part near $x = t$, and splitting the term explicitly proportional to $\Phi(x)$ in (7.7). This leaves the contribution

$$-\text{P} \int_{-\infty}^{+\infty} dt \frac{\Phi(t)(x+t)}{(x-t)[1+(x+t)^2]},$$

where again the integral is evaluated as a principal part. The analytic extension of this principal part is found by using the formula:

$$\text{P} \int_{-\infty}^{+\infty} dt \frac{f(t)}{(x-t)} = \lim_{x \to \text{real}} \left[i\pi f(x) + \int_{-\infty}^{+\infty} dt \frac{f(t)}{(x-t)} \right],$$

where $f(\cdot)$ is a smooth function, and where we mean by $x \to$ real that the argument x goes from complex to real values by tending continuously to the real axis from above, and where the last integral is a standard Cauchy integral. As announced, the use of this formula allows one to write in a compact way the complex extension of the principal part operator defined on the real axis. The contribution from the Cauchy integral when computed with $\Phi(\cdot)$ does not add any important contribution near the singularity. As one expects, this integral remains well defined when computed on the real axis, because it is obtained as the integral over the real axis of a smooth function. Whence, near the singularity, the dominant contribution comes from the first term on the r.h.s. of the above formula and one gets finally:

$$K[\xi, x] \sim \int_{-\infty}^{+\infty} dt \frac{[\Phi(x) - \Phi(t)](x+t)}{(x-t)[1+(x+t)^2]} \sim \kappa \Phi(x) \tag{7.8}$$

near $x = \frac{1}{2}i$. Putting all this together, we see that the inner equation (i.e. a reduction of the original equation (7.2) near $x = \frac{1}{2}i$ in the large C limit) has the form:

$$4\left(-2 + \frac{d^2 F}{dv^2}\right) = -F\left(4iv - 2i\frac{dF}{dv}\right)^{3/2}. \tag{7.9}$$

This parameterless form is obtained with $\beta = -\frac{2}{7}$. Note that this inner equation is an ordinary differential equation, although the starting point was an integro-differential equation. This is because of the analytical continuation of the principal part is dominated by the local contribution. A very similar reduction is true for the WKB part of the perturbative solution. In the present problem, there is a supplementary complication (besides the one due to the fact that one deals with one integral, instead of differential equations), because the inner equation is not uniformly valid near $x = \frac{1}{2}i$: the solutions of this equation behave at small v as $F \sim v^{-1/3}$. This algebraic divergence may be retraced to the neglect of higher order terms, of order $C(\cdot)\mathrm{d}^2F/\mathrm{d}v^2$ in the denominator of the metric term for the curvature. This means that there is a second inner region, nested inside the one where the inner equation (7.7) holds and in which another inner equation must be used. We refer the interested reader to the original publication (Ben Amar and Pomeau [1986, a, b]), where a few more details are given.

Till now we have followed a similar path to the one which led from the equation (6.3) of the geometrical model to the inner equation (6.13) for the singular behavior of the solution of this model in the complex plane. The kind of theory developed in section 6 continues to apply to the analysis of the integro-differential equations (7.2–7.3) in the large C limit.

We have already seen that this expansion in $1/C$ is potentially troublesome, because the small term $(1/C)$ is in front of the highest derivative [the curvature expressed in Cartesian coordinates on the left hand side of (7.2)]. This makes it natural to seek a WKB contribution to the pertubation, with a small amplitude and a variation fast enough to make this highest derivative of the same order as the r.h.s. linearized near Ivantsov's parabola. The linearization of the integral term yields in general a rather complicated linear integral operator, as written on the r.h.s. of (7.6), but this takes a much simpler form if the perturbation is a rapidly varying function, a remark already used when deriving (7.8).

From all these remarks, one gets that the WKB part of Φ is the solution of a homogeneous linear differential equation. It is useful to write this equation in a form which allows the easiest comparison with the inner equation (7.9). For that purpose, we introduce the imaginary part of $\Phi(\cdot)$, $\mathrm{Im}(\Phi)$, and define the function $h(t) = \mathrm{Im}[\Phi(t)]/C$, where t is real on the imaginary axis and defined as $t = -ix$, so that the singularity is now at $t = \frac{1}{2}$. Restricting ourselves to the WKB part of the perturbation, one finds that $h(t)$ is the solution of:

$$\frac{d^2h}{dt^2}(1 - 4t^2)^{-1/2} + 12\frac{dh}{dt}(1 - 4t^2)^{-3/2} - C\frac{h}{2(1 - 2t)} = 0. \quad (7.10)$$

At large C, this equation has a WKB like solution because, once again, the small term $(1/C)$ is in front of the highest derivatives. Its solution in the large C limit can be computed by standard application of the eikonal method (note that the first derivative dh/dt has been retained in order to compute consistently the Van Vleck prefactor):

$$h(t) = 2^{1/4}B(1 - 4t^2)^{3/8}(\tfrac{1}{2} + t)^{1/4}\exp[C^{1/2}S(t)],$$

where the phase $S(t)$ is defined as:

$$S(t) = \int_0^t dt'(\tfrac{1}{2} - t')^{1/2}(1 - 4t'^2)^{1/4}. \quad (7.11)$$

Again, the coefficient B is left undefined by this analysis. It can be computed by matching this WKB solution with a similar contribution to the asymptotic behavior of the solution of the inner equation, this matching taking place for large values of the inner variable v, and for small values of the variable t. As for the geometrical model, the calculation of this constant cannot be done exactly, it needs either a numerical solution of the inner equation or some approximate solution of it. We give below a numerical estimate of a quantity related to B.

Again, as for the geometrical model, this WKB contribution yields a nonzero first derivative of the solution of the integral equation (7.3) at $x = 0$ and in the absence of anisotropy ($\varepsilon = 0$ in 7.2–7.3). This first derivative is

$$\left[\frac{d\Phi}{dx}\right]_{x=0} = B_1 C^{1/28}\exp-[C^{1/2}S(\tfrac{1}{2})], \quad (7.12)$$

where B_1 is a numerical constant, proportional to B and about -0.85 ± 0.05. This completes the analysis and shows that there is no solution of the integral equation in the large C limit, with $[dF/dx]_{x=0} = 0$ in the absence of crystal anisotropy.

This result carries over to finite values of C, as shown numerically, as well as to three dimensional axisymmetric needle crystals. The physical meaning of this is that other physical effects are at play for explaining the growth of real needle crystals. The one we shall consider below is the effect of anisotropic surface tension, always present in real crystals (although one could consider it to be suppressed in smectic liquid crystals and in particular geometries).

Other phenomena may also play an important rôle in real experiments: one might think of kinetic effects (although they are generally believed to be negligible for the non-faceted crystals we are considering here), of the effect of flows in the melt, either due to thermal convection or to other sources, and finally it is also likely that the secondary branching should play a rôle, as it makes the solution non-stationary in any frame of reference, contrary to what we assumed. If anisotropy of surface tension is included (that is ε is different from zero but less than one) one finds a discrete set of possible profiles (Ben Amar and Pomeau [1986, a, b] and Ben Amar and Moussallam [1987]). In the limit of large eigenvalues C_n, we have established that :

$$C_n(\varepsilon) = \frac{(\theta_0 + n\pi)^2}{T(\varepsilon)^2}. \tag{7.13a}$$

θ_0 is a nonlinear phase shift, $T(\varepsilon)$ is given by a path integral on the imaginary axis between two singularities the previous one $\frac{1}{2}i$ and it_0, so:

$$T(\varepsilon) = \int_{t_0}^{1/2} dt \frac{S'(t)}{\left(4\alpha t^2/(1-4t^2)^2 - 1\right)^{1/2}} \quad \text{with} \tag{7.13b}$$

$$\alpha = \frac{8\varepsilon}{1-\varepsilon} \quad \text{and} \quad t_0 = \left[-\alpha^{1/2} + (a+4)^{1/2}\right]^{1/4}. \tag{7.13c}$$

Although valid for small ε, (7.13a–c) reproduce conveniently the eigenvalues found numerically by solving (7.3). By a partly numerical linear stability analysis (it is rather easy to prove that all but the fastest solution are linearly unstable, but much more difficult to prove the linear stability of the quickest one), Kessler and Levine [1986] show that only the smallest one (which gives the quickest needle crystal) can be observed in an experiment so we deduce the growth rate:

$$v = \frac{16\Delta^4}{\pi^2 C_0(\varepsilon)} \frac{\Delta}{\Gamma} \quad \text{while} \quad a^{-1} = 2\rho = \tfrac{1}{4}\pi C_0(\varepsilon) \frac{1}{\Delta^2} \Gamma. \tag{7.14}$$

7.2 Solution of the integral equation for arbitrary undercooling

In this subsection, we shall consider the same problem as before, that is the large C limit for the integro-differential equation giving the shape of a needle crystal, but now at finite undercooling and Péclet number. Our starting point will be the equation (7.1).

a) *Analytic continuation in the complex plane*

We will follow exactly the steps of the previous analysis. If C is large enough, (7.1) can be linearized around the Ivantsov solution in part of the complex z plane where ξ is not singular. On the real axis, we get:

if $\qquad \xi(x) = -x^2 + \dfrac{1}{C'}u(x) \quad \text{with} \quad C' = \dfrac{C}{1-\varepsilon}$ $\qquad$ (7.15)

then $\qquad \dfrac{-2}{(1+4x^2)^{3/2}} + \dfrac{1}{C'}\dfrac{\mathrm{d}}{\mathrm{d}x}\dfrac{u_x G(x)}{(1+4x^2)^{3/2}} =$ $\qquad$ (7.16a)

$$- \frac{\mathrm{Pe}}{4\pi}\int_{-\infty}^{+\infty}\left[u(x) - u(t)\right] H(x,t)\exp\left[\tfrac{1}{2}\,\mathrm{Pe}(x^2 - t^2)\right]\mathrm{d}t,$$

with $\qquad H(x,t) = K_0\left[\tfrac{1}{2}\,\mathrm{Pe}\left|R_{\mathrm{Iv}}(x) - R_{\mathrm{Iv}}(t)\right|\right]$ $\qquad$ (7.16b)

$$- \frac{x^2 - t^2}{\left|R_{\mathrm{Iv}}(x) - R_{\mathrm{Iv}}(t)\right|}K_1\left[\tfrac{1}{2}\,\mathrm{Pe}\left|R_{\mathrm{Iv}}(x) - R_{\mathrm{Iv}}(t)\right|\right],$$

$$G(x) = 1 + \frac{4\alpha x^2}{(1+4x^2)^2}.$$ $\qquad$ (7.16c)

Written in this way, (7.16a–c) are not suitable for analytical continuation, mainly because of the absolute value. Recalling that

$$K_0(x \pm ia) = K_0\left(|x|\right) \mp i\pi I_0(x)\theta_H(-x) \quad \text{if} \quad a \to 0^+,$$

$$K_1(x \pm ia) = K_1\left(|x|\right) \pm i\pi I_1(x)\theta_H(-x) \quad \text{if} \quad a \to 0^+,$$ $\qquad$ (7.16d)

I_0 and I_1 are the modified Bessel functions, regular at the origin, $\theta_H(x)$ is the Heaviside distribution, equal to 1 when ξ is positive and zero elsewhere. We have chosen the Riemann sheet which makes symmetric the upper and lower half complex planes. Hence, we will focus on the behavior of u along the positive imaginary axis ($z = iv$), we write only the analytical continuation of (7.16) on this axis, for sake of simplicity:

$$\frac{-2}{(1-4v^2)^{3/2}} - \frac{1}{C'}\frac{d}{dv}\frac{G(iv)}{(1-4v^2)^{3/2}} = \tag{7.17a}$$

$$-\frac{u(iv)}{2(1-4v^2)} + \frac{\mathrm{Pe}}{4\pi}\mathrm{P}\int_{-\infty}^{+\infty} u(t)L(iv,t)\exp\left[-\tfrac{1}{2}\mathrm{Pe}(v^2+t^2)\right]dt$$

$$+\frac{i\,\mathrm{Pe}}{4}\int_{\Gamma} u(z)l(iv,z)\exp\left[-\tfrac{1}{2}\mathrm{Pe}(v^2+z^2)\right]dz$$

with:

$$L(iv,t) = K_0\left[\tfrac{1}{2}\mathrm{Pe}(iv-t)\left(1+(iv+t)^2\right)^{1/2}\right] \tag{7.17b}$$

$$-\frac{iv+t}{\left(1+(iv+t)^2\right)^{1/2}}K_1\left[\tfrac{1}{2}\mathrm{Pe}(iv-t)\left(1+(iv+t)^2\right)^{1/2}\right]$$

$$l(iv,t) = I_0\left[\tfrac{1}{2}\mathrm{Pe}(iv-t)\left(1+(iv+t)^2\right)^{1/2}\right] \tag{7.17c}$$

$$-\frac{iv+t}{\left(1+(iv+t)^2\right)^{1/2}}I_1\left[\tfrac{1}{2}\mathrm{Pe}(iv-t)\left(1+(iv+t)^2\right)^{1/2}\right].$$

The first integral in (7.17a) has to be performed on the real axis, while the Γ contour involves two parts: Γ_1 and Γ_2. The contour Γ_1 begins at iv on the imaginary axis, and goes to zero along this axis. The contour Γ_2 takes place on the positive real axis. The notation P means the Cauchy principal part, as before. Notice that $l(iv,z)$ is purely real on Γ_1. Note that the second integral in (7.17a) was pointed out first by Barbieri and Langer [1989] and Tanveer [1989]. As shown later, it is responsible for the Péclet dependence of the eigenvalues and cannot be neglected. The Kruskal and Segur method [1985] rests on a solution of (7.17a–c) in the complex plane: it assumes the validity of the ordinary perturbation expansion of $\xi(x)$ when x goes to $+\infty$. The first term of this expansion u_1 is a solution of (7.17a–c) if one formally makes $C' = \infty$.

Contrary to the vanishing Péclet number limit (Ben Amar and Moussallam [1987]), no analytical solution of this linear equation is known to exist, but after numerical investigation which will be explained later, we can claim nevertheless that u_1 has no derivative at the tip (with or without any crystalline anisotropy). So as before, the ordinary perturbation expansion does not give any information on the selection mechanism. The linearization of (7.1a–c) breaks down in the part of the complex plane where each term of (7.17a–c) becomes singular: that is $\pm\frac{1}{2}i$. However, in the close neighborhood of $\pm\frac{1}{2}i$, we can take into account the most singular terms to simplify the original nonlinear equation a little.

b) *The internal region and the WKB analysis on the imaginary axis*

We assume the validity of the scaling laws found at $Pe = 0$, since we are interested in an arbitrary but finite Péclet number, contrary to the limit studied by Barbieri [1987]. So, if $\varepsilon = 0$

$$\frac{1}{C}u(\tfrac{1}{2}i - iv) = C^{-4/7}F(C^{2/7}v). \tag{7.18}$$

F obeys the following local but nonlinear equation:

$$4(2 + F_{tt}) = \left[F + P_1\textstyle\int F\right](4t + 2F_t)^{3/2} \quad \text{with} \quad P_1 = Pe\, C^{-2/7}. \tag{7.19}$$

t is the natural variable of F, that is: $C^{2/7}v$. After inspection of the second integral in (7.17a), one can deduce that the main contribution comes from $x \approx t$, so $l(iv, z)$ is equal to one, and (7.19) involves only an integral of F denoted: $P_1\int F$.

If α is not equal to zero: let us define $C_1 = C/((1-\varepsilon)\alpha)$ and a similar definition of F

$$\frac{1}{C'}u(\tfrac{1}{2}i - v) = C_1^{-4/11}F(C_1^{2/11}v) \quad \text{if} \quad t = C_1^{2/11}v, \tag{7.20}$$

$$(2 + F_{tt})16 = -\left[F + P_2\textstyle\int F\right](4t + 2F_t)^{7/2} \quad \text{with} \quad P_2 = Pe\, C_1^{-2/11}. \tag{7.21}$$

Equations (7.19) and (7.21) are valid in a region of the complex plane around $\frac{1}{2}i$ of very small extent in the first case: $C^{-2/7}$, in the second: $C_1^{-2/11}$. So we need the asymptotic behavior of F as soon as we leave this area, in order to match to the ordinary WKB solution, valid on the imaginary axis between $\frac{1}{2}i$ and zero. At first, let us focus on the case without anisotropy:

i) isotropic case

When t is large, the leading behavior of F is purely real and given by:

$$\int F \approx \frac{1}{P_1 t^{3/2}} + \cdots \quad \text{when} \quad t \to +\infty. \tag{7.22}$$

Let us recall that only odd terms contribute to the first derivative at zero. They are purely imaginary on the imaginary axis. So we focus only on the asymptotic expansion of the imaginary part of F. Its main contribution is obtained by linearization of (7.21) around the leading real behaviour (7.19) which gives

$$h_{tt} - \frac{3}{2}\frac{h_t}{t} - 2h(t)t^{3/2} - 2P_1 t^{3/2} \int h = 0 \qquad (7.23a)$$

so we obtain:

$$h(t) = 2^{1/2} A(P_1) t^{3/8} \exp\left(-2^{5/2} t^{7/4}/7\right) \exp\left(\tfrac{1}{2}P_1 t\right) \qquad (7.23b)$$

A is a nonlinear eigenvalue which obviously is a function of P_1. Equation (7.20) gives in variable v:

$$\mathrm{Im}\left[u(\tfrac{1}{2}i - iv)\right] = u^*(\tfrac{1}{2}i - iv) =$$
$$2^{1/2} A(P_1) C^{-13/28} v^{3/8} \exp\left(-\tfrac{1}{7}C^{1/2}2^{5/2}v^{1/4}\right) \exp\left(\tfrac{1}{2}\mathrm{Pe}\,v\right).$$

On the imaginary axis, in the outer region, $\mathrm{Im}(u)$ obeys an ordinary differential equation:

$$\frac{u_{vv}^*}{(1 - 4v^2)^{3/2}} + \frac{12u_v^*}{(1 - 4v^2)^{5/2}} - \frac{1}{2}C\frac{u^*(v)}{1 + 2v} + \frac{1}{4}\mathrm{Pe}\int u^* = 0. \qquad (7.24)$$

We can solve this equation by a WKB approximation since the small parameter $1/C$ is in front of the highest derivative u_{vv}^*. Assuming this approximation for u^*, we get simply a integral of $\int u^*$ by a simple asymptotic evaluation of the second integral in (7.17a) and we derive

$$u^*(v) = B(1 - 4v^2)^{3/8}(\tfrac{1}{2} + v)^{1/4} \qquad (7.25)$$
$$\times \left[\exp\left(-\tfrac{1}{4}\mathrm{Pe}(v + v^2)\right)\exp\left(C^{1/2}S(v)\right)\right.$$
$$\left. - \exp\left(\tfrac{1}{4}\mathrm{Pe}(v + v^2)\right)\exp\left(-C^{1/2}S(v)\right)\right]$$

with $S(v) = \int_0^v S'(t)\mathrm{d}t$ as in (7.13a–c). By matching the two expressions of u in their common domain of validity, we can calculate B in terms of the nonlinear eigenvalue A and finally obtain for the first derivative at zero:

$$u_v(0) = 2^{1/4} BC^{1/2} \qquad (7.26)$$
$$= A\left(\mathrm{Pe}\,C^{-2/7}\right)C^{1/28}\exp\left(\tfrac{3}{16}\mathrm{Pe}\right)\exp\left(-C^{1/2}S(\tfrac{1}{2})\right).$$

We shall assume that A is not a singular function of P_1, so it admits a regular series expansion in terms of P_1:

$$A(\mathrm{Pe}\,C^{-2/7}) = A_0 + A_1(\mathrm{Pe}\,C^{-2/7}) + A_2(\mathrm{Pe}\,C^{-2/7})^2 + \cdots. \qquad (7.27)$$

When compared to (7.12), Eq. (7.26) proves that the mismatch at the tip is larger than at non-zero Péclet number. Although derived at rather small Péclet number (Pe $< C^{2/7}$), it indicates that there is no chance of observing a needle crystal without anisotropy, at any Péclet number.

ii) Anisotropic case

Let us come back to the local nonlinear equation with anisotropy (7.21). As in the former case, we do not need to solve this equation completely, valid only around $\frac{1}{2}i$, but we need its asymptotic expansion. The leading behavior is real :

$$\int F \approx -\frac{1}{P_2 t^{7/2}} + \cdots \quad \text{when} \quad t \to +\infty \qquad (7.28)$$

and the imaginary part is the solution to:

$$h_{tt} - \frac{7}{2}\frac{h_t}{t} + 8h(t)t^{7/2} + 8P_2 t^{7/2}\int h = 0. \qquad (7.29)$$

Choosing $h(t) \approx e^{iS(t)}$ we finally find :

$$\begin{aligned} h(t) = D(P_2)t^{7/8}\exp\left(\tfrac{1}{11}i2^{9/2}t^{11/14} + i\Phi(P_2)\right) \\ \times \exp(\tfrac{1}{2}P_2 t) + \text{C.C.} \end{aligned} \qquad (7.30)$$

$\Phi(P_2)$ is a nonlinear eigenphase-shift, and for it we will assume a regular series expansion in powers of P_2. So we get

$$\Phi(P2) = \Phi_0 + \Phi_1 P_2 + \Phi_2 P_2^2 + \cdots \quad \text{with} \quad P_2 = \text{Pe}\, C^{-2/11}.$$

$h(t)$ has to be matched to the WKB outer solution u^* which is given by:

$$u^*(v) = r(v)\exp\left[C'^{1/2}S_1(v)\right]\exp\left[-\tfrac{1}{4}\text{Pe}(v+v^2)\right], \qquad (7.31a)$$

with

$$C' = \frac{C}{1-\varepsilon},$$

$$r(v) = E\frac{(1-4v^2)^{3/8}(\tfrac{1}{2}+v)^{1/4}}{\left(1-4\alpha v^2/(1-4v^2)^2\right)^{1/2}}, \qquad (7.31b)$$

$$S_1(v) = \int_0^v \frac{S'(t)\,dt}{\left(1-4\alpha t^2/(1-4t^2)^2\right)^{1/2}}.$$

Notice that the Péclet contribution to u^* is real along the imaginary axis, so does not modify the selection rule. This WKB expansion breaks down at t_0 which has been given earlier (7.13c) where a new internal region has to be defined and a new local equation has to be established, taking care only of the most singular terms. If we neglect terms of order $\text{Pe}\,C^{-1}$, we derive the same equation as for Pe equal to zero (see Ben Amar and Pomeau [1986, a, b]). So the eigenvalues are fixed by a path integral between t_0 and $\frac{1}{2}i$:

$$C'_n(\varepsilon, \text{Pe})^{1/2}T(\varepsilon) = n\pi + \theta_0 + \Phi_1 P_2.$$

θ_0 takes into account the two eigenphase shifts arising from the treatment of the two singularities. We can estimate it numerically by solving either the local boundary layer equations around t_0 and $\frac{1}{2}i$ or by solving the original Nash and Glicksman equation. Note that neither θ_0 nor Φ_1 depend on the anisotropy. If P_2 is small, that is Pe smaller than $C^{2/11}$, then we get for the Péclet dependence of the discrete set of eigenvalues:

$$C_n(\varepsilon, \text{Pe}) = C_n(\varepsilon) + \frac{2\,\text{Pe}\,\Phi_1}{T(\varepsilon)}(1 - \varepsilon)^{15/22}\left[\frac{8\varepsilon}{1 - \varepsilon}\right]^{2/11} C_n(\varepsilon)^{7/22}. \quad (7.32)$$

$C_n(\varepsilon)$ is the eigenvalue at Pe equal to zero, found earlier. Notice that the slope of the linear effect in (7.32) is not small : it is given by $\varepsilon^{-5/4}$, while the C eigenvalues (at Pe $= 0$) are scaled by $\varepsilon^{-7/4}$ at small anisotropy. It is important to realize that the Péclet contribution to the eigenvalues $C_n(\varepsilon, \text{Pe})$ arises from the internal region around $\pm\frac{1}{2}i$ and not from the WKB expansion along the imaginary axis. The relation (7.32) is valid within the WKB approximation, i.e. for large C.

To do so, we can choose either very small anisotropy ε or nonphysical eigenvalues C_n corresponding to high n values. They are nonphysical since they cannot be observed in an experiment. For usual anisotropy values (Bouissou [1989]), we can compare this asymptotic relation to numerical eigenvalues, solution of the real nonlinear, nonlocal integro-differential equation (7.1). By direct comparison with the numerics (see Ben Amar [1990]), estimation of the validity of our treatment can be made. In any case, the WKB treatment always explains the existence of the eigenvalues but a precise comparison with experimental data requires a numerical solution of the integro-differential equation (7.1). This is the question considered below.

To our knowledge, the experiment of Willnecker *et al.* [1989] is the only one which covers such a wide domain of undercooling between 0 and 0.75, that is, in Péclet between 0 and 10. The authors have measured the growth rate of pure nickel and of an alloy ($Cu_{70}Ni_{30}$) using a rapid solidification technique. We can discuss here only the experiment where solidification is induced by undercooling since our model, the symmetric one, assumes equality of the diffusion coefficients between the liquid and solid phases. For the alloy solidification, the one-sided model (with vanishing diffusion coefficient in the solid phase) is more adapted and results will be published elsewhere.

The authors insist on a break in the experimental data around $D = 0.440$. It is clear that the curve representing the growth rate versus the undercooling shows a break in curvature for this value. They explain this by kinetic effects and indicate a coefficient μ of order 1.6m/sK. Unfortunately, they do not indicate the anisotropy of surface tension which is so important for selection. It is time to discuss now the importance of the kinetic effects which indicate that the interface presents a small departure from thermodynamical equilibrium. To measure their strength, it is necessary to write down the proper equation with dimensionless parameters that is for us C and Pe. If one restricts to the linear expansion, in (7.1), one must add on the left hand side of (7.1) the following term:

$$-\beta_K \, \text{Pe} \, \frac{1}{(1 + \xi_r^2)^{1/2}}.$$

Here, we assume dendritic growth in three dimensions and axisymmetry. Note that β_K is also anisotropic if surface tension is chosen to be anisotropic. Kinetic effects are not important at low undercooling for two reasons: first, in the equation we have to solve, they are multiplied by Pe so they modify the growth rate only for Pe $> 1/\beta_K$. Second, they include only the first derivative, so within the WKB approximation, they modify only the Van-Vleck prefactor which is not important for selection. The phase factor (called S in the previous section) depends on the highest derivative, that is the curvature, and so the selection is mostly induced by surface tension. When the Péclet number increases, as shown by Lemieux *et al.* [1987], it is necessary to incorporate these effects, at least for calculating numerically the growth rates.

We want to stress the especially high value of the β_K coefficient for nickel when written with our units

$$\beta_K = \frac{c}{L\mu} \frac{D}{\Gamma} \approx 15.65, \quad \Gamma = \frac{c\gamma}{L^2 \rho^*} T_m \approx 6.53.$$

For the definition of each thermodynamical constant, see section 2. We have taken into account experimental data collected in Willnecker *et al.* [1989]. Note that γ is the interface energy and ρ^* the density of nickel, that is $8.88 \times 10^3; \text{kg/m}^3$. After some trials, we decide to calculate the 3D growth rate for a four fold anisotopy coefficient $\delta_3 D$ of 0.0266 ($\delta_3 D \approx \varepsilon/15$). In fact here we have two adjustable parameters: anisotropy of free energy and the kinetic coefficient. We fix the latter to the numerical value indicated above. Fig. 3 proves that the agreement between experimental and numerical data is rather good as soon as Pe is greater than 1.4 ($\Delta > 0.4$). For $\Delta < 0.4$, our calculation seems to underestimate the C eigenvalues so to overestimate the growth rate.

We recall that the growth rate is an increasing function of the anisotropy and a decreasing function of the β_K coefficient. However, surface tension acts more at low undercooling and the kinetic effects at large undercooling. To obtain a better agreement, it will be necessary to reduce both the anisotropy of surface tension and the β_K coefficient or to increase the sophistication of the calculation by introducing anisotropy in β_K. When β_K is neglected, the growth rate is too high and never fits the experimental data, as shown in Fig. 3.

Conclusion

Crystal growth is sometimes considered more an art than a science when it comes to making real crystals in the laboratory. This is because the process of growth may depend on many different physical effects: diffusion of impurities, thermal convection, crystal defects, etc.. In these notes, we have concentrated mostly on the effects that might be called 'macroscopic', probably the easiest type to control in experiments. We believe that we have reached clearcut conclusions concerning, for instance, the rôle of crystal anisotropy and the size of facets. Because the reader could be worried by the experimental relevance of all this mathematical machinery, we only briefly alluded to this, except at the end of section 7. Generally, the agreement between theory and experiment is encouraging, if not good for the effect of anisotropy on crystal growth (Bouissou [1989]). Some recent experiments have also adressed the question of growth in flows (Gill and collaborators [1988], Bouissou et al [1989]), but then, the agreement with the theory is not completely clear, perhaps because growth is limited by various impurities

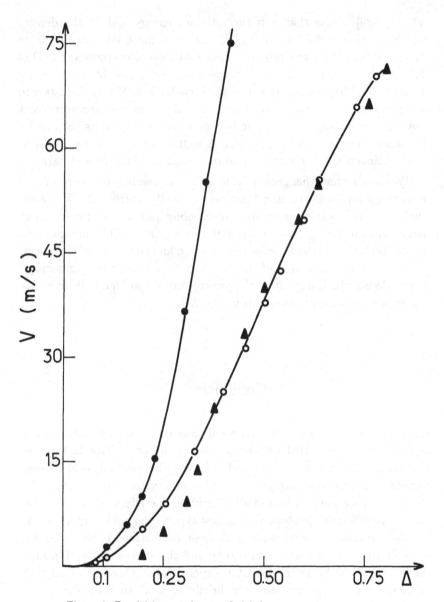

Figure 3: Dendritic growth rate of nickel.

●—●—● Calculated growth rate assuming thermodynamical equilibrium for the interface.

○—○—○ Calculated growth rate with anisotropy of surface tension and kinetics effets.

▲—▲—▲ Experimental points of Willnecker *et al.* [1989].

Dendritic growth and related topics 429

with widely different diffusion coefficients. As far as we know, there is no direct experimental check of the relation fixing the size of the facets of growing crystals (section 4), although the theory developed in section 4 explains well experiments on faceted needle crystals (Ben Amar and Pomeau [1988, a, b]). A difficulty in the subject of crystal growth is the large variety of possible experimental situations. However, we believe that the ideas and methods presented here can be relevant in a non-negligible class of experiments in this large domain.

References

R. Ananth and W.N. Gill: *Chem. Eng. Comm.*, **68**, 1, 1988, a.

R. Ananth and W.N. Gill: *Journ. of Cryst. Growth*, **91**, 587, 1988, b.

A. Barbieri, *Phys. Rev.*, **A36**, 5353, 1987.

A. Barbieri and J. Langer, *Phys. Rev.*, **A39**, 5314, 1989.

G.I. Barenblatt, *Similarity, Self Similarity and Intermediate Asymptotics*, Consultants Bureau, New York, 1979.

M. Ben Amar, *Physica*, **31D**, 409, 1988.

M. Ben Amar and Y. Pomeau, *Euro. Phys. Lett.*, **2**, 307, 1986, a.

M. Ben Amar and Y. Pomeau, *Physico-chemical Hydrodynamics, Interfacial Phenomena*, Ed. M. Velarde, Nato Asi Series, 1986.

M. Ben Amar and B. Moussallam, *Physica*, **25D**, 155, 1987.

M. Ben Amar, P. Bouissou and P. Pelcé, *J. of Crystal Growth*, **92**, 97, 1988.

M. Ben Amar and Y. Pomeau, *Euro. Phys. Lett.*, **6**, 609, 1988, a.

M. Ben Amar and Y. Pomeau, *C.R.A.S*, Paris, **307**, 1055, 1988, b.

M. Ben Amar, *Phys. Rev.*, **A41**, 2080, 1990.

P. Bouissou, *Influence d'un Écoulement en Croissance Dendritique: Aspects Expérimentaux et Théoriques*, Ph. D. Thesis, Université P. et M. Curie, 1989.

P. Bouissou, B. Perrin and P. Tabeling, *Phys. Rev.*, **A40**, 509, 1989.

R. Bowley, B. Caroli, C. Caroli, F. Graner, P. Nozières and B. Roulet, *Journ. de Phys.*, **50**, 1377, 1989.

430   *Y. Pomeau and M. Ben Amar*

W.K. Burton, N. Cabrera and F.C. Franck, *Philos. Trans. Roy. Soc. London*, **A243**, 299, 1951.

B. Caroli, C. Caroli, B. Roulet and J.S. Langer, *Phys. Rev.*, **A33**, 442, 1986.

R. Combescot, V. Hakim, T. Dombre, Y. Pomeau and A. Pumir, *Phys. Rev.*, **A37**, 1270, 1988.

M.E. Glicksman, S.R. Coriell and G.B. McFadden, *Ann. Rev. Fluid Mech.*, **A307**, 1986.

C. Herring, *Phys. Rev.*, **92**, 87, 1951, a.

C. Herring, Chapter 8 in *The physics of powder metallurgy*, McGraw-Hill, New York, Ed. W.E. Kingston, 1951, b.

C. Herring, *Structure and Properties of Solid Surfaces*, Ed. R. Gomer and C.S. Smith, University of Chicago Press, 1953.

G.P. Ivantsov, *Dokl. Akad. Nauk SSSR*, **58**, 567, 1947.

D. Kessler, J. Koplik and H. Levine, *Phys. Rev.*, **A33**, 3352, 1986.

D. Kessler, J. Koplik and H. Levine, *Adv. in Phys.*, **37**, 255, 1988.

D. Kessler and H. Levine, *Phys. Rev. Lett.*, **A57**, 3069, 1986.

M. Kruskal and H. Segur, *Asymptotics Beyond all Orders in a Model of Dendritic Crystals*, Aero. Res. Ass. of Princeton Tech. Memo, 1985.

G. Lamé and B.P. Clapeyron, *Ann. de Chimie Phys.*, **47**, 250, 1831.

J. Langer, *Rev. Mod. Phys.*, **52**, 1, 1980.

J. Langer, *Chance and Matter*, Compte-Rendus des Houches, North Holland, 1987.

L. Landau and I.M. Lifshitz, *Mécanique Quantique*, Ed. Mir, 1967, a.

L. Landau and I.M. Lifshitz, *Physique Statistique*, Ed. Mir, 1967, b.

M.A. Lemieux, J. Liu and G. Kotliar, *Phys. Rev.*, **36**, 1849, 1987.

H.M. McConnel, D. Keller and H. Gaub, *J. Chem. Phys.*, **90**, 1717, 1986.

D. Meiron, *Phys. Rev.*, **A33**, 2704, 1986.

W.W. Mullins and R.F. Sekerka, *Journ. of Appl. Phys.*, **35**, 444, 1964.

G.E. Nash and M.E. Glicksman, *Acta Metall.*, **22**, 1283, 1974.

P. Pelcé, *Dynamics of Curved Fronts*, in Perspective in Physics, Ed. H. Araki, A. Libchaber and G. Parisi, Academic Press, New York, 1988.

P. Pelcé and Y. Pomeau, *Stud. in Appl. Math.*, **74**, 245, 1986.

Y. Pomeau, *Euro Phys. Lett.*, **3**, 1201, 1987.

P.G. Saffman and G.I. Taylor, *Proc. Roy. Soc. London*, **A234**, 312, 1958.

B. Shraiman and D. Bensimon, *Phys. Rev.*, **A30**, 2840, 1984.

S. Tanveer, *Phys. Rev.*, **A40**, 4756, 1989.

R.M. Weis and H.M. McConnell, *Nature*, **310**, 47, 1984.

R. Willnecker, D.M. Herlach and B. Feuerbacher, *Phys. Rev. Lett.*, **62**, 2707, 1989.

CHAPTER 5

GROWTH AND AGGREGATION FAR FROM EQUILIBRIUM

L.M. Sander

Introduction

In this chapter we will give a picture of growth by using simple models which show scaling behavior. We will emphasize the relationship between growth and morphology in situations which are very far from equilibrium. The result of many recent investigations has been that some of the remarkable properties of the models which are related to scaling are widely applicable to natural systems.

Much of the activity in this area was inspired by the work of Witten and Sander (1981, 1983) on one of the models that we will discuss below, the diffusion-limited aggregation (DLA) process. In recent years several reviews have treated the subjects that we will discuss from various points of view. The student is particularly encouraged to consult the recent book of Vicsek (1989).

1

Growth and fractals; fractal geometry

1.1 Disorderly growth and fractals

We will be dealing with objects which grow in such a way as to have the property of scaling, that is, such that a part of the object looks like the whole. The purpose of this section is to give the elementary mathematical notions necessary to describe this property precisely. Much of the original work which underlies this section is due to B. Mandelbrot, and is described in his beautifully illustrated book (Mandelbrot, 1982).

As an example, Figure 1 gives an artificial model for the sort of thing we are interested in. It is a figure generated by an obvious rule such that one part of it (one fifth, in this case) is identical to the entire object. The object is *scale-invariant*: it possesses symmetry under changes of length scale. (Another term often used to designate objects with this symmetry is *self-similarity*.) Objects that are self-similar are called fractals (Mandelbrot, 1982). In this section we will be concerned with the mathematical description of scale-invariance, i.e. with the geometry of fractals.

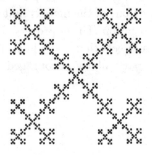

Figure 1: The Vicsek snowflake, a regular fractal.

The kind of growth that we will model does not have a determined pattern like that of Figure 1. Dendritic solidification (which is treated elsewhere in this book) is a case of growth far from equilibrium which has a determined steady-state pattern with many symmetries, but *not* scaling; snowflakes are probably not fractals. Rather we will concern ourselves with *disorderly* growth dominated by random processes and instabilities where scaling does occur. Objects which are formed by processes very similar to those that form dendritic crystals, but where no steady-state is possible, *are* scale-invariant fractals. In fact, in a great number of cases of disorderly growth far from equilibrium, fractals are produced: we will see examples taken from the physics of crystals, colloids, rough surfaces, the patterns of fluids in porous media, and dielectric breakdown of gases. (For completeness, we must add that certain objects in equilibrium, such as polymers in a good solvent, are fractals as well. These will not concern us here.) Thus the statements that we will make about the morphology of the objects grown will be, in most cases, *statistical* in nature: they will be about the ensemble average of certain correlation functions that characterize the shape of the objects formed.

1.2 Scale-invariance

In the case of a regular object such as the Vicsek snowflake of Figure 1, the simplest way to define fractal symmetry is to consider a mapping of the object in question into a number, N, of identical objects each reduced in size by λ, where λ is less than unity. The object is scale-invariant if it is identical to the union of these N scaled replicas. For example, for the Vicsek snowflake, if we take $\lambda = (\frac{1}{3})^k$ and $N = 5^k$, we clearly have scale invariance for any k.

In Figure 2 we show the most famous example of a regular fractal, the Sierpinski 'gasket'. In this case scalings by $(\frac{1}{2})^k$ will give the same structure and the corresponding $N = 3^k$.

For a disorderly structure, however, the method of mapping onto scaled replicas is not convenient, because the object mapped is defined only by a statistical distribution. It is convenient to consider the set of density correlation functions:

$$C_n(\mathbf{r}_1, \mathbf{r}_2, \ldots, \mathbf{r}_n) = \langle \rho(\mathbf{r}_1)\rho(\mathbf{r}_2)\ldots\rho(\mathbf{r}_n) \rangle \tag{1}$$

which define the relationships between the different pieces of the object and $\rho(\mathbf{r})$ is the density at position $\mathbf{r}$. The symbol $\langle \ \rangle$ indicates an ensemble average. If the object is scale-invariant, changing the distances

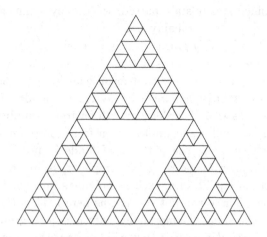

Figure 2: The Sierpinski gasket.

by a factor of λ cannot change C_n in an essential way: only an overall constant is possible:

$$C_n(\lambda \mathbf{r}_1, \lambda \mathbf{r}_2, \ldots, \lambda \mathbf{r}_n) = A_n(\lambda) C_n(\mathbf{r}_1, \mathbf{r}_2, \ldots, \mathbf{r}_n). \qquad (2)$$

For example, consider the two-point function, C_2. If the object is homogeneous on the average we can write:

$$\langle \rho(\mathbf{r}_1)\rho(\mathbf{r}_2) \rangle = C_n(\mathbf{r}_1, \mathbf{r}_2) = C_n(\mathbf{r}_1 - \mathbf{r}_2) = c(\mathbf{x}), \qquad (3)$$

which can be interpreted as the probability of finding a small volume around $\mathbf{r} + \mathbf{x}$ occupied, provided that the volume at $\mathbf{r}$ is occupied.

Clearly we have:

$$c(\lambda \mu \mathbf{x}) = A(\lambda) A(\mu) c(\mathbf{x}) = A(\lambda \mu) c(\mathbf{x}). \qquad (4)$$

This implies in turn that:

$$A(\lambda) = \lambda^{-\alpha} \qquad (5)$$

for some α. Finally, we see that if c depends only on the magnitude of $\mathbf{x}$, we can take $\lambda = 1/x$ to find

$$c(1) = x^\alpha c(x), \quad c(x) \sim x^{-\alpha}. \qquad (6)$$

Two-point correlation functions with power law behavior characterize fractal scaling in a simple and convenient way.

1.3 The fractal dimension

In the case of the two regular fractals illustrated above, as the size of the object goes to infinity, it becomes more and more *tenuous*. That is, if we surround the structure by a box the fraction of the interior occupied becomes smaller and smaller as the object gets larger. This is because, as we will see, the fractals of Figures 1 and 2 are intermediate between curves (dimension 1) and surfaces (dimension 2). In particular, they do not have a well-defined area or length.

The standard description of what is happening is to follow the usual methods of measure theory: we attempt to cover the fractal with boxes of diameter ξ, as in Figure 3. (Of course we could use 'balls', that is circles or spheres, instead of boxes, i.e. squares or cubes, and, indeed, it is more conventional to do so.)

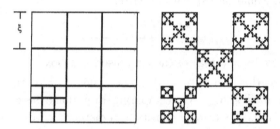

Figure 3: Covering a square and covering a fractal.

The area of an ordinary object is measured by counting the number, $N(\xi)$, of boxes needed to cover it in the limit $\xi \to 0$: for a surface, $\xi^2 N(\xi)$ approaches a constant (the area). For a curve, $\xi N(\xi)$ approaches the length. In both cases the power of ξ is the dimensionality of the object to be measured.

For the fractals above, the length defined in this manner is infinite, and the area is zero. However, the fractals can be measured by generalizing the definition of dimensionality: the (similarity) fractal dimension, D, is defined (by analogy with the statements above) to be:

$$N(\xi) \longrightarrow C\xi^{-D}, \quad C = \text{constant},$$

$$D = \lim_{\xi \to 0} \frac{\log\left[N(\xi)\right]}{\log[1/\xi]}. \tag{7}$$

It is easy to calculate the fractal dimension for the fractals of Figures 1 and 2. For the snowflake we have, from Figure 3, for $\xi = (\frac{1}{3})^k$, $N = 5^k$, so that $D = \log(5)/\log(3) = 1.46$. The fractal dimension need not be an integer.

For an object which grows, it is convenient to use another method to measure D, that of measuring the *mass dimension*. We note that if we take ξ large enough, it will cover the entire object. Let us call this value (that of a circumscribing box or ball) R. Then $N(R) = 1$. However, from Equation (7) we then have $C = R^D$. Now suppose that the object is made up of 'particles', that is, units (such as the smallest triangles in Figure 2) each of mass m and size a. The total mass, M, of the object then grows with R as:

$$M = m\left[\frac{R}{a}\right]^D \sim R^D. \tag{8}$$

Now the average density, ρ, of these tenuous objects decreases with the size, R. From Equation (8), we can write:

$$\rho = \frac{M}{V} \sim \frac{R^D}{R^d} = \frac{1}{R^{d-D}}. \tag{9}$$

Here V is the volume (area) of the circumscribing box.

We can use this result to calculate the behavior of the two-point function described above. By comparing Equations (9) and (6), and noting that $c(x)$ is the average density at a distance x from an occupied point, we have $\alpha = d - D$.

1.4 Fractal surfaces

The scale-invariant object of Figure 1 admits no real distinction between the object and its surface. Every point of the object is on the surface. However, it is rather easy (Mandelbrot, 1982) to give examples of fully scale-invariant objects that bound regions of finite area. The classic random example of this sort of behavior is found in the statistical study of coastlines: the apparent length of a coastline depends on the resolution of the map on which it is measured, because, over a certain range of scales, coastlines have dimension larger than that of a smooth curve: many studies find dimensions ≈ 1.15.

1.5 Self-affine fractals

In the pages that follow we will discuss models for the growth of random rough surfaces. As we will see, the surfaces produced are not scale-invariant in the sense that we described above. Nevertheless, they have well-defined, interesting scaling properties, because they are *self-affine* fractals.

For a surface, there is a clear distinction between the growth direction, perpendicular to the substrate, and the directions along the substrate. It should come as no surprise that scaling can occur in which the two directions are treated differently. To define self-affine symmetry consider, as above, a mapping of the object in question into a number, N, of identical objects each reduced in size by λ, in all the directions except the growth direction, and λ^H *in the growth direction*, where H is less than unity. The object is self-affine if it is identical to the union of these N scaled replicas.

The most famous example of a (random) self-affine curve is the graph of a noise signal. It is well-known that for an ordinary gaussian noise, $h(x)$, the typical change of the signal depends on the time of observation as:

$$\langle (\delta h)^2 \rangle \sim \delta x. \tag{10}$$

Here we have a self-affine curve with $H = \frac{1}{2}$. As above, we can express this sort of statement as a scaling law for the correlation function of the curve:

$$\langle [h(\lambda x) - h(\lambda y)]^2 \rangle = \lambda^{2H} \langle [h(x) - h(y)]^2 \rangle,$$
$$\langle [h(x) - h(y)]^2 \rangle \sim |x - y|^{2H}. \tag{11}$$

The expressions x and y include all the directions perpendicular to the growth direction, h. An example of a random self-affine fractal is given in Figure 4.

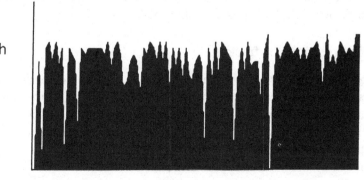

Figure 4: A self-affine surface generated by ballistic aggregation.

It is rather easy to show that a self-affine fractal has a fractal dimension which depends on the scale of the observation: i.e. scaling in the strict sense only exists over a certain range of lengths. We can see this most easily if we re-write Equation (11) with the units restored:

$$(\delta h) = a\left[\frac{\delta x}{b}\right]^{H}. \tag{12}$$

Here a and b are two lengths. It is simple algebra to show that:

$$\begin{aligned} \delta h &\gg \delta x \quad \text{if} \quad \delta x \ll b_c, \\ \delta h &\ll \delta x \quad \text{if} \quad \delta x \gg b_c, \\ b_c &= b\left[\frac{a}{b}\right]^{1/(1-H)}. \end{aligned} \tag{13}$$

The curve looks steep (in fact, as we will see, fractal) for increments of distance less than the cutoff b_c and flat for large scales. For small scales we can calculate the fractal dimension by an argument that often appears in the literature (see Vicsek (1989) for example). We try to cover the curve with boxes of size $\delta x \ll b_c$. The curve is steep and we need to 'stack up' boxes: we cover the plane, of size L perpendicular to h, with $[L/\delta x]^{d-1}$ boxes, where d is, as usual, the dimension of the space in which the curve or surface is drawn, and then stack up to a typical height $s = \delta x^H$ on each box in this base. Then we need:

$$N(\delta x) \sim \left[\frac{s}{\delta x}\right]\left[\frac{L}{\delta x}\right]^{d-1} \sim \delta x^{-(d-H)}. \tag{14}$$

Thus from Equation (7) we have $D = d - H$. However, for the models that we will consider in the following, this relation is only of academic interest, because in physical situations the cutoff b_c is a microscopic lower cutoff below which scaling is no longer valid.

1.6 Intersections of fractals

In the following sections we shall often have occasion to consider the intersection of two fractal objects. It is relatively simple, in the case of two irregular fractals, of dimensions D_1 and D_2 to estimate the mean number of intersections between them when they occupy the same region of space, of size R. This will be, roughly, the mean density of the first, ρ_1, multiplied by the number of points of the second:

$$M_{1,2} = \rho_1 N_2 \sim R^{-(d-D_1)}R^{D_2} \sim R^{D_1+D_2-d}. \tag{15}$$

Thus, if $D_1 + D_2 > d$, intersections are frequent and grow with R, and if $D_1 + D_2 < d$, intersections become more and more unlikely as the tenuous objects get larger. Also, from the dependence of the number of common points in the region, we conclude, by comparison with Equation (8) that the intersection scales with fractal dimension

$$D_{12} = D_1 + D_2 - d. \qquad (16)$$

Two very useful applications of this idea are to cross-sections and to projections. For a cross-section, we can take $D_2 = d - 1$. Then the fractal dimension of the cross-section is $D_1 - 1$.

If we project a fractal, say, from three dimensions into two, we must consider the intersection of the fractal with a line, $D_2 = 1$. If each such line encounters many points, i.e., if $D_1 + 1 > d$ (in the usual case, if $D_1 > 2$) then the projection will be 'black' and points will obscure each other. Otherwise, if $D_1 < 2$, each projection line will encounter on the order of one point of the fractal, and the scaling of the projection will be the same as that of the original object.

1.7 Multifractals

Physical considerations often lead us to define a function on a fractal with scaling properties that are different at different points. We can consider, in particular, a probability, i.e., a function, μ, such that:

$$0 \le \mu \le 1 \qquad \int d\mu = 1. \qquad (17)$$

For example, consider a rough electrode such as is formed in electro-deposition. As we will see this is sometimes a fractal of the DLA type. The charge, q, which resides on the surface of the deposit is very unequally distributed, because the surface has many points. At each point the normal electric field, E_n, is magnified (this is the principle of the lightning rod), and since the surface charge is proportional to the field, it will be much more irregular than the original irregular surface. In this case we can take:

$$d\mu = \frac{dq}{Q} = \frac{E_n}{4\pi Q}.$$

Here Q is the total charge. In two dimensions this is an example of the harmonic measure which is the boundary value of the normal derivative of a harmonic function (the electrostatic potential, ϕ) with Neumann boundary conditions (i.e., $\phi = 0$).

In the case of the example just cited, we can expect many singularities of the function μ. From ordinary electrostatics, we know that at each point we will have a power law singularity which depends on how sharp the point is. We will attempt to classify this sort of singularity.

As above, we will cover the fractal support for μ with boxes of diameter ξ. Set $\varepsilon = \xi/R$, where R is the total size. Now if we integrate μ over the i-th box, we expect

$$\int_i d\mu = p_i \sim \varepsilon^\alpha. \tag{18}$$

Note that $\sum p_i = 1$. Now we simply count how many boxes have the same singularity strength, α, and call this $n(\alpha)$. We should expect:

$$n(\alpha) \sim \varepsilon^{-f(\alpha)} \rho(\alpha). \tag{19}$$

Here ρ is a smooth function which is not singular as ε goes to zero. If $f(\alpha) > 0$ we can interpret it as a fractal dimension of the set on which singularity α occurs. (The case $f(\alpha) < 0$, means that the singularity becomes less common as, say, $R \to \infty$. This case of latent singularities will not concern us further.) Now a plot of $f(\alpha)$ versus α will give a histogram of the information about the singularities. Note that $f(\alpha) \leq D$. It can be shown that the curve of $f(\alpha)$ is always convex. A function for which Equations (18) and (19) hold is called multifractal.

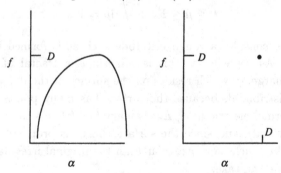

Figure 5: $f(\alpha)$ for a multifractal and an ordinary fractal.

For the uniform measure, μ =constant, on an ordinary fractal, the $f(\alpha)$ curve reduces to the single point $f(D) = D$. There are geometric objects, such as the strange attractors of chaos theory, such that the uniform measure has a non-trivial $f(\alpha)$. These are sometimes called geometric multifractals.

It is useful to consider the generating function of the probabilities, p_i,

$$Z(q) = \sum_i p_i^q, \qquad -\infty < q < \infty. \qquad (20)$$

Note that

$$Z(0) = \sum_I 1 = N(\varepsilon) \sim \varepsilon^{-D}, \quad Z(1) = 1.$$

Now, if the p_i scale, $Z(q)$ will also. Define:

$$Z(q) \sim \varepsilon^{(q-1)D_q}. \qquad (21)$$

The D_q are called *generalized dimensions*.

Now we can write, using (18) and (19):

$$Z(q) = \sum_i p_i^q = \int \mathrm{d}\alpha\, \rho(\alpha) \varepsilon^{[\alpha q - f(\alpha)]}. \qquad (22)$$

This integral can be evaluated by the method of steepest descents as ε approaches zero. For each q there exists an $\alpha(q)$ such that the exponent $\alpha q - f(\alpha)$ attains a maximum, i.e. $q = \mathrm{d}f/\mathrm{d}\alpha$ at $\alpha(q)$, and the value of the integrand at that point will dominate the integration. That is:

$$Z(q) \sim \varepsilon^{[\alpha(q)q - f(\alpha)]}. \qquad (23)$$

By comparison with Equation (21) we have:

$$(q-1)D_q = \alpha(q)q - f(\alpha),$$
$$q = \frac{\mathrm{d}f[\alpha(q)]}{\mathrm{d}\alpha}. \qquad (24)$$

This is a Legendre transform of the usual sort, whose inverse is:

$$\frac{\mathrm{d}}{\mathrm{d}q}[(q-1)D_q] = \alpha,$$
$$q\frac{\mathrm{d}}{\mathrm{d}q}[(q-1)D_q] - (q-1)D_q = f. \qquad (25)$$

These relationships show how to reconstruct the $f(\alpha)$ curve from the generating function. In practice this is usually what is done.

In Figure 6 an often quoted example of a multifractal is given. In this case (see Vicsek 1989, for more discussion) the fractal that carries

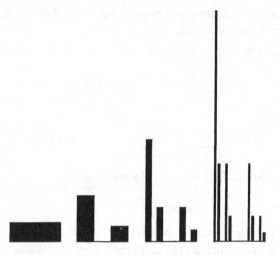

Figure 6: Four stages in the production of a multifractal.

the function μ is a Cantor set, constructed by successively omitting the middle third of the unit interval. The function itself is constant on each third, and equal to $\frac{3}{4}$ and $\frac{1}{4}$ on the right and left subinterval, respectively, at the first stage of the construction. Then the same pattern is repeated by dividing up the weight in each subinterval. Clearly, this pattern scales in a way which depends on the point in the interval.

2

Model for rough surfaces; the Eden model and ballistic aggregation

In this section, and the two which follow, we will consider growth from a very simple point of view, in the approximation of aggregation. That is, we imagine that the process of development of a cluster of material can be idealized by the addition of matter from outside, one unit at a time (these units are called 'particles' in what follows) in such a way that each particle added sticks to the cluster, and then retains its position. Thus all of the complex processes that characterize equilibrium physics, the constant rearrangements, relaxations, and explorations of configurations, are completely neglected. The clusters that are formed have no way to attain equilibrium.

This strong approximation can be considered to be an idealization of the case, which often occurs in nature, when the rate of growth greatly exceeds the rate of relaxation. There are three things to remark about this approximation:

(i) The clusters formed are almost always determined by the fluctuations in the incident flux. In the cases we will consider, the aggregates are irregular, and often have scaling properties.

(ii) The models, as we will soon see, often correspond very closely to physical reality. Though aggregation is a brutal approximation, it seems to often be the case that the features that we describe are preserved whenever the growth dominates relaxation.

(iii) The models are ideally suited for computer simulation, but, in most cases, very difficult to analyze by the traditional methods of theoretical physics.

2.1 Eden growth

The first model that we will discuss is due to Eden (1961) and was introduced in the context of mathematical biology. It deals with the growth of a cell colony in the situation where the major bottleneck to growth (say by cell splitting) is the availability of room to grow. That is, every site at the edge of the present colony has an equal probability of growth.

We can most easily imagine the simulation of the process in the following terms: imagine a lattice with one site occupied by a 'particle' (which in this case represents a cell). Then introduce a new particle at any of the adjacent sites with equal probability. Then, at the next step, choose again one of the perimeter sites (including the new ones created at the previous step) and so forth. The result of many steps of growth is a roughly circular cluster with rough edges. Another geometry which is often studied is a growth from substrate in a channel of width L with periodic boundary conditions at the edges. In this case the deposit grows with a rough upper surface. For the two cases, which are often referred to as *aggregation* and *deposition*, respectively, see Figure 7.

Despite its simplicity, the Eden model has not been solved. We can, however, prove certain things about the general nature of the growth, and refer to the extensive numerical simulations which have been done for a more complete understanding.

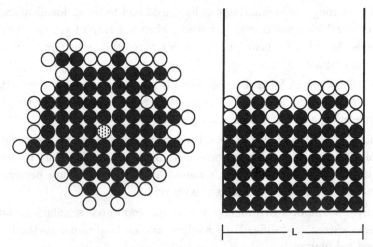

Figure 7: Eden growth in aggregation and deposition geometries.
The white circles are the sites that can grow at the next step.

We will first prove, using a method adopted from Leyvraz (1985), that Eden aggregates are not fractals, despite their random and porous appearance. (The result considerably predates Leyvraz; see Richardson, 1973.) The proof of Leyvraz starts by considering the smallest convex set which encloses the Eden cluster, the 'convex hull'. This surface has a number of polygonal faces. Its total surface area will be called S_N, and its volume V_N, both of which depend on the number of particles currently present, N. The ratio of these two quantities has the units of length, and is an average radius of the hull:

$$R = \frac{V_N}{S_N}. \tag{26}$$

The probability of growth at any of the perimeter sites is equal to $1/N_s$, where N_s is the number of such sites.

Now consider the change in V_N when one particle is added. If the particle is on the outer reaches of the cluster it increases V_N, but it need not do so. On the average, then, the increase in V_N is bounded above by the sum over the i polygonal faces of the hull, S_i, multiplied by the probability to grow at any one of these faces:

$$\left\langle \frac{\partial V_N}{\partial N} \right\rangle \leq \sum_i \frac{S_i}{N_s} = \frac{S_N}{N_s}. \tag{27}$$

Now we can integrate the above with respect to N, assuming that both S_N and N_s are power laws in N:

$$\langle V_N \rangle \leq CN\frac{S_N}{N_s}, \quad N_s \leq N\frac{C}{R}. \tag{28}$$

Here, C is a constant. However, N_s is surely bounded *below* by the value for a compact cluster. Thus:

$$N_s \geq C'R^{d-1}. \tag{29}$$

Now combining Equations (28) and (29) and taking $N \sim R^D$ with D the assumed fractal dimension, we have:

$$C''R^{d-1} \leq C'''R^{D-1},$$
$$R^{d-D} \leq \text{constant}. \tag{30}$$

However, Equation (30) implies at once that $d \leq D$, because otherwise the exponent of the increasing quantity R would be positive, and we must have $d \geq D$. Hence $d = D$ and the cluster is dense, not fractal. Simulations verify this point.

Eden aggregates are not fractal in the bulk, but simulations show that their surfaces are self-affine fractal curves. These show scaling properties which are identical for this model and the next one that we will discuss, ballistic aggregation; both models are described by the same continuum equation. The nature of this scaling will be discussed below.

2.2 Ballistic growth

Thin films are commonly grown by the process of vapor deposition: a chamber is prepared with a furnace or other source of particles which are allowed to fall on a surface (the substrate) and build up the film. In order to visualize this process, many materials scientists (see Leamy *et al.* (1980) and references therein) have employed a simple deposition model for the process: particles drop along straight lines from far away and attach to the substrate or to the other particles at any perimeter site. The particles sometimes stick to the *sides* of the deposit. Once more the deposits have a rough upper surface. For an example, see Figure 4.

The deposition geometry is the closest to the physical process being considered, namely the growth of a deposit from a substrate. However we can equally well consider (and we will find it useful to do so) the growth of a cluster from a center, say a point in our evaporation chamber. In this aggregation case we can take the trajectories to be random in direction.

Ballistic aggregates are also not fractal. Careful numerical simulations verify this point (Meakin, 1983a). A proof of this fact similar to the one above can be given, but it is simpler to use an argument derived from the work of Ball and Witten (1984). Consider a ballistic cluster growing in aggregation geometry. If there is a fixed flux of particles from far away at random angles, clearly the growth rate of the mass will be proportional to the projected surface area. In d dimensions this means:

$$\frac{\mathrm{d}M}{\mathrm{d}t} \sim R^{d-1}, \tag{31}$$

where R is a mean radius of the cluster. Now

$$\frac{\mathrm{d}M}{\mathrm{d}t} = \frac{\mathrm{d}M}{\mathrm{d}R}\frac{\mathrm{d}R}{\mathrm{d}t} \sim R^{D-1}v. \tag{32}$$

The speed, v, is the rate of increase of the mean radius, and D is the fractal dimension. Now we can surely assume that v is bounded. Thus:

$$R^{d-D} \leq \text{const.} \tag{33}$$

Once more this implies that $d = D$. The bulk of a grown cluster is solid and the density achieves a constant value. The scaling properties are present only in the surface.

2.3 Surface scaling

Surface scaling for ballistic aggregation was first discussed by Family and Vicsek (1985). For the Eden model similar properties were noticed by Jullien and Botet (1985). A very useful way to state the scaling (our notation is closest to that of Family and Vicsek) is to consider the deposition geometry with width L; cf. Figures 7, 8. The roughness, σ (the mean fluctuation of the height averaged over the surface) is found to depend on the mean height, h_0. Or, since h_0 is proportional to time (for a non-fractal deposit the density is constant) σ depends on time. The simulations then can be summarized by saying that σ increases for a while and then saturates at a time that depends on the width of the deposit. The exact form is:

$$\sigma = L^\alpha f(h_0 L^{-z}) = \begin{Bmatrix} h_0^\beta & ; & h_0 \ll L^z \\ L^\alpha & ; & h_0 \gg L^z \end{Bmatrix}. \tag{34}$$

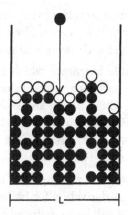

Figure 8: Ballistic growth in deposition geometry. The white circles are the sites that can grow at the next step. Note that the holes in the interior are screened from further growth.

That is, there is power law growth with power β at early times. Later, when roughness fluctuations begin to sample the system boundaries the roughness attains a steady-state. The scaling function, f, interpolates between the regimes. In order for the roughness to be independent of L at early times, when the surface is made up of independently fluctuating pieces:

$$f(y) \sim y^{\beta}, \quad \beta z = \alpha. \tag{35}$$

The scaling exponents α, z (or α, β) characterize the scaling. The surface is not fractal and the roughness grows more slowly than the height or the width. Thus α, $\beta < 1$. We will see that $z > 1$. Note that the late time behavior is characteristic of a self-affine fractal with affinity exponent $H = \alpha$.

It can be numerically verified that the surfaces satisfy a stronger assumption which implies the above, and which can be motivated as follows. The roughness grows by fluctuations in the addition of particles at different places. From the above, there is a kind of 'diffusion' of roughness which moves $|\delta x|^z$ in a 'time' h_0. This same idea can be applied locally. We can look at the growth of the mean square difference between the height fluctuations, $s(\mathbf{x}, t) = h(\mathbf{x}, t) - h_0$ at different points (h_0 is taken at time t). It is natural to assume the following:

$$\left\langle [s(\mathbf{x}', t') - s(\mathbf{x}, t)]^2 \right\rangle \sim |\mathbf{x}' - \mathbf{x}|^{2\alpha} G\left(|t' - t||\mathbf{x}' - \mathbf{x}|^{-z}\right). \tag{36}$$

The scaling function G must be taken to be constant for small argument y and approach $y^{2\beta}$ for large y, as above. For $t = t'$ Equation (36) has the same form as Equation (11). Growing surfaces are self-affine.

There is a considerable literature which involves numerical simulations of these models in various versions in order to find the exponents α, z (or α, β). For the Eden model see Wolf and Kertesz (1987) and for ballistic deposition Meakin et al. (1986a), Meakin and Jullien (1987), Kim and Kosterlitz (1989). The best current evidence is that these exponents are the *same* for the two models.

In two dimensions $\alpha = \frac{1}{2}$, $\beta = \frac{1}{3}$; these are known to be exact (Kardar et al., 1986). The simplest derivation for ballistic aggregation is given in Meakin et al. (1986a) and will be reproduced below. In three dimensions $\alpha \approx 0.35$ and $\beta \approx 0.21$ (see the references cited above).

2.4 Continuum models

2.4.1 The field theory of Kardar, Parisi, and Zhang

It is reasonable to assume that scaling behavior should still be given correctly after coarse-graining and passing to the continuum limit. Kardar, Parisi and Zhang (1986) have proposed a simple non-linear Langevin equation which is intended to describe these models. They proposed

$$\frac{\partial s}{\partial t} = \eta \nabla^2 s(\mathbf{x}, t) + \lambda |\nabla s|^2 + \zeta(\mathbf{x}; t) \tag{37}$$

as the simplest equation consistent with the symmetry of the problem. The non-linear term, $\lambda |\nabla s|^2$, builds into the physics the fact that in both kinds of deposition slopes grow at a different rate from flat surfaces. For example, in Eden growth (Vicsek, 1989) growth in a coarse-grained description is, on the average, normal to the surface. The vertical growth is then given by:

$$\delta h \approx \sqrt{1 + (\nabla h)^2}\, v \delta t,$$
$$\delta s \approx \tfrac{1}{2} (\nabla s)^2 v \delta t, \tag{38}$$

for small slopes; thus $\lambda = \frac{1}{2} v$. In ballistic growth the 'sideways' nature of the growth is built into the model and is even faster than $\frac{1}{2} v$. It is surprising that the addition of this feature in this simple way should be adequate to induce scaling in the same form as the lattice models, but all our current evidence is that it is so.

The first term in Equation (37) describes coupling and smoothing processes. The last term in Equation (37), $\zeta(\mathbf{x}; t)$, is a random noise, usually taken to be 'white', which describes the fluctuations in the arrival of the particles.

If we admit that the scaling behavior which we have described above arises only from the large-scale, coarse-grained behavior, which is correctly given by Equation (37), then we can understand why Eden and ballistic growth have the same scaling; they are both in the universality class of Equation (37).

2.4.2 *Polymers in a random medium*

There is another advantage in using the continuum formulation of Kardar *et al.*, namely that we can make a connection to another model (not otherwise related to aggregation) which has its own literature (see, for example, Kardar and Zhang, 1987). It also seems to be in the same universality class. This connection is obtained by making the non-linear transformation:

$$W = \exp(\lambda s/\eta). \tag{39}$$

Then Equation (37) becomes:

$$\frac{\partial W}{\partial t} = \eta \nabla^2 W + \frac{2\zeta W}{\lambda}. \tag{40}$$

This is a diffusion equation with a random distribution of sources and sinks. In order to simulate the equation, one considers a *directed polymer*, i.e. a random walk which is required to advance in one direction (time) but can wander freely in the transverse directions. The probabilities to walk to each site of the lattice are taken to be random variables, corresponding to random energies, and the statistical distribution is taken at finite temperature. W plays the role of a partition function and the height fluctuations are related to fluctuations of the free energy, from Equation (39).

2.4.3 *A scaling law*

Meakin *et al.* (1986a) pointed out that, if Equation (37) is valid, the exponents α, β are not independent. In fact,

$$\frac{2}{\alpha} = 1 + \frac{1}{\beta}, \quad \text{or} \quad \alpha + z = 2. \tag{41}$$

The derivation is simple. If we admit that the new, non-linear term dominates the large scale behavior, then we can find, from Equation (36), that a typical excursion of the surface, $s(x,t)$ behaves as $x^{\alpha}G(tx^{-z})$. Putting this into Equation (37) and matching the behavior of $\partial s/\partial t$ and the non-linear term gives the result at once. All known lattice simulations satisfy this rule, except, as one would expect, ones which correspond to $\lambda = 0$.

2.4.4 *Weak coupling*

It is perfectly simple to solve Equation (37) explicitly if the non-linear term is neglected (Edwards and Wilkinson, 1982, Sander, 1986). A Fourier transform gives:

$$\alpha = \tfrac{1}{2}(3-d), \quad \beta = \tfrac{1}{4}(3-d), \quad z = 2. \tag{42}$$

Note that the scaling law is not obeyed in this case because $\lambda = 0$. In three dimensions the surface thickens logarithmically, $\alpha = \beta = 0$.

The linear solution was used by Kardar *et al.*, as the starting point in a perturbation expansion followed by a renormalization. In two dimensions the exact result is obtained, but in higher dimensions the situation is complex, and we will return to it below.

2.5 The castle-wall model

Meakin *et al.* (1986a) showed how to derive the exact result $\alpha = \tfrac{1}{2}$, $\beta = \tfrac{1}{3}$ in two dimensions in a much simpler way than that of the section above. They accomplished this by introducing yet another model of the ballistic type, which is called the 'castle-wall' or 'single-step' model. Simulations show that it is in the same universality class as the others described above, and it is soluble in the long-time regime.

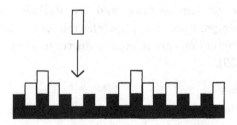

Figure 9: The castle-wall model for deposition.

Suppose we start our surface with the configuration of Figure 9, like that of a medieval fortification. Now the surface is allowed to grow by deposition *only at local minima* of the current surface, and only by dropping 'bricks' of height 2. Some of the stages of growth are shown in the figure. Clearly, the surface is formed, at any stage, by steps of ± 1 unit. (The related work of Kim and Kosterlitz (1989) also allows steps of size 0.)

The model can be mapped onto a model of spins: spin up represents a step up, and spin down a step down. The growth rule (Figure 10) corresponds to spin exchange with up spins moving left and down spins right, but not the reverse.

Figure 10: Mapping the castle-wall model to a spin system. One growth stage and the corresponding spin flip is shown.

If we are dealing with a system with periodic boundary conditions, we can understand the onset of the steady-state regime as the time required for each spin to make enough transits around the system so that the initial correlations are completely lost, i.e.

$$\langle \sigma_k \sigma_m \rangle = \delta_{km}. \tag{43}$$

Suppose that this is true. Then since the height difference is the sum of the steps we have:

$$h_i - h_j = \sum_{k=i}^{j} \sigma_k,$$

$$\langle (h_i - h_j)^2 \rangle = \sum_{k,m} \langle \sigma_k \sigma_m \rangle = |i - j|. \tag{44}$$

Now comparing Equations (44) and (36) gives $\alpha = \frac{1}{2}$, and the scaling law gives $\beta = \frac{1}{3}$.

It remains to show that Equation (43) is valid. This is a subtle point, which can be proved by considering the distribution of probabilities among the different configurations. It is surely the case that we will get Equation (43) if all configurations are equally likely in the steady state. Let us prove that such a statistical state is, in fact, the steady state. (This proof is due to C. Caroli, private communication.)

In the steady state the distribution of probabilities obeys detailed balance, i.e., for each configuration the probability to destroy it by growth is equal to the probability to create it. However the number of ways to destroy a configuration is the number of minima, and the number of ways to create it is the number of maxima (which are minima in the parent configuration). But for a periodic curve on a one dimensional substrate the number of minima is always equal to the number of maxima. This simple topological fact is enough to complete the proof. No comparable exact solution is known for dimensions above 2.

2.6 Higher dimensions

Many groups are now working on growing surfaces for dimensions above 2. This is a difficult problem for which the perturbative renormalization group of Kardar *et al.* gives ambiguous answers. The difficulty may arise from the fact that for $d \geq 3$ this class of models shows a *phase transition* from strong coupling to weak coupling as a function of the parameters of the model. For large 'smoothing' term (large η) the system reverts to weak coupling. Halpin-Healy (1989) speculated that this should be true for dimensions 4 and above, and very recently, Yan, Kessler, and Sander (1990) showed numerically that the transition exists in 4 *and* in 3 dimensions. Amar and Family (1990) and Pelligrini and Jullien (1990) have results of the same sort for somewhat different models. A similar transition has been found for directed polymers in dimensions 4 and above (Cook and Derrida, 1989) and in three dimensions (Derrida and Golinelli, 1991) as a function of the temperature of the polymer.

In the case of real thin films the purpose of much of the technology is to produce smooth deposits. That is, relaxation processes are encouraged in order to make layer-by-layer crystal growth possible. It is not clear where this kind of growth fits in to our picture. Perhaps the practical situation is always represented by a very large η. In fact, the scaling behavior described here which is on a fairly sound theoretical basis lacks experimental confirmation of any kind.

3

Diffusion limited aggregation

In ballistic deposition particles arrive at the deposit in a straight line. However, it is easy to imagine another extreme: growth controlled by the arrival of particles with a short mean free path — which perform many steps of a random walk before sticking. This is growth controlled by diffusion. A model appropriate to this case was introduced by Witten and Sander (1981, 1983). It is called diffusion-limited aggregation, DLA.

3.1 Simulations

In the model for deposition, a substrate is prepared as before, and particles are allowed to wander in from far away, one at a time, by a random walk and attach to the substrate or the other particles, whichever they encounter first. Once more, we can also think about the growth of a cluster from a point, aggregation, and indeed, it is most common in DLA studies to discuss the aggregation case. Figure 11 is a DLA cluster with about 2100 particles. The current record for simulations is in the millions. (See for example, Meakin *et al.*, 1987.) As we will see, the simulation represents many natural processes rather well.

DLA deposits and clusters are not merely rough, as ballistic deposits are; they seem to be *fractals* with $D \approx 1.7$ for $d = 2$ and ≈ 2.5 for $d = 3$. There are two popular methods to measure fractal scaling: one is to look for power laws in the measured correlation function, and the other to look at the increase of a typical radius with M, cf. Equations (6), (8).

The essential point to note about DLA is that the particles tend to stick at the outside of the cluster or near the top of the deposit. This is a direct result of Equation (15). The random walker that feeds the cluster has fractal dimension $D = 2$ (this follows at once from Equation (8) and the well-known fact that $R^2 \approx M$ where M is the number of steps for a random walk). The number of intersections will be large if $D_{DLA} + 2 > d$. However, this is so for the two values we have quoted, and we will see later that it is true in all dimensions. Since the mean number of intersections with the cluster of the random walker

Figure 11: A DLA cluster grown on a square lattice.

is large, the probability of attaching near the outer boundary is large. Qualitatively, this corresponds to the fact that a random walker has a very small probability of wandering very far down one of the 'fiords' in Figure 11.

There are a series of tricks which have been developed in order to make DLA simulations more tractable. Consider the aggregation geometry. One trick is that the random walker need not actually start far away from the aggregate. It can start at a random point on a circle of size R_{max} which just encloses the cluster since its probability of arrival on this circle is random. (This observation is due to M.E. Sander.) However, the walker may wander away from the aggregate. In that case, it is necessary to allow a free walk until it is quite far away (many aggregate radii). This is not a serious problem since it is possible to allow the walker to take large steps when it is outside R_{max}. It cannot encounter any matter, so it can take a step as large as the distance to the nearest point on the cluster, but in a random direction. (This trick was invented by P. Meakin.) The most refined modern programs use techniques that allow large steps even inside holes in large aggregates, and in this way very large clusters can be grown (Ball and Brady, 1985).

In the following we will often find it useful to consider a DLA cluster growing in a radial geometry with a fixed probability density, u_0, to find a random walker far away. The probability density elsewhere, $u(\mathbf{r}, t)$, satisfies the diffusion equation (here written on a lattice):

$$u(\mathbf{r}, t + \tau) - u(\mathbf{r}, t) = \frac{1}{q} \sum_{\delta} [u(\mathbf{r} + \boldsymbol{\delta}, t) - u(\mathbf{r}, t)].$$

This equation simply says that the walker has an equal probability to jump from any q neighboring sites located at $\mathbf{r} + \boldsymbol{\delta}$ to the site in question at any time. This is the discrete version of the continuum diffusion equation:

$$\frac{\partial u}{\partial t} = \eta \nabla^2 u, \tag{45}$$

with diffusion constant η. In the case of a single walker at a time, we can neglect the right-hand side of the equation because the behavior we are interested in is independent of the time it takes to attach a single particle: the average over realizations of the random walk depends only on position, not on time. Thus:

$$\nabla^2 u = 0. \tag{46}$$

Since the particles are absorbed

$$u_s = 0, \tag{47}$$

on the surface. Also, the probability to grow at a point on the surface, P_s, is given by the probability that a particle arrives there. Thus:

$$P_s \sim u(\mathbf{r} + \boldsymbol{\delta}) \sim \frac{\partial u}{\partial n_s}. \tag{48}$$

The last equation, which relates the growth probability to the normal derivative on the surface, is to be used in the continuum limit. The average growth velocity, v_s, is proportional to $\partial u / \partial n_s$.

Using the formulation of the last few paragraphs, it is possible to give an extension of the type of reasoning given above that showed that ballistic aggregates are not fractal (Ball and Witten, 1984). However, the result for DLA is different: we will find that DLA can be fractal.

Consider a large sphere which is far away from the cluster and a small sphere whose radius R is of order R_{max} and which is defined as the radius at which the average absorption takes place. The solution to Equation (46) near the outer sphere is:

$$u = u_0\left[1 - \left(\frac{R}{r}\right)^{d-2}\right]. \tag{49}$$

(Near the outer sphere, all of the angle-dependent terms can be taken to have died out. In two dimensions the reasoning must be altered slightly, but the final result is the same.) The total amount of matter added, dM/dt, is the integral over the large sphere of P_s:

$$\frac{dM}{dt} \sim R^{d-2}. \tag{50}$$

Now we repeat the argument of Equation (32):

$$R^{d-2} \sim \frac{dM}{dR}\frac{dR}{dt} \sim R^{D-1}v. \tag{51}$$

The speed, v, is bounded. Thus we find:

$$R^{d-D-1} \leq \text{const}. \tag{52}$$

The only way for this to happen is for

$$d - 1 \leq D \leq d. \tag{53}$$

This bound on the fractal dimension has been checked up to $d = 6$ (Meakin, 1983a).

3.2 Universality

There has been a substantial effort in looking at different features of universality of the scaling of DLA, that is, its lack of dependence on details of the growth. Witten and Sander (1983) and Meakin (1983a) showed that the fractal scaling properties were independent of the sticking probability. That is, if we allow a particle to sometimes bounce off the aggregate before sticking, the branches thicken but the fractal nature at large scales is preserved. In effect, the particle size has been increased.

The model can be reformulated by directly solving Eqs. (46)–(48) without changing the scaling behavior. Numerical simulations of these equations (Niemeyer et al., 1984) using a relaxation method to solve the Laplace equation lead to DLA-type fractals. This is sometimes referred to as the dielectric breakdown version of DLA because of the particular experimental realization that the authors had in mind (see below). Niemeyer *et al.* also gave an interesting extension of the model to allow the growth probability to be given by:

$$P_s \sim \left[\frac{\partial u}{\partial n_s} \right]^\eta, \tag{54}$$

where the growth exponent, η, represents possible non-linearities in breakdown, also gives rise to fractals whose scaling is different from DLA: their fractal dimension depends on η. In fact, $D(\eta = 0) = d$ since in this case the growth probability is independent of the perimeter site and we are back to the Eden model. D approaches 1 for large η. Of course, $D = 1.7$ for $\eta = 1$, because then we recover DLA, Equation (48).

Voss (1984) considered the possibility of a finite density of aggregating particles. In this case there is a crossover from DLA to compact aggregation: on small scales the cluster scales like ordinary DLA but on larger sizes it achieves a finite density. This is reasonable when one realizes that the density of the aggregate can hardly fall below the original ambient density. Any particles trapped between two arms will certainly be caught. A generalization of Equations (46)–(48) was given for this case by Nauenberg, Richter and Sander (1983) who pointed out that the relevant equation in this case is the diffusion equation, Equation (45). We will see below how this gives rise to a finite scale (known as the diffusion length) in a problem which is otherwise scale-invariant.

For aggregates up to about 10^6 particles the pattern is independent of the lattice on which the growth takes place. For larger clusters the pattern does depend on the lattice (Brady and Ball, 1984, Meakin *et al.*, 1987, Meakin 1988). In fact, for large aggregates the pattern deforms and takes on a shape for which the individual branches are aligned with the lattice axes. The lattice anisotropy acts to distort the cluster. Another anisotropy effect was given by Ball *et al.* (1985) who showed that if the sticking probability is anisotropic on a square lattice, so that it is more likely to stick right and left than up or down, then the cluster distorts very quickly into an elongated shape where the one radius increases as $M^{1/3}$ and the other as $M^{2/3}$. The universality of the model is not preserved in the presence of anisotropy. This sensitivity to anisotropy is shared with models for orderly growth by diffusion.

3.3 Experimental manifestations

One remarkable thing about the DLA model is that it is possible to find real systems that appear to be well described by it. DLA is not merely a computer game, and its surprising ubiquity gives extra interest to its study.

3.3.1 *Crystallization*

In the original work of Witten and Sander (1981) it was pointed out that Equations (46)–(48) are very similar to the standard description of crystal growth. (See the chapters by Caroli *et al.* and by Pomeau and Ben Amar.)

However, there are several things that we must point out. Normally, in studies of dendritic solidification (say from a solution) the concentration of matter, referred to the equilibrium concentration near a flat surface, is given by the diffusion equation, Equation (45), not the Laplace equation, (46). We can, however, estimate the size of the term $\partial u/\partial t$ by noting that if there is a typical velocity of growth, v, then $\partial u/\partial t \sim v \partial u/\partial x$. Now:

$$|\nabla^2 u| \approx \frac{1}{L_d}\left|\frac{\partial u}{\partial x}\right| \qquad (55)$$

where $L_d = \eta/v$, the diffusion length, sets the scale for the diffusion field. It can be interpreted as the scale over which correlations are communicated by changes in the diffusion field; we can hardly imagine a fractal over distances larger than L_d. However, in practical cases L_d is often very much larger than the other scales in the problem, so that the right hand member of Equation (55) can be neglected. In these cases only we might get something like DLA in crystallization.

The Gibbs-Thompson boundary condition of crystallization theory:

$$u_s \sim \gamma\kappa, \qquad (56)$$

where γ is related to the surface tension and κ is the curvature, is different from Equation (47), and controls the formation of orderly patterns when they exist. This condition means that very curved surfaces act as if the supersaturation is smaller and thus grow more slowly. We believe that this effect acts as a cutoff quite analogous to the finite particle size in the DLA model. However the student should be warned that this point (that the two sorts of cutoff act in a similar manner) is quite controversial, and not accepted by many experts in the field.

The final boundary condition in the standard formulation of dendritic growth says that matter arrives at the interface by diffusion:

$$v \sim \frac{\partial u}{\partial n_s}. \tag{57}$$

This is a version of Equation (48) as we pointed out above.

The usual interpretation of Equations (55)–(57) is as a set of *deterministic* equations to be solved for a steady-state shape, given some suitable initial conditions. An example of a steady-state solution is a dendrite, a pattern with a stable tip, for example one of the branches of a snowflake. The DLA model produces something quite different, namely disorderly fractal growth, with tips which split repeatedly; however, the two cases arise from essentially the same physics. The feature that we have left out, and which seems to distinguish between orderly dendritic growth and fractal growth, is *anisotropy*. As is explained in detail in chapter by Pomeau and Ben Amar, a stable solution to Equations (55)–(57) exists only if another effect is added, namely the *anisotropy* of the surface tension arising from the crystalline *anisotropy*. This small effect is necessary for an orderly steady-state solution: otherwise tip-splitting occurs (Ben-Jacob *et al.*, 1984).

When, then, do we see DLA-like behavior? We believe that the following qualitative picture is consistent with the present evidence. First, we must be in a situation with diffusion length long compared to the size of the deposit, as we have pointed out. Also, since some kind of anisotropy is necessary to stabilize diffusive growth into a steady-state pattern, and suppress tip-splitting, anisotropy must either be absent, or too small to overcome the averaging effects of extrinsic noise present in the experiment. If that is the case, tip-splittings proliferate and produce fractals. An analytic treatment to support this picture does not yet exist. The evidence that we have is based on experiment and simulations.

For example, suppose the crystal is polycrystalline, with no long-range correlation for the crystalline axes. One case that has been studied in detail (Radnoczi *et al.* (1987), see Figure 12), is the formation of a polycrystalline deposit of $GeSe_2$ when an amorphous film rapidly crystallizes. In this case well-defined DLA-like growths are observed with fractal dimension ≈ 1.67. In another experiment, Honjo *et al.* (1986) grew NH_4Cl between rough plates separated by 5 μm. Once more DLA scaling was observed even though the analogous experiment with smooth plates gives rise to ordered dendrites.

Figure 12: A fractal growth of GeSe$_2$ of the type studied
by Radnoczi *et al.* (1987). (Courtesy of T. Vicsek.)

These and related experiments lead us to believe that DLA patterns
will develop, at least on scales less than the diffusion length, whenever the
intrinsic instabilities in crystal growth are not controlled by anisotropy.

3.3.2 *Electrochemical deposition*

Electrochemical deposition of metals is a special case of crystallization
which has been studied in detail in recent years in the context of DLA.
Suppose we consider an experiment in which metal is deposited on a
cathode from an electrolytic solution, and the motion of the ions in
solution is limited by diffusion. In this case we are dealing with an
almost literal realization of the DLA model provided the electric field in
the solution is not important for the motion of the ions over most of the
bulk. This is the case in the experiments of Brady and Ball (1984) who
considered the electrodeposition of copper from CuSO$_4$ and found fractal
deposits. In order to ensure that the electric force is not important in
this case they added a supporting electrolyte (i.e. another inert set of
ions in addition to the Cu$^+$ and SO$_4^-$) to the solution to screen the fields,
and a powder to suppress convection. The resulting deposits on the end
of a fine wire are of the DLA type in $d = 3$.

Brady and Ball were able to measure the fractal dimension of their
deposits using electrical measurements alone. They did this by noting
that by Equation (50), the instantaneous electrical current in $d = 3$ is
proportional to the radius of the cluster. Also, the integrated current
gives the total mass. Now by plotting $\log(R)$ versus $\log(M)$ (cf. Equation (8)) the fractal dimension follows at once.

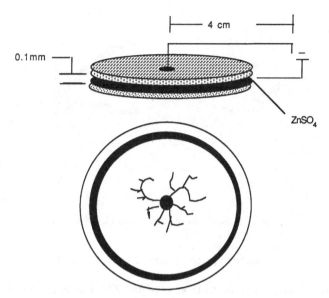

Figure 13: The experimental configuration of Grier *et al.* (1986) for producing two dimensional electrodeposits.

In another set of experiments (Matsushita *et al.* (1984), Grier *et al.* (1986), Sawada *et al.* (1986)) the driving force for the motion of the ions *is* the electric field. The ions follow field lines on large scales. However, in this case we are once more in a regime that can be treated by the same model equations, Equations (46)–(48), because we can reinterpret the u to be the electrostatic potential which obeys the Laplace equation for a conductor taken to be an equipotential. The deposit grows when an ohmic current arrives at the surface. The current, by Ohm's law, is proportional to $\partial u / \partial n$ (the electric field) at the surface, cf. Equation (48). The process is DLA-like (and can produce patterns very like DLA clusters) but is really more similar to dielectric breakdown, see below, than to diffusion-limited electrodeposition. Here, as in crystallization, there is the possibility of making DLA-type fractals, *and* stable, dendritic crystals, depending on the experimental conditions. An example of these two patterns is shown in Figure 14 (a) and (b): DLA patterns arise for slow growth and dendritic crystals for fast growth. The way in which the experimental patterns are controlled (presumably) by anisotropy, and how anisotropy can be related to growth rate, is a complex question in electrochemistry which is currently being studied.

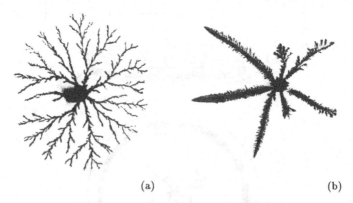

(a)   (b)

Figure 14: Some results of Grier et al. (1986) for
the electrodeposition of zinc.
(a) Slow growth, DLA-like pattern.
(b) Dendritic pattern produced by rapid growth.

In the original work (Sawada *et al.*, 1986, Grier *et al.*, 1986) for intermediate growth rates we find a pattern that we have not yet discussed,
Figure 15. It has many tip-splittings like a fractal, but its overall outline
is stably round in a round cell, and its average density is constant. We
call this the dense radial pattern. How does this fit in to the overall
picture?

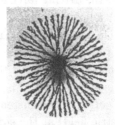

Figure 15: The dense radial pattern, Grier *et al.* (1986, 1987),
in the electrodeposition of zinc at intermediate growth rates.

Our explanation of the effect (Grier *et al.*, 1987) is that the metal
deposit has an appreciable resistance compared to the electrolyte. In this
case Equation (47) is no longer valid. A detailed analysis shows that in
this case the pattern is stable on large scales (corresponding to the overall
outline) and unstable on small scales, which presumably has something
to do with the filamentary structure. A very recent experiment (Melrose
and Hibbert, 1989) finds that the onset of the dense radial pattern is

associated with a peak in the resistivity of the deposit, but then the apparent resistance decreases. It is clear that this fascinating subject is still open.

3.3.3 *Viscous fingering*

Viscous fingering is the unstable motion of the interface of two fluids, one more viscous than the other when the inviscid one is injected into the more viscous (see Bensimon *et al.* (1986) for a review). The fingering instability is a problem for oil recovery processes in which air or water is injected into an oil field in order to drive the oil to a distant well; if the pattern of the interface is complex, it is difficult to know where to put the well.

The phenomenon occurs for flow in a porous medium like an oil-bearing rock, and for flow between two parallel plates with a small gap, a so-called Hele-Shaw cell. The pattern of the relatively inviscid injected fluid plays the role of the crystal in the sections above, or the DLA cluster.

In order to relate this process to what we have discussed so far, we need only point out that flow in porous media or a Hele-Shaw cell is described by an empirical rule called D'Arcy's law which relates the fluid velocity, v, to the gradient of the pressure, u, in the viscous fluid:

$$v \sim \nabla u. \tag{58}$$

The proportionality constant depends on the viscosity. Since most fluids are almost incompressible

$$\nabla \cdot v = \nabla^2 u = 0. \tag{59}$$

We take the zero of pressure to be that in the inviscid fluid (since it has small viscosity, its pressure there is approximately constant). We have:

$$u_s = \gamma \kappa, \tag{60}$$

due to the pressure drop due to the curvature of the interface. Equation (58), applied at the interface gives:

$$v_s \sim \frac{\partial u}{\partial n}. \tag{61}$$

These are formally identical to the equations for crystallization in the limit of large diffusion length (Paterson, 1984). Thus we have another example of the kind of diffusion-limited growth that should be describable using DLA.

Experimental work in this area has been very extensive. We will cite a partial set of references. Paterson (1981) was among the first to note the complexity of a viscous fingering pattern in a radial cell. For a recent example see Figure 16 (see Couder, 1988 and Rauseo *et al.* 1987). In this case, the fractal dimension has been measured, and appears to be near, but slightly larger than that of DLA.

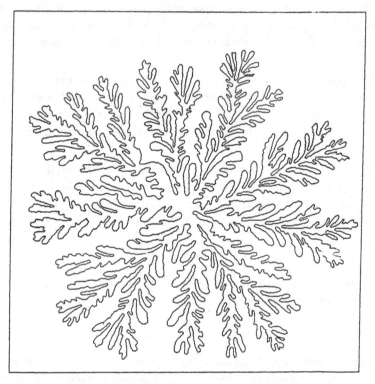

Figure 16: A viscous fingering pattern (courtesy of Y. Couder).

These results are in contradiction with the observations of Ben-Jacob *et al.* (1986) who saw the analogue of the dense radial pattern in viscous fingering. This is very hard to explain in the framework we have developed here: the analogue of finite resistivity (finite viscosity of the injected fluid) is far too small here to account for the effect. In our view the only way to make sense of the situation, and to account for the contradiction between the experimental results of Ben-Jacob *et al.* and those quoted above is to assume that in the high-pressure experiments of Ben-Jacob *et al.*, other physical effects (partial wetting, mechanical deformations, varying pressure at the boundaries) intervene and make Equations (59)–(61) inapplicable. In a direct numerical solution of Equations (59)–(61) (Sander *et al.*, 1985) there is no sign of dense behavior over a small range of sizes, but there does seem to be the beginning of fractal scaling.

Some groups intentionally introduce randomness into their experiments. For example, Maloy *et al.* (1985) and Nittmann *et al.* (1985) studied fingering in artificial porous media by trapping glass beads in a cell and injecting fluid. In these cases well-developed DLA patterns are easily observed.

The classic work on viscous fingering, Saffman and Taylor (1958), is in another regime entirely, that of very large anisotropy. They studied a channel which is very long compared with its width. In this case a single orderly finger occupies the center of the channel. An experiment which showed in a qualitative manner the transition between the two regimes was performed by the Michigan group (Ben-Jacob *et al.*, 1985). In this case a set of grooves was inscribed on the plates. At low flow rates something like Figure 16 was observed, but at high flow rates a dendritic pattern occurred. In a way that is not entirely clear, increasing the overall flow rate enhanced the effect of the imposed anisotropy. A spontaneous appearance of anisotropy in viscous fingering was given by Buka *et al.* (1986), by producing viscous fingering in liquid crystals.

3.3.4 *Dielectric breakdown*

Another DLA-type process is the breakdown of a dielectric material when it is exposed to a very large electric field. Niemeyer *et al.* (1984) showed that the pattern formed in some cases is exactly that of DLA.

The connection to the model is once more through Eqs. (46)–(48). Suppose we have a material with two electrodes with a large potential between them. Let u be the electrostatic potential between them. We have the Laplace equation for u. The breakdown channel where the material is highly ionized has a large conductivity, and thus a constant potential. And, it is reasonable to suppose (though it may depend on the material in question) that the probability of breakdown is linear in the electric field, ∇u. Thus we have recovered Eqs. (46)–(48). One possible way to describe non-linear breakdowns was described above, Eq. (54).

3.4 Theory of DLA

The theory of DLA is not well developed. We have no analytic proof that the scaling which is observed in numerical simulations is really present in the asymptotic case, and no way (apart from measurements of numerically generated patterns) to find the fractal dimension. Many groups have attempted to formulate a theory. We will give here some ideas about the approaches which appear interesting.

There are two aspects of the problem which make the theory difficult, far more difficult than that of ballistic aggregation, for example. They are the fact that there is no 'upper critical dimension' in the problem, and the fact that the growth probabilities are multifractal, so that unlike ballistic aggregation, where there is only one important exponent, in DLA there appear to be an infinite number.

3.4.1 *Upper critical dimension and mean-field theory*

We have shown above that the fractal dimension of a DLA cluster is bounded below by $(d-1)$. Witten and Sander (1983) pointed out that this makes DLA quite different from other statistical physics problems for which there exists a dimension above which the fractal dimension ceases to depend on d, and the theory becomes trivial. For example, it is thought (Halpin-Healy, 1989) that above $d = 5$ ballistic deposits are always in the trivial weak-coupling state. In this sense DLA has no upper critical dimension, which makes it impossible to apply methods of expansion about the upper critical dimension.

It is interesting to compare DLA with an equilibrium fractal which has an upper critical dimension. This is the self-avoiding walk (see de Gennes, 1979) the standard model of polymers in a good solvent. In this case the fractal dimension is equal to 2 for $d \geq 4$, which is the upper critical dimension of the model.

We can derive this fact from the simple considerations of fractal intersections outlined earlier. A self-avoiding walk is defined to be a trajectory which is forbidden to cross itself. To see the significance of $d = 4$, let us see when the constraint of self-avoidance is actually unimportant. This could happen if the fractal were so tenuous that it would be unlikely to cross itself simply from statistical considerations. However, if two parts of a fractal are to simply not see each other, we have, from Equation (15), that

$$2D < d. \tag{62}$$

Since a free walk has dimension 2, above $d = 4$ it will not cross itself.

For DLA this clearly does not happen: for all dimensions the constraint of self-avoidance is important. In fact, above we have extended the reasoning to show why the intersection of a DLA with the random walks that feed it occurs at the surface.

It is thus very probable that no simplified theory of DLA will be exact in any dimension. However, it is interesting to inquire whether there is a mean-field theory of some sort which could, in principle, be used as a starting point for understanding. There is, indeed such a theory, which was suggested by Witten and Sander (1983) and developed by Ball, Nauenberg, and Witten (1984) and Nauenberg (1983). Unfortunately, though this approach does give some results, it has led to no further development to date.

In this approach we coarse grain the density of the aggregate, and average over angles. Thus we seek the average profile of a cluster. The density is replaced by a function $\rho(\mathbf{r}, t)$, which is coupled to the random walker probability function, u, which we have discussed up to now. Since the walker can be absorbed inside the coarse-grained profile, we treat the change in the density as a sink for particles. Thus we have instead of the Laplace equation with a boundary condition:

$$\nabla^2 u = \frac{\partial r}{\partial t}. \tag{63}$$

We now must describe the probability of absorption. Since a particle is absorbed only where there is matter, we are tempted to set this to be $u\rho$. However, a bit of thought will show that this will not do: the finite particle size is nowhere present here: we must allow the absorption to allow growth into a region where ρ is initially zero. The simplest way to represent this is to write:

$$\frac{\partial r}{\partial t} = u[\rho + a^2 \nabla^2 \rho]. \tag{64}$$

These two equations are to be solved for the profile. We expect the two profiles to look like the sketch of Figure 17.

Well inside the aggregate we can define an average screening length, ξ, from dimensional analysis. Using the first term of Equation (54) together with Equation (63), we have:

$$\xi^{-2} \sim \frac{\nabla^2 u}{u} = \rho. \tag{65}$$

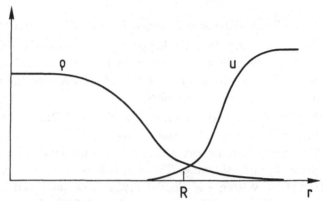

Figure 17: Expected behavior of the averaged aggregate density ρ and the random-walker density u, in the mean-field description of DLA.

Now the approximate solution of Ball, Nauenberg, and Witten (1984) is based on the observation that the solution to Equation (64) in the exterior region (where we can take $u \approx u_0$) is, up to power laws in t:

$$\rho \sim \exp(t - r^2/4a^2 t). \tag{66}$$

This is to be matched to the interior solution in the surface region, i.e., at some average radius, R. A reasonable definition of R is the point where the profile becomes steep enough so that it changes within a screening length, ξ, so that we can no longer solve Equation (64) with constant u. Thus:

$$\frac{1}{\rho}\frac{\partial \rho(R)}{\partial r} \sim \xi^{-1} \sim \sqrt{\rho(R)}. \tag{67}$$

Combining Equations (66) and (67) gives:

$$\frac{R^2}{4a^2 t^2} \sim \exp(t - R^2/4a^2 t). \tag{68}$$

Now this matching can only be consistent if $R = 2at$. Thus in this model the velocity is constant. Now return to Equation (51) which says $R^{d-2} \sim R^{D-1}v$. This is only consistent if $D = d - 1$. This mean-field theory result for D is equal to the lower bound on the fractal dimension. It seems that for large d (Meakin, 1983a) the lower bound is approached.

3.4.2 *Multifractal scaling*

Many authors have tried to use the results of the previous section to give an approximate theory of DLA. For example Tokuyama and Kawasaki

(1984) gave a kind of Flory theory based on the notion that DLA is a kind of polymer with ξ as a basic unit of length. Unfortunately, all theories of this sort seem very doubtful because of the extreme inhomogeneity of the growth. Theories based on a running scale like ξ neglect the fact that there are many relevant scales because the growth probability, P_s, is *multifractal*, the tips of the growth advance much, much faster than the fiords.

The numerical proof of this statement was given by Halsey *et al.* (1986), Amitrano *et al.* (1987), Meakin *et al.* (1986b), and most recently, by Hayakawa *et al.* (1963). P_s is found by either sending many random walkers to a fixed cluster and recording where they land, or directly solving the Laplace equation around a cluster, and using $P_s \sim [\partial u / \partial n_s]$. In all cases a multifractal distribution is found with an $f(\alpha)$ like that of Figure 5.

A theoretical framework for this array of singularities was given by Turkevich and Scher (1986). They pointed out that the solution of the Laplace equation near a sharp point naturally gives rise to a singularity in $\partial u / \partial n_s$: the fact that sharp points on grounded conductors have large electric fields is the essential point of the ordinary theory of the lightning rod. A DLA cluster has many sharp points.

For example, if we consider the leading tip (which grows fastest) on a DLA cluster to be more or less a wedge of included angle β, then in two dimensions standard electrostatic theory gives:

$$\nabla u \sim X^{A-1}, \quad A = \frac{\pi}{[2\pi - \beta]}. \tag{69}$$

Here X is the distance of the observation point from the tip. For example, for a flat conductor, $\beta = 0$, we have an inverse square-root singularity.

Now suppose we integrate Equation (69) over some small distance near the tip. The total probability, which gives the tip velocity, is then:

$$v \sim \left(\frac{a}{R}\right)^A, \tag{70}$$

where we have made the natural assumption that the probability must be made dimensionless by the only available length in the problem, the current radius, R. Now repeating Equation (51), we have, for $d = 2$:

$$C \sim \frac{\mathrm{d}M}{\mathrm{d}R}\frac{\mathrm{d}R}{\mathrm{d}t} \sim R^{D-1}v \sim R^{D-1-A}, \quad D = A + 1. \tag{71}$$

Thus an arbitrary 'wedge angle' gives rise naturally to a non-integer dimension for a DLA cluster. Unfortunately, no one has yet given a geometrical argument for the average angle at the leading tip. Working backwards from Equations (71) and (69) and $D = 1.7$ gives a β about equal to the interior angle of a regular pentagon.

An example of these ideas is in the development of the anisotropic DLAs of Ball *et al.* (1985). Since the angles approach those of a 'wire' we should expect for the 'fast' direction, say the x-direction, from Equation (70),

$$v \sim \frac{dR_x}{dM} \sim \frac{1}{\sqrt{R_x}}, \tag{72}$$

assuming (for this purpose only) that we add mass uniformly in time. Integrating this we get $R_x \sim M^{2/3}$, as quoted above.

The best way to interpret A is as the strongest singularity of the multifractal array (i.e. the smallest α, cf. Equations (18), (69) and (70)): $A = \alpha_0$. Then Equation (71) is a constraint on the form of the $f(\alpha)$ curve for DLA since D is the top of the curve, and α_0 is the left-hand intersection with the axis. This constraint is obeyed by the numerical calculation of Hayakawa *et al.* with reasonable precision.

Formulating a real theory of DLA will probably require a renormalization of the whole $f(\alpha)$ curve, or alternately, of all of the D_q. This is beyond our theoretical capability at the moment.

4

Cluster-cluster aggregation

In an atmosphere of particles which can diffuse and combine, forming multiparticle clusters which continue to diffuse, we have a different sort of physics, the *aggregation of aggregates* or cluster-cluster aggregation. A model for the process (Meakin, 1983b, and Kolb, Botet and Jullien, 1984) produces fractal aggregates. This model is directly applicable to many situations in colloid and aerosol chemistry. We will not discuss the experimental aspects of the subject: Vicsek (1989) and Meakin (1988) give full discussions with many references. Our purpose here is to illustrate the parts of the subject which are related to the kinds of geometric scaling that we have seen for single-particle aggregation.

4.1 Simulations

In the model, one begins with a large collection of particles, each of which is allowed to diffuse until it encounters another. Then the resulting cluster continues to move until large clusters result. Cluster-cluster aggregates have a fractal dimension which is much smaller than that of DLA. Some results (Meakin, 1988) are $D(2) \approx 1.43$, $D(3) \approx 1.75$ (compare DLA with 1.7, 2.5 respectively). The reason is relatively simple to grasp: however much difficulty particles have in wandering down fiords, clusters will have much more trouble and will stick near to the surface. This is true if the kinetics is such that at any time the cluster size distribution has a mean that increases in time so that a cluster does not encounter many individual particles. Any 'normal' kinetics turns out to have this property, as we will now discuss.

We noted that we must continue to let aggregates diffuse. In order to completely define the model, we must decide on how fast they are to diffuse, namely what the diffusion coefficient of an aggregate is to be. If we assume the friction to be given by Stokes's law, the retarding force is proportional to the inverse radius of the cluster. In the simulations it is usual to take the diffusion coefficient of the clusters to be a power law in the mass (Meakin, 1983b, Kolb *et al.*, 1983):

$$\eta_M \sim M^\gamma. \tag{73}$$

For Stokes's law $\gamma = -D^{-1}$. In the simulations it is usual to take γ to be a free parameter. In fact, the fractal dimensions are independent of γ (and equal to the values above) provided γ is less than 1. In this case the mean cluster size increases. However, if γ exceeds 1, that is large clusters move much faster than small ones (not 'normal' kinetics at all — quite non-physical in fact) one large cluster begins to 'eat' the smaller ones who never grow (Jullien and Kolb, 1984). Thus we are back to particle aggregation of the DLA type.

Two variations of the model have been studied in detail. We can look at ballistic cluster-cluster aggregation. In this case we *still* produce fractals, because of the exclusion effect mentioned above, but the fractal dimension is higher. (Jullien and Kolb, 1984.) We have $D(2) = 1.55$. And one can consider 'reaction-limited' aggregation (Jullien and Kolb, 1984, Brown and Ball, 1985) where the probability of sticking is small at each try. This is relevant to real situations. Once again we expect that after many tries the clusters will penetrate more than for the diffusive case. In fact $D(2) \approx 1.53$, $D(3) \approx 2.11$.

4.2 Cluster size distributions and kinetic equations

The cluster size distribution is clearly a matter of great interest in this type of aggregation study. For diffusion-limited cluster aggregation the cluster size distribution, $n(s,t)$, the number of clusters of size s at time t, has a scaling form (Vicsek and Family, 1984, Kolb, 1984):

$$n(s,t) \sim s^{-2} f\left(\frac{s}{t^z}\right). \tag{74}$$

The form of the scaling function f depends on the value of γ.

The cluster size distribution is the object treated by the famous kinetic equations of Smoluchowski (1915). These relate the cluster size distribution to the so-called reaction kernel, K_{ij}, the probability of reaction of two clusters of size i, j, to make a cluster of size $i + j$. Using these it is possible, for example, to find a relation between z, γ, and the fractal dimension.

In what follows we will neglect all of these very interesting details, which are very relevant, however, to real kinetics. We will restrict ourselves to the simplified case where the cluster size distribution is narrowly peaked. This is realistic, for example, for diffusion-limited cluster aggregation for small γ.

4.3 Theory; upper critical dimension

There is a remarkable theoretical result that one can derive for cluster-cluster aggregation: this process, unlike DLA, has an upper critical dimension. This result follows simply from arguments about fractal geometry (Ball and Witten, 1984, see also Witten and Cates, 1986).

In this discussion, we assume that, as in the case of the random walk in high dimensions, the clusters are so tenuous that they can freely interpenetrate. Then we see what value of d is necessary for this to be true.

If the clusters are tenuous, they can stick to one another anywhere. If two clusters A, B, of the same size, M, interact, then choose two sites at random. With probability $\frac{1}{4}$ they will both be on the same cluster, but with probability $\frac{1}{2}$ one of the two sites will be on each cluster.

Now measure the mean distance, $b(2M)$, between the two sites *along the cluster* (the so-called chemical distance). If the two sites came from the same parent cluster, the distance will be $b(M)$, but if they come from different parents, the distance will be $2b(M)$, where the 2 comes from the fact that in order to get to the random joining point a distance b will have to be travelled on each cluster. Now combining the possible values with the proper probabilities we have:

$$b(2M) = \tfrac{1}{4}b(M) + \tfrac{1}{4}b(M) + \tfrac{1}{2}2b(M) = \tfrac{3}{2}b(M). \tag{75}$$

The solution to this recursion relation is $b \sim M^{\ln(3/2)/\ln(2)}$.

Now the physical distance is simply related to the chemical distance by $R^2 \sim b$, because the cluster arms are randomly coiled. Thus:

$$M \sim R^{2\ln(2)/\ln(3/2)}. \tag{76}$$

Thus the limiting fractal dimension is 3.4. In order to have consistency, we need (cf. Equation (62)) a dimension greater than 2D. The upper critical dimension is 6.8.

In fact, this argument is applicable to the reaction-limited case since each site is taken to be equally likely to be the attachment point. For diffusion limited and ballistic cluster-cluster aggregation similar arguments give larger values for d_c.

5

Acknowledgements

The work of the present author reported here was supported by NSF Grants DMR 82-03698, 85-05474, and 88-15908, and by DOE Grant DEFG-02-85ER54189. I would like to thank the Groupe de Physique des Solides, Université Paris 7, for hospitality while these notes were being prepared.

References

Amar, J. and Family, F., 1990, *Phys. Rev. Lett.*, **64**, 543.

Amitrano, C., Coniglio, A. and diLiberto, F., 1987, *Phys. Rev. Lett.*, **57**, 1016.

Ball, R.C. and Brady, R., 1985, *J. Phys.*, **A18**, L809.

Ball, R.C., Brady, R., Rossi, G. and Thompson, B., 1985, *Phys. Rev. Lett.*, **55**, 1406.

Ball, R.C., Nauenberg, M. and Witten, T., 1984, *Phys. Rev.*, **A29**, 2017.

Ball, R. C. and Witten, T.A., 1984, *Phys. Rev.*, **A29**, 2966.

Ben-Jacob, E., Deutscher, G., Garik P., Goldenfeld, N.D. and Lareah, Y., 1986, *Phys. Rev. Lett.*, **57**, 1903.

Ben-Jacob, E., Godbey, R., Goldenfeld, N.D., Koplik, J., Levine, H., Mueller, T. and Sander, L.M., 1985, *Phys. Rev. Lett.*, **55**, 1315.

Ben-Jacob, E., Goldenfeld, N.D., Langer, J.S. and Schon, G., 1984, *Phys. Rev.*, **A29**, 330.

Bensimon, D., Kadanoff, L.P., Liang, S., Shraiman, B. and Tang, C., 1986, *Rev. Mod. Phys.*, **58**, 977.

Botet, R. and Jullien, R., 1984, *J. Phys.*, **A17**, 2517.

Brady, R. and Ball, R. C., 1984, *Nature* (London), **309**, 225.

Brown, W. and Ball, R.C., 1985, *J. Phys.*, **A18**, L517.

Buka, A., Kertesz, J. and Vicsek, T., 1986, *Nature*, **323**, 424.

Cook, J. and Derrida, B., 1989, *J. Stat. Phys.*, **57**, 89.

Couder, Y., 1988, in *Random Fluctuations and Pattern Growth*, H. E. Stanley and N. Ostrowsky eds. (Kluwer).

Derrida, B. and Golinelli, O., 1990, *Phys. Rev.*, **A41**, 4160.

Eden, M., 1961, *Proc. of the Fourth Berkeley Symp. on Math. Stat. and Prob.*, **4**, 223.

Edwards, S.F. and Wilkinson, D.R., 1982, *Proc Roy. Soc. London*, **A381**.

Family, F. and Vicsek, T., 1985, *J. Phys.*, **A18**, L75.

de Gennes, P., 1979, *Scaling Concepts in Polymer Physics* (Cornell).

Grier, D., Ben-Jacob, E., Clarke, R. and Sander, L.M., 1986, *Phys. Rev. Lett.*, **56**, 1264.

Grier, D., Kessler, D. and Sander, L.M., 1987, *Phys. Rev. Lett.*, **59**, 2315.

Halpin-Healy, T., 1989, *Phys. Rev. Lett.*, **63**, 442.

Halsey, T., Meakin, P. and Procaccia, I., 1986, *Phys. Rev. Lett.*, **56**, 854.

Hayakawa, Y. Sato, S. and Matsushita, M., 1987, *Phys. Rev.*, **A36**, 1963.

Honjo, H., Ohta, S. and Matsushita, M., 1986, *J. Phys. Soc. Japan*, **55**, 2487.

Jullien, R. and Botet, R., 1985, *J. Phys. A.*, **18**, 2279.

Jullien, R. and Kolb., M., 1984, *J. Phys. A.*, **17**, L639.

Kardar, M., Parisi, G. and Zhang, Y., 1986, *Phys. Rev. Lett.*, **56**, 889.

Kardar, M. and Zhang, Y., 1987, *Phys. Rev. Lett.*, **58**, 2087.

Kim, J. and Kosterlitz, J., 1989, *Phys. Rev. Lett.*, **62**, 2289.

Kolb, M., 1984, *Phys. Rev. Lett.*, **53**, 1653.

Kolb, M., Botet, R. and Jullien, R., 1983, *Phys. Rev. Lett.*, **51**, 1123.

Leamy, H., Gilmer, G. and Dirks, A., 1980, in *Current Topics in Materials Science*, Vol. **6** (North-Holland).

Leyvraz, F., 1985, *J. Phys. A*, **18**, L941.

Maloy, K., Feder, J. and Jossang, J., 1985, *Phys. Rev. Lett.*, **55**, 2681.

Mandelbrot, B.B., 1982, *The Fractal Geometry of Nature*, W.H. Freeman.

Matsushita, M. Sano, M., Hayakawa Y., Honjo, H. and Sawada, Y., 1984, *Phys. Rev. Lett.*, **52**, 286.

Meakin, P., 1983a, *Phys. Rev.*, **A27**, 604.

Meakin, P., 1983b, *Phys. Rev. Lett.*, **51**, 1119.

Meakin, P. 1988, in *Phase Transitions and Critical Phenomena*, Vol. **12**, C. Domb and J. Lebowitz, eds, Academic.

Meakin, P., Ball, R., Ramanlal, P. and Sander, L., 1986a, *Phys. Rev.*, **A34**, 5091.

Meakin, P., Coniglio, A., Stanley, H.E. and Witten, T., 1986b, *Phys. Rev.*, **A34**, 3325.

Meakin, P., Ball, R., Ramanlal, P. and Sander, L., 1987, *Phys. Rev.*, **A35**, 5233.

Meakin, P. and Jullien, R., 1987, *J. Phys. (Paris)*, **48**, 1651.

Melrose, J. and Hibbert, D., 1989, *Phys. Rev.*, **A40**, 1727.

Nauenberg, M., 1983, *Phys. Rev.*, **B28**, 449.

Nauenberg, M., Richter, R. and Sander, L., 1983, *Phys. Rev.*, **B28**, 1649.

Niemeyer, L., Pietronero, L. and Weismann, H., 1984, *Phys. Rev. Lett.*, **52**, 1033.

Nittman, J., Daccord, G. and Stanley, H., 1985, *Nature*, **314**, 141.

Paterson, L., 1981, *J. Fluid Mech.*, **113**, 513.

Paterson, L., 1984, *Phys. Rev. Lett.*, **52**, 1621.

Pellegrini, Y.P. and Jullien, R., 1990, *Phys. Rev. Lett.*, **64**, 1745.

Radnoczi, G., Vicsek, T., Sander, L. and Grier, D., 1987, *Phys. Rev.*, **A35**, Rapid Communications, 4012.

Rauseo, S., Barnes P. and Maher, J., 1987, *Phys. Rev.*, **A35**, 5686.

Richardson, D., 1973, *Proc. Camb. Phil. Soc.*, **74**, 515.

Saffman, P. and Taylor, G.I., 1958, *Proc. Roy. Soc. London*, **A245**, 312.

Sander, L., Ramanlal, P. and Ben-Jacob, E., 1985, *Phys. Rev.*, **A32**, Rapid Communications, 3160.

Sander, L., 1986, *Proceedings of Symposium on Multiple Scattering of Waves and Random Rough Surfaces*, (College Station, PA, July, 1985). Published in *Multiple Scattering of Waves in Random Media and Random Rough Surfaces*, V.V. Varadan and V.K. Varaden eds. (Technomic).

Sawada, Y., Dougherty, A. and Gollub, J.P., 1986, *Phys. Rev. Lett.*, **56**, 1260.

Smoluchowski, M., 1915, *Z. Phys. Chem.*, **92**, 129.

Tokuyama, M. and Kawasaki, K., 1984, *Phys. Lett.*, **100A**, 337.

Turkevich, L.A. and Scher, H., 1986, *Phys. Rev. Lett.*, **55**, 1026.

Vicsek, T., 1989, *Fractal Growth Phenomena* (World Scientific).

Vicsek, T. and Family, F., 1984, *Phys. Rev. Lett.*, **52**, 1669.

Voss, R., 1984, *Phys. Rev.*, **B30**, 334.

Witten, T. A. and Cates, M., 1986, *Science*, **232**, 1067.

Witten, T.A. and Sander, L.M., 1981, *Phys. Rev. Lett.*, **47**, 1400.

Witten, T.A. and Sander, L.M., 1983, *Phys. Rev.*, **B27**, 5686.

Wolf, D.E. and Kertesz, J., 1987, *Europhys. Lett*, **4**, 651.

Yan, H., Kessler, D. and Sander, L.M., 1990, *Phys. Rev. Lett.*, **64**, 926.

CHAPTER 6

KINETIC ROUGHENING OF GROWING SURFACES

J. Krug and H. Spohn

1

Introduction

Solids form through growth processes which take place at the surface. Physically, there is a huge variety of growth mechanisms depending on the materials involved, their temperature, composition, phases, etc. In this chapter we will discuss a very particular growth mechanism: We imagine an already formed nucleus to which further material sticks from the ambient atmosphere. The process of attachment is

▷ reaction limited (there is a good supply of material, but a permanent link to the nucleus is formed only after many attempts);

▷ far away from equilibrium (we consider time scales, on which the surface has not yet relaxed through surface diffusion and not yet reached a state of local thermal equilibrium with the surrounding gas/fluid phase).

We follow the tradition of Statistical Mechanics in studying oversimplified models which nevertheless attempt to capture some of the essential physics. It will turn out that these models also describe other physical processes of interest. Some of them will be explained in Section 4. The idea behind the most basic model is to focus on the two properties just mentioned and to ignore all other details. We disregard the ambient atmosphere and assume that particles stick randomly at the surface of an already formed cluster. Once a particle sticks, it remains there forever. Such a model was first proposed by Murray Eden (1958) in a biological context. The Eden model is one of the simplest growth processes.

Let us make the effort to define the Eden model (better one version of it) more precisely. To simplify the geometry we let the cluster grow on an underlying square lattice. (The generalization to higher dimensions will be obvious.) We start with a single seed at the origin. At each one of the four available perimeter sites we add an extra particle at a random time with an exponential distribution. We continue this process. At some time t we have a cluster of sites, A_t. (A_t is a random set because it depends on the particular growth history.) The sites adjacent to it are the growth sites. Each one of the growth sites is filled independently with a particle after an exponentially distributed random time. In Figure 1 we show a cluster grown this way consisting of 2.5×10^6 particles. If we let the aggregation process run for a while, then the cluster will take a definite shape with some fuzziness — the shape fluctuations, compare with the enlargement in Figure 1.

$S = 2.5 \times 10^6$

01

2000 LATTICE UNITS

200 LATTICE UNITS

Figure 1: Eden cluster on a square lattice. The right hand figure shows the top part of the cluster enlarged by a factor of ten. Courtesy of P. Meakin.

What can be learned from such a model? Most basically we want to understand the interplay between the microscopic growth rule and the macroscopic shape. Under what conditions does the cluster form facets, edges, or corners, as observed for many real materials? From a statistical mechanics point of view the immediate question is how to understand the properties of the shape fluctuations. In fact, we will argue that they are universal, i.e. essentially independent of the particular growth rule. If so, this leads to the difficult problem of determining the universality classes and the general characteristics of the growth rules defining them.

For readers not familiar with critical phenomena an example may be useful here. For a diffusing particle the microscopic motion will differ from material to material. Still the mean square displacement is always proportional to t. In this sense diffusion is universal. The only condition needed is that the velocities of the diffusing particle are statistically essentially independent when separated by a long time. This assumption breaks down e.g. in a turbulent fluid, where the mean square displacement grows as t^3.

By definition universality means that the large scale properties of the fluctuations are independent of microscopic details (within the given class). Thus real systems must have the same behavior. In this way the study of simplified models leads to predictions on real materials.

To give a guide through our undertaking: In Section 2 we consider a scale where fluctuations are negligible. We explain the link between the macroscopic shape and the inclination dependent growth velocity. In Section 3 our resolution is increased and we focus on a mesoscopic scale. On this scale there is still a well-defined and fairly smooth surface. Atomic roughness and overhangs are ignored. The scale is fine enough however to capture the stochastic nature of the growth process. We develop a scaling theory for shape fluctuations based on the notion of statistical self-similarity. Our discussion emphasizes generality. Up to then the only guiding example of a concrete growth process we have is the Eden model introduced above. This situation is rectified in Section 4 where we list and discuss a large variety of growth processes. The literature on the subject is rather ramified because of different interests and times. Since apparently not available we take the space to systematize somewhat. As an additional bonus, so to speak by example, we delineate more sharply the physical domain of applicability of our theory.

At this point serious business has to start. Even if only approximately, we want to compute on the basis of microscopic models the inclination dependence of the growth velocity, determine universality classes and their critical exponents, understand faceting transitions, etc. As far as one can go, these topics are covered in the remaining four sections. In Section 5 we develop and analyse the continuum theory of Kardar, Parisi and Zhang (1986). In particular, we exploit the mapping to a directed polymer in a random medium, a model rather close to spin glasses. Thus methods from the theory of disordered systems come into play. Section 6 deals with two-dimensional models ($\hat{=}$ one-dimensional surface). They are closely related to one-dimensional lattice gases driven by an external force. Probabilistically growth has been studied mostly through first passage percolation, Section 7. This approach leads, in particular, to a proof that the cluster takes a definite shape after a long time. In the final section we develop a simple theory for the average cluster shape by neglecting correlations. Such an approach cannot deal properly with surface fluctuations, but it is a useful tool for studying the macroscopic shape.

2

Macroscopic shape

The shape of a cluster growing from a seed is related to the direction dependent growth velocity through a simple geometric construction that was known to crystallographers a long time ago (Wulff 1901, Gross 1918). Here we give a derivation based on an effective equation of motion for the cluster surface, and discuss some of the shapes which may occur.

2.1 Derivation

We fix a d-dimensional plane of reference which contains the seed at the origin and measure the height of the cluster surface at time t perpendicular to the plane by a function $h_t(\mathbf{x})$. $\mathbf{x}$ is a vector in the plane. To obtain the full cluster shape several coordinate systems may have to be glued together. We work on such a large scale that fluctuations in the height can be neglected. In spirit $h_t(\mathbf{x})$ is to be compared to the hydrodynamic fields of a fluid. $\mathbf{x}$ refers to a cell which is small on a macroscopic scale but contains so many lattice points that upon spatial averaging the surface has a well-defined non-fluctuating height. Our basic assumption is that the local growth velocity $v = \partial h_t(\mathbf{x})/\partial t$ in the h-direction is uniquely determined by the local surface gradient $\mathbf{u}_t = \nabla h_t$

through a known function,

$$\frac{\partial}{\partial t} h_t = v(\nabla h_t). \tag{2.1}$$

For an isotropic system (the cluster grows as a ball) we have $v(\mathbf{u}) = c\sqrt{1 + \mathbf{u}^2}$ because of our particular choice of the coordinate system with c the normal growth velocity. Due to anisotropies in the aggregation process, in general v has a more complicated dependence on the surface gradient and as a consequence there will be more interesting macroscopic shapes. For a microscopic model $v(\mathbf{u})$ is determined by growing from a flat substrate orthogonal to $(\mathbf{u}, -1)$. After some transient time the surface will grow parallel to the substrate with normal velocity $v(\mathbf{u})/\sqrt{1 + \mathbf{u}^2}$.

To simplify our presentation we work in two dimensions, so $h_t(x)$ is a curve. For definiteness we take the underlying lattice to possess fourfold symmetry and choose the x–axis along one of the symmetry directions. Then $v(u)$ is even and needs to be specified only for $|u| \leq 1$, since the large u behavior is fixed by

$$v(u) = |u| v\left(\frac{1}{u}\right). \tag{2.2}$$

Starting from a seed the stationary growth shape is a solution to (2.1) which is of the scaling form

$$h_t(x) = t g\left(\frac{x}{t}\right). \tag{2.3}$$

Hence the shape function $g(y)$ satisfies

$$g(y) = y g'(y) + v\big(g'(y)\big). \tag{2.4}$$

Any solution of (2.4) has a definite curvature in the sense that either $g''(y) \geq 0$ or $g''(y) \leq 0$ everywhere. To see this, suppose that $g'(y_1) = g'(y_2)$ for some $y_1 < y_2$. It then follows from (2.4) that

$$g(y_2) - g(y_1) = \int_{y_1}^{y_2} g'(y) \, \mathrm{d}y = g'(y_1)(y_2 - y_1). \tag{2.5}$$

This is possible only if either $g'(y) \equiv g'(y_1)$ on $[y_1, y_2]$, or if $g'(y_3) = g'(y_1)$ for some point $y_3 \in (y_1, y_2)$. In the latter case the argument is repeated for the interval $[y_1, y_3]$ to show that $g'(y)$ must be constant on $[y_1, y_2]$. Thus $g'(y)$ is monotone. Trivial solutions of (2.4) are the straight lines $g_u(y) = uy + v(u)$. The cluster shape is obtained from these through the following familiar construction: We draw the lines $g_u(y)$ for all possible slopes $u, -\infty < u < \infty$. Equation (2.4) requires $g(y)$ for every y to be tangent to one of the $g_u(y)$'s. We conclude that $g(y)$ is the envelope of the family of lines $g_u(y)$, which may be written as

$$g(y) = \min_u [v(u) + uy]. \qquad (2.6)$$

The cluster shape is the Legendre transform of the growth velocity. For convex $v(u), v''(u) > 0$, the minimum in (2.6) is unique and given by $v'(u) = -y$. The curvatures of v and g are related through

$$g''(y)v''(g'(y)) = -1. \qquad (2.7)$$

The physics behind (2.6) is clarified by considering a somewhat different growth geometry. Suppose that the initial condition for (2.1) is an infinitely extended corner, $h_0(x) = -u_0|x|$ for some $u_0 > 0$. Taken literally, the corner will propagate unchanged at velocity $v(u_0)$. However, physically we expect the initial corner to be rounded in some tiny neighborhood $-\epsilon \le x \le \epsilon$ of $x = 0$. If $v(u)$ is convex, we may as well use its Legendre transform (2.6) to model the rounded part, i.e. $h_0(x) = (\epsilon/y_0)(g(xy_0/\epsilon) - g(y_0)) - u_0\epsilon$ for $-\epsilon \le x \le \epsilon$ where $g'(y_0) = -u_0$. But then, by (2.3) the rounded part expands linearly under the growth, and the stationary growth shape of the corner is given by an expression similar to (2.6),

$$g(y) = \min_{|u| \le u_0} [v(u) + uy]. \qquad (2.8)$$

The convexity of $v(u)$ implies that the inclinations $|u| \le u_0$, which are initially present close to $x = 0$, propagate at a slower rate than $v(u_0)$. Hence they lag behind the 'ideal' corner solution $h_t(x) = -u_0|x| + v(u_0)t$ and thereby spread laterally along the x–axis. The surface inclination at fixed y is chosen by minimizing the local growth rate $v(u) + uy$ among the inclinations that were present initially. For cluster growth, the seed contains all possible inclinations and hence (2.8) reduces to (2.6).

A similar discussion applies to the related problem of a dissolving corner. Within the framework of (2.1) this is modeled by an initial condition $h_0(x) = u_0|x|$, for some $u_0 > 0$, and $v(u)$ is the inclination dependent dissolution velocity. In this case a rounding of the corner occurs if the inclinations $|u| \le u_0$ propagate $faster$ than $v(u_0)$, i.e. if $v''(u) < 0$. Then the shape function for dissolution is

$$g_{\text{diss}}(y) = \max_{|u| \le u_0} [v(u) + uy].$$ (2.9)

For convex $v(u)$ the maximum in (2.9) is always attained at $u = \pm u_0$, which implies that the dissolving corner remains sharp.

2.2 Edges, facets and other singularities

Singularities in the growth velocity translate into corresponding non-analyticities in the growth shape. It is clear from (2.6) that the convex envelope $\hat{v}(u)$ of $v(u)$ determines the shape, rather than $v(u)$ itself. In particular, the extension of the cluster shape along the symmetry axes is given by $g(0) = \hat{v}(0)$, which also determines the domain of definition of $g(y)$ as $[-\hat{v}(0), \hat{v}(0)]$. $\hat{v}(u)$ is obtained from $v(u)$ by removing all nonconvex parts using the double tangent (Maxwell) construction. Unless $v(u)$ is convex, $\hat{v}(u)$ contains linear pieces where $\hat{v}''(u) \equiv 0$. We consider such a piece located between u_1 and u_2, $\hat{v}'(u) \equiv \hat{v}'(u_1) = \hat{v}'(u_2)$ for $u_1 \le u \le u_2$. As y in (2.6) passes through $y = -\hat{v}'(u_1)$, the inclination where the minimum is attained jumps discontinuously from u_1 to u_2. Thus $g(y)$ develops an edge at $y = -\hat{v}'(u_1)$ and the range of inclinations $u_1 < u < u_2$ disappears from the growth shape. This is quite analogous to two-phase coexistence in a fluid where a range of densities is thermodynamically unstable (Rottman and Wortis 1984).

While nonconvex parts of $v(u)$ are irrelevant to the growth shape, they may appear in the shape of a dissolving corner, cf. Equation (2.9), which is determined by the *concave* envelope of $v(u)$ in the range $|u| \le u_0$. Figure 2 demonstrates the construction of growth and dissolution shapes for nonconvex $v(u)$. An example of a microscopic growth model where such a shape actually occurs will be given in Section 6.

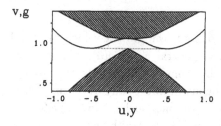

Figure 2: Growth shape $g(y)$ (bottom) and dissolution shape $g_{\text{diss}}(y)$ (top) for nonconvex growth velocity $v(u)$. The dotted line is the convex envelope $\hat{v}(u)$.

Next we consider the case where $v(u)$ itself has a cusp at $u = 0$, say,

$$v(u) = v(0) + \lambda|u| + O(|u|^\delta), \tag{2.10}$$

where $\delta > 1$ and both λ and the next to leading term are positive. Inserting this into (2.6) we must minimize $\lambda|u| + uy$ with respect to u. For $-\lambda \leq y \leq \lambda$ the minimum is attained at $u = 0$ and hence $g(y) \equiv v(0)$ in this range. The growing cluster develops a facet of size $2\lambda t$. The next to leading term in (2.10) determines how the rounded part of the cluster shape joins the facet. Applying (2.6) yields

$$g(\lambda) - g(\lambda + \epsilon) \sim \epsilon^{\delta/(\delta-1)} \tag{2.11}$$

for $\epsilon \to 0$. Growing crystals are often faceted due to the slow (nucleation dominated) growth rate at singular faces (Chernov 1984). In Section 7 we will discuss a class of growth models which show a faceting transition.

The occurrences of edges and facets in the growth shape are limiting cases of the curvature relation (2.7), provided $v(u)$ is replaced by $\hat{v}(u)$. Along linear pieces of $\hat{v}(u)$, $\hat{v}''(u) = 0$ and thus $g''(y)$ is forced to diverge, whereas a cusp in $\hat{v}(u)$, $\hat{v}''(u) = \infty$, leads to $g''(y) = 0$. For a general singularity of $v(u)$ of the form

$$v(u) \approx v(0) + \lambda|u|^\alpha, \qquad \alpha \geq 1, \tag{2.12}$$

for $u \to 0$, (2.7) yields the singular growth shape

$$\begin{aligned} g(y) &\approx v(0) - \lambda'|y|^{\alpha/(\alpha-1)}, \\ \lambda' &= ((\alpha-1)/\alpha)(\lambda\alpha)^{-1/(\alpha-1)}, \end{aligned} \tag{2.13}$$

for $y \to 0$. This kind of singularity occurs at the critical point of the faceting transition discussed in Section 7.

Finally we mention a type of singularity in $g(y)$ which is characteristic of the ballistic deposition models to be introduced below in Section 4. The deposition flux singles out one direction and hence fourfold symmetry, cf. Equation (2.2), does not hold. One finds a convex growth velocity $v(u)$ with the asymptotic behavior (Krug and Meakin 1989)

$$v(u) \approx v_\perp(a + |u|), \quad u \to \pm\infty. \tag{2.14}$$

Here $a > 0$ and $v_\perp$ denotes the growth velocity in the lateral direction (along the x–axis); in general $v_\perp < v(0)$. The shape function $g(y)$ is defined on $[-v_\perp, v_\perp]$. From (2.6) one gets $g(y) \to v_\perp a$ as $y \to \pm v_\perp$, i.e. $g(y)$ is discontinuous at the boundaries (since $g(\pm v_\perp) = 0$). The discontinuity is due to the nonlocal shadowing effects in these models, which produce fan-shaped clusters with an opening angle of $2\arctan(1/a)$ (cf. Section 4). The derivative $g'(v_\perp)$ can be finite or infinite depending on the corrections to (2.14).

2.3 The Wulff construction

Summarizing our main results, Equations (2.6), (2.8) and (2.9), we may conclude that a growing cluster (or crystal) attempts to minimize its total growth rate, while a dissolving cluster (crystal) maximizes the rate of dissolution. These general principles were stated by Wulff and Gross at the beginning of the century. Wulff's geometric treatment of faceted growth was extended by Gross to include curved growth shapes and was given an analytic formulation by Chernov (1963) and Wolf (1987).

The celebrated Wulff construction in its textbook form determines the *equilibrium* shape of a crystal by minimizing the surface free energy at fixed volume. While such a general variational principle cannot be used as a starting point for a theory of growth shapes, we shall see below that the kinetically determined prescription (2.6) is completely equivalent to the Wulff construction with the direction dependent growth velocity replacing the surface free energy. Wulff (1901) originally assumed growth shapes and equilibrium shapes to be the same, and hence concluded that surface free energies could be derived from growth rate measurements. However, since growth rates are determined by kinetic as well as thermodynamic requirements, growth and equilibrium shapes of crystals are generally quite different (Métois *et al.* (1982)).

In the context of equilibrium shapes it is well known (Andreev 1982, Rottman and Wortis 1984) that the Wulff construction can be recast in terms of Legendre transforms similar to (2.6), with the surface inclination u playing the role of a thermodynamic variable. To show the equivalence in the present case, we introduce the normal growth velocity $w(\vartheta)$ as a function of the angle ϑ formed by the growth direction with the vertical, $\vartheta = \arctan(-u)$, through

$$w(\vartheta) = v(-\tan\vartheta)\cos\vartheta. \qquad (2.15)$$

The growth shape is given in polar coordinates by a function $r(\varphi)$,

$$g(y) = r(\varphi)\cos\varphi, \quad y = r(\varphi)\sin\varphi, \qquad (2.16)$$

where $-\frac{1}{2}\pi < \vartheta$, $\varphi < \frac{1}{2}\pi$. Once the angle of inclination of the cluster surface at a given polar angle φ, $\vartheta(\varphi)$, is known, an elementary geometric construction gives

$$r(\varphi) = \frac{w\big(\vartheta(\varphi)\big)}{\cos\big(\varphi - \vartheta(\varphi)\big)}. \qquad (2.17)$$

The relation between φ and ϑ follows from (2.6). Since $\tan\varphi = y/g(y)$, we have

$$\tan\varphi = \left[u - \frac{\hat{v}(u)}{\hat{v}'(u)}\right]^{-1} \tag{2.18}$$

which is rewritten using (2.15) as

$$\tan\big(\varphi - \vartheta(\varphi)\big) = \frac{\widehat{w}'\big(\vartheta(\varphi)\big)}{\widehat{w}\big(\vartheta(\varphi)\big)} \tag{2.19}$$

where $\hat{v}$ and $\widehat{w}$ are the convex envelopes of v and w. Taking the derivative of (2.17) with respect to ϑ it follows that the angle determined by (2.19) is the one which minimizes (2.17), hence

$$r(\varphi) = \min_{\vartheta}\left(\frac{w(\vartheta)}{\cos(\varphi - \vartheta)}\right). \tag{2.20}$$

This is the analytic representation of the Wulff construction: For each value of ϑ a line of inclination $-\tan\vartheta$ is drawn at a distance $w(\vartheta)$ from the origin. Then the growth shape is the inner envelope of all lines. An example of a $w(\vartheta)$–plot and the corresponding growth shape is shown in Figure 3.

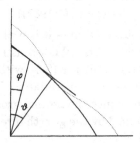

Figure 3: Wulff construction of one quadrant of the growth shape for the PNG model (see Section 4.2). The dotted line is the direction dependent growth velocity $w(\vartheta)$ and the full line is the growth shape $r(\varphi)$.

For future reference we record some general properties of the growth angle $\varphi(\vartheta)$ defined by (2.18). The convexity of $\hat{v}$ implies that φ is a monotonously increasing function of ϑ. Note however that the sign of $\varphi - \vartheta$ in (2.19) is not fixed, since convexity of $\hat{v}(u)$ only requires that $\widehat{w}(\vartheta) + \widehat{w}''(\vartheta) > 0$. A cusp in $v(u)$ (Equation (2.10)) leads to a discontinuity in $\varphi(\vartheta)$, and a general singularity of the form (2.12) yields

$$\varphi(\vartheta) \approx \frac{\lambda\alpha}{v(0)}\vartheta^{\alpha-1}, \qquad \vartheta \to 0. \tag{2.21}$$

The asymptotically linear behavior (2.14) implies that $\varphi \to \arctan(1/a)$ for $\vartheta \to \frac{1}{2}\pi$.

Before closing this section we should note the intrinsic significance of $\varphi(\vartheta)$. In the context of the macroscopic evolution equation (2.1), $\varphi(\vartheta)$ determines the direction $\mathbf{m} = (\sin\varphi, \cos\varphi)$ in the (x, t)–plane along which a small surface segment with inclination $u = -\tan\vartheta$ translates locally. The trajectories in the (x, t)–plane of such segments of constant inclination form the *characteristics* of (2.1). A microscopic characterization of $\varphi(\vartheta)$ is obtained considering the growth of a planar surface of fixed inclination $u = -\tan\vartheta$. The growing film is decomposed into clusters of connected particles which share the same ancestor substrate site (Meakin 1987b). For thick films the clusters are elongated in the direction determined by φ. While usually the clusters are hidden in the bulk of the film (Meakin 1987a), they become visible to the naked eye in the case of oblique incidence ballistic deposition, where they form the ubiquitous columnar microstructure (cf. Section 4.3).

3

Scaling Theory of Shape Fluctuations

Having determined (at least in principle) the macroscopic shape of the growing cluster, we may turn our attention to more refined aspects of the growth process. A striking feature already noted by Eden in his 1961 paper is the roughness of the cluster surface. We introduce and employ the notion of statistical scale invariance to characterize the roughness of growing surfaces. Statistical scale invariance is really at the heart of the more widely promoted concept of fractal geometry (Mandelbrot 1982) and shares with it certain limits of applicability to the real world of natural processes and computer simulations. We return to this further below, but start out with a discussion of 'ideal' kinetic roughness.

3.1 Statistical scale invariance

Since in general the average cluster shape is not explicitly known, it is impractical to study shape fluctuations in the cluster geometry (Zabolitzky and Stauffer 1986). We therefore use the substrate geometry instead and take the substrate to be an infinite, d-dimensional hyperplane in

$(d + 1)$-dimensional space. After some local coarse-graining the surface configuration at time t can be described by a single valued, continuous function $h_t(\mathbf{x})$ which measures the height of the surface perpendicular to the substrate above the substrate point $\mathbf{x}$. $h_t(\mathbf{x})$ should not be confused with the deterministic (macroscopic) surface profile discussed before in Section 2. Here we work on a 'mesoscopic' scale which is fine enough to capture the stochastic nature of the growth process, but sufficiently coarse to allow us to ignore the discrete lattice structure, overhangs and other microscopic details (for a more detailed discussion of microscopic length scales see Section 3.2).

The initial condition is $h_0(\mathbf{x}) \equiv 0$. Being interested in fluctuations, we subtract the average height at time t. Hence $h_t(\mathbf{x})$ is a random function with zero mean. We call the growth process statistically scale invariant, if typical surface configurations can be made to 'look the same' by suitable simultaneous rescaling of space $(\mathbf{x})$, time (t), and height (h). More precisely, we require that for an arbitrary rescaling factor $b > 0$ the statistical properties of the rescaled process

$$h_t'(\mathbf{x}) = b^{-\zeta} h_{b^z t}(b\mathbf{x}) \tag{3.1}$$

coincide with those of $h_t(\mathbf{x})$. For a given growth process, this requirement fixes the scaling exponents ζ and z, which therefore carry the central information about the scaling properties of the surface fluctuations.

We illustrate the significance of ζ and z using as an example the height difference correlation function

$$G(|\mathbf{x} - \mathbf{x}'|, t) = \left\langle \left| h_t(\mathbf{x}) - h_t(\mathbf{x}') \right| \right\rangle. \tag{3.2}$$

For long times $(t \to \infty)$ we expect G to become stationary (time independent). Equation (3.1) then implies

$$\xi_\perp(r) := \lim_{t \to \infty} G(r, t) = a r^\zeta \tag{3.3}$$

with $a > 0$ some constant. $\xi_\perp(r)$ measures the transverse 'wandering' of the surface over the horizontal distance r. This is why ζ has been termed the *wandering* or *roughness exponent* (M.E. Fisher 1986, Lipowsky 1988, 1990). Rough surfaces have $0 < \zeta < 1$. Marginal roughness with $\zeta = 0$ is characteristic of two dimensional surfaces in thermal equilibrium but also occurs in some nonequilibrium situations such as critical faceting (cf. Section 7) and diffusion-limited annihilation (Meakin and Deutch 1986). In these cases the power law (3.3) is replaced by

$\xi_\perp(r) = a'(\log r)^{\zeta'}$. The opposite limit $\zeta = 1$ implies that the large scale surface orientation differs from the substrate orientation since $\xi_\perp(r)/r$ does not vanish for $r \to \infty$. Depending on the situation, $\zeta = 1$ may be the signature of a fractal (Meakin and Jullien 1989, 1990), crumpled (Lipowsky 1988, 1990) or discontinuous (Krug and Meakin 1989) surface. For equilibrium models $\zeta = 1$ has been associated with the breakdown of two-phase coexistence at the lower critical dimensionality (Huse *et al.* 1985b, M.E. Fisher 1986). Whatever happens in such cases, one must be prepared to abandon the picture of a single valued, continuous height function and look for other means of description.

Next we consider the approach of $G(r, t)$ to the stationary limit (3.3). Inserting (3.1) into (3.2), we obtain the homogeneity relation

$$G(r, t) = b^{-\zeta} G(br, b^z t). \tag{3.4}$$

Choosing $b = 1/r$ this may be rewritten as

$$G(r, t) = \xi_\perp(r) g\left(\frac{r}{t^{1/z}}\right). \tag{3.5}$$

with $g(0) = 1$ from (3.3). The interpretation of (3.5) is as follows: For finite t there exists a correlation length

$$\xi_\parallel(t) \sim t^{1/z} \tag{3.6}$$

such that on scales $r \ll \xi_\parallel(t)$ the surface is stationary and rough, whereas on scales $r \gg \xi_\parallel(t)$ the surface looks smooth in the sense that the transverse wandering does not further increase with r. Requiring that $G(r, t)$ becomes independent of r for $r \gg t^{1/z}$ implies that the scaling function in (3.5) vanishes as $g(x) \sim x^{-\zeta}$ for $x \gg 1$ and hence

$$G(r, t) \sim t^{\zeta/z} \sim \xi_\parallel^\zeta, \qquad r \gg \xi_\parallel(t). \tag{3.7}$$

Thus the *dynamic exponent* z describes the temporal spread of surface fluctuations and, via (3.7), the increase of surface roughness in time. If the spread of fluctuations were purely diffusive, we would have $\xi_\parallel(t) = \sqrt{Dt}$ and $z = 2$. However, as we will demonstrate below, growth conditions usually lead to a superdiffusive $z < 2$. Figure 4 shows a numerical example of the correlation function $G(r, t)$. In this case $\zeta = \frac{1}{2}$ and $z = \frac{3}{2}$, compare with Section 5.

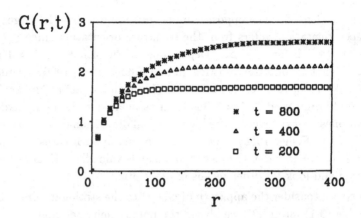

Figure 4: Height difference correlation function $G(r,t)$ for the one dimensional discrete time PNG model (the growth rule is defined in Section 4.2). Simulations were carried out on a lattice of size $L = 5000$ with a nucleation rate $p = 0.01$. The data are an average over 50 runs.

3.2 Corrections to scaling

In real systems and computer models, simple power laws like (3.3) are often obscured by the presence of additional length scales. Since experimental results in this field are scarce, we give a discussion appropriate for the standard type of computer simulation.

Let us first explore the consequences of taking (3.3) literally for all r, $0 < r < \infty$. The prefactor a then introduces a crossover scale r_c such that $\xi_\perp(r_c) = r_c$,

$$r_c = a^{1/(1-\zeta)}. \tag{3.8}$$

On scales $r \ll r_c$, $\xi_\perp(r)/r \gg 1$, and the surface is wildly agitated. In fact, in this regime it looks fractal. To understand this, we recall the definition of the fractal dimensionality D of a surface (Mandelbrot 1982). The height variables are averaged over horizontal patches of linear size ϵ. The area $A(\epsilon)$ of the averaged surface (the surface area 'on the scale ϵ') is measured and compared to the projected (substrate) area A_0. D is then defined through

$$\frac{A(\epsilon)}{A_0} \sim \epsilon^{-(D-d)}, \tag{3.9}$$

where d denotes the substrate dimension. For a fractal surface $(D > d)$ $A(\epsilon)$ increases indefinitely as $\epsilon \to 0$. For a rough surface simple arguments show that (Wong and Bray 1987, Burkhardt 1987)

$$A(\epsilon) = A_0\left[1 + \frac{\xi_\perp(\epsilon)}{\epsilon}\right] = A_0\left[1 + \left(\frac{\epsilon}{r_c}\right)^{-(1-\zeta)}\right]. \qquad (3.10)$$

Comparing with (3.9) we find that the fractal dimension is scale dependent,

$$D = \begin{cases} d + 1 - \zeta & (\epsilon \ll r_c), \\ d & (\epsilon \gg r_c), \end{cases} \qquad (3.11)$$

so $D > d$ below the crossover scale r_c.

Mandelbrot (1986) has coined the term 'self-affine fractals' for geometrical objects with the property (3.11), 'affinity' replacing 'similarity' because such objects are (statistically) invariant under *anisotropic* rescaling of space. The standard example of a self-affine fractal (with $d = 1$ and $\zeta = \frac{1}{2}$) is the record of a one dimensional Brownian motion (Wiener process). In this case the time axis plays the role of the spatial coordinate and the particle position corresponds to the surface height. The notion of self-affinity has become quite popular. It should be noted however that in contrast to the mathematically constructed Wiener process, for physical surfaces the extrapolation to arbitrarily fine length scales implicit in (3.11) is impeded by various small-scale cutoffs, which we now discuss.

An obvious small-scale limit to scale invariance is the lattice constant r_0. More subtle corrections to scaling often arise from the local surface structure. Following Kertész and Wolf (1988) these are summarized in the intrinsic width ξ_i, which is the r–independent part of $\xi_\perp(r)$ and is introduced through

$$\xi_\perp(r)^2 = \xi_i^2 + a^2 r^{2\zeta}. \qquad (3.12)$$

The pure scaling form (3.3) is recovered only on length scales $r \gg r_i := (\xi_i/a)^{1/\zeta}$. The intrinsic width is built up from high steps (i.e. nearest neighbor height differences exceeding one lattice constant), overhangs and holes. Its significance is similar to that of the bulk correlation length in the theory of equilibrium interfaces between coexisting phases (Huse *et al.* 1985b). In particular, any continuum description of the surface must start from a level of coarse-graining which is large compared to r_c. In computer simulations the intrinsic width can be reduced either by using a modified growth algorithm (Kertész and Wolf 1988) or by restricting the set of admissible surface configurations to exclude holes, overhangs and

high steps as in the solid-on-solid models, cf. Section 4. The crossover
scale (3.8) usually turns out to be comparable to or smaller than r_i
(or r_0), and hence the self-affine fractal regime $r \ll r_c$ in (3.11) does not
exist.

The second severe limitation to the simple scaling form (3.5) is due to
the finite size of the system. Once the correlation length $\xi_\parallel(t)$ becomes
comparable to the linear size L of the system, i.e. after a time of order L^z,
the growth turns stationary. This can be expressed through the finite
size scaling form for the overall surface roughness

$$W(L,t) = \left[L^{-d} \int d^d x \left(h_t(\mathbf{x}) - \bar{h}_t(L) \right)^2 \right]^{1/2}, \tag{3.13}$$

where the integral runs over the substrate coordinates and $\bar{h}_t(L)$ is the
spatial average

$$\bar{h}_t(L) = L^{-d} \int d^d x \, h_t(\mathbf{x}). \tag{3.14}$$

Note that since the ensemble average of $h_t(\mathbf{x})$ has already been subtracted, (3.14) is the difference between the spatial and the ensemble
averages. The two coincide only in the limit $L \to \infty$. For long times the
surface diffuses as a rigid object and one expects that $\langle \bar{h}_t(L)^2 \rangle \sim t$.

$W(L,t)$ should have a scaling form similar to $G(r,t)$ with $r = L$: It is
a function of $L/\xi_\parallel(t)$ which saturates at a value proportional to L^ζ for
$\xi_\parallel(t) \gg L$, and grows as $t^{\zeta/z}$, independent of L, for $\xi_\parallel(t) \ll L$, compare
with Equations (3.3), (3.5) and (3.7). Thus we may write (Family and
Vicsek 1985)

$$W(L,t) = L^\zeta f\left(\frac{t}{L^z}\right) \tag{3.15}$$

where $f(x \to \infty) = \text{const.}$ and $f(x \to 0) \sim x^{\zeta/z}$. This form has been
used in many simulations to determine the exponents ζ and z.

3.3 Scaling relations

Static and dynamic surface fluctuations are coupled by the growth
process itself. This leads to a scaling relation between ζ and z which
has been derived in a number of more or less formal ways (Huse and
Henley 1985, Meakin *et al.* 1986b, Krug 1987, Kardar and Zhang 1987,
Medina *et al.* 1989, Krug and Meakin 1989, Wolf and Kertész 1989).

Here we show how it arises as a natural and immediate consequence of the general scaling picture.

Looking at the growing surface at time t, we observe bulges of all sizes up to the correlation length $\xi_\parallel(t)$. Let us focus our attention on one of the largest bulges and watch how it evolves in time. Its height is proportional to $\xi_\perp(\xi_\parallel(t))$ and its slopes have an inclination of the order $\xi_\perp(\xi_\parallel(t))/\xi_\parallel(t) \ll 1$. In default of any specific knowledge about the growth process, we may assume that everywhere the direction of growth is normal to the surface (Figure 5).

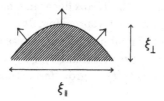

Figure 5: Widening of a surface fluctuation through normal growth.

Then the bulge widens at a rate proportional to the horizontal projection of the normal growth velocity of its slopes, i.e. proportional to the slope inclination. Thus

$$\frac{d}{dt}\xi_\parallel \sim \frac{\xi_\perp(\xi_\parallel)}{\xi_\parallel} \sim \xi_\parallel^{\zeta-1} \tag{3.16}$$

and

$$\xi_\parallel(t) \sim t^{1/(2-\zeta)}. \tag{3.17}$$

Comparing (3.17) to the definition (3.6) of the dynamic exponent z, we conclude that

$$z = 2 - \zeta. \tag{3.18}$$

This scaling relation looks very universal: It is independent of both the surface dimension and the details of the growth process. Still there are some instances where it fails. To explore its range of validity, we must scrutinize the suppositions inherent in the above argument. Treating the growth of the bulge as a deterministic process, we have in fact assumed that the macroscopic theory applies also at the mesoscopic level of (large-scale) fluctuations. But we know already that the direction of growth does not generally coincide with the surface normal. Using the expression (2.21) for the growth angle in the case of a general nonlinearity of the inclination dependent growth velocity, cf. Equation (2.10), we find that (3.16) generalizes to

$$\frac{d}{dt}\xi_\| \sim \left(\frac{\xi_\perp(\xi_\|)}{\xi_\|}\right)^{\alpha-1} \tag{3.19}$$

and hence $\xi_\|(t) \sim t^{1/z}$ with (Krug and Spohn 1988)

$$z = \alpha + \zeta(1-\alpha). \tag{3.20}$$

The relation between ζ and z is determined by the leading nonlinear term of $v(u)$ in the limit $u \to 0$. Terms linear in u only shift the bulge as a whole and do not contribute to its spreading. It should be noted however that even in the absence of nonlinearities the noise in the growth process causes the fluctuations to spread diffusively, $\xi_\| \sim t^{1/2}$. The nonlinearity is relevant only if it causes $\xi_\|(t)$ to grow *faster* than diffusively, i.e. if

$$\alpha < \frac{2-\zeta}{1-\zeta}. \tag{3.21}$$

4

Growth models

At this point we have to provide some examples of the growth phenomena covered by our theory. The general physical setup we have in mind is that of two phases, one of them stable and the other unstable, separated by an interface. The stable phase grows at the expense of the unstable phase and the interface moves at a constant speed. The interface is sharp and globally flat on a macroscopic scale, and we are interested in its mesoscopic roughness. This explicitly excludes the kind of large-scale interfacial instabilities often encountered in solidification processes, such as dendritic or cellular growth (Langer 1987). While in real systems the transition from stable to unstable growth (these terms referring now to the state of the interface) usually occurs upon changing the growth parameters rather than the underlying mechanism (Saito and Ueta 1989), the 'microscopic' models we shall consider are constructed to describe stable growth only. Clearly this limits their applicability to physical growth processes.

The basic stochastic model for *unstable* growth, which in fact triggered most of the present-day interest in growth processes, is the celebrated Witten-Sander model of diffusion-limited aggregation (DLA) (Witten and Sander 1981). The model is extremely easy to define: Given a cluster of N particles, the $(N + 1)$th particle is launched anywhere on the surface of a sphere enclosing the cluster and is allowed to diffuse until it touches the cluster and sticks to it. This procedure generates the intriguing, ramified patterns which now decorate the covers of countless conference proceedings. Despite considerable numerical and theoretical efforts, a satisfactory understanding of DLA still eludes us (Meakin 1988a). The difficulty is due to the *nonlocal* nature of the diffusion field which governs most pattern forming interfacial instabilities. To determine the probability for the next particle to stick at some given point on the surface, one has to solve the stationary diffusion equation (Laplace equation) including the whole cluster surface as a zero-field boundary condition. In contrast, the growth models of interest here are *local* in the sense that the probability of adding a particle depends only on the local environment of the respective surface site. Physically, this implies that the transport mechanisms which carry new material to the growing surface and expel impurities or latent heat are, to a large extent, neglected, and the growth is assumed to be dominated by the aggregation kinetics.

Even within the restricted class of local models there is an abundance of varieties which can be (and have been) studied. Our aim here is to describe the major representatives and to discuss some of their features within the conceptual framework of the previous sections. Before doing so, it may be useful to list a few general properties of local growth models in order to provide a first classification.

(i) *Lattice-/off-lattice models:*

For reasons of computational efficiency, simulations of growth processes are usually carried out on a discrete lattice of sites, which are either vacant or occupied by a particle. Off-lattice models, in which the particles are represented by hard discs or spheres with real spatial coordinates, are more realistic but harder to program. While the macroscopic cluster shape obviously reflects the symmetry of the underlying lattice, the scaling properties of the shape fluctuations are generally expected to be insensitive to the lattice structure. A notable exception to this rule occurs for a class of deposition models to be described below.

(ii) *Fully/partly irreversible models:*

Although of course any growth process is irreversible, we may distinguish between fully irreversible models, which only allow the addition of particles to the aggregate, and partly irreversible models, which also include disaggregation, albeit at a smaller rate. In high spatial dimensionalities ($d \geq 3$) there is the possibility of a surface roughening transition as the degree of irreversibility is varied (cf. Section 5).

(iii) *Flux-limited/reaction-limited models* (Krug 1989a):

This distinction is best illustrated by two idealized physical situations. For the flux-limited case, consider vapor deposition onto a cold substrate. A dilute flux of particles impinges on the surface and the particles stick irreversibly where they hit. Reaction-limited growth is typified by molecular-beam epitaxy on vicinal faces of a crystalline semiconductor. In this case there is a constant density of atoms on the terraces which diffuse along the steps, looking for a growth site (e.g. a kink site) where they can be incorporated in the crystal. Although, as noted above, local growth models treat transport in a very summary fashion, they can still be distinguished according to whether the rate of growth is limited by the supply of new material (flux-limited) or by the availability of growth sites (reaction-limited). Among the models to be discussed in the following, the Eden- and SOS-models are reaction-limited, while the deposition models are flux-limited. It should be noted that the reaction-limited case appears quite naturally in the context of biological applications, in which the Eden-type models were first formulated: If the growth proceeds through cell division, there is no need to transport new material (new cells) to the cluster surface.

(iv) *Sequential/synchronous models:*

In most simulations particles are added to the aggregate sequentially, i.e. one by one, and 'time' is counted in terms of the aggregate mass. This procedure can be shown to generate the same ensemble of configurations as the following continuous time random process: In an infinitesimal time interval dt, growth occurs independently at every growth site (site at which a particle can be added) with probability Γdt, where Γ is a rate constant. In contrast, in a synchronous (discrete time) process a randomly chosen finite fraction or possibly all of the growth sites are simultaneously updated in a single time step. One characteristic feature of synchronous models is the occurrence of a faceting phase transition, cf. Sections 7 and 8.

4.1 Eden models

The basic lattice model for cluster growth was devised by Murray Eden in 1956 as a minimal description of biological morphogenesis (Eden 1958). It is so simple that it hardly needs an explanation: Given a cluster of N particles, the $(N+1)$th particle is added at a randomly chosen perimeter site of the cluster. A perimeter site is a vacant lattice site which has at least one occupied neighbor. The occupied neighbors of the perimeter sites will be referred to as surface sites. Jullien and Botet (1985) noted three different, equally natural ways of choosing among the perimeter sites: Either the perimeter sites themselves (version A), or the bonds connecting perimeter and surface sites (version B), or the surface sites (version C) may be assigned equal probabilities. In version C, an additional random choice is necessary if the chosen surface site is adjacent to several perimeter sites. Eden originally studied version B on the two dimensional square lattice. Our introductory model (Section 1) is the continuous time process corresponding to version A.

Large Eden clusters contain growth (perimeter) sites only in a thin surface layer which occupies a negligible fraction of the cluster volume as $N \to \infty$. This implies that clusters attain a well-defined macroscopic shape as was proved by Richardson (1973).

On the basis of simulations up to $N = 2^{15}$, Eden (1961) noted that the clusters are 'essentially circular in outline' (Figure 1). It was necessary to increase the cluster sizes by more than three orders of magnitude to firmly establish a slight deviation from the circular shape (Freche *et al.* 1985), corresponding to a weak angular dependence of the normal growth velocity $w(\vartheta)$ (cf. Section 2). Keeping in mind the corresponding continuous-time process, it is easily seen that $w(\vartheta)$ is proportional to the number of perimeter sites (version A), the number of open bonds (version B) or the number of surface sites (version C) per unit area of the tilted substrate. These quantities were measured for versions A (Hirsch and Wolf 1986) and C (Meakin *et al.* 1986a), and the growth was found to be slower by about 2% along the lattice diagonal, as compared to the lattice axes. The anisotropy is expected to become more pronounced in higher dimesions, since rigorous bounds show that the growth velocity along the lattice axes is of the order $d/\log(d)$ while along the space diagonal it is only $O(\sqrt{d})$ (Kesten 1986).

Eden also observed that the cluster perimeter, defined as the number of open bonds, is larger than that of a perfect circle by a factor of 1.8.

The 'excess perimeter' reflects the local crinkliness of the cluster surface
(Mollison 1972). It consists of holes, overhangs and high steps, and
constitutes the major contribution to the intrinsic surface width ξ_i
introduced in (3.12) (Kertész and Wolf 1988). The excess perimeter is
sensitive to the local growth rule: Going from version A to version C
it decreases by a factor of 6 (Jullien and Botet 1985). The *noise
reduction* algorithm originally developed for DLA (Szép *et al.* 1985)
allows to systematically reduce the excess perimeter, and thereby ξ_i.
This algorithm introduces an integer valued noise reduction parameter m
and requires a perimeter site (for version A) to be selected m times
before it is occupied. As a consequence of this local averaging, the excess
perimeter density and the intrinsic width both decrease as $1/m$ (Kertész
and Wolf 1988).

In the limiting case $m = \infty$ all growth sites are occupied simultaneously and the growth becomes deterministic (no noise) (Krug and Spohn
1988). The resulting growth shape is a diamond. Since the shape would
be expected to depend continuously on m there must be a transition
toward the diamond shape with increasing m (Figure 6). Noise reduc-

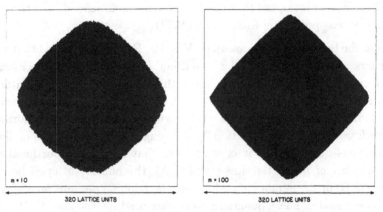

Figure 6: Eden clusters on a square lattice grown with
noise reduction parameters $m = 10$ and $m = 100$.
Courtesy of P. Meakin.

tion enforces the lattice anisotropy. The surprising (and unexplained)
feature is that without noise reduction $(m = 1)$, the noise *almost* suc-
ceeds in turning the diamond into a circle. Clusters which are even
less anisotropic can be generated by extending the definition of growth
(perimeter) sites to include the next nearest neighbors of a surface site.

Since the deterministic ($m = \infty$) shape is a square, the growth is now expected to be *faster* along the diagonal. This was confirmed and the anisotropy (for $m = 1$) was found to be only 1.2% (Garmer 1989).

A similar shape transition occurs in the synchronous (discrete time) Eden model introduced by Richardson (1973). In one time step $t \rightarrow t + 1$, all perimeter sites of the cluster are filled simultaneously and independently with probability p. Again, the growth is deterministic for $p = 1$ and in the limit $p \rightarrow 0$, combined with a rescaling of time, the original Eden model is recovered. However the sequence of shapes which interpolate between (almost) circular and diamond are qualitatively different. While, in the case of noise reduction, the cluster edges remain slightly curved even for large m (Meakin 1988b), the Richardson model shows a faceting transition at some critical value $0 < p_c < 1$. For $p > p_c$ the cluster shape coincides with the diamond close to the diagonal, cf. Sections 7 and 8. Synchronous models in which a finite lifetime is attributed to the growth sites have also been considered (Savit and Ziff 1985). In these cases there is a second (bulk percolation) threshold $p_c^b < p_c$, such that for $p < p_c^b$ clusters die out eventually. At p_c^b the clusters (conditioned on survival) are fractal.

In an attempt to describe tumor induction, Williams and Bjerknes (1972) introduced a partly reversible variant of the Eden model which has motivated much of the subsequent mathematical work on growth models. The basic idea is that the rate of division of cancer cells exceeds that of normal cells by some factor $k > 1$ (the 'carcinogenic advantage'). Each time a cell (normal or cancer) divides, the daughter cell displaces one of the neighbors of the mother cell. Clearly the configuration changes only along the perimeter of the tumor. (Initially, there is a single cancer cell at the origin.) For $k \rightarrow \infty$ one recovers the (B-version of) the Eden model. Note that due to partial reversibility there is a finite probability ($= 1/k$) for the cluster to ultimately disappear from the lattice.

Based on their computer simulations Williams and Bjerknes originally conjectured the cluster surface to be fractal, which would correspond to an infinite excess perimeter density. This was disproved by Mollison (1972) who showed that the excess perimeter density has a limit which is bounded from above by $(6k + 1)/(k - 1)$. Using results from the equivalent biased voter model, Bramson and Griffeath (1980, 1981) later proved that for $k > 1$ the cluster (provided it survives) attains a well-defined macroscopic shape. Mollison's bound suggests (as do computer simulations) that the excess perimeter density, and thereby the intrinsic surface width, increases without limit as $k \rightarrow 1$. The most striking mechanism which contributes to this intrinsic roughening is the creation of islands separate from the original cluster (Williams and Bjerknes 1972). Nothing appears to be known about the way in which the surface dissolves in the limit $k \rightarrow 1$.

All kinds of modified Eden models, including e.g. anisotropic growth rules (Sawada *et al.* 1982), directed lattices (Chernoutsan and Milošević 1985, Botet 1986) and an off-lattice version (Meakin 1988c) have been introduced in the literature (for a survey see Meakin 1986). As they do not add much to the general picture, we refrain from discussing them here. Likewise we do not consider other theoretical approaches to Eden growth, such as $1/d$ expansions (Parisi and Zhang 1984, Friedberg 1986), Cayley trees (Vannimenus *et al.* 1984) and field theory (Parisi and Zhang 1985, Cardy 1983), since these methods focus on bulk properties rather than surface fluctuations of clusters.

4.2 SOS models

The major contributions to the intrinsic width in (3.12) — holes, overhangs and high steps — can be eliminated from the outset by using a solid-on-solid (SOS) model with a restriction on the nearest-neighbor height differences. The prototype model in this class is the single step model (Meakin *et al.* 1986b, Plischke *et al.* 1987), which we explain here for the one-dimensional case (the two-dimensional model is discussed in Section 6).

By definition, SOS-models require the substrate geometry. The surface position at time t above site i of the one dimensional substrate lattice is given by an integer valued height variable $h_t(i)$ which is subject to the single step constraint $|h_t(i+1) - h_t(i)| = 1$. This is satisfied by choosing odd (even) values for $h_t(i)$ on odd (even) sites i. Initially $h_0(i) = 1(0)$ on odd (even) sites. In order to stay within the prescribed set of configurations, growth / evaporation events $h_t(i) \rightarrow h_t(i)+2$ / $h_t(i) \rightarrow h_t(i)-2$ occur only at local minima $(h_t(i+1) - 2h_t(i) + h_t(i-1) = 2)$ / local maxima $(h_t(i+1) - 2h_t(i) + h_t(i-1) = -2)$ of the surface. In the continuous time setting we assign the rate $\Gamma_+(\Gamma_-)$ to a growth (evaporation) event. As for the Eden models the growth velocity v is determined by the density of growth sites, i.e. the process is reaction-limited. More precisely, $v = \Gamma_+ \times$ (density of local minima) $-\Gamma_- \times$ (density of local maxima). Due to a symmetry particular to one dimensional surfaces, the number of local minima is equal to the number of local maxima in every configuration. Using the lattice gas representation to be introduced in Section 6 for a general class of SOS-models, one easily finds the growth velocity for a surface of average inclination u,

$$v(u) = \tfrac{1}{4}(\Gamma_+ - \Gamma_-)(1 - u^2), \qquad |u| \leq 1. \tag{4.1}$$

Inclinations $|u| > 1$ clearly cannot occur in a single step configuration. Hence growth shapes can only be discussed in the corner geometry of Section 2. Since $v''(u) < 0$, we conclude from Equations (2.8) and (2.9) that a growing corner remains sharp, while a dissolving corner attains a rounded (parabolic) shape (Rost 1981, Marchand and Martin 1986).

A synchronous version of the single step model (with $\Gamma_- = 0$) is obtained by filling all local surface minima simultaneously (Krug and Spohn 1988). The growth is deterministic and a flat surface remains flat. However the relaxation of an initially rough surface shows interesting scaling properties which will be discussed below in Section 5.4.

Our next example is the polynuclear growth (PNG) model, a classic in the theory of crystal growth (Frank 1974, Gilmer 1980, Goldenfeld 1984). It is a continuous time, off-lattice model which contains both random and deterministic events. Starting from a flat surface, monolayer islands nucleate, at a rate Γ per unit area, at random times and random positions. Once created, an island spreads laterally with constant speed c in all directions until it merges with another island in the same layer. Thus the basic constituents of the PNG model are layers rather than individual particles. Dimensional arguments show that the growth velocity of a d-dimensional horizontal ($u = 0$) surface is (van Saarloos and Gilmer 1986)

$$v(0) \sim (\Gamma c^d)^{1/(d+1)}. \tag{4.2}$$

For comparison with real crystal growth rates the dependence of Γ and c on temperature and chemical potential must be known. Since growth is limited by the density of surface steps (= edges of islands), tilting the surface increases the growth rate, so $v''(u) > 0$. This is confirmed by the exact calculation of $v(u)$ for $d = 1$ in Section 6.

A synchronous lattice version of the one dimensional PNG model is obtained as follows (Krug and Spohn 1989). In one unit of time ($t \rightarrow t + 1$) up/down surface steps are shifted one lattice unit to the left/right, and nucleation centers (pairs of up-down steps) are created with probability p at each site. The dynamics of the integer valued height variables $h_t(i)$ can be written as

$$h_{t+1}(i) = \max\left\{h_t(i-1), h_t(i), h_t(i+1)\right\} + \delta_t(i), \qquad (4.3)$$

where $\delta_t(i) = 1$ (resp. 0) with probability p (resp. $1 - p$). For small p (4.3) mimics the continuous time model with $\Gamma = p$, $c = 1$. At $p = p_c = 0.539$ there is a faceting transition analogous to that of the Richardson model (Kertész and Wolf 1989). The transition is suppressed if nucleation events which create steps of more than unit height are discarded. For $p = 0$ (4.3) reduces to a deterministic growth model which is equivalent to the synchronous single step model, cf. Section 5.4.

As our final example we mention the restricted solid-on-solid (RSOS) model (Kim and Kosterlitz 1989). This is a sequential lattice model. A site $\mathbf{x}$ of the d-dimensional substrate lattice is chosen at random and a particle is added $(h_{t+1}(\mathbf{x}) = h_t(\mathbf{x}) + 1)$ *provided* this does not generate nearest neighbor height differences of more than one lattice constant; otherwise no growth takes place and a new site is selected. This simple procedure is surprisingly effective in suppressing corrections to scaling, cf. Section 5.3.

4.3 Ballistic deposition models

The first simulations of ballistic deposition were concerned with the structural properties of random packings of hard spheres. In 1959, Marjorie Vold introduced a model for the sedimentation of moist glass spheres in a nonpolar solvent. Spheres are dropped sequentially above randomly chosen positions of the horizontal substrate, move towards the surface along linear (ballistic) trajectories and stick permanently at the point of first contact with a previously deposited sphere (or the substrate). This procedure generates a chainlike structure of very low density (0.128 volume fraction from a simulation of 155 spheres). Having in mind stacks of ball bearings, Visscher and Bolsterli (1972) simulated a related model in which the dropped sphere is allowed to roll downhill, in contact with previously deposited spheres, until it reaches a three sphere contact which is stable under gravity. The density obtained this way is much higher than for the Vold model (0.58 volume fraction), but significantly lower than the close packed h.c.p. density (0.74 volume fraction). A sequential off-lattice model intermediate between these two was employed by Henderson *et al.* (1974) in their study of structural anisotropy and void formation in vapor deposited thin films. In this model, a deposited sphere comes to rest at its first three sphere contact, irrespective of whether it is stable or not. Large-scale simulations of all three models have been performed recently by Jullien and Meakin (1987).

The reader will have inferred from our description that ballistic deposition is *not* strictly local, since protruding parts of the surface can shadow others from the incoming flux. Nevertheless the *accessible* (*active*) surface of the deposit has a local dynamics. To see this we consider a two dimensional square lattice version of the Vold model (Family and Vicsek 1985, Meakin *et al.* 1986b). The one dimensional substrate is oriented along the x-axis of the lattice and particles move in the y-direction along randomly chosen lattice columns, sticking permanently at the first perimeter site (as defined above in the context of Eden models) they encounter. Denoting by $h_t(i)$ the maximum y-coordinate for any of the occupied sites in column i, this height variable evolves upon deposition in column i according to the local rule

$$h_{t+1}(i) = \max\Big\{ h_t(i-1), h_t(i)+1, h_t(i+1) \Big\}. \qquad (4.4)$$

A detailed numerical investigation of the surface configurations generated by (4.4) shows that the nearest neighbor height differences have an exponential distribution with mean $\langle |h_t(i+1) - h_t(i)| \rangle \simeq 1.136$ for $t \to \infty$ (Meakin *et al.* 1986b). Hence, although the bulk of the deposit is highly porous (Figure 7), with a density $\rho_0 \simeq 0.4684$, the accessible surface is well defined and has a rather small intrinsic width. The *internal* surface (the set of all perimeter sites) is much larger and scales like the deposit mass.

In the lattice model (4.4) there is always exactly one growth site per lattice column. Unlike the reaction-limited models discussed above, the density of growth sites is independent of surface inclination. To see how the inclination dependent growth velocity $v(u)$ arises in this case, we consider a general ballistic deposition process and fix the coordinate system such that the particles fall along the vertical (negative h-) direction. The deposit mass per horizontally projected substrate area increases at a constant rate J which is equal to the mass flux through a unit area perpendicular to the particle trajectories. The deposit thickness is related to its mass through the density ρ, hence

$$v(u) = \frac{J}{\rho(u)}. \qquad (4.5)$$

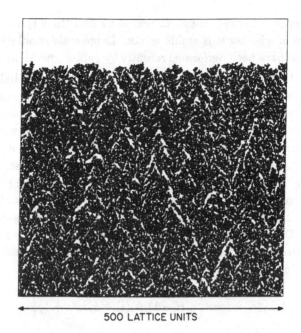

500 LATTICE UNITS

Figure 7: Square lattice simulation of ballistic deposition at normal incidence. Courtesy of P. Meakin.

Any variation of the growth velocity with surface inclination is therefore due to a corresponding variation of the deposit density (Krug 1989a). In general, deposition onto an inclined substrate (or equivalently deposition at oblique particle incidence) increases the porosity and lowers the density of the deposit, thereby leading to a (quadratic) minimum in $v(u)$ at $u = 0$. Clearly this effect is absent in models which presuppose an ordered (crystalline) deposit structure and thus do not allow for voids and defects. In such cases (4.5) implies that v is independent of u. An example is the surface diffusion model studied by Family (1986) and by Liu and Plischke (1988).

These considerations shed some light on recent simulations of a lattice version of the Visscher-Bolsterli model (Meakin and Jullien 1987, Jullien and Meakin 1987). Since an exact implementation of the deposition rules for this model leads to a regular close packed deposit structure, one might think that the introduction of a lattice makes no difference. However, in off-lattice simulations or experiments the regular packing is never realized, because any amount of fluctuations leads to a defective structure of lower density which presumably shows a nontrivial $\rho(u)$

dependence (Figure 8). As the nonlinearity of $v(u)$ determines the scaling properties of the shape fluctuations (cf. Section 5), it follows that these are *different* for the lattice- and off-lattice versions of the Visscher-Bolsterli model, the reason being a difference in the *bulk* structure (Krug 1989a).

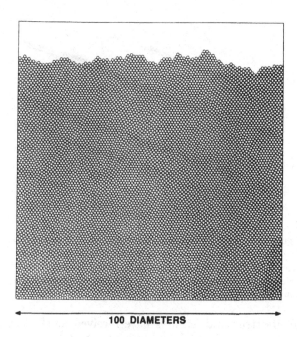

100 DIAMETERS

Figure 8: Off-lattice ballistic deposition with relaxation in two dimensions. Discs are dropped at random. From its first point of contact a disc rolls downhill until it reaches a position locally stable under gravity. Courtesy of P. Meakin.

For the Vold model and its variants the deposit density $\rho(u)$ vanishes in the limit of grazing particle incidence, $u \to \infty$ (Meakin and Jullien 1987, Jullien and Meakin 1987, Krug and Meakin 1989). This is due to the formation of a macroscopic network of voids which separate the columns of the 'columnar' deposit microstructure (Figure 9). As a consequence the accessible deposit surface acquires macroscopic discontinuities which can be viewed as a divergent contribution to the intrinsic surface width.

While the scaling properties of the columnar microstructure have only recently been elucidated (Krug and Meakin 1989, Meakin and

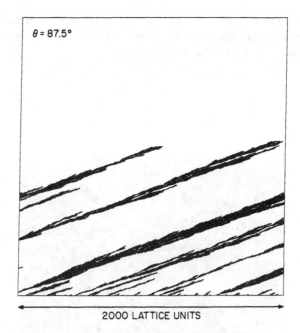

Figure 9: Off-lattice ballistic deposition at near-grazing incidence in two dimensions. The substrate is horizontal. Particles
enter from the right along randomly chosen linear trajectories
which form an angle $\theta = 87.5°$ with the substrate normal.
Courtesy of P. Meakin.

Krug 1990), its structural features have received much attention both
from the experimental and the theoretical side (Leamy *et al.* 1980).
Looking at Figure 9, an immediate question arises which concerns the
relationship between the column orientation and the angle of particle
incidence. In our geometry the orientation of the columns and of the
substrate normal relative to the direction of incidence are measured by
the angles ϑ and φ, respectively, which were introduced in Section 2. For
some time φ and ϑ were believed to be related by the empirical 'tangent
rule' (Nieuwenhuizen and Haanstra 1966)

$$\tan(\vartheta - \varphi) = \tfrac{1}{2}\tan\vartheta. \tag{4.6}$$

This is ruled out however by the general considerations of Section 2,
since it would imply a nonmonotonic dependence of φ on ϑ. Indeed (4.6)
has been refuted by large scale simulations of various deposition models
(Meakin *et al.* 1986b, Meakin 1988d). One finds instead a monotonic
increase of $\varphi(\vartheta)$ which appears to saturate at a constant value $\varphi_{\max}$ for

$\vartheta \to \frac{1}{2}\pi$ (near-grazing incidence). This is expected to be characteristic of columnar growth, since simple arguments show that the growth velocity of a column structure is a *linear* function of inclination, as in (2.14), with $a = \cotan\varphi$ and $v_\perp = 1$ $(\frac{1}{2})$ for the lattice (off-lattice) version of the two dimensional Vold model (Krug and Meakin 1989, Limaye and Amritkar 1986). The theory of Section 2 then predicts that $\varphi = \varphi_{max}$ independent of ϑ.

Clusters can be grown through ballistic deposition by exposing a point seed to the unidirectional particle flux (Bensimon *et al.* 1984a, Liang and Kadanoff 1985). (Note that this is different from the ballistic aggregation model also due to Vold (1963), in which the particle trajectories have *random* orientations.) The shadowing effects lead to a fan-shaped cluster with a domed upper surface (Figure 10). The upper surface is related to the nonlinear part of $v(u)$, while the opening angle is determined by the asymptotically linear behavior (2.14), cf. Section 2. The opening angle is $2\varphi_{max}$ as would be expected intuitively.

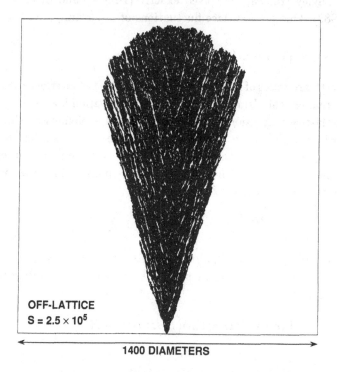

OFF-LATTICE
S = 2.5 × 10⁵

◄──────────────────────────────►
1400 DIAMETERS

Figure 10: Cluster grown by off-lattice ballistic deposition onto a point seed. Particles rain down vertically. Courtesy of P. Meakin.

An interesting shape transition occurs in the synchronous version of the Vold lattice model, in which the growth rule (4.4) is applied simultaneously to a finite fraction p of sites (Baiod *et al.* 1988). At the threshold value p_c ($p_c \simeq 0.7058$ in two dimensions) facets appear at the corners of the fan and spread towards the top as $p \to 1$. At the same time the opening angle of the fan increases according to $\varphi_{\max}(p) = \arctan[p/(1 - p)]$ (Krug and Meakin 1991). The origin of the transition is the same as for the other synchronous models discussed previously, and will be explained in Section 7.

Several modifications of ballistic deposition have been introduced, including e.g. spatial correlations in the particle flux (Meakin and Jullien 1989, 1990), particle diffusion superimposing the ballistic motion (Meakin 1983, Jullien *et al.* 1984, Nadal *et al.* 1984), and a finite density of depositing particles (Jullien and Meakin 1989). Long range correlations in the particle flux modify the surface scaling properties in a highly nontrivial way (Medina *et al.* 1989). A ballistic deposition model on the Cayley tree can be solved exactly (Bradley and Strenski 1985, Krug 1988). One finds surface fluctuations of order unity, i.e. $\zeta = 0$.

4.4 Low temperature Ising dynamics

The growth processes of interest here describe also the interface dynamics of a low temperature Ising model without conservation law. We consider a square lattice (extension to higher dimensions is obvious) with a spin $\sigma_{\mathbf{x}} = \pm 1$ at each lattice site $\mathbf{x} = (i, j)$. The spins interact through a ferromagnetic nearest neighbor coupling $J > 0$ and are subject to an external field $\mu > 0$ which prefers the $+$ phase. The energy of a configuration $\sigma = \{\sigma_{\mathbf{x}} \mid \mathbf{x} \in \mathbf{Z}^2\}$ is

$$H(\sigma) = -J \sum_{\langle \mathbf{xy} \rangle} \sigma_{\mathbf{x}} \sigma_{\mathbf{y}} - \mu \sum_{\mathbf{x}} \sigma_{\mathbf{x}} \tag{4.7}$$

where the first sum is over pairs $\langle \mathbf{xy} \rangle$ of nearest neighbors. The configurations evolve through single spin flips which occur according to the Metropolis rates,

$$\Gamma(\sigma \to \sigma^{(\mathbf{x})}) = \Gamma_0 \phi \Big(\beta \big(H(\sigma^{(\mathbf{x})}) - H(\sigma) \big) \Big),$$
$$\phi(\lambda) = \min(1, e^{-\lambda}), \tag{4.8}$$

where $\sigma^{(\mathbf{x})}$ denotes the configuration σ with the spin at site $\mathbf{x}$ flipped, and $\beta = 1/k_B T$ is the inverse temperature. We choose a temperature below the critical temperature and prepare an initial configuration with a flat interface.

The Ising dynamics differs from other growth dynamics in one important respect. By random events deep inside the bulk of the unstable − phase + droplets of sufficient size may nucleate and start to grow themselves. Once there is a finite density of such droplets, the interface loses its geometrical meaning. These considerations determine the time scale over which our theory is applicable to Ising interfaces. It depends strongly on the temperature and the external field.

To understand the equivalence with growth processes we orient the initial configuration with an interface along the $(1,-1)$ direction, i.e. $\sigma_{ij} = 1$ for $i + j \leq 1$ and $\sigma_{ij} = -1$ for $i + j > 1$. In the limit of zero temperature ($\beta \to \infty$) and if the field is not too strong ($0 < \mu < 2J$) the only allowed processes are those in which a − spin with two − and two + neighbors flips at rate Γ_0. The interface advances into the unstable phase while remaining single valued with respect to the line $i+j = 0$, and on average parallel to the $(1,-1)$ direction. We introduce a coordinate system relative to the lines $i+j = 0$ and $i-j = 0$, $(k,m) = (i-j, i+j)$, and define the position $h(k)$ of the interface above point k by

$$h(k) = \max_{(i,j)}\Big\{ i+j \mid \sigma_{ij} = 1, \ i - j = k \Big\}. \tag{4.9}$$

Then by construction $|h(k+1) - h(k)| = 1$ and when the spin at $(i,j) = \frac{1}{2}(h(k) + k, h(k) - k)$ flips, $h(k)$ increases by 2. The interface evolves according to the growth rule of the single step model with $\Gamma_+ = \Gamma_0$ and $\Gamma_- = 0$ (Marchand and Martin 1986). A partly reversible dynamics ($\Gamma_- > 0$) is obtained by letting $\mu \to 0$ as $\beta \to \infty$ such that $\beta\mu = \text{const.} > 0$. The reversible case ($\mu = 0$) is discussed by Kandel and Domany (1990).

Similar considerations apply to an interface initially oriented along the $(1,0)$ direction. In order to have anything at all happening at zero temperature, we must choose the field $\mu > 2J$. On the other hand the suppression of droplet formation in the − phase requires $\mu < 4J$. Taking $2J < \mu < 4J$ and $\beta \to \infty$ the only allowed spin flips are those of a − spin which has at least one + neighbor. Since all allowed processes occur at equal rate Γ_0, we recover the continuous time (version A) Eden model.

Although the exact equivalence holds only in the limiting cases discussed, we expect that the interface fluctuations of a kinetic Ising model are governed by our theory, provided $\mu \neq 0$ and times are so short that droplet formation in the bulk can be neglected.

5

Continuum Theory

5.1 The Kardar-Parisi-Zhang equation

Our starting point is the time evolution of the macroscopic height profile
as governed by

$$\frac{\partial}{\partial t} h_t = v(\nabla h_t), \qquad (5.1)$$

compare with Section 2. v is the growth velocity in the h-direction. We
assume that v depends smoothly on ∇h_t. In fact, this assumption may
be violated. (An example will be given in Section 7.2.) But in general
this will happen only at isolated values in parameter space.

In the spirit of the scaling theory developed in Section 3 we assume
now that (5.1) remains valid even on finer scales. On a finer scale we
will see more details, in particular random increases in the surface height
and local smoothing. Using a conventional description of both processes
we obtain the equation

$$\frac{\partial}{\partial t} h_t = v(\nabla h_t) + \nu \Delta h_t + \zeta_t. \qquad (5.2)$$

Since growth events occur independently when separated in space-time,
the noise ζ_t is chosen to be Gaussian white noise with mean zero and
covariance

$$\langle \zeta_t(\mathbf{x}) \zeta_{t'}(\mathbf{x}') \rangle = \gamma \delta(t - t') \delta(\mathbf{x} - \mathbf{x}'). \qquad (5.3)$$

Note that we really have defined a *new* growth model. ν can
be thought of as an effective surface tension. Physically the relaxation term $\nu \Delta h_t$ may arise through evaporation-condensation processes
(Mullins 1959) or as a consequence of gravity induced restructuring (Edwards and Wilkinson 1982). On the basis of Equation (5.2), at least in
principle, an effective inclination dependent growth velocity, $v_{\text{eff}}(\nabla h)$,
may be computed which determines then the macroscopic shape accord-
ing to the rules explained in Section 2. Of course, because of noise, in
general v_{eff} will be different from the 'bare' growth velocity v.

We expect that the large scale behavior of the growth model (5.2) is again governed by our scaling theory. If the large scale properties of growing surfaces are universal, then we may as well rely for their prediction on Equation (5.2). As a common experience in statistical physics, for continuum models more and better developed techniques are available. Thus a continuum growth model may be of advantage. We lose thereby the precise specification of the microscopic processes of adhesion and relaxation. But the real physics is anyhow more complicated than our simplified microscopic models. If only for this reason, we have to concentrate on those properties which are independent of microscopic details (of course within the class of local growth rules).

From our remarks it is clear that it makes no sense to 'derive' (5.2) from a microscopic model. It would not add to the credibility of this equation. A notable exception is the Ginzburg-Landau model A at low temperatures. In this case h_t is the position of the interface between the $+$ and $-$ phases. Its surface free energy is

$$H_0 = \sigma \int d^d x \left[1 + (\nabla h_t)^2\right]^{1/2} \qquad (5.4)$$

with σ an effective surface tension (Buff *et al.* 1965, Diehl *et al.* 1980). As usual, H_0 acts as a potential for the dynamics. Keeping in mind our particular choice of the coordinate system we obtain

$$\frac{\partial}{\partial t} h_t = -L_0 \left[1 + (\nabla h_t)^2\right]^{1/2} \frac{\delta H_0}{\delta h_t} + L_0^{1/2} \left[1 + (\nabla h_t)^2\right]^{1/4} \zeta_t, \qquad (5.5)$$

where L_0 is a kinetic coefficient and the noise strength follows from detailed balance (Bausch *et al.* 1981, Kawasaki and Ohta 1982). Now under an applied external field μ, the system can gain energy by translating the interface into the unstable phase. Then H_0 is changed to the total energy

$$H = H_0 - \mu \int d^d x \, h_t(\mathbf{x}) \qquad (5.6)$$

and the equation of motion becomes

$$\frac{\partial}{\partial t} h_t = L_0 \sigma \Delta h_t \qquad (5.7)$$

$$- L_0 \sigma \left[1 + (\nabla h_t)^2\right]^{-1} \sum_{i,j=1}^{d} \frac{\partial h_t}{\partial x_i} \frac{\partial^2 h_t}{\partial x_i \partial x_j} \frac{\partial h_t}{\partial x_j}$$

$$+ L_0 \mu \left[1 + (\nabla h_t)^2\right]^{1/2} + L_0^{1/2} \left[1 + (\nabla h_t)^2\right]^{1/4} \zeta_t,$$

which is of the form (5.2) with $v(\nabla h) = L_0\mu\sqrt{1 + (\nabla h)^2}$ corresponding to isotropic growth.

We proceed with the analysis of Equation (5.2) in order to see what it can teach us about growth. We are interested in shape fluctuations around a surface which is flat on the average. The surface gradients are then small and it is natural to expand $v(\nabla h_t)$. As we will show below terms of third and higher order are irrelevant for the large scale behavior. Therefore it suffices to consider

$$\frac{\partial}{\partial t}h_t = v(\mathbf{0}) + \nabla v'(\mathbf{0}) \cdot \nabla h_t \tag{5.8}$$
$$+ \tfrac{1}{2}\nabla h_t \cdot \nabla\nabla v(\mathbf{0}) \cdot \nabla h_t + \nu\Delta h_t + \zeta_t.$$

$v(\mathbf{0})$ and $\nabla v(\mathbf{0})$ can be absorbed through the Galilei transformation

$$\tilde{h}_t(\mathbf{x}) = h_t(\mathbf{x} - \nabla v(\mathbf{0})t) - v(\mathbf{0})t. \tag{5.9}$$

For simplicity we assume isotropy in the sense that $\partial^2 v/\partial u_i\partial u_j = \lambda\delta_{ij}$. Denoting $\tilde{h}_t$ again by h_t we arrive at the KPZ equation (Kardar, Parisi and Zhang 1986)

$$\frac{\partial}{\partial t}h_t = \tfrac{1}{2}\lambda(\nabla h_t)^2 + \nu\Delta h_t + \zeta_t. \tag{5.10}$$

The equation has the three parameters λ, ν and γ. λ is the strength of the nonlinearity. We record that

$$\lambda = \frac{1}{d}\Delta v(\mathbf{0}). \tag{5.11}$$

5.1.1 *Linear theory*

In some of the microscopic models discussed in Section 4 v is independent of ∇h_t and therefore $\lambda = 0$. Thus the linear theory will govern the large scale behavior. Since linear, the details are worked out easily. Of particular interest is the stationary two-point function. For $d = 1, 2$ only the height differences become stationary with the result

$$\left\langle \left[h_t(\mathbf{x}) - h_t(\mathbf{x}')\right]\left[h_s(\mathbf{x}) - h_s(\mathbf{x}')\right]\right\rangle \tag{5.12}$$
$$= (2\pi)^{-d}\int d^d k\left[1 - e^{i\mathbf{k}\cdot(\mathbf{x}-\mathbf{x}')}\right]\frac{\gamma}{2\nu}\frac{1}{k^2}e^{-\nu k^2|t-s|}.$$

For $d \geq 3$ the surface is smooth with correlations

$$\langle h_t(\mathbf{x})h_s(\mathbf{x}')\rangle = (2\pi)^{-d}\int d^d k\, e^{i\mathbf{k}\cdot(\mathbf{x}-\mathbf{x}')}\frac{\gamma}{2\nu}\frac{1}{k^2}e^{-\nu k^2|t-s|}. \tag{5.13}$$

Independent of dimension the dynamical exponent is

$$z = 2. \tag{5.14}$$

For $d = 1$ $h_t(x)$ is a random walk with respect to x, i.e.

$$\left\langle \left[h_t(x) - h_t(x')\right]^2 \right\rangle = \frac{\gamma}{2\nu}|x - x'|.$$

For $d = 2$ there are logarithmic fluctuations, whereas for $d \geq 3$ the static correlations decay as

$$\langle h_t(\mathbf{x})h_t(\mathbf{x}')\rangle \sim |\mathbf{x} - \mathbf{x}'|^{-(d-2)}.$$

We summarize this behavior by the wandering exponent

$$\zeta = \frac{2 - d}{2}. \tag{5.15}$$

5.1.2 *Scaling*

Henceforth we assume $\lambda > 0$ (if $\lambda < 0$, then we change h_t to $-h_t$). We look for a statistically self-similar solution to (5.10). Therefore we define the rescaled surface

$$\tilde{h}_t(\mathbf{x}) = b^{-\zeta} h_{b^z t}(b\mathbf{x}), \tag{5.16}$$

b large, and insert in (5.10) to obtain the following equation for $\tilde{h}_t$,

$$\frac{\partial}{\partial t}\tilde{h}_t = b^{(\zeta+z-2)}\tfrac{1}{2}\lambda(\nabla\tilde{h}_t)^2 + b^{(z-2)}\nu\Delta\tilde{h}_t + b^{(z-2\zeta-d)/2}\zeta_t. \tag{5.17}$$

Here we used the scale invariance

$$\zeta_{b^z t}(b\mathbf{x}) = b^{-(z+d)/2}\zeta_t(\mathbf{x}) \tag{5.18}$$

which follows from (5.3).

If $\zeta > 0$, then the first term dominates as $b \to \infty$. To ensure scale invariance we have to set

$$\zeta + z = 2. \tag{5.19}$$

Thus we recovered the scaling relation (3.18). Note that higher orders in the expansion of $v(\nabla h_t)$, like $(\nabla h_t)^3$, are irrelevant compared to the second order term. For $b \to \infty$ the diffusion and noise terms vanish. Nevertheless they are needed to single out the invariant distribution of physical relevance.

If $\zeta < 0$ the nonlinearity is irrelevant. z and ζ are then given by the linear theory, cf. Equations (5.14) and (5.15). At this point we have no tool for computing ζ. Taking the ζ of the linear theory suggests that in dimensions $d > 2$, at least for small λ, the nonlinearity is of no importance.

5.1.3 *1+1 dimensions, noisy Burgers equation*

In $1+1$ dimensions the height differences, $h_t(x') - h_t(x)$, become stationary as $t \to \infty$. Writing them as $\int_x^{x'} dy \, \partial h_t(y)/\partial y$ we may as well consider the surface slope $u_t = \partial h_t/\partial x$. If h_t is governed by the KPZ equation, then u_t satisfies

$$\frac{\partial}{\partial t} u_t = \frac{\partial}{\partial x}\left[\tfrac{1}{2}\lambda u_t^2 + \zeta_t + \nu \frac{\partial}{\partial x} u_t\right], \qquad (5.20)$$

which is the noise driven Burgers equation (Burgers 1974). (Burgers equation has $\lambda = -1$, which can be achieved through substituting u_t by $-(1/\lambda)\tilde{u}_t$.) For Burgers u_t is the velocity field of a one dimensional fluid. Equation (5.20) is then the Navier-Stokes equation with random forcing. Burgers investigated in great detail the deterministic equation ($\zeta_t \equiv 0$) with random initial data (cf. Section 5.4).

Since u_t is locally conserved we are free to still fix its average value, which is taken $\langle u_t \rangle = 0$ by our choice of initial data. Noise and diffusion alone (i.e. Equation (5.20) with $\lambda = 0$) determine then a unique invariant distribution. It is the Gaussian white noise with covariance

$$\langle u_t(x) u_t(x') \rangle = \frac{\gamma}{2\nu}\delta(x - x'). \qquad (5.21)$$

It so happens that this measure is also invariant under the flow generated by the solutions of

$$\frac{\partial u_t}{\partial t} = \frac{\lambda}{2}\frac{\partial u_t^2}{\partial x}, \qquad (5.22)$$

(Forster *et al.* 1977, Huse *et al.* 1985a). To prove it we consider an interval of length L with periodic boundary conditions. Formally, the right side of Equation (5.22) is divergence free, since

$$\int_0^L dx \frac{\partial}{\partial x} u_t(x) = 0 \qquad (5.23)$$

because of periodic boundary conditions. Therefore we only have to check the time invariance of the density

$$\exp\left[-\frac{\nu}{\gamma}\int_0^L dx \, u_t(x)^2\right]. \qquad (5.24)$$

Differentiating in time yields

$$-\frac{\lambda\nu}{2\gamma}\left[\int_0^L dx \, u_t(x)\frac{\partial}{\partial x}u_t(x)^2\right]\exp\left[-\frac{\nu}{\gamma}\int_0^L dx \, u_t(x)^2\right]. \qquad (5.25)$$

By partial integration the prefactor vanishes. Note that in higher dimensions the prefactor is $\int d^d x \, \mathbf{u}_t(\mathbf{x})(\mathbf{u}_t(\mathbf{x}) \cdot \nabla) \mathbf{u}_t(\mathbf{x})$ which does not vanish, in general.

The stationarity of white noise is a little bit of a surprise because a profile $u_t(x)$ which is smooth initially converges as $t \to \infty$ to a constant profile. Profiles typical for white noise are fairly rough and may not settle down as $t \to \infty$. These considerations provoke the question of how well defined Equation (5.20) is mathematically. Of course, physically a short distance cutoff should be introduced. If one discretizes Equation (5.20) as

$$\frac{d}{dt} u_t(j) = \frac{\lambda}{6} \Big[u_t(j+1)\big(u_t(j) + u_t(j+1)\big) \tag{5.26}$$
$$- u_t(j-1)\big(u_t(j-1) + u_t(j)\big) \Big]$$
$$+ \nu \Big[u_t(j+1) - 2u_t(j) + u_t(j-1) \Big]$$
$$+ \zeta_t(j+1) - \zeta_t(j),$$

then in the steady state the $u_t(j)$'s are distributed as independent Gaussians with variance $\gamma/2\nu$ (Nieuwenhuizen 1989). Also, as will be discussed in Section 6, for several lattice growth models the stationary distribution can be computed explicitly. For it the slopes are independent at large separation. These results strengthen our trust in (5.21).

We conclude that for the stationary growth process

$$\big\langle \big(h_t(x) - h_t(x')\big)^2 \big\rangle = \frac{\gamma}{2\nu} |x - x'|. \tag{5.27}$$

Therefore

$$\zeta = \tfrac{1}{2} \quad \text{and} \quad z = \tfrac{3}{2}. \tag{5.28}$$

Our argument gives no handle on the scaling function. Renormalization (Janssen and Schmittmann 1986) shows that there is a universal function, g, such that

$$\int dx \, e^{ikx} \langle u_t(x) u_0(0) \rangle = \frac{\gamma}{2\nu} g\Big[\Big(\frac{\lambda^2 \gamma}{2\nu}\Big)^{1/3} k |t|^{2/3} \Big] \tag{5.29}$$

for small k and large t, average in the stationary measure (5.21). The scaling function g is not known explicitly. By symmetry $g(x) = g(-x)$ and $g'(0) = 0$. Janssen and Schmittmann (1986) have computed $g(x)$ for small x and find $g''(0) \simeq -4.5$. Approximations suggest that g

decays as $\exp(-c|x|^{3/2})$ for large x (van Beijeren *et al.* 1985, Yakhot and She 1988, Zaleski 1989). The crucial point in (5.29) is that only macroscopic parameters of the growth model appear: λ is determined by the growth velocity, compare with (5.11), and $\gamma/2\nu$ measures the strength of the static fluctuations. The scaling form (5.29) governs the large scale behavior of any two dimensional growth process (provided the growth rules are sufficiently local and $\lambda \neq 0$, cf. Equation (5.11)).

Since with some luck we have guessed the stationary measure, the effective growth velocity $v_{\text{eff}}(\nabla h)$ can be computed (we refer to the discussion below Equation (5.3)). Actually, the Gaussian (5.24) is the steady state for any bare growth velocity $v(\nabla h)$. Then, on the basis of (5.20) with $(\lambda/2)u_t^2$ replaced by $v(u_t)$, the effective growth velocity is given by

$$v_{\text{eff}}(\nabla h) = \int du \left(\frac{\nu}{\pi\gamma}\right)^{1/2} e^{-\nu u^2/\gamma} v(\nabla h + u). \qquad (5.30)$$

5.1.4 *Renormalization*

The method of dynamic renormalization has been applied to the noise driven Burgers equation by Forster, Nelson and Stephen (1977). Without alteration their results carry over to the KPZ equation (see Kardar *et al.* 1986, Medina *et al.* 1989). We rewrite this equation in terms of the height Fourier coefficients. They satisfy the integral equation

$$\hat{h}_t(\mathbf{k}) = e^{-\nu k^2 t}\, \hat{h}_0(\mathbf{k}) + \int_0^t ds\, e^{-\nu k^2(t-s)} \qquad (5.31)$$

$$\times \left[\hat{\zeta}_s(\mathbf{k}) - \frac{\lambda}{2}(2\pi)^{-d}\int d^dq\, \mathbf{q}\cdot(\mathbf{k}-\mathbf{q})\hat{h}_s(\mathbf{k}-\mathbf{q})\hat{h}_s(\mathbf{q})\right].$$

Equation (5.31) is solved perturbatively. The first terms in the perturbation expansion for $\langle\hat{h}_t(\mathbf{k})\hat{h}_t(-\mathbf{k})\rangle$ diverge for $d < 2$. The perturbation series is then reorganized into a renormalization by integrating out only the modes in the shell $e^{-l}\Lambda \leq |\mathbf{k}| \leq \Lambda$. Wave vector and time are rescaled as $\mathbf{k}' = e^l \mathbf{k}$, $t' = e^{zl}t$ and the remaining height modes as $\hat{h}'_{t'}(\mathbf{k}') = e^{-(d+\zeta)l}\hat{h}_t(\mathbf{k})$. The rescaled Fourier coefficients obey then, in approximation, again (5.31) provided the coefficients ν, λ and γ are adjusted according to the flow equations

$$\frac{d}{dl}\nu = \left[z - 2 + A_d\frac{2-d}{4d}\bar{\lambda}^2\right]\nu,$$

$$\frac{d}{dl}\gamma = \left[z - d - 2\zeta + A_d\frac{\bar{\lambda}^2}{4}\right]\gamma, \qquad (5.32)$$

$$\frac{d}{dl}\lambda = [z + \zeta - 2]\lambda$$

with $A_d = (2^{d-1}\pi^{d/2}\Gamma(d/2))^{-1}$ and $\bar{\lambda}^2 = \lambda^2\gamma/2\nu^3$. For $A_d = 0$ (5.32) just expresses the scaling already found in (5.17). The terms proportional to A_d are the result of a first order expansion in $\bar{\lambda}$.

To discuss the renormalization group flow (5.32) z and ζ are adjusted such that $d\nu/dl = 0 = d\gamma/dl$. The effective coupling constant evolves then according to

$$\frac{d}{dl}\bar{\lambda} = \frac{2-d}{2}\bar{\lambda} + A_d\frac{2d-3}{4d}\bar{\lambda}^3. \qquad (5.33)$$

Since obtained in an expansion around $\bar{\lambda} = 0$, we can trust (5.33) only for small $\bar{\lambda}$. For $d < 2$ the fixed point $\bar{\lambda} = 0$ is repulsive. Thus the large scale behavior is governed by the nonlinearity $(\nabla h_t)^2$ and $\bar{\lambda}$ should flow to a strong coupling fixed point. On the other hand, for $d > 2$, $\bar{\lambda} = 0$ is an attractive fixed point. There is a weak coupling regime where the nonlinearity is irrelevant. The large scale behavior is governed by the linear theory.

These findings agree with the scaling analysis in (5.17). Equation (5.33) has a strong coupling fixed point but only for $0 < d < \frac{3}{2}$. It predicts $z = (8 - 4d - d^2)/2(3 - 2d)$ and $\zeta = (d-2)^2/2(3 - 2d)$ satisfying $z + \zeta = 2$. ζ increases for $d > 1$ and ζ becomes even larger than one for $d > 1.415$, both non-physical results. For dimension one the exact exponents $\zeta = \frac{1}{2}$, $z = \frac{3}{2}$ are reproduced. Dimension $d = 0$ corresponds to a single growing column. Its height fluctuations increase as $\sqrt{t}$, i.e. $\zeta/z = \frac{1}{2}$ and $\zeta = \frac{2}{3}$, $z = \frac{4}{3}$. This is also reproduced correctly by (5.33). For $d > 2$ a separatrix emerges. Above it $\bar{\lambda}$ flows to infinity, below to zero. This suggests that for $d > 2$ there is a weak and strong coupling regime. In the strong coupling regime the nonlinearity dominates. At this stage we cannot understand the mechanism causing the transition. We will return to this point once we have achieved the mapping to the directed polymer. The harvest from the renormalization is fairly meagre. In particular no tool is offered which would allow the computation of either the dynamic or the static exponent.

5.2 Directed polymer representation

Hopf (1950) and Cole (1951) observed that the Burgers equation can be
mapped to a linear diffusion equation. The Hopf-Cole transformation
extends to the KPZ equation in arbitrary dimension. The price to be
paid is that additive noise is turned into multiplicative noise. We gain
however a rather different perspective. Surprisingly enough the physics
of disordered systems in thermal equilibrium enters the play.

We define

$$Z_t(\mathbf{x}) = \exp\left[-\frac{\lambda}{2\nu} h_t(\mathbf{x})\right]. \qquad (5.34)$$

Then $Z_t(\mathbf{x})$ is governed by the 'diffusion' equation

$$\frac{\partial}{\partial t} Z_t = -\left(-\nu\Delta + \frac{\lambda}{2\nu}\zeta_t\right)Z_t. \qquad (5.35)$$

(Note that $\int \mathrm{d}^d x\, Z_t(\mathbf{x})$ is not conserved.) We may think of (5.35) also
as an imaginary time Schrödinger equation. $\zeta_t(\mathbf{x})$ is then a space-time
random potential. Through the Feynman-Kac formula the solution to
(5.35) can be written in the form of the path integral

$$Z_t(\mathbf{x}) = \int_{\substack{\mathbf{y}(0)=\mathbf{0} \\ \mathbf{y}(t)=\mathbf{x}}} \mathcal{D}\mathbf{y}(\cdot) \exp\left[-\frac{1}{2\nu}\int_0^t \mathrm{d}s\, \tfrac{1}{2}\dot{\mathbf{y}}(s)^2\right] \qquad (5.36)$$

$$\times \exp\left[-\frac{1}{2\nu}\int_0^t \mathrm{d}s\, \lambda\zeta_s\big(\mathbf{y}(s)\big)\right].$$

We have split up the path integral into two factors. The first one is the
'free' (unperturbed) part. It is Brownian motion starting at the origin
ending at $\mathbf{x}$ at time t, normalized to one when being integrated over all
final points. A Brownian path, $\mathbf{y}(t)$, is then weighted with the exponential of an 'energy', which is given simply by summing the 'potential'
$\zeta_t(\mathbf{x})$ along the path. Note that $Z_t(\mathbf{x})$ is random, because $\zeta_t(\mathbf{x})$ is.

The following physical interpretation is suggestive: $Z_t(\mathbf{x})$ is the partition function of a polymer chain in $d+1$ dimensions with length t and
constrained endpoints. The chain is directed (no U–turn in time) and
sits in an external random potential. 2ν plays the role of temperature.
By (5.34) the height is the free energy, in particular the average height is

$$\langle h_t(\mathbf{x})\rangle = -\frac{2\nu}{\lambda}\langle \log Z_t(\mathbf{x})\rangle. \qquad (5.37)$$

Thus our interest is quenched disorder (as in spin glasses).

The statistical mechanics problem (5.36) can be posed also in discrete form. To illustrate the method we present one example for simplicity in $1+1$ dimension. We choose the square lattice $\mathbf{Z}^2$ and label the horizontal axis by $t = 0, 1, \ldots$ and the vertical axis by $x = 0, \pm 1, \ldots$. A walk (polymer) is a sequence of connected bonds in $\mathbf{Z}^2$. Admissible walks are directed and make at most one step, i.e. $y(t+1) - y(t) = 0, \pm 1$. (These walks are just the restricted SOS interface configurations of a two dimensional Ising model at low temperatures.) To each bond of the lattice we associate a potential energy. The vertical bonds have the energy V_0. The horizontal bonds, b, have a random energy V_b. The V_b's are independent and have identical distribution. To each walk we assign the energy $E(y(\cdot)) = \sum$ *energies along the walk* $y(s)$, $0 \le s \le t$.

Our discrete version of (5.36) reads then

$$Z_t(x) = \sum_{\substack{\text{admissible } y(\cdot) \\ y(0)=0,\ y(t)=x}} e^{-\beta E(y(\cdot))} \tag{5.38}$$

with β the inverse temperature, $\beta > 0$. We have to distinguish now between the thermal average corresponding to (5.38), denoted by E (expectation), and the disorder average, denoted by $\langle \cdot \rangle$ as before. There are two control parameters. βV_0 regulates the diffusion coefficient of the free walk: for $V_b \equiv 0$ one has

$$\mathsf{E}\big(y(t)^2\big) = \frac{\sum_x x^2 Z_t(x)}{\sum_x Z_t(x)} = \big[\tfrac{1}{2}\, e^{\beta V_0} + 1\big]^{-1} t. \tag{5.39}$$

β controls the strength of the random potential. For large β a walk tries to minimize its energy by taking advantage of deep potential minima. This is counteracted by the entropic term which wants to maintain a mean square displacement as in (5.39). As for other problems in statistical mechanics there is a fight between energy and entropy. In fact our previous analysis shows that for $d = 1$ energy always wins, disorder dominates, whereas for $d \ge 3$ entropy wins for small disorder and energy for large disorder.

At this stage the reader may worry that we have lost touch completely with growing surfaces. This is not the case. $Z_t(x)$ satisfies the recursion relation

$$Z_{t+1}(x) = Z_t(x-1)\exp\Big[-\beta(V_0 + V_{(t,x-1)})\Big] \qquad (5.40)$$
$$+ Z_t(x)\exp\Big[-\beta V_{(t,x)}\Big]$$
$$+ Z_t(x+1)\exp\Big[-\beta(V_0 + V_{(t,x+1)})\Big],$$

where in the random potential we indicated only the left endpoint of the bond. Following (5.34) we define

$$h_t(x) = -\frac{1}{\beta}\log Z_t(x). \qquad (5.41)$$

Then

$$h_{t+1}(x) = -\frac{1}{\beta}\log\Big\{\exp\Big[-\beta\big(h_t(x-1) + V_0 + V_{(t,x-1)}\big)\Big] \qquad (5.42)$$
$$+ \exp\Big[-\beta\big(h_t(x) + V_{(t,x)}\big)\Big]$$
$$+ \exp\Big[-\beta\big(h_t(x+1) + V_0 + V_{(t,x+1)}\big)\Big]\Big\}.$$

For $\beta \to \infty$ the iteration becomes

$$h_{t+1}(x) = \min\Big\{h_t(x-1) + V_0 + V_{(t,x-1)}, \qquad (5.43)$$
$$h_t(x) + V_{(t,x)}, h_t(x+1) + V_0 + V_{(t,x+1)}\Big\}.$$

The iterations (5.42), (5.43) combine two steps: First to the current height configuration, $h_t(x)$, one adds independently at each site a random amount V_b. This corresponds to the noise term of the KPZ equation. In the second step one chooses at each site the minimum of the heights at the site itself and its two neighbors. The neighbors receive a penalty V_0. (For finite β the minimum rule is washed out somewhat.) This operation is a discrete version of $\nu\Delta h_t + \frac{1}{2}\lambda(\nabla h_t)^2$, i.e. a combination of lateral spreading and smoothing of the height profile. Note the similarity between (5.43) and the PNG growth rule (4.3).

Before entering into a more detailed analysis of the partition function (5.36) we should translate the objects of interest into the language of directed polymers. From (5.37) we conclude that the growth velocity in the h-direction is just the quenched free energy/length of the polymer. There are two obvious fluctuation quantities: (1) the fluctuations in the free energy and (2) the end-to-end distance (mean square displacement).

ad (1): The free energy is the height of the surface. If starting from a flat surface, the height at $\mathbf{x}$ at time t corresponds to summing over all walks with endpoint $\mathbf{x}$ and arbitrary starting point. We may as well sum over all endpoints of walks starting at the origin. Thus if we define

$$Z_t = \int \mathrm{d}^d x\, Z_t(\mathbf{x}), \qquad (5.44)$$

then from the scaling theory of the growing surface

$$\left\langle \left[-\frac{1}{\beta} \log Z_t - \left\langle -\frac{1}{\beta} \log Z_t \right\rangle \right]^2 \right\rangle^{1/2} \sim t^{\zeta/z} \qquad (5.45)$$

for large t. One expects the same behavior with both endpoints fixed.

ad (2): The mean square displacement is defined by

$$\left\langle \mathsf{E}\left[\mathbf{y}(t) - \mathsf{E}(\mathbf{y}(t))^2 \right] \right\rangle \qquad (5.46)$$
$$= \left\langle \int \mathrm{d}^d x\, \mathbf{x}^2 Z_t(\mathbf{x}) \Big/ \int \mathrm{d}^d x\, Z_t(\mathbf{x}) \right\rangle$$
$$- \left\langle \left[\int \mathrm{d}^d x\, \mathbf{x} Z_t(\mathbf{x}) \Big/ \int \mathrm{d}^d x\, Z_t(\mathbf{x}) \right]^2 \right\rangle.$$

Using the statistical self-similarity of the height profile, one obtains

$$\left\langle \left[\mathsf{E}(\mathbf{y}(t)^2) - \mathsf{E}(\mathbf{y}(t))^2 \right]^{1/2} \right\rangle \sim t^{1/z} \qquad (5.47)$$

for large t. If $z < 2$, the walk is superdiffusive. The disorder roughens the walk. It makes larger excursions than an ordinary random walk.

5.2.1 *Weak coupling*

To estimate the importance of disorder a standard criterion is to consider the ratio

$$\frac{\langle Z_t^2 \rangle}{\langle Z_t \rangle^2}. \qquad (5.48)$$

If the ratio remains bounded for large t, then $Z_t \simeq \langle Z_t \rangle$ and the quenched and annealed free energies are equal. Disorder is irrelevant. The t-dependence of (5.48) is grasped most easily for the continuum directed polymer. To have a well defined path integral we have to allow for spatial correlations in the noise (white noise is too rough). Thus ζ_t is taken to be Gaussian with covariance

$$\langle \zeta_t(\mathbf{x}) \zeta_{t'}(\mathbf{x}') \rangle = \delta(t - t') V(\mathbf{x} - \mathbf{x}'), \qquad (5.49)$$

where V has a good decay at infinity. Performing the Gaussian average we obtain

$$\langle Z_t \rangle = \int_{\mathbf{y}(0)=\mathbf{0}} \mathcal{D}\mathbf{y}(\cdot) \exp\left[-\frac{1}{2\nu} \int_0^t ds \left(\tfrac{1}{2}\dot{\mathbf{y}}(s)^2 - \frac{\lambda^2}{4\nu}V(0) \right) \right] \quad (5.50)$$

and

$$\langle Z_t^2 \rangle = \int_{\mathbf{y}_1(0)=\mathbf{0}} \mathcal{D}\mathbf{y}_1(\cdot) \int_{\mathbf{y}_2(0)=\mathbf{0}} \mathcal{D}\mathbf{y}_2(\cdot) \quad (5.51)$$

$$\times \exp\left[-\frac{1}{2\nu} \int_0^t ds \left\{ \tfrac{1}{2}\dot{\mathbf{y}}_1(s)^2 + \tfrac{1}{2}\dot{\mathbf{y}}_2(s)^2 \right.\right.$$

$$\left.\left. - \frac{\lambda^2}{2\nu}V(0) - \frac{\lambda^2}{2\nu}V\big(\mathbf{y}_1(s) - \mathbf{y}_2(s)\big) \right\} \right].$$

In order to translate back to a Schrödinger equation we define

$$\begin{aligned} H_1 &= -\nu\Delta_1, \\ H_{12} &= -\nu\Delta_1 - \nu\Delta_2 - \frac{1}{2}\left(\frac{\lambda}{2\nu}\right)^2 V(\mathbf{x}_1 - \mathbf{x}_2). \end{aligned} \quad (5.52)$$

Then, writing the operator e^{-tH} as a kernel in position space,

$$\langle Z_t^2 \rangle / \langle Z_t \rangle^2 = \quad (5.53)$$

$$\int d^d x_1 \int d^d x_2 \, e^{-tH_{12}}(\mathbf{0}, \mathbf{0} \mid \mathbf{x}_1, \mathbf{x}_2) \Big/ \left(\int d^d x \, e^{-tH_1}(\mathbf{0} \mid \mathbf{x}) \right)^2.$$

H_{12} has zero as the continuum edge. Therefore the ratio of partition functions remains bounded provided H_{12} has no bound state. Now in one and two dimensions a Schrödinger equation with a short range attractive potential always binds, whereas in three and more dimensions it binds only if λ is sufficiently large. (Note that, because of (5.49), $\widehat{V}(\mathbf{k}) \geq 0$ and $V(0) = (2\pi)^{-d/2} \int d^d k \widehat{V}(\mathbf{k}) < \infty$. The binding/no binding property is a theorem of Simon 1976.) Thus for $d > 2$ the directed polymer has a weak coupling regime. The same conclusion holds for lattice versions. In this case there are even rigorous proofs (Imbrie and Spencer 1988, Bolthausen 1989) guaranteeing that the free energy equals with probability one the annealed free energy and that the large scale behavior of the walk agrees with the free walk.

Our result provides a lower bound on the critical coupling for $d > 2$. To show that there must be a transition to a disordered phase we use an argument familiar from spin glasses. It applies only to the lattice version. One notes that the entropy is always positive. For small coupling the free energy agrees with the annealed free energy. We compute the entropy through the Legendre transform of the annealed free energy. It becomes negative at sufficiently large β. Thus there must be a transition.

To work out numbers one has to solve the lattice analogue of the two particle Schrödinger equation (5.52). This has been carried out for the Gaussian site case (Cook and Derrida 1989a). The random potential, $V_{\mathbf{x}}$, is assigned to each site of a $(d+1)$ dimensional hypercubic lattice. The $V_{\mathbf{x}}$'s are independent Gaussians with variance one. A walk jumps forward in the one-direction ($=$ time direction) and to one of the nearest neighbors in the remaining coordinates. Following the strategy explained one obtains

$$\begin{aligned} \beta_c &\leq 1.67 & \text{for} \quad d = 2, \\ 1.03 \leq \beta_c &\leq 1.90 & \text{for} \quad d = 3, \\ (\log 2d)^{1/2} \leq \beta_c &\leq (2\log 2d)^{1/2} & \text{for} \quad d \to \infty. \end{aligned} \tag{5.54}$$

For $d = 2$ our arguments do not exclude a strictly positive β_c. The leading large d behavior is expected to coincide with the upper bound.

5.2.2 *Replica*

Computing the n-th moment of the partition function leads, as in (5.51), to the n-particle Schrödinger operator

$$H_n = -\nu \sum_{j=1}^{n} \Delta_j - \left(\frac{\lambda}{2\nu}\right)^2 \sum_{i,j=1}^{n} V(\mathbf{x}_i - \mathbf{x}_j). \tag{5.55}$$

Let E_n be the ground state energy of H_n. Then

$$\langle Z_t^n \rangle \approx e^{-E_n t} \tag{5.56}$$

for large t. The replica trick consists in the hope of obtaining

$$\langle \log Z_t \rangle = \lim_{n \to 0} \frac{1}{n} (\langle Z_t^n \rangle - 1). \tag{5.57}$$

For the Sherrington-Kirkpatrick model of spin glasses, E_n corresponds to the free energy of the n-replicated system. It is defined through a variational problem. The famous Parisi ansatz chooses a solution which breaks the symmetry in replica space. No direct analogue of this phenomenon is expected here (Derrida 1990).

If in $d = 1$ we replace $-V(x)$ by an attractive δ-potential, $-\delta(x)$, and ignore the self-interactions (Kardar and Zhang 1987), then the ground state wave function of H_n is

$$\psi = Z^{-1} \exp\left[-\frac{\kappa}{2} \sum_{i \neq j = 1}^{n} |x_i - x_j|\right] \qquad (5.58)$$

with $\kappa = \lambda^2/16\nu^3$. It has the energy

$$E_n = -\frac{1}{3}\nu\kappa^2 n(n^2 - 1) \qquad (5.59)$$

(Lieb and Liniger 1963, Kardar 1987). The cumulant expansion for $\langle Z_t^n \rangle$ is given by

$$\langle Z_t^n \rangle = \exp\left[\sum_{j=1}^{\infty} \frac{n^j}{j!} C_j(\log Z_t)\right], \qquad (5.60)$$

where C_j is the j-th cumulant of $\log Z_t$ ($C_1 = \langle \log Z_t \rangle$, $C_2 = \langle (\log Z_t)^2 \rangle - \langle \log Z_t \rangle^2$, etc.). Using (5.56) and taking the limit $n \to 0$ in (5.59) yields then (Kardar 1987)

$$\lim_{t\to\infty} \frac{1}{t} C_1 = \frac{1}{3}\nu\kappa^2, \quad \lim_{t\to\infty} \frac{1}{t} C_2 = 0, \quad \lim_{t\to\infty} \frac{1}{t} C_3 = -\frac{1}{18}\nu\kappa^2. \qquad (5.61)$$

Thus we recover again $\zeta/z = \frac{1}{3}$, compare with (5.45).

The argument is slightly more subtle than it appears. If we take the Hamiltonian (5.55), then, for large n, particles pile on top of each other and $E_n \sim -n^2$. We expect then an intermediate regime where (5.59) holds and which determines the behavior near $n = 0$. This discussion signals the difficulties for $d \geq 2$. Even for model potentials the exact E_n's are not known. The large n asymptotics, $\sim -n^2$, does not reflect the limit $n \to 0$. Furthermore a non-rational exponent ζ/z is hard to accommodate (McKane and Moore 1988, Zhang 1989).

5.2.3 *Functional renormalization*

Halpin-Healy (1989a) proposed a renormalization which follows the flow of the spatial part of the disorder under rescaling and integrating out large fluctuations. As in (5.49) he chooses a general covariance

$$\langle \zeta_t(\mathbf{x}) \zeta_{t'}(\mathbf{x}') \rangle = \delta(t - t') R(|\mathbf{x} - \mathbf{x}'|). \qquad (5.62)$$

He computes the effective action to one loop order followed by differential time $t \to (1 + z\delta l)t$ and length $\mathbf{x} \to (1 + \delta l)\mathbf{x}$ rescalings. The covariance R is then approximately governed by (Halpin-Healy 1990b)

$$\frac{\partial}{\partial l} R = \left(3 - \frac{4}{z}\right) R + \frac{r}{z} R' + \frac{1}{2}(R'')^2 - R'' R''(0) \qquad (5.63)$$

$$+ \frac{(d-1)}{2}\left(\frac{R'}{r}\right)^2 - (d-1)\frac{R'}{r} R''(0),$$

where ' denotes differentiation with respect to the radial coordinate, r. We search for fixed points of Equation (5.63). For large r the nonlinearities can be neglected. The linear equation has two independent solutions: the short range fixed point $R^*_{SR}(r) \approx r^{(3z-4-d)} \exp[-r^2/2zR''(0)]$ and the Flory fixed point $R^*_F(r) \approx r^{(4-3z)}$ for large r. Equating the algebraic decays yields (Nattermann 1989, Halpin-Healy 1990a)

$$z = \frac{8+d}{6}. \tag{5.64}$$

Equation (5.64) reproduces the known exponent for $d = 1$ and gives $z = \frac{5}{3}$ for $d = 2$ close to numerical results (cf. Section 5.3). However for large d, z diverges in contradiction to the known asymptotic value $z = 2$. Halpin-Healy (1990b) argues that this implies an upper critical dimension $d_u = 4$ above which $z = 2$ in the strong and weak coupling phase. Even if (5.63) is a reasonable approximation, the difficulty remains of how to extract from it the exponent relevant for short range disorder. The matching rule alluded to above is *ad hoc* and a better understanding of the flow generated by (5.63) is needed.

There is one further point of interest: Halpin-Healy (1989a) attempts to view the directed polymer in a disordered medium and interface fluctuations in a system with bond disorder (D.S. Fisher 1986) in a unified framework. Equation (5.36) is generalized in the obvious way to the case where $\mathbf{y}(t)$ is an n-component vector field over a d'-dimensional base space. The directed polymer is then the particular case of $n = d$ and $d' = 1$ whereas the interface between the + and − phases of a bond-disordered d-dimensional ferromagnet corresponds to the case $n = 1$ and $d' = d - 1$.

5.2.4 *Real space renormalization, hierarchical lattices*
For spin systems in thermal equilibrium the Migdal-Kadanoff real space renormalization scheme becomes exact for hierarchical lattices (Berker and Ostlundv 1979, Griffiths and Kaufmanv 1982, Kaufman and Griffiths 1984). We follow these ideas and define the directed polymer on a disordered hierarchical lattice (Derrida and Griffiths 1989, Cook and Derrida 1989a, Derrida 1990). For the construction of the lattice we refer to Figure 11.

At generation $(n + 1)$ each bond of generation n is substituted by the starting motif. As an analogue for dimension the starting motif has b branches. In the n-th generation lattice we assign to each bond

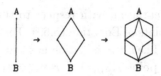

Figure 11: Construction of a hierarchical lattice with branching ratio $b = 2$.

independently a random energy V_b. A walk, w, is directed from A to B. It has the energy

$$E(w) = \sum_{b \in w} V_b. \tag{5.65}$$

The partition function is then given by

$$Z_t = \sum_w e^{-\beta E(w)}, \tag{5.66}$$

where the sum is over all walks from A to B in the n-th generation lattice and $t = 2^n$. We have lost the spatial structure. But through the fluctuations in the free energy we can still identify the exponent ζ/z, compare with Equation (5.45).

Without modification the method developed in Section 5.2.1 can be applied to the b-branch hierarchical lattice. If V_b has a Gaussian distribution of width one, then one obtains the bounds

$$\bigl(\log(b-1)\bigr)^{1/2} \le \beta_c(b) \le (2 \log b)^{1/2}. \tag{5.67}$$

Although not a consequence of (5.67) it is suggestive to set the lower critical branching at $b = 2$.

Next we note that for branching $b = 1$ a one-dimensional lattice is generated. This is the $d = 0$ directed polymer. There is only one walk and the free energy is a sum of independent random variables. Therefore its fluctuations increase as $\sqrt{t}$ and $\zeta/z = \frac{1}{2}$. Together with the scaling relation (3.18) we obtain $\zeta = \frac{2}{3}$, $z = \frac{4}{3}$. It is natural to try an expansion around $b = 1$. This is not at all straightforward but has been accomplished to second order (Cook and Derrida 1989a). As to be expected ζ/z is decreasing in b.

It may be of interest to understand why the directed polymer on a hierarchical lattice is difficult to attack analytically. If one is only interested in the strong coupling ζ/z it suffices to consider the ground state energy (the minimal energy walk). Let E_n be the ground state energy for the n-th generation. Because of the hierarchical structure

$$E_{n+1} = \min\Bigl\{ E_n^{(1)} + E_n^{(2)}, \ldots, E_n^{(2b-1)} + E_n^{(2b)} \Bigr\}, \tag{5.68}$$

where the $E_n^{(i)}$'s are the ground state energies of the sublattices corresponding to generation n. As these are independent random variables, (5.68) implies the following recursion relation for the probability distribution P_n of E_n:

$$
Q_n(E) = \int dE' P_n(E - E') P_n(E'),
$$
$$
P_{n+1}(E) = b Q_n(E) \left[\int_E^\infty dE' Q_n(E') \right]^{b-1}. \tag{5.69}
$$

Clearly (5.69) are meaningful also for noninteger b. To obtain ζ/z one has to follow how the distribution P_n scales for large n. The branching ratio plays the role of dimension. For spin models the usual identification is $b = 2^d$ (Melrose 1983). This relation overestimates b. For example, the exact $d = 1$ exponent $\zeta/z = 1/3$ would be reached at $b = 2$ whereas it occurs in fact already at $b = 1.65$. Also the large d behavior is not reproduced, compare Equations (5.54) and (5.67). We have only a qualitative link between b and d.

The recursion relation (5.69) has been iterated numerically up to $b = 12$ (Halpin-Healy 1989b). One finds a continuous decrease of ζ/z. There is no indication of an upper critical dimension. Extrapolating beyond the known, a rather simple picture is suggested. There are weak and strong coupling regimes. In the strong coupling regime ζ shows a steady decrease from $\zeta = \frac{2}{3}$ at $b = 1$ to $\zeta = 0$ at $b = \infty$. Correspondingly z increases from $\frac{4}{3}$ to 2. The nature of the transition has not been explored yet.

5.2.5 1/d-expansion, trees

Many problems of statistical mechanics simplify in large dimensions. For a directed polymer the limit $d \to \infty$ leads to a disordered tree. Again we assign to each bond of the tree independently a random potential V_b. A walk is directed branching outwards. The energy of a walk is as in (5.65) and the partition function is given by (5.66), where the sum is over all walks of length t (b^t walks for a tree of branching ratio b). Using an analogy with traveling fronts, the directed polymer on a tree can be solved fairly explicitly (Derrida and Spohn 1988), even for complicated branching structures. The main point here is that there is a transition from a high temperature phase to a low temperature 'spin glass' phase. The transition is not drastic. The fluctuations in the free energy are always of order one. They increase with increasing β and freeze into a

β-independent distribution above the critical point β_c. Thus $\zeta = 0$ in both phases. Cook and Derrida (1989b, 1990) develop a $1/d$-expansion which is based on a sequence of approximations consisting of more and more complicated trees. In particular, they compute the mean square displacement for large d and find it to be proportional to t, i.e. $z = 2$ in (5.47). If not an artifact of the method, this result indicates an upper critical dimension d_u. The picture implied is that in the strong coupling phase $\zeta = 0$ and $z = 2$ for $d > d_u$. The prefactor of the mean square displacement (the 'diffusion coefficient') diverges as $d \to d_u$. Below d_u the walk is superdiffusive, $z < 2$.

5.3 Numerical results for the KPZ exponents

To illustrate the present state of affairs regarding the values of the KPZ (strong coupling) scaling exponents, Tables 1, 2 and 3 give an up-to-date summary of simulation results obtained from a variety of growth models (see Section 4) as well as from the directed polymer. In the simulations various combinations of ζ and z are actually measured. We have chosen to express all results in terms of ζ and z, using the known scaling relations when necessary. In cases where two exponents were determined independently, this provides a check of the relation $\zeta + z = 2$, cf. Equation (3.18), which is shown then in the corresponding column of the Tables. We refrain from citing error bars, as the different assessments of systematic errors by different authors renders a comparison difficult.

The results for $1 + 1$ dimensional growth have been included mainly to demonstrate the difficulty in obtaining reliable results even in this now theoretically well understood case. Convincing agreement with the theoretical prediction $\zeta = \frac{1}{2}$, $z = \frac{3}{2}$ was achieved only using algorithms (Wolf and Kertész 1987a) and models (Meakin *et al.* 1986b, Krug and Spohn 1989, Kim and Kosterlitz 1989) specifically devised to suppress corrections to scaling.

As discussed in the preceding sections of this chapter, no exact results are available in higher dimensions ($d \geq 2$). Based on their simulations of the noise reduced Eden model, Wolf and Kertész (1987b) conjectured the following dimension dependence of the scaling exponents,

$$\zeta_d = \frac{1}{d+1}, \quad z_d = \frac{2d+1}{d+1} \tag{5.70}$$

which gives the exact result for $d = 1$ (but not for $d = 0$). In contrast to the prediction (5.64), Equation (5.70) does not contain a finite

Model	ζ	z	$\zeta + z$	Reference
polymer, $T = 0$	0.50	1.52	2.02	Huse and Henley (1985)
polymer, $T = 0$	0.41	1.59	—	Kardar (1985)
polymer, $T > 0$	0.51	1.49	—	Kardar (1985)
polymer, $T > 0$	0.42	1.58	—	Bovier *et al.* (1986)
polymer, $T = 0$	0.50	1.50	2.00	Nattermann and Renz (1988)
ballistic	0.42	1.40	1.82	Family and Vicsek (1985)
Eden	0.45	1.55	—	Plischke and Rácz (1985)
Eden	0.50	1.70	2.20	Jullien and Botet (1985)
Eden	0.50	1.67	2.17	Meakin *et al.* (1986a)
Eden	0.51	1.57	2.08	Zabolitzky and Stauffer (1986)
ballistic	0.48	1.54	2.02	Meakin *et al.* (1986b)
single step	0.50	1.51	2.01	Meakin *et al.* (1986b)
single step	0.57	1.43	—	Plischke *et al.* (1987)
Eden	0.50	1.50	—	Wolf and Kertész (1987a)
Eden	0.50	1.50	—	Meakin (1987b)
ballistic	0.50	1.50	—	Meakin (1987b)
PNG	0.50	1.50	—	Krug and Spohn (1989)
restricted SOS	0.50	1.50	2.00	Kim and Kosterlitz (1989)

Table 1: Numerical results for the strong coupling KPZ exponents in $d = 1$.

upper critical dimension, i.e. $\zeta_d \to 0$ and $z_d \to 2$ only asymptotically for $d \to \infty$.

Later work by Kim and Kosterlitz (1989) on a restricted solid-on-solid model (see Section 4.2) led the authors to conjecture

$$\zeta_d = \frac{2}{d+3}, \quad z_d = \frac{2(d+2)}{d+3} \qquad (5.71)$$

which is exact for $d = 1$ and $d = 0$, and also predicts no upper critical dimension. We note that both (5.70) and (5.71) give integer values for z_d/ζ_d in all dimensions, a suggestive prospect in the framework of the replica approach outlined in Section 5.2.2 (McKane and Moore 1988, Zhang 1989). However both conjectures (5.70) and (5.71) have

Model	ζ	z	$\zeta + z$	Reference
polymer, $T = 0$	0.53	1.61	2.14	Kardar and Zhang (1987)
polymer, $T = 0$	0.37	1.63	2.00	Renz (1990)
Eden	0.20	1.80	—	Jullien and Botet (1985)
ballistic	0.33	1.39	1.72	Meakin *et al.* (1986b)
single step	0.36	1.58	1.94	Meakin *et al.* (1986b)
Eden	0.33	1.50	1.83	Wolf and Kertész (1987b)
Eden	0.28	1.72	—	Meakin (1987b)
ballistic	0.41	1.59	—	Meakin (1987b)
single step	0.38	1.63	2.01	Liu and Plischke (1988)
ballistic	0.36	1.64	—	Baiod *et al.* (1988)
restricted SOS	0.40	1.60	2.00	Kim and Kosterlitz (1989)
Eden	0.40	1.60	—	Devillard and Stanley (1989)
single step	0.39	1.60	1.99	Forrest and Tang (1990)
KPZ equation	0.18	1.80	1.98	Chakrabarti and Toral (1989)
KPZ equation	0.24	1.85	2.09	Guo *et al.* (1990)
KPZ equation	0.38	1.58	1.96	Amar and Family (1990b)
ballistic	0.35	1.67	2.02	Family (1990)

Table 2: Numerical results for the strong coupling KPZ exponents in $d = 2$.

Model	ζ	z	$\zeta + z$	Reference
polymer, $T = 0$	0.64	1.56	2.20	Kardar and Zhang (1987)
polymer, $T = 0$	0.28	1.71	1.99	Renz (1990)
Eden	0.08	1.92	—	Jullien and Botet (1985)
Eden	0.24	1.64	1.88	Wolf and Kertész (1987b)
restricted SOS	0.33	1.67	2.00	Kim and Kosterlitz (1989)
Eden	0.33	1.67	—	Devillard and Stanley (1989)
single step	0.30	1.67	1.97	Forrest and Tang (1990)

Table 3: Numerical results for the strong coupling KPZ exponents in $d = 3$.

been questioned by the very recent results of Forrest and Tang (1990). Using a variant of the single step model and exceptionally large lattices (11520 × 11520 sites in $d = 2$), they find $\zeta_2/z_2 = 0.240 \pm 0.001$, slightly but significantly less than the simple fraction $\frac{1}{4}$.

Renz (1990) has performed simulations of the $T = 0$ directed polymer up to dimension $d = 5$. He finds $\zeta_4 = 0.26$, $z_4 = 1.75$ and $\zeta_5 = 0.21$, $z_5 = 1.79$. In particular the results in $d = 5$ seem to rule out the upper critical dimension $d_u = 4$ conjectured by Halpin-Healy (1990b), cf. Section 5.2.3.

5.4 KPZ type equations without noise

On an abstract level we may regard the KPZ equation (5.10) as a nonlinear mechanism which transforms the correlations of the noise ζ_t into those of the field h_t. An alternative way to gain insight into this mechanism is to investigate the deterministic equation ($\zeta_t \equiv 0$) subject to noisy initial data. This approach has been pursued in the hydrodynamic context of Burgers' equation (Burgers 1974, Kida 1979 and references therein). For surface growth, the physical picture behind the statistical initial value problem is the flattening of an initially rough surface due to growth. We show here that such processes can be described by a scaling theory quite similar to that presented in Section 3 for the stochastic case. A more detailed account can be found in Krug and Spohn (1988).

5.4.1 *The Kuramoto-Sivashinsky equation*

Before turning to surface growth we briefly discuss a close cousin of the deterministic KPZ equation, the Kuramoto-Sivashinsky equation of chemical turbulence (Kuramoto and Yamada 1976)

$$\frac{\partial}{\partial t}\theta(\mathbf{x}, t) = \omega_0 + \nu\Delta\theta + \mu(\nabla\theta)^2 - \lambda\Delta^2\theta. \tag{5.72}$$

Here $\theta(\mathbf{x}, t)$ is the phase of a complex Ginzburg-Landau type order parameter $w(\mathbf{x}, t)$. Physically, $w = a + ib$ where a and b are local concentrations of two reacting species in an autocatalytic reaction. ω_0 is the local oscillation frequency in the absence of spatial coupling. Equations similar to (5.72) arise in combustion theory, with $\theta(\mathbf{x}, t)$ describing the position of a flame front moving at average velocity ω_0 (Sivashinsky 1977).

The coefficients of the gradient terms depend on the system parameters. Clearly for $\nu > 0$ the fourth order derivative is irrelevant and (5.72) reduces to the deterministic KPZ equation. In contrast, for $\nu < 0$, the system becomes linearly unstable for small wave numbers $k^2 < -\nu/\lambda$, and a turbulent stationary state evolves. Using a momentum shell renormalization Yakhot (1981) argues that the short wavelength components of θ in the turbulent state effectively generate a stochastic force with short range correlations. Moreover the viscosity ν acquires a correction which is positive in $d = 1$. Hence the turbulent state appears to be described by the *stochastic* KPZ equation. From the exact results of KPZ for $d = 1$ one expects then the stationary power spectrum to scale as $\langle |\hat{\theta}(k)|^2 \rangle \sim k^{-2}$ for small k. This has been confirmed by numerical integration of (5.72) (Yamada and Kuramoto 1976). The verification of the dynamic exponent $z = \frac{3}{2}$ is hampered by strong crossover effects (Hyman *et al.* 1986, Zaleski 1989, and references therein).

5.4.2 *General nonlinearity*

For the KPZ equation there is no linear instability, so all reasonable initial profiles flatten in the course of time. We consider deterministic growth with a general nonlinearity $v(\mathbf{u}) \approx v(\mathbf{0}) + \lambda |\mathbf{u}|^\alpha$ in the inclination dependent growth velocity,

$$\frac{\partial}{\partial t} h_t(\mathbf{x}) = \nu \Delta h_t + \lambda |\nabla h_t|^\alpha \qquad (5.73)$$

where a small 'viscosity' $\nu > 0$ has been added to suppress non-physical solutions. The ensemble of random initial profiles $h_0(\mathbf{x})$ is characterized by $\langle h_0(\mathbf{x}) \rangle = 0$ and the covariance

$$\left\langle \left[h_0(\mathbf{x}) - h_0(\mathbf{x}') \right]^2 \right\rangle \approx A_0 |\mathbf{x} - \mathbf{x}'|^{2\zeta} \qquad (5.74)$$

for $|\mathbf{x} - \mathbf{x}'| \gg 1$, $\zeta > 0$, compare with Equation (3.3). Here a crucial difference appears between the stochastic evolution (5.10) and its deterministic counterpart: While in the former case the static roughening exponent ζ is fixed by the steady state of the dynamics, for the deterministic problem it is an input parameter which is introduced through the initial condition (5.74). Since the dynamics (5.73) is purely relaxational, it does not generate a proper steady state but merely transforms the static scaling properties of the initial data into a dynamically scale invariant process.

Having initialized the deterministic growth at time $t = 0$, we follow the evolution of dynamic correlation functions of the type introduced in Section 3.1. Consider in particular the height-height correlation function

$$G(|\mathbf{x} - \mathbf{x}'|, t) = \left\langle \left[h_t(\mathbf{x}) - h_t(\mathbf{x}') \right]^2 \right\rangle. \qquad (5.75)$$

Assuming statistical scale invariance for the relaxation process, as in Equation (3.1), this may be written

$$G(r, t) = A\left(\frac{r}{\xi_\|(t)}\right) r^{2\zeta} \tag{5.76}$$

just as in the stochastic case. However, the amplitude scaling function A is expected to have a different form: At $t = 0$ the surface is rough on all scales, so $A(y) \to A_0$ for $y \to \infty$. For $t > 0$ the surface is smoothed on scales $r \ll \xi_\|(t)$, hence $A(y) \to 0$ for $y \to 0$.

As in the stochastic case the growth of the correlation length $\xi_\|(t) \sim t^{1/z}$ defines a dynamic exponent z. Since here both the non-linearity exponent α and the roughness exponent ζ can be fixed at will, z is uniquely determined by the scaling relation

$$z = \alpha + \zeta(1 - \alpha) \tag{5.77}$$

cf. Equation (3.20). For $\alpha > (2 - \zeta)/(1 - \zeta)$ the nonlinearity is irrelevant and $z = 2$.

At our present level of generality not much is known about the amplitude scaling function. Burgers (1974) investigates the particular case of dimension $d = 1$, $\alpha = 2$ and $\zeta = \frac{1}{2}$. Since he regards $u_t(x) = -2\lambda \partial h_t / \partial x$ as the velocity field of a one dimensional fluid, $\zeta = \frac{1}{2}$ means that the initial velocity field has short range correlations. Using the Hopf-Cole transformation (5.34), Burgers shows that the asymptotic ($t \to \infty$, $\nu \to 0$) velocity profile consists of linear pieces separated by shock discontinuities with an average distance $l \sim t^{2/3}$. The corresponding surface profile $h_t(x)$ is then composed of parabolic arcs of typical extension l, so l can be identified with $\xi_\|$ and $z = \frac{3}{2}$ as expected from the scaling relation (5.77). He also calculates the short distance behavior of A. The full amplitude scaling function is determined numerically by Kida (1979).

In the context of turbulence one would also like to know how the mean kinetic energy decays in the course of time. In the surface picture this corresponds to a typical slope squared, which we estimate as

$$\langle u_t^2 \rangle \sim \frac{G\big(\xi_\|(t), t\big)}{\xi_\|(t)^2} \sim t^{-2(1-\zeta)/z}. \tag{5.78}$$

Indeed Burgers obtains a $t^{-2/3}$ decay of the kinetic energy.

The case $\alpha = 1$ (still retaining $d = 1$, $\zeta = \frac{1}{2}$) is more tractable (Krug and Spohn 1988). Details will be given in the following section.

As explained already, the deterministic KPZ equation ($\alpha = 2$) is an approximation to isotropic growth,

$$\frac{\partial}{\partial t} h_t = \lambda \sqrt{1 + (\nabla h_t)^2}, \tag{5.79}$$

compare with Equation (5.7). The solution to (5.79) can be obtained from a simple construction, known as Huygens principle in optics. Given $h_0(\mathbf{x})$, for every $\mathbf{x}$ one draws a sphere of radius λt with center $h_0(\mathbf{x})$. $h_t(\mathbf{x})$ is then the envelope function. Since the surface flattens in the course of time, the large t solution to (5.79) agrees with the one of the deterministic KPZ equation.

We now choose the initial height $h_0(\mathbf{x})$ to be a stochastic process stationary in $\mathbf{x}$ and with a finite correlation length. Then the height-height correlation function, cf. Equation (5.74), decays exponentially. Naïvely, this would indicate a roughness exponent $\zeta = 0$. However, the spheres which determine the surface at time t must emanate from local maxima of $h_0(\mathbf{x})$ on the appropriate scale. Therefore we *define* the roughness exponent ζ by

$$\left\langle \max_{|\mathbf{x}| \leq l} h_0(\mathbf{x}) \right\rangle \sim l^\zeta \tag{5.80}$$

for large l. As before the dynamic exponent is then determined through the scaling relation $\zeta + z = 2$.

If $h_0(\mathbf{x})$ is a self-similar Gaussian field with asymptotics as in Equation (5.74), $\zeta > 0$, then the definitions (5.80) and (5.74) for the roughness exponent agree. If $h_0(\mathbf{x})$ is independent at different sites (we imagine here a spatial discretization with the correlation length as unit), then ζ depends on the tail of the distribution of $h_0(\mathbf{x})$. Let $P(h)$ denote this single site distribution. For $P(h) \sim \exp(-B|h|^\beta)$, $\beta > 0$, the roughness exponent $\zeta = 0$ with logarithmic corrections depending on β. In this case $z = 2$. On the other hand for an algebraic decay as $P(h) \sim h^{-\tau}$, $\tau > 1$, one finds $\zeta = d/(\tau - 1)$ and therefore $z = (2\tau - 2 - d)/(\tau - 1)$. For a single site distribution of finite support the left hand side of Equation (5.80) tends to a constant for large l with a correction $-O(l^{-d})$. Thus we set $\zeta = -d$ and $z = 2 + d$. These predictions are supported by an explicit computation of the average density of spheres forming the surface at time t and by numerical simulations (Kida 1979, Tang *et al.* 1990). For dimension $d = 1$ and a single site distribution with exponential decay Kida (1979) verifies the scaling form (5.76) and obtains the amplitude scaling function. As in other one dimensional models, his technique is to follow the dynamics of the surface cusps.

5.4.3 *An exactly solved case*

As an example of a deterministic growth process for which all details can be worked out, we study the one dimensional PNG model on the lattice in the limit of zero nucleation rate. As initial conditions we choose the height differences independently at each bond,

$$u_0(i) = h_0(i+1) - h_0(i) = \pm 1 \qquad (5.81)$$

with equal probability, $\langle u_0(i) \rangle = 0$. The initial profile is the record of a one-dimensional random walk with unit step length, i.e. $\zeta = \frac{1}{2}$. The dynamics takes the simple form (see Equation (4.3))

$$h_{t+1}(i) = \max\Big\{ h_t(i-1), h_t(i), h_t(i+1) \Big\}. \qquad (5.82)$$

Iterating this recursion t times yields the formal solution

$$h_t(i) = \max_{i-t \leq j \leq i+t} \{ h_0(j) \}. \qquad (5.83)$$

Hence the statistics of the surface configurations is related to the statistics of maxima of one dimensional random walks.

To see what to expect in terms of the scaling picture, we note that, since downwards (upwards) slopes propagate with velocity one (minus one), the continuum limit of (5.82) is

$$\frac{\partial}{\partial t} h_t(x) = \left| \frac{\partial}{\partial x} h_t(x) \right| \qquad (5.84)$$

supplemented by the condition that $h_t(x)$ remains continuous at discontinuities of $\partial h_t/\partial x$. Thus the inclination dependent growth velocity is $v(u) = |u|$ and $\alpha = 1$ in the scaling relation (5.77), which gives the dynamic exponent $z = 1$ (fluctuations spread at a finite velocity).

The clue to solving the problem posed in Equations (5.81) and (5.82) is to consider the dynamics of surface steps. (This point of view will be further developed in Section 6.) Under (5.82) a downward step, $u_t(i) = h_t(i+1) - h_t(i) = -1$, moves one lattice spacing to the right and an upward step, $u_t(i) = 1$, moves one lattice spacing to the left. Steps of opposite sign annihilate when they collide. Hence the problem is reduced to a one dimensional gas of particles and antiparticles with velocities ± 1 which initially have an ideal gas distribution (cf. Elskens and Frisch 1985).

To illustrate the method of computation we consider the decay of the step density $\rho(t) = \langle |u_t(i)| \rangle$. This is the probability for a step to survive up to time t. Suppose the step starts from the origin at $t = 0$ and moves to the right $(u_0(0) = -1)$. This step has a left-going partner which starts at a site $j > 0$ at $t = 0$. The two will annihilate at time $t_1 = \frac{1}{2}j$, hence our chosen step survives up to time t iff $j > 2t$. Obviously the two steps have to be at the same height, i.e. j is the first site to the left of the origin for which $h_0(j) = h_0(0)$. The condition for survival is therefore

$$h_0(0) - h_0(i) > 0 \quad \text{for} \quad 0 \leq i \leq 2t. \tag{5.85}$$

Keeping in mind that $h_0(i)$ is the record of a random walk, it follows that $\rho(t)$ is the probability for a symmetric, one dimensional random walk not to return to its starting position in $2t$ steps. For large t (see e.g. Feller 1950)

$$\rho(t) \approx \frac{1}{\sqrt{\pi t}} \tag{5.86}$$

in accord with the scaling law (5.78) for $\zeta = \frac{1}{2}$ and $z = 1$.

The same approach can be used to compute second and higher order, spatial and temporal step correlation functions (Krug and Spohn 1988, Krug 1989b). By integration this gives explicit expressions for the surface correlations. As an illustrative example we quote the result for the height-height correlation

$$\left\langle [h_t(0) - h_t(r)]^2 \right\rangle \approx \begin{cases} A\big(r/\xi_{\parallel}(t)\big)r & r < \xi_{\parallel}(t), \\ r - (4/\pi - 1)\xi_{\parallel}(t) & r \geq \xi_{\parallel}(t), \end{cases} \tag{5.87}$$

for large r and t, with the correlation length

$$\xi_{\parallel}(t) = 2t \tag{5.88}$$

and the amplitude scaling function

$$A(y) = \frac{4}{\pi}\left(\frac{1+y}{y}\arctan\sqrt{y} - \frac{1}{\sqrt{y}}\right). \tag{5.89}$$

This confirms the scaling picture and the dynamic exponent $z = 1$: the roughness is suppressed on scales $r < \xi_{\parallel}(t)$, $A(y) \sim \sqrt{y}$ for $y \to 0$, while on scales $r > \xi_{\parallel}(t)$ it still grows with the exponent $\zeta = \frac{1}{2}$ of the initial data, merely reduced by a constant. We note that (5.88) can be immediately read off from (5.83): for $|i - j| > 2t$, $h_t(i)$ and $h_t(j)$ depend on disjoint portions of the initial condition.

Higher order correlations contain information about the asymmetry of the growing surface with respect to the direction of growth. Specifically, we have computed the local surface skewness (cf. Wolf 1987, Wolf and Kertész 1987a, b)

$$s(r,t) = \frac{\langle \Delta(r,t)^3 \rangle}{\langle \Delta(r,t)^2 \rangle^{3/2}},$$

$$\Delta(r,t) = h_t(0) - \tfrac{1}{2}\big(h_t(r) + h_t(-r)\big),$$

(5.90)

with the result (Krug and Spohn 1988)

$$s(r,t) \approx -\tilde{s}\Big(\frac{r}{\xi_{\parallel}(t)}\Big)$$

(5.91)

where $\tilde{s}(y)$ is a positive, single-humped function which decays as $y^{-3/2}$ for $y \to \infty$.

It is also of interest to consider correlations between events occurring at the same site at different times. Due to the special rôle of $t = 0$, there is no temporal stationarity. For example, the two-point correlation of the step current $|u_t(i)|$ is given by

$$g(t,\tau) := \big\langle |u_t(i)u_{t+\tau}(i)| \big\rangle \approx \frac{1}{\pi\sqrt{t\tau}} F\Big(\frac{t}{\tau}\Big)$$

(5.92)

where $F(y)$ varies monotonously between $F(0) = 1$ and $F(\infty) = 1/\sqrt{2}$. For fixed t and $\tau \to \infty$ the truncated current correlation vanishes as

$$g(t,\tau) - \rho(t)\rho(t+\tau) \approx \frac{1 - 1/\sqrt{2}}{\pi\tau}.$$

(5.93)

The curious statistical properties of the train of steps passing a fixed site emerge more clearly from the conditional probability $C_t(\tau)$ of observing a step at time $t + \tau$, given a step at time t. From Equations (5.86) and (5.92) we have

$$C_t(\tau) = \frac{g(t,\tau)}{\rho(t)} = \frac{1}{\sqrt{\pi\tau}} F\Big(\frac{t}{\tau}\Big)$$

(5.94)

which is practically independent of t. Hence, *conditioned* on the event $\{|u_t(i)| = 1\}$ at time t, the subsequent history has the same appearance as if one had started at $t = 0$, although the *total* density of events vanishes according to (5.86). Similar behavior occurs in the context of intermittent dynamical systems with a non-normalizable invariant measure, where it has been termed 'sporadic' (Gaspard and Wang 1988, Wang 1989). In the temporal sequences $|u_t(i)|$ this intermittency appears as a strong bunching of events, which is also implied by (5.93).

Coming back to surface properties, (5.92) can be integrated to obtain the temporal fluctuations of the local height increase

$$\left\langle \left[h_t(i) - h_0(i) - \left\langle h_t(i) - h_0(i) \right\rangle \right]^2 \right\rangle \approx Dt. \qquad (5.95)$$

While the analytic expression for the diffusion constant D is difficult to evaluate, we have the bounds $\sqrt{2} - 4/\pi \leq D \leq 2 - 4/\pi$ and the numerical estimate $D \simeq 0.354$. Again, the exact result (5.95) supports the scaling theory which predicts this quantity to grow as $t^{2\zeta/z}$ in general.

6

Driven lattice gases

To introduce the microscopic model we imagine a hexagonal packing of discs with surface, cf. Figure 12 (Gates 1988, Gates and Westcott 1988). No overhangs are allowed. Further discs attach to the surface (respecting the hexagonal packing and no-overhang rules) according to the rate α_n if the disc has n neighbors, $n = 2, 3, 4$. We also want to allow for processes where discs detach from the surface and evaporate into the ambient atmosphere (respecting the hexagonal packing and no-overhang rules). The corresponding rates are γ_n if the disc has n neighbors.

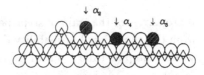

Figure 12: Edge of a two dimensional hexagonal crystal. The zig zag line is the instantaneous surface configuration. Hatched discs attach at the indicated rates.

For the KPZ equation we noticed already that it pays to consider the surface slope rather than the height. For our model let us follow then how the slope changes in time. We draw a line connecting the center of the discs along the top layer, such that only slopes $\pm 60°$ occur, compare with Figure 12. To the j-th line segment we assign the variable $\eta_j = 0, 1$ depending on whether the slope is $+60°$ or $-60°$. We think of η_j, $j = 0, \pm 1, \ldots$, as a lattice gas configuration. There is at most one particle per site. $\eta_j = 1$ if site j is occupied and $\eta_j = 0$ if site j is vacant. η is our shorthand for a whole configuration.

The dynamics of the lattice gas is easily obtained. An attachment of a disc results in a jump to the right and a detachment in a jump to the left. The jump rates between j and $j+1$ are given by

$$\eta_j = 1, \ \eta_{j+1} = 0 : \tag{6.1}$$
$$\quad \alpha_2 \ \text{for} \ \eta_{j-1} = 0, \ \eta_{j+2} = 1,$$
$$\quad \alpha_3 \ \text{for} \ \eta_{j-1} = 0, \ \eta_{j+2} = 0 \ \text{and} \ \eta_{j-1} = 1, \ \eta_{j+2} = 1,$$
$$\quad \alpha_4 \ \text{for} \ \eta_{j-1} = 1, \ \eta_{j+2} = 0,$$
$$\eta_j = 0, \ \eta_{j+1} = 1 :$$
$$\quad \gamma_2 \ \text{for} \ \eta_{j-1} = 0, \ \eta_{j+2} = 1,$$
$$\quad \gamma_3 \ \text{for} \ \eta_{j-1} = 0, \ \eta_{j+2} = 0 \ \text{and} \ \eta_{j-1} = 1, \ \eta_{j+2} = 1,$$
$$\quad \gamma_4 \ \text{for} \ \eta_{j-1} = 1, \ \eta_{j+2} = 0.$$

We denote these jump rates by $c(j, j+1, \eta)$. The probability distribution, $p_t(\eta)$, of the lattice gas at time t is then governed by the master equation

$$\frac{\mathrm{d}}{\mathrm{d}t} p_t(\eta) = \sum_j \Big[c(j, j+1, \eta^{jj+1}) p_t(\eta^{jj+1}) \tag{6.2}$$
$$- c(j, j+1, \eta) p_t(\eta) \Big].$$

Here η^{jj+1} stands for the configuration η with the occupancies at j and $j+1$ interchanged.

Before turning to specific properties of (6.2) let us see how the basic quantities of interest for the large scale structure of the surface are reexpressed in terms of the lattice gas.

(i) By changing the density of the lattice gas the inclination of the surface is varied. Density one corresponds to a slope of $-60°$, density zero to $60°$. Clearly, the particle number, N, is conserved. This implies that the average inclination does not change in time, as it should be.

(ii) The growth velocity is related to the average current. For given density ϱ, the lattice gas settles in a steady state which is spatially uniform. We denote this steady state, in the infinite volume limit, by $\langle \cdot \rangle_\varrho$. In principle it can be obtained as the stationary solution of (6.2) for a ring of L sites with periodic boundary conditions in the limit $N \to \infty$, $L \to \infty$, $N/L \to \varrho$. The steady state current is then given by

$$j(\varrho) = \big\langle c(0, 1, \eta) (\eta_0 - \eta_1) \big\rangle_\varrho. \tag{6.3}$$

If a is the disc diameter, then the growth velocity is $v(u) = \sqrt{3/4}\,aj(\varrho)$ with slope $\partial h/\partial x = u = \sqrt{3}(1 - 2\varrho)$.

(iii) Surface correlations translate to density correlations. If we study the lattice gas for its own sake, then from a statistical mechanics point of view the basic quantity is the time dependent density-density correlation in the steady state. Most conveniently one defines its Fourier transform (the so-called structure function or intermediate scattering function)

$$S(k,t) = \sum_j e^{ikj}\left(\langle \eta_{t,j}\,\eta_{0,0}\rangle_\varrho - \varrho^2\right), \qquad (6.4)$$

where we used that by space-time stationarity $\langle \eta_{t,j}\rangle_\varrho = \varrho$. The term in the round brackets is the correlation of the occupation at the origin at time $t = 0$ and at the site j at time t. If $P_t(\eta, \eta')$ denotes the transition probability of the lattice gas to the configuration η' at time t given the configuration η at time $t = 0$ (in principle, computable from the master equation (6.2)), then

$$\langle \eta_{t,j}\,\eta_{0,0}\rangle_\varrho = \sum_\eta \sum_{\eta'} \bar{p}(\eta)\,\eta_0\,P_t(\eta, \eta')\,\eta'_j \qquad (6.5)$$

with $\bar{p}$ the steady state solution of (6.2) at $N/L = \varrho$. (It is understood that expectations are in the infinite volume limit at constant density.)

Since the density is the only conserved quantity, $S(k,t)$ should scale for small k and large t as

$$S(k,t) = \chi\,e^{ikc(\varrho)t}\,e^{-\nu k^2|t|}. \qquad (6.6)$$

χ is the compressibility of the lattice gas,

$$\chi(\varrho) = \sum_j \left(\langle \eta_j\,\eta_0\rangle_\varrho - \varrho^2\right). \qquad (6.7)$$

$c(\varrho)$ is the sound velocity, which is determined through

$$\frac{1}{\chi}\sum_j j\left(\langle \eta_{t,j}\,\eta_{0,0}\rangle_\varrho - \varrho^2\right) = tj'(\varrho) = tc(\varrho). \qquad (6.8)$$

This result should be no surprise. If we consider a situation where the density varies slowly, then locally the lattice gas is stationary and its density is governed by

$$\frac{\partial}{\partial t}\varrho_t(x) + \frac{\partial}{\partial x}j\bigl(\varrho_t(x)\bigr) = 0. \qquad (6.9)$$

Since correlations correspond to a small deviation from the steady state, the sound velocity is obtained by linearizing (6.9) around a uniform density. Finally, ν is the usual sound damping coefficient.

Equation (6.6) is the result of linearized hydrodynamics, just as for fluids, only with the simplification of a single conservation law. For fluids it has been recognized for a long time that the nonlinearities in the hydrodynamic equations cannot be completely neglected (Pomeau and Résibois 1975). In three dimensions the nonlinearities give rise to a slow decay in the current-current correlations (the so-called long time tails), but they do not modify the scaling form of the structure function. For a one dimensional fluid the effects are more drastic. For these we are already well prepared from our discussion of the KPZ equation. Thus (6.6) is to be replaced by the correct scaling form (van Beijeren *et al.* 1985)

$$S(k,t) \approx \chi \, e^{ikc(\varrho)t} g\big((\lambda^2\chi)^{1/3}k|t|^{2/3}\big) \qquad (6.10)$$

for small k and large t, compare with (5.29). According to (5.11) the coupling constant is

$$\lambda = v''(u) = \frac{1}{8\sqrt{3}} a j''(\varrho). \qquad (6.11)$$

Of course, we assumed here that $0 < \lambda$, $\chi < \infty$. If $\lambda = 0$, then the linear theory of Section 5.1.1 applies and the conventional scaling form (6.6) is the correct one.

Having said all this about the structure function of the lattice gas, we still owe to the reader the link to surface fluctuations. As one relevant example we work out how, starting from a flat substrate, the surface width increases, compare with (3.13). Let h_t be the height of the growing surface at time t and let $j_t(i, i+1)$ be the actual particle current through the bond $(i, i+1)$. Then

$$h_t(i) - h_0(i) = \sqrt{3}a \int_0^t ds \, j_s(i-1, i), \qquad (6.12)$$

where we used that, because of the hexagonal structure, the height increases in units of $\sqrt{3}a$. Therefore the average width, $W(t)$, of the surface is given by

$$W(t)^2 = \left\langle \left[h_t(1) - h_0(1) - \langle h_t(1) - h_0(1) \rangle \right]^2 \right\rangle \qquad (6.13)$$

$$= 3a^2 \left\langle \left[\int_0^t ds \, \{ j_s(0,1) - \langle j_s(0,1) \rangle \} \right]^2 \right\rangle.$$

To compute the average on the right hand side we use the conservation law,

$$\left\langle \left[\sum_{i=-l}^{l} (\eta_{t,i} - \eta_{0,i}) \right]^2 \right\rangle = \tag{6.14}$$

$$\left\langle \left[\int_0^t ds \left\{ j_s(-l-1,-l) - j_s(l,l+1) \right\} \right]^2 \right\rangle.$$

Since t is fixed, the integrated currents through the bonds $(-l-1,-l)$ and $(l,l+1)$ are independent for sufficiently large l. Therefore, in the limit $l \to \infty$ and using translation invariance, we obtain

$$W(t)^2 = 3a^2 \frac{1}{2\pi} \int_{-\pi}^{\pi} dk \left(4(1 - \cos k) \right)^{-1} \tag{6.15}$$

$$\times \sum_j e^{ikj} \left\langle (\eta_{t,j} - \eta_{0,j})(\eta_{t,0} - \eta_{0,0}) \right\rangle.$$

At this point it may be tempting to insert the stationary structure function (6.4) in its scaling form (6.10). This would lead to a long time behavior as $3a^2 [\frac{1}{2}\chi c(\varrho)t + \text{Const} \cdot t^{2/3}]$. Only for vanishing sound velocity does the surface width increase as $t^{1/3}$. However, since we start from a flat substrate, the initial configuration is alternating: $\eta_j = 1$ for even j, $\eta_j = 0$ for odd j and with equal weight the same configuration shifted by one lattice unit. Therefore in (6.15) the non-stationary structure function appears. It is not immediately obvious how to relate it to the scaling form (6.10). To obtain a clue we consider the KPZ equation (5.8) with nonlinearity neglected. Setting $\varrho_t = -\partial/\partial x \, h_t$ and $c = -v'(0)$ it reads

$$\frac{\partial}{\partial t} \varrho_t + c \frac{\partial}{\partial x} \varrho_t = \nu \frac{\partial^2}{\partial x^2} \varrho_t - \frac{\partial}{\partial x} \zeta_t \tag{6.16}$$

with initial conditions $\varrho_0(x) = 0$. Then

$$\int dx \, e^{ikx} \left\langle \left[\varrho_t(x) - \varrho_0(x) \right] \left[\varrho_t(0) - \varrho_0(0) \right] \right\rangle \tag{6.17}$$

$$= \int dx \, e^{ikx} \left\langle \varrho_t(x) \varrho_t(0) \right\rangle = \chi(1 - e^{-2\nu k^2 t}).$$

Note that the contributions from the sound velocity precisely cancel. If we assume (6.17) to continue to hold with the correct scaling form, then for large t

$$W(t)^2 = 3a^2 \frac{\lambda^{2/3} \chi^{4/3}}{2\pi} t^{2/3} \int_{-\infty}^{\infty} dk \frac{1}{k^2} (1 - g(k)). \tag{6.18}$$

The surface width increases as $t^{1/3}$ with a prefactor determined by the scaling function.

6.1 Steady states

We turn to the master equation (6.2). The problem posed is to compute the structure function $S(k,t)$. Clearly this is an impossible task. More modestly we may ask for the static $(t=0)$ structure function. We have to know then first the steady state solution to (6.2). Most commonly the steady state is found by means of detailed balance. Since we assumed no particular relationship between the α's (forward jumps) and the γ's (backward jumps), in general detailed balance cannot hold. Thus we are left with guesswork. Since the jump rates depend only on the two nearest neighbors, it is natural to try the ansatz

$$\frac{1}{Z}\exp\left[-\beta H(\eta) + h\sum_j \eta_j\right] \tag{6.19}$$

with

$$H(\eta) = -\sum_j \eta_j\eta_{j+1}. \tag{6.20}$$

h just fixes the density. Inserting (6.19), (6.20) in (6.2) one has indeed a stationary solution provided

$$\alpha_2 = e^\beta\alpha_4, \quad \alpha_2 + \alpha_4 = 2\alpha_3, \tag{6.21}$$

and

$$\gamma_2 = \alpha_2\, e^{-\beta}\, e^{-E}, \quad \gamma_3 = \alpha_3\, e^{-E}, \quad \gamma_4 = \alpha_4\, e^\beta\, e^{-E}. \tag{6.22}$$

Following (6.19) β is interpreted as 'inverse temperature' (times a unit binding energy). E is regarded as the external driving field causing biased jumps. Since the corresponding potential is linear, E is the energy gained in a jump to the right. Thus (6.22) can be written in the form of a local detailed balance condition,

$$c(j, j+1, \eta) = c(j, j+1, \eta^{jj+1})$$
$$\times \exp\left[-\beta\big(H(\eta^{jj+1}) - H(\eta)\big) + E(\eta_j - \eta_{j+1})\right]. \tag{6.23}$$

Of course, there are other choices of the jump rates satisfying (6.23). We introduced the parameter E also because it regulates the growth velocity.

If $E > 0$, then the attachment dominates and the surface grows. If $E < 0$, then the crystal dissolves. For $E = 0$ both processes balance and we have the model of an equilibrium interface. In equilibrium $j(\varrho) = 0$ which implies $c(\varrho) = 0$, $j''(\varrho) = 0$. The dynamic exponent is $z = 2$. The structure function is of the standard form (6.6) with ν the bulk diffusion coefficient.

Considering the local detailed balance condition (6.23) just by itself it is tempting to try the stationary solution $Z^{-1} \exp[-\beta H + h \sum_j \eta_j]$ with

$$H(\eta) = -\sum_j \eta_j \eta_{j+1} - \frac{E}{\beta} \sum_j j \eta_j. \qquad (6.24)$$

Physically, this is the energy of a lattice gas in a uniform gravitational field. If $E < 0$, then particles pile up at the bottom of the 'box'. Unfortunately, the periodic boundary conditions cannot be satisfied. Of course, the infinite system does not know about boundary conditions. In fact, there are then two stationary solutions for (6.2): the homogeneous solution (6.19), (6.20) and the inhomogeneous solution (6.19), (6.24). Since for the inhomogeneous state detailed balance holds, the average current vanishes. Clearly, as applied to surfaces, (6.19), (6.20) is the state of interest.

Returning to the rates satisfying (6.21), (6.22), physically the processes of attachment become faster for increasing numbers of neighbors. Therefore $\alpha_4 > \alpha_2$ and β should be negative. In terms of a spin system, the steady state is antiferromagnetic.

At least for particular rates we have found explicitly the steady state. Its two-point function $\langle \eta_0 \eta_j \rangle_\varrho - \varrho^2$ decays exponentially. Statically, the statistics of surface configurations is an ensemble of random walks with a one-step memory and hence $\zeta = \frac{1}{2}$. Furthermore the average current $j(\varrho)$ can be computed (Brandstetter 1991). We have then one of the few growth models for which the macroscopic inclination dependent growth velocity can be determined explicitly. In Figure 13 we show three representative examples. Note that there are exceptional densities at which $j''(\varrho) = 0$. If we adjust the average inclination to that particular value, then the surface fluctuations are governed by the linear theory, i.e. $z = 2$.

As β decreases the curvature at $\varrho = \frac{1}{2}$ changes its sign. Qualitatively this can be understood as a switching between two different growth mechanisms. If β is very negative, then $\alpha_2 \ll \alpha_3$. The dominant growth

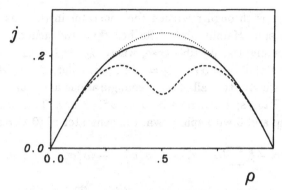

Figure 13: Average current as a function of the density for $\beta = 0$ (dotted line), $\beta = \beta_c = -2\log 3$ (full line) and $\beta = -4$ (dashed line).

is at surface steps, corresponding to a nucleation limited regime. Since a small tilt of the surface increases the step density, we expect the growth velocity to increase and hence $\lambda > 0$. On the other hand for $\beta = 0$ nucleation on flat portions of the surface and step motion occur at the same rates. The steady current of the lattice gas is proportional to $\varrho(1 - \varrho)$. Hence the growth rate is *maximal* at zero tilt and $\lambda < 0$. Such a phenomenon occurs also in higher dimensions (Amar and Family 1990a, 1990c, Krug and Spohn 1990, Kim *et al.* 1990, Huse *et al.* 1990). In two dimensions the effect is even more spectacular because for vanishing λ the surface fluctuations are only logarithmic.

The particular role of the rates satisfying (6.21) and (6.22) has been recognized in an equilibrium context by Singer and Peschel (1980), Zwerger (1981), for driven lattice gases by Katz, Lebowitz and Spohn (1984), and for crystal growth by Gates (1988). Since the steady state is independent of the driving force, the conductivity is frequency independent. The response of the lattice gas is instantaneous. Also the bulk diffusion coefficient is given by a static average. We refer to Spohn (1990) for further details.

For general rates the steady state is not known. Monte Carlo simulations of the lattice gas indicate still a rapid decay of correlations (Katz *et al.* 1984). No expansion around the exactly solved cases has been tried.

Why is it so difficult to extract any kind of dynamical information from (6.2)? After all we have something like a $1 + 1$ dimensional field theory which ought to be 'exactly solvable'. A way to explain the difficulty is through rewriting the generator in (6.2) as a quantum mechanical spin Hamiltonian. Rather than the general case let us consider the simplest possible case, where $\gamma_2 = \gamma_3 = \gamma_4 = 0$ ($E = \infty$, growth only) and $\alpha_2 = \alpha_3 = \alpha_4 = 1$. In particular, $\beta = 0$ which means that in the steady state all allowed configurations are equally likely. We denote the Pauli spin matrices by $\sigma = (\sigma^x, \sigma^y, \sigma^z)$. Then, identifying 1 with spin up and 0 with spin down, the generator of (6.2) reads

$$H = -\tfrac{1}{4}\sum_j (\sigma_j \cdot \sigma_{j+1} + i\sigma_j^x \sigma_{j+1}^y - i\sigma_j^y \sigma_{j+1}^x - 1) \qquad (6.25)$$

on a ring with periodic boundary conditions. The transition probability from σ to σ' is

$$\langle \sigma | e^{-tH} | \sigma' \rangle \qquad (6.26)$$

in the σ^z-representation. It is already correctly normalized as

$$\sum_{\sigma'} \langle \sigma | e^{-tH} | \sigma' \rangle = 1. \qquad (6.27)$$

H is a Heisenberg Hamiltonian with complex couplings. Because of the biased jumps, H is not self-adjoint. The ground states of H (energy zero) are the same as for the ferromagnetic Heisenberg chain, namely with a factorized wave function.

Given our experience with one dimensional spin chains, it is natural to try the Bethe ansatz for the eigenfunctions of (6.25). It is claimed (Dhar 1987) that the dynamical exponent $z = \tfrac{3}{2}$ follows from considering the low lying excitations. No details are available yet.

For the symmetric case $E = 0$ and jump rates satisfying (6.21), (6.22), the scaling form (6.6) is proved by DeMasi et al. (1986). A lower bound on $S(k,t)$ of the form $\exp[-\nu k^2 t]$ follows easily from Jensen's inequality. To actually establish the limit $k \to 0$, $t \to \infty$ with $k^2 t$ fixed is difficult.

6.2 Other one dimensional models

The mapping to driven lattice gases also works for other models. For example, the single step model maps to a lattice gas as above with $\alpha_2 = \alpha_3 = \alpha_4 = \Gamma_+$, $\gamma_2 = \gamma_3 = \gamma_4 = \Gamma_-$. The driven lattice gas corresponding to the PNG model is a little bit more complicated (Krug and Spohn 1989). One has a collection of 'particles' (step down) and 'antiparticles' (step up). The particles move with velocity c, the antiparticles with velocity $-c$. At a collision particle and antiparticle annihilate each other.

The nucleation of steps corresponds to the creation of a pair consisting of a particle and an antiparticle with rate Γ. In the steady state particles and antiparticles have an ideal gas (uniform Poisson) distribution. The conserved quantity is $\phi = \varrho_+ - \varrho_-$, the density of particles minus the density of antiparticles. The steady state current is $j = c(\varrho_+ + \varrho_-)$. Stationarity requires $2c\varrho_+\varrho_- = \Gamma$. Thus the steady current is

$$j(\phi) = c\left(\phi^2 + \frac{2\Gamma}{c}\right)^{1/2}. \qquad (6.28)$$

and $j(0) = (2\Gamma c)^{1/2}$ in agreement with (4.2). As before ϕ is proportional to the average surface inclination. The corresponding growth shape is shown in Figure 3.

A further illuminating example is the Visscher-Bolsterli model for ballistic deposition of discs. The mapping to the lattice gas proceeds as for the crystal edge in Figure 12. Discs are dropped down vertically above randomly chosen positions and are allowed to roll downward along the surface until they reach a local surface minimum. Clearly then the growth rate at a given minimum depends on its basin of attraction. In terms of the lattice gas, the jump rate of a particle at site j, which is surrounded by the local configuration

$$\eta_{j-n+1} = \eta_{j-n+2} = \ldots = \eta_j = 1,$$
$$\eta_{j+1} = \eta_{j+2} = \ldots = \eta_{j+m} = 0 \qquad (6.29)$$

is given by

$$\Gamma_{nm} = \Gamma_0(m + n), \qquad (6.30)$$

where Γ_0 is the number of discs deposited per unit time and unit substrate length. As shown in Section 4, using a more general argument, the peculiar jump rates (6.30) lead to a current (growth velocity) which is *independent* of density (inclination). To see this, we pick an arbitrary configuration η on a ring of N sites and subdivide it into $M(\eta)$ local configurations of type (6.29), containing m_j neighboring particles and n_j holes, respectively, $j = 1, \ldots, M(\eta)$. Then the total current is

$$J(\eta) = \sum_{j=1}^{M} \Gamma_0(m_j + n_j) = \Gamma_0 N \qquad (6.31)$$

independent of η. The steady state corresponding to the rates (6.30) does not have a simple structure, as was erroneously claimed in Krug and Spohn (1989). Inspection of the solution for small rings ($N = 6$) indicates an effectively repulsive interaction between the lattice gas particles.

Finally, we note that the space-time histories of a particular version of an asymmetric lattice gas can be mapped onto the two dimensional six vertex model in *equilibrium* (Kandel and Domany 1990, Lebowitz 1990). Thereby one dimensional growth is related to a two dimensional equilibrium model. The dynamics is given by a parallel, discrete time algorithm. At $t = 0$ we block the particle configuration into pairs of neighboring sites. Each pair is updated independently according to the transition probabilities $T(\eta_1, \eta_2 \mid \eta_1', \eta_2')$. Clearly, by particle number conservation,

$$T(0,0 \mid 0,0) = T(1,1 \mid 1,1) = 1.$$

In the other cases exchanges may occur with probabilities

$$T(1,0 \mid 0,1) = p = 1 - T(1,0 \mid 1,0),$$
$$T(0,1 \mid 1,0) = q = 1 - T(0,1 \mid 0,1),$$

$0 < q, p < 1$. At the next time step the pairing is shifted by one lattice unit and pairs are updated according to the same rule as before. The asymmetric choice $q \neq p$ gives a growth model.

To see the connection to the six vertex model on the square lattice, we choose the diagonal as space axis and the line $x_1 = -x_2$ as time axis. The bonds carry arrows which point either right, left or up, down. Each time slice $x_2 = x_1 + t$ maps onto a lattice gas configuration: Arrows pointing up and left correspond to particles, arrows pointing down and right correspond to holes. Pairs of bonds meet at vertices in the forward time direction. Each vertex corresponds to a 'collision' event in which a particle and a hole can be exchanged. Particle number conservation leads to the ice rule for the allowed vertices: At each vertex two ingoing and two outgoing arrows have to meet. The transition matrix T translates to the Boltzmann weights of the six vertex model as $e^{-\beta \varepsilon_1} = p$, $e^{-\beta \varepsilon_2} = q$, $e^{-\beta \varepsilon_3} = 1$, $e^{-\beta \varepsilon_4} = 1$, $e^{-\beta \varepsilon_5} = 1-q$, $e^{-\beta \varepsilon_6} = 1-p$ (we use the standard labeling, see Lieb and Wu 1972, Baxter 1982). The symmetric choice $q = p$ lies on the boundary between the disordered and ferroelectric phases at zero electric field. The asymmetry induces an external electric field pointing along the space axis. As in the other models, one can still determine the steady state: The occupation variables at different sites are independent, but the steady state has a period of two. Only for $q = p$ does the superstructure vanish.

6.3 Higher dimensions

In $1 + 1$ dimensions we are able to obtain fairly detailed information about the steady state because surface gradients are essentially independent. As an additional bonus the dynamics of the surface gradients can be viewed as a many-particle system with a single conservation law. This certainly helps physical intuition. Although we did not present any details, it also allows one to use theoretical methods as developed in the context of kinetic theory, e.g. mode coupling equations (van Beijeren *et al.* 1985, Krug 1987). Such simplifications are lost in higher dimensions. We have no theoretical result to report here, but we regard it as instructive to understand from a somewhat different perspective why higher dimensions are more difficult.

As a starting point we use the KPZ equation in the real world of $(2 + 1)$ dimensions. The surface gradient $\mathbf{u} = \nabla h$ satisfies

$$\frac{\partial}{\partial t}\mathbf{u}_t + \nabla\left(-\tfrac{1}{2}\lambda\mathbf{u}_t^2 - \nu\nabla\cdot\mathbf{u}_t - \zeta_t\right) = 0. \qquad (6.32)$$

We may regard $\mathbf{u}_t$ as the velocity field of a two dimensional fluid. Instead of the incompressibility condition $\nabla\cdot\mathbf{u}_t = 0$, however it satisfies the potential condition

$$\frac{\partial}{\partial x_1}u_{t,2} = \frac{\partial}{\partial x_2}u_{t,1}. \qquad (6.33)$$

We may interpret (6.32) also as a two-component particle system where the α-current points along the α-direction. Equation (6.33) couples then the two components. To come from a surface the gradients must satisfy an integrability condition: Line integrals over closed loops have to vanish. Even in the linear theory ($\lambda = 0$) this constraint causes a slow decay of the steady state correlations in the surface gradients. The static structure function for the surface is $(\gamma/2\nu)k^{-2}$. The gradients are then correlated in the position space as

$$\langle u_1(x_1, x_2)u_1(0,0)\rangle = \frac{\gamma}{4\pi\nu}\frac{x_2^2 - x_1^2}{(x_1^2 + x_2^2)^2},$$

$$\langle u_1(x_1, x_2)u_2(0,0)\rangle = \frac{\gamma}{2\pi\nu}\frac{x_1 x_2}{(x_1^2 + x_2^2)^2}. \qquad (6.34)$$

The correlations decay as $|\mathbf{x}|^{-2}$ without definite sign.

To illustrate the microscopic gradient dynamics we consider the two-dimensional single step model (Meakin *et al.* 1986b). The height variables $h_t(\mathbf{x})$ at sites $\mathbf{x} = (i,j)$ take only integer values, which are even/odd for $i + j$ even/odd. They also satisfy the single step condition $|h_t(\mathbf{x}) - h_t(\mathbf{y})| = 1$ for $|\mathbf{x} - \mathbf{y}| = 1$. As in one dimension each allowed height configuration can be mapped onto a spin configuration (van Beijeren 1977). To each bond in the dual lattice we give a direction (arrow) in such a way that looking along the arrow the higher point of the surface is to the right. The continuum integrability condition (6.33) translates to the ice rule. There can be only closed loops, i.e. no sinks and sources. Thus at each site of the dual lattice there must be two incoming and two outgoing arrows. The allowed vertices are those of the six vertex model.

The surface grows by filling up local minima with the rate Γ_+: if $h_t(i + 1, j) + h_t(i, j + 1) + h_t(i - 1, j) + h_t(i, j - 1) - 4h_t(i, j) = 4$ (local minimum), then $h_t(i, j)$ increases to $h_t(i, j) + 2$ with rate Γ_+. Correspondingly, we also introduce evaporation processes: if $h_t(i+1, j) + h_t(i, j + 1) + h_t(i - 1, j) + h_t(i, j - 1) - 4h_t(i, j) = -4$ (local maximum), then $h_t(i, j)$ decreases to $h_t(i, j) - 2$ with rate Γ_-. The translation to the six vertex model is straightforward. A local minimum corresponds to a closed four loop (plaquette) oriented counterclockwise. As the height increases, it changes its orientation (with rate Γ_+). The reverse process (with rate Γ_-) corresponds to a height decrease.

If $\Gamma_+ = \Gamma_-$, then in the steady state all allowed configurations are equally likely. This is the situation of an equilibrium surface above the roughening transition. Sutherland (1968) computed the arrow-arrow correlations. Their large distance behavior is as in (6.34). (For the finite temperature six vertex model we refer to Lieb and Wu (1972). It models ferroelectrics and the equilibrium roughening transition, cf. van Beijeren and Nolden 1987.) For the growing surface one has $\Gamma_+ > \Gamma_-$. At present no theoretical method is available for the prediction of steady state correlations. Numerically one finds $\zeta \simeq 0.37$ (Meakin *et al.* 1986b, Liu and Plischke 1988), corresponding to a static height structure function $S(k) \sim |k|^{-2.74}$ for small k and hence a decay of the arrow-arrow correlation as $|\mathbf{x}|^{-1.26}$. Thus, in contrast to the driven one dimensional lattice gases, the steady state correlations depend on the bias. The steady states for $\Gamma_+ \neq \Gamma_-$ should all be in the same universality class with a crossover to the equilibrium behavior as $\Gamma_+ \to \Gamma_-$.

The difference between the one dimensional and the two (and higher) dimensional cases can be traced back to a basic topological property: In one dimension the number of local maxima is equal to the number of local minima for *any* surface configuration. This property, which is no longer true for higher dimensions, is the reason for the simplicity of the steady states in $d = 1$.

6.4 Shock fluctuations

Shock formation and shock propagation are traditional subjects of fluid dynamics. As in growth processes there is a traveling front. We may inquire then about fluctuations in the position of the front. Properly speaking, our investigation should be carried out in the context of nonlinear hydrodynamics with fluctuating currents. To our knowledge, such a program has not been achieved. Instead we discuss a simplified model. Our main aim is yet another illustration of the theory developed.

As in the previous sections we consider driven lattice gases, for simplicity the infinite temperature case. The lattice gas is now in d dimensions and the driving force points along the positive 1-axis. Subject to the constraint of single occupancy, the jump rates are $1 \pm \alpha$ along the ± 1-axis, $0 < \alpha \le 1$, and 1 along all other directions. Before we were in $d = 1$ with a uniform density of particles. For a shock we need an inhomogeneous density distribution. We impose therefore an average density, ϱ, where $\varrho(x) = \varrho_-$ for $x_1 < 0$ and $\varrho(x) = \varrho_+$ for $x_1 \ge 0$, $\varrho_- < \varrho_+$. For visualization, the extreme case $\varrho_+ = 1$ is helpful. Since the particles in the right half space are stuck, the particles in the left half space are pushed against a solid and pile up. With fluctuations the front moves to the left. Our usual picture of a stable phase growing at the expense of an unstable phase is not applicable here. Rather we have two stable, spatially homogeneous steady states separated by a front moving through external driving.

Without further information, we would expect naïvely the front to smear out diffusively. Certainly this happens when particles jump independently. However, the interaction due to the single occupancy constraint stabilizes the front. We will argue that the shock fluctuations are governed by the linear ($\lambda = 0$) KPZ equation. In particular, in three dimensions this implies that the front has only logarithmic fluctuations.

The steady state current of the lattice gas is

$$\mathbf{j}(\varrho) = 2\alpha\varrho(1 - \varrho)\mathbf{e_1}. \qquad (6.35)$$

If we assume local equilibrium, then the Euler equation for the density reads

$$\frac{\partial}{\partial t}\varrho_t + \frac{\partial}{\partial x_1}\left[2\alpha\varrho_t(1-\varrho_t)\right] = 0. \tag{6.36}$$

(Note that the density is the only locally conserved field.) Equation (6.36) is a text book example of a nonlinear hyperbolic equation leading to shock formation: even if the initial data are smooth, the solution of (6.36) may develop discontinuities. To determine the location of the shock (6.36) has to be supplemented then by the entropy condition (for an introduction see Chorin and Marsden 1979). For the initial condition from above the solution to (6.36) is a sharp shock which travels with velocity

$$v_s = 2\alpha(1 - \varrho_+ - \varrho_-). \tag{6.37}$$

If $\varrho_- + \varrho_+ > 1$, then particles pile up and the shock moves to the left, whereas for $\varrho_+ + \varrho_- < 1$ the shock travels to the right. Phenomenologically it is natural to add a viscosity term, as $\nu\Delta\varrho_t$, to (6.36). This smears out the shock over a length $\sqrt{\nu/\alpha}$. The shock velocity remains unchanged.

Since the dynamics is local, we can use the KPZ equation as the large scale theory. To determine λ the inclination dependent growth velocity is then needed. If one solves (6.36) with an initial step along a plane tilted relative to the $\{x_1 = 0\}$ plane, then the growth velocity along the 1-axis is again v_s, also if the viscosity term $\nu\Delta\varrho_t$ is added. Even without computation, this result follows from mass conservation together with the fact that the densities away from the shock are determined by the initial conditions and do not change in time. We conclude that $\lambda = 0$ and that the linear theory governs the shock fluctuations. (The one-dimensional problem is studied by Boldrighini *et al.* 1989, Gärtner and Presutti 1990, Ferrari *et al.* 1991.)

7

Growth and percolation

7.1 First passage percolation

The connection between growth processes and percolation has been cultivated mostly in the probabilistic camp. Besides proving some basic properties, like the existence of an asymptotic shape, this approach could deepen our understanding since growth is viewed as a sort of optimization problem. Somewhat unexpected, we will find a close relation to directed polymers.

In standard percolation on the simple hypercubic lattice $\mathbf{Z}^d$ a bond is open (occupied, present) with probability p and closed (vacant, absent) with probability $(1-p)$. (We restrict our attention to bond percolation with independent bond probabilities. There is also the essentially equivalent site percolation.) One investigates such problems as the size of a connected cluster containing the origin, the probability for the origin to be connected to infinity, etc. To a large extent the interest in percolation stems from the theory of second order phase transitions (Stauffer 1985).

The term 'percolation' signals the picture of a fluid being pushed into solid material (like a filter or a rock). The fluid crosses a given bond if open. With this background it is natural to assume that it takes some time for the fluid to spread across a bond. To model such a physical situation we assign, independently, to each bond b, $b = \{\mathbf{x}, \mathbf{y}\}$, $|\mathbf{x} - \mathbf{y}| = 1$, a random variable, τ_b, $\tau_b \geq 0$. τ_b is the amount of time the fluid needs to cross b. A path, w, in our lattice is a sequence of connected bonds. The *passage time* from $\mathbf{x}$ to $\mathbf{y}$ along the path w, w starts at $\mathbf{x}$ and ends at $\mathbf{y}$, is then given by

$$\tau_{\mathbf{x}\mathbf{y}}(w) = \sum_{b \in w} \tau_b. \tag{7.1}$$

The fluid is injected at the origin. One would like to know the time at which a certain site is first reached by the fluid. Therefore we define the *first passage time* from $\mathbf{0}$ to $\mathbf{x}$ by

$$\tau_{\mathbf{x}} = \min\left\{\tau_{\mathbf{0}\mathbf{x}}(w) \mid w \text{ is a path from } \mathbf{0} \text{ to } \mathbf{x}\right\}. \tag{7.2}$$

First passage percolation studies the asymptotics of the first passage times. Besides the point to point first passage, the point to line first passage is also of interest. We characterize a hyperplane, $H(\mathbf{r})$, through a vector $\mathbf{r}$: The plane is orthogonal to $\mathbf{r}$ and its smallest distance from the origin is $r = |\mathbf{r}|$. The first passage time from the origin to the hyperplane $H(\mathbf{r})$ is then

$$\tau_{\mathbf{r}} = \min\Big\{ \tau_{\mathbf{x}} \mid \mathbf{x} \text{ lies beyond } H(\mathbf{r}) \text{ as seen from the origin} \Big\}. \quad (7.3)$$

Standard percolation can be understood as the particular case where $\tau_b = 1$ with probability p and $\tau_b = \infty$ with probability $(1 - p)$.

The connection to growth is immediate. We simply consider the set of sites which are reached by the fluid at time t. This is a random set $A_t \subset \mathbf{Z}^d$ with $A_0 = \{0\}$. Clearly A_t is defined by

$$A_t = \big\{ \mathbf{x} \mid \tau_{\mathbf{x}} \le t \big\}. \quad (7.4)$$

A_t is a cluster growing from a single seed.

It may help to illustrate the connection of first passage percolation and growth by a simple example. Let us consider a random walk, on a one dimensional lattice, starting at the origin. We may study then the transition probability $p_t(j)$ to find the walker at site j at time t. This corresponds to growth. On the other side we may introduce a boundary at L and ask for the time the walker first hits the boundary. This is the first passage problem. To amplify even further: For the one dimensional lattice let τ_b have an exponential distribution with mean α. A_t is then simply an interval expanding at both ends. The probability for the edge to be at site j at time t is $p_t(j) = (1/j!)(t/\alpha)^j \, e^{-t/\alpha}$. In particular the growth velocity is $1/\alpha$. On the other hand the first passage time to the site j has the distribution $(1/j\alpha) \, e^{-t/j\alpha}$, because the minimum of a sum of independent exponentials is again exponential. As a consequence, $(1/j)\tau_j \to \alpha$ as $j \to \infty$.

Actually, we have introduced already two growth models defined through first passage percolation.

(i) Let τ_b have an exponential distribution with mean 1. This gives the bond version of the Eden model of the Introduction: A perimeter site becomes part of the growing cluster with a rate equal to the number of nearest neighbor sites already belonging to the cluster.

(ii) Let τ_b have the discrete distribution $\sum_{n=1}^{\infty} p(1 - p)^{n-1}\delta(t - n)$. This is the Richardson model (discrete time version of the Eden model).

In one time step (here set equal to one) an already infected site infects a neighboring site with probability p and does not infect it with probability $(1 - p)$.

Some rigorous results

It is impossible to do justice to a beautiful probabilistic development. Good reviews are available (Smythe and Wierman 1978, Kesten 1986, Durrett 1988a, b). We 'explain' only a few results of relevance in our context.

The most basic result is the existence of an asymptotic cluster shape, in the sense that the scaled down random set $(1/t)A_t = \{(1/t)\mathbf{x} \mid \mathbf{x} \in A_t\}$ tends to a deterministic limit, $\overline{A}$, with probability one as $t \to \infty$. Of course, $\overline{A}$ is the macroscopic form discussed in Section 2. Let p_c be the critical bond percolation probability. If $\text{Prob}(\{\tau_b = 0\}) < p_c$, then $\overline{A}$ is a compact and convex subset of $\mathbf{R}^d$. (Otherwise $\overline{A}$ would be all of $\mathbf{R}^d$. For the behavior close to p_c cf. Chayes *et al.* (1986).) One also knows that $\overline{A}$ depends continuously on τ_b. This means the following: Let $F(t)$ be the distribution function for τ_b, $F(t) = \text{Prob}(\{\tau_b \leq t\})$. The first passage percolation with distribution $F(t)$ defines a macroscopic shape $\overline{A}$. If $F_n(t) \to F(t)$ as $n \to \infty$ with the exception of the jump points of F, then also the corresponding shapes converge. Note that no recipe to compute $\overline{A}$ is given.

The strategy of the proof is similar to the existence of the free energy for systems in equilibrium. The first passage time from the origin to a plane is subadditive, which ensures the existence of the limit

$$\lim_{\lambda \to \infty} \frac{1}{\lambda r} \tau_{\lambda \mathbf{r}} = c\left(\frac{\mathbf{r}}{r}\right) \qquad (7.5)$$

with probability one. Thus the first passage time scales linearly with the distance of the plane from the origin. $c(\mathbf{r}/r)$ is called the time constant. It depends on the orientation of the plane. As in our one dimensional example above, the time constant is inverse to the growth velocity,

$$c\left(\frac{\mathbf{r}}{r}\right)v\left(\frac{\mathbf{r}}{r}\right) = 1. \qquad (7.6)$$

Using (7.5) the macroscopic shape $\overline{A}$ is then built up from planes.

Very little is known about the shape on a rigorous level. One has conditions on the distribution of τ_b, which imply that the form has flat pieces (Kesten 1986). One example is the Richardson model, which will be discussed in detail below. Another result concerns the asymptotics in large dimensions. For the continuous time Eden model we consider the growth along one of the lattice axes, say $e_1 = (1, 0, \ldots, 0)$. Lower and upper bounds on the growth velocity imply

$$v(e_1) \approx \frac{2d}{\log d} \qquad (7.7)$$

for large d (Dhar 1988).

Fluctuations

For first passage percolation the macroscopic object is the time constant $c(\mathbf{r}/r)$ of (7.5) and, associated with it, the Wulff constructed shape $\bar{A}$. Two fluctuating quantities are of obvious interest: (1) $c(\mathbf{r}/r)$ gives the mean first passage time, $\langle \tau_\mathbf{r} \rangle \approx rc(\mathbf{r}/r)$ for large r. What is then the typical width of the first passage time distribution? (2) Instead of the time we may also consider the *location* of the first passage in the plane $H(\mathbf{r})$. Denoting this location by $\mathbf{x}(\mathbf{r}) \in H(\mathbf{r})$, how does $\mathbf{x}(\mathbf{r})$ then scatter typically around its average?

We blow up $\bar{A}$ self similarly as $t\bar{A}$. Let $\mathbf{x}_0$ be the point of first contact with $H(\mathbf{r})$ and let t_0 be the time of contact. Then $\langle \tau_\mathbf{r} \rangle \simeq t_0$ and $\langle \mathbf{x}(\mathbf{r}) \rangle \simeq \mathbf{x}_0$ for sufficiently large r. We consider the surface of the cluster A_t at times slightly less than t_0 and close to $\mathbf{x}_0$. Let h_t be the height of the surface measured relative to the hyperplane $H(\mathbf{r})$. We decompose h_t into a deterministic part and a fluctuating part, denoted by $\tilde{h}_t$. Then $h_t(\mathbf{x}) = v(\mathbf{r}/r)(t - t_0) - \lambda(\mathbf{x} - \mathbf{x}_0)^2/2t + \tilde{h}_t(\mathbf{x})$, where we assumed that $\bar{A}$ has a nonvanishing curvature λ at $\mathbf{x}_0/t_0$. From the general scaling theory we know that $\langle \tilde{h}_t(\mathbf{x})^2 \rangle^{1/2}$ grows as $t^{\zeta/z}$. The cluster surface crosses the plane $H(\mathbf{r})$ with a finite velocity. Therefore the fluctuations in $\tilde{h}_t(\mathbf{x})$ translate linearly to fluctuations in $\tau_\mathbf{r}$ and

$$\left\langle \left[\tau_\mathbf{r} - \langle \tau_\mathbf{r} \rangle \right]^2 \right\rangle^{1/2} \sim r^{\zeta/z}. \qquad (7.8)$$

Typical scatters of $\mathbf{x}(\mathbf{r})$ originate in the events where the fluctuations in the surface just reach $H(\mathbf{r})$. Thus

$$\left\langle \left[\mathbf{x}(\mathbf{r}) - \langle \mathbf{x}(\mathbf{r}) \rangle \right]^2 \right\rangle \sim t \langle \tilde{h}_t(\mathbf{x})^2 \rangle^{1/2} \sim t^{1+\zeta/z} = t^{2/z}, \qquad (7.9)$$

where in the last step we used the scaling relation (3.18). For example in two dimensions the first passage time distribution has a width of the order $r^{1/3}$ whereas the first passage location scatters as $r^{2/3}$.

Directed first passage percolation

In directed first passage percolation the fluid spreads across a bond only along a preassigned direction. Again for simplicity, let us only consider the two dimensional case. In the first quadrant all bonds are positively oriented. We study the first passage from the origin to the line $\{x_1 + x_2 = n\}$. The reader will have noticed immediately that we have described nothing else than the ground state problem of a directed polymer. The walks are directed along the (1,1) direction. Each walk has n steps. Adding the passage times, τ_b, along the walk corresponds to adding up the random potentials V_b. The minimal energy is the first passage time. The time constant is the ground state energy per length. As for undirected first passage percolation the growing cluster is the set of sites reached by the fluid at time t. This cluster is described by the standard theory. No surprise then that the fluctuations of directed first passage percolation are again governed by (7.8), (7.9), which of course coincide with (5.45) and (5.47). (Large d bounds for the time constant are proved by Cox and Durrett 1983.)

7.2 Facets and directed percolation

We noted in Section 4 that several synchronous growth models exhibit a faceting transition. Here we study this transition in more detail, using as an example the two dimensional Richardson model (Richardson 1973). Time runs in discrete steps, $t = 0, 1, 2, \ldots$. Initially the cluster consists only of the origin. A perimeter site at time t becomes part of the cluster at time $t + 1$ with probability p. By symmetry it suffices to consider only the cluster in the upper right quadrant of the square lattice. Clearly the growth velocity v_1 along the diagonal cannot exceed $v_{\max} = 1/\sqrt{2}$ in units of the lattice spacing. For small p, $v_1(p)$ is linear in p. At some critical value p_c, $v_1(p_c) = v_{\max}$ and remains constant up to $p = 1$. For $p > p_c$ there is a sector enclosing the diagonal in which the cluster edge is faceted and propagates at the maximal speed $1/\sqrt{2}$. At $p = 1$ the sector contains the whole quadrant and the cluster forms a diamond. The surprise is to have facets already at values of p with $p_c < p < 1$.

This can be understood through a mapping to directed percolation (Durrett and Liggett 1981, Savit and Ziff 1985, Kertész and Wolf 1989, Krug *et al.* 1990). We draw the line $x_1 + x_2 = t$ which marks the range of influence of the origin. The cluster at time t cannot extend beyond this line. We record all sites of the cluster located on the line at time t and call this set B_t. B_t is determined through a graphical rule: Each site of the quadrant is open with probability p and closed with probability $(1 - p)$, just as in ordinary (site) percolation. However, we orient the bonds positively in the direction of the diagonal. A site $\mathbf{x}$ in $\{x_1 + x_2 = t\}$ belongs to B_t if $\mathbf{x}$ is connected to the origin by a path of adjacent open sites and bonds respecting the orientation. This is the so-called directed percolation problem, in our case the symmetric site version (for an introduction see Kinzel 1983). For $p < p_c$, there are too many closed sites and the set B_t will be empty for large t. The infection spreading from the origin dies out. This means that $v_1 < 1/\sqrt{2}$. However, for $p > p_c$ the infection survives and B_t is a set expanding linearly, of course with some holes. Close to the diagonal the surface sticks then to the line $\{x_1 + x_2 = t\}$ with random excursions of a few lattice spacings. The surface width is O(1) and purely intrinsic, cf. Section 3.2. In two dimensions the percolation threshold is at $p_c \simeq 0.705489$ (Essam *et al.* 1988).

To proceed we need some concepts from directed percolation theory. We focus first on the *subcritical case*, $p < p_c$. The infected set B_t has a typical survival time ξ_t and a typical (maximal) spatial extension ξ_r. Both lengths diverge as $p \rightarrow p_c$, defining the correlation length exponents ν_t and ν_r,

$$\xi_t \sim |p - p_c|^{-\nu_t}, \quad \xi_r \sim |p - p_c|^{-\nu_r} \tag{7.10}$$

with $\nu_t > \nu_r$. In two dimensions $\nu_t \simeq 1.733$ and $\nu_r \simeq 1.097$ (Essam *et al.* 1988). When viewed as a cluster on the square lattice, the infection history is club-shaped with an extension ξ_t along the diagonal and ξ_r perpendicular to it. As $p \rightarrow p_c$ typical clusters become increasingly elongated with an opening angle φ_0 relative to the diagonal, $\varphi_0 \sim \xi_r/\xi_t \sim |p - p_c|^{\nu_t - \nu_r}$. Let us now choose a ray forming some angle φ with the diagonal and ask for the typical extension $\xi(\varphi)$ of the directed percolation cluster along that ray. Obviously $\xi(\varphi) = 0$ for $\varphi > \varphi_0$ and $\xi(0) = \xi_t$. This suggests the scaling form

$$\xi(\varphi) = \xi_t f\left(\frac{\varphi}{\varphi_0}\right), \tag{7.11}$$

where $f(x) = 0$ for $x > 1$, $f(0) = 1$, and $f(x)$ has a quadratic maximum at $x = 0$, $\xi(\varphi) \approx \xi_t(1 - a(\xi_t/\xi_r)^2\varphi^2)$ for $\varphi \ll \varphi_0$.

To see how (7.11) relates to the growth shape of the Richardson cluster (not to be confused with the directed percolation clusters), we put a coordinate axis through the origin perpendicular to the diagonal and consider the scaled shape function $g(y)$ relative to this axis ($g(y)$ was defined in Equation (2.3)). Close to $\varphi = 0$ we have $y \approx g(0)\varphi$ ($g(0) = v_1$). For a fixed angle φ, the distance between the cluster edge and the line $\{x_1 + x_2 = t\}$ is $t(v_{\max} - g(y))$ at time t. This distance becomes observable (of the order of a few lattice spacings) after a time of the order $\xi(\varphi)$, hence

$$v_{\max} - g(y) \sim \xi(\varphi)^{-1}. \tag{7.12}$$

Inserting the expansion of (7.11) for $\varphi \ll \varphi_0$ it follows that $v_{\max} - v_1 \sim \xi_t^{-1}$ and that the curvature of the cluster edge vanishes as

$$g''(0) \sim -\frac{\xi_t}{\xi_r^2} \sim -(p_c - p)^{2\nu_r - \nu_t}. \tag{7.13}$$

At the critical point $p = p_c$, $\xi(\varphi)$ is finite for any $\varphi \neq 0$. The scaling form (7.11) then requires $\xi(\varphi)$ to diverge for $\varphi \to 0$ as $\xi(\varphi) \sim \varphi^{-\nu_t/(\nu_t - \nu_r)}$, and therefore using (7.12) we obtain the singular growth shape

$$g(0) - g(y) \sim |y|^{\nu_t/(\nu_t - \nu_r)}, \quad p = p_c. \tag{7.14}$$

The above considerations apply equally well to d-dimensional surfaces, corresponding to $(d+1)$-dimensional directed percolation, although the exponents ν_t and ν_r depend on dimension of course. The upper critical dimension of directed percolation is $d_c + 1 = 5$. For $d \geq 4$, $\nu_t = 1$ and $\nu_r = \frac{1}{2}$ independent of d, while $\nu_t/\nu_r < 2$ for $d < 4$ (Kinzel 1983). In the mean field regime $d > 4$, the power law (7.13) is replaced by a logarithmic behavior, $g''(0) \sim 1/\log(p_c - p)$, as will be shown explicitly in Section 8.

The critical behavior of the inclination dependent growth velocity $v(u)$ is obtained by simply inverting the Legendre transform (2.6). From (2.7) we conclude that the curvature (and hence the coupling constant λ in the KPZ equation, cf. (5.11)) diverges for $p \to p_c$ as $v''(0) \sim (p_c - p)^{\nu_t - 2\nu_r}$, and logarithmically for $d > 4$. At criticality $v(u)$ is of the singular form (2.12), $v(u) - v(0) \sim |u|^\alpha$ with

$$\alpha = \frac{\nu_t}{\nu_r}. \tag{7.15}$$

Via (3.20) this determines the relation between the exponents ζ and z of the shape fluctuations. The dynamic exponent z can be obtained by noting that at p_c the spatial spread of the infection (conditioned on survival) grows with time as $\xi_r(t) \sim t^{\nu_r/\nu_t}$. This is identified with the surface correlation length $\xi_\parallel(t)$ in Equation (3.6) and leads to (Kertész and Wolf 1989)

$$z = \frac{\nu_t}{\nu_r}. \tag{7.16}$$

From (7.15) and (7.16) we conclude, using (3.20), that

$$\zeta = 0 \tag{7.17}$$

independent of the surface dimension. Numerical simulations at p_c indicate that the surface width increases logarithmically, $W(t) \sim (\log t)^{\zeta'}$, where $\zeta' \simeq 0.4 - 0.5$ in two dimensions (Kertész and Wolf 1989, Krug and Meakin 1991).

In the *supercritical regime* $(p > p_c)$ the set B_t of infected sites spreads linearly at some speed $c(p)$. The size of the facet of the Richardson cluster is $2ct$ at time t, so the leading behavior of the growth velocity is $v(u) \approx v_{\max} + c|u|$ (see Equation (2.10)). The exponent in (2.12) is $\alpha = 1$, hence the scaling relation (3.20) predicts that fluctuations spread at a finite velocity, $z = 1$ (Krug and Spohn 1988). As $p \to p_c$ from above, c vanishes as $(p-p_c)^{\nu_t - \nu_r}$ (Kinzel 1983). The directed percolation clusters extend to infinity within the angle $\varphi_c(p) = \arctan(c(p))$ from the diagonal (the 'percolation cone'). Thus the direction dependent correlation length $\xi(\varphi)$ is infinite for $\varphi < \varphi_c$ and finite for $\varphi > \varphi_c$. The behavior of the shape function $g(y)$ close to the facet can be obtained from (7.12), once we know how $\xi(\varphi)$ diverges as $\varphi \to \varphi_c$ from above. We define the *supercritical correlation length exponent* ν through

$$\xi(\varphi) \sim (\varphi - \varphi_c)^{-\nu}. \tag{7.18}$$

This exponent is related to the fuzziness of the boundary of the set B_t (Krug *et al.* 1990). We pick a ray at an angle $\varphi = \varphi_c + \Delta\varphi$ outside the percolation cone. At time t the distance between the ray and the cone boundary is $t\Delta\varphi$. For $t \sim \xi(\varphi)$ this distance is comparable to the width $W_B(t)$ of the boundary. To determine $W_B(t)$, we note that the growth of the supercritical cluster B_t is yet another example of local growth. Hence its edge fluctuations are governed by the general theory of a $(d-1)$-dimensional growing surface (recall that d is the surface dimension of the Richardson cluster) and $W_B(t) \sim t^{\zeta_{d-1}/z_{d-1}}$. We conclude then that

$$\nu_d = \frac{z_{d-1}}{z_{d-1} - \zeta_{d-1}}. \tag{7.19}$$

In particular, $\nu_1 = 2$ and $\nu_2 = \frac{3}{2}$. The result for $d = 1$ has been noted previously (Grassberger and de la Torre 1979, Domany and Kinzel 1981, Harms and Straley 1982). It should be emphasized that the exponent ν arises from purely kinetic considerations and is in no way related to the critical point of directed percolation.

Using (7.12), Equation (7.18) is rewritten in terms of the shape function $g(y)$ as

$$g(c) - g(c + \epsilon) \sim \epsilon^\nu \qquad (7.20)$$

(recall that the infection speed c determines the location of the facet). Comparing with (2.10) we obtain the exponent δ of the next to leading term of the growth velocity $v(u) = v_{\max} + c|u| + O(|u|^\delta)$,

$$\delta = \frac{\nu}{\nu - 1} = \frac{z_{d-1}}{\zeta_{d-1}}. \qquad (7.21)$$

There are several surprising features of (7.20) and (7.21). Firstly, the shape fluctuations of a $(d - 1)$-dimensional surface show up in the d-dimensional macroscopic cluster shape. Secondly, the result $\nu = \frac{3}{2}$ for a three dimensional cluster $(d = 2)$ happens to coincide with the behavior of equilibrium crystals below the roughening temperature (van Beijeren and Nolden 1987). Needless to say, the $\frac{3}{2}$-power law has a totally different origin in that case. Thirdly, the correction exponent δ takes integer values ($\delta = 2$ and 3 resp.) in the exactly solved cases $d = 1$ and 2. If one could show generally that only integer powers of $|u|$ appear in the expansion of $v(u)$, this would lend support to the conjecture that z/ζ is an integer for any d (Wolf and Kertész 1987b, Kim and Kosterlitz 1989, Zhang 1989). Finally we note that in high dimensions one expects $\zeta \to 0$, $z \to 2$ and hence $\delta \to \infty$, $\nu \to 1$. This is supported by the mean field calculation in Section 8, but it contradicts the Cayley tree result of Harms and Straley (1982), who find $\nu = 2$.

8

An approximation of mean field type

One of the most useful concepts in equilibrium statistical mechanics is the mean field approximation. Of course, we now understand that fine details, such as the critical exponents at a second order phase transition,

cannot be correctly predicted by this method. But the overall phase
diagram, the free energy and susceptibilities are reproduced qualitatively
by mean field theory, if applied with the appropriate caution.

We want to explain that for growth processes there is an approximation in a similar spirit. Mean field type approaches have repeatedly
appeared in the literature, both in discrete (Bensimon *et al.* 1984b,
Savit and Ziff 1985, Cheng *et al.* 1987) and continuum (Nauenberg
1983, Nauenberg *et al.* 1983, Ball *et al.* 1984, Parisi and Zhang 1985)
formulations. However it has not been commonly recognized what kind
of useful information these theories contain, and how it can be extracted.
Here we give a general treatment and apply our method to the problem
of growth shapes in the Eden and Richardson models. No information
about kinetic roughening is obtained, since the exponent ζ/z always
takes its $d \to \infty$ value of zero. The mean field approach to ballistic deposition (Bensimon *et al.* 1984b) is discussed in detail elsewhere (Krug
and Meakin 1991).

As an explanatory example we choose the continuous time Eden
model in its bond version. This time we need a little bit of notation.
We let $\eta_{\mathbf{x}}$ be the occupation variable at the site $\mathbf{x} \in \mathbf{Z}^d$, $\eta_{\mathbf{x}} = 0$ if site
$\mathbf{x}$ is vacant and $\eta_{\mathbf{x}} = 1$ if site $\mathbf{x}$ is occupied. In the course of time $\eta_{\mathbf{x}}$
will change from zero to one (the reverse process is forbidden). Such an
event happens with rate $c_{\mathbf{x}}(\eta)$, which is proportional to the number of
'infected' neighbors, i.e.

$$c_{\mathbf{x}}(\eta) = (1 - \eta_{\mathbf{x}}) \sum_{\mathbf{e}, |\mathbf{e}|=1} \eta_{\mathbf{x}+\mathbf{e}}, \qquad (8.1)$$

where for simplicity we fixed the time scale. The master equation
then reads

$$\frac{\mathrm{d}}{\mathrm{d}t} f_t(\eta) = L f_t(\eta) \qquad (8.2)$$

with

$$L f(\eta) = \sum_{\mathbf{x}} c_{\mathbf{x}}(\eta) [f(\eta^{\mathbf{x}}) - f(\eta)]. \qquad (8.3)$$

Here $\eta^{\mathbf{x}}$ is the configuration η with the occupancy at $\mathbf{x}$ changed from
zero to one. The formal solution to (8.2), $e^{Lt}(\eta, \eta')$, is the probability
to have the configuration η' at time t given the initial configuration η.
Now the average occupation is governed by

$$\frac{d}{dt}\langle \eta_x \rangle_t = \langle L\eta_x \rangle_t \tag{8.4}$$

$$= \Big\langle (1 - \eta_x) \sum_{\mathbf{e}, |\mathbf{e}|=1} \eta_{\mathbf{x+e}} \Big\rangle_t.$$

The mean field approximation consists in neglecting correlations on the right hand side. If we define $\tilde{\varrho}_t(\mathbf{x}) = \langle \eta_x \rangle_t$, then, in this approximation,

$$\frac{\partial}{\partial t} \tilde{\varrho}_t(\mathbf{x}) = (1 - \tilde{\varrho}_t(\mathbf{x})) \sum_{\mathbf{e}, |\mathbf{e}|=1} \tilde{\varrho}_t(\mathbf{x} + \mathbf{e}). \tag{8.5}$$

A representative initial condition for (8.5) is $\tilde{\varrho}_0(\mathbf{0}) = 1$ and $\tilde{\varrho}_0(\mathbf{x}) = 0$ for $\mathbf{x} \neq 0$. We are interested in the macroscopic shape. There is no easy way to solve the nonlinear equation (8.5). However, we really need only the inclination dependent macroscopic growth velocity. If the growth direction is characterized by the unit vector $\mathbf{n}$, then the appropriate initial condition for (8.5) is $\tilde{\varrho}_0(\mathbf{x}) = 1$ for $\mathbf{n} \cdot \mathbf{x} < 0$ and $\tilde{\varrho}_0(\mathbf{x}) = 0$ for $\mathbf{n} \cdot \mathbf{x} > 0$. With the solution ansatz $\tilde{\varrho}_t(\mathbf{x}) = \varrho_t(\mathbf{n} \cdot \mathbf{x})$, Equation (8.5) reduces to the *one dimensional* equation

$$\frac{\partial}{\partial t} \varrho_t(x) = (1 - \varrho_t(x)) \sum_{\mathbf{e}, |\mathbf{e}|=1} \varrho_t(x + \mathbf{n} \cdot \mathbf{e}). \tag{8.6}$$

(Properly speaking, we should take for $\mathbf{n}$ a vector with integer entries and produce the general case through approximation.) We have to solve (8.6) with the initial condition $\varrho_0(x) = 1$ for $x < 0$, $\varrho_0(x) = 0$ for $x \geq 0$. Physically we expect as solution a front traveling with velocity $v(\mathbf{n})$ for large t. This leads to a:

Digression on traveling fronts and minimal speed

The simplest and best understood equation with a traveling front is the Fisher-Kolmogorov equation (R.A. Fisher 1937, Kolmogorov *et al.* 1937). One version of it reads

$$\frac{\partial}{\partial t} \varrho_t = \frac{\partial^2}{\partial x^2} \varrho_t + \varrho_t(1 - \varrho_t) \tag{8.7}$$

with initial condition $\varrho_0(x) = 1$ for $x \leq 0$ and $\varrho_0(x) \to 0$ for $x \to \infty$. To obtain a traveling front we make the ansatz

$$\varrho_t(x) = w(x - ct). \tag{8.8}$$

w satisfies then

$$w'' + cw' + w(1 - w) = 0. \qquad (8.9)$$

We interpret (8.9) as the equation of motion of a mechanical particle with friction coefficient c rolling down the potential hill $\frac{1}{2}w^2(1 - \frac{2}{3}w)$. The boundary conditions are $w(-\infty) = 1$ and $w(\infty) = 0$. Clearly (8.9) does not fix c. If we assume an exponential decay as e^{-qx} for $x \to \infty$, then the corresponding velocity is

$$c(q) = q + \frac{1}{q}. \qquad (8.10)$$

Thus any $c(q) \geq 2$ is allowed. For $c < 2$ the solution to (8.9) overshoots at $w = 0$ and becomes negative, which is not admissible. Note that the front is exponentially sharp, i.e. $\zeta/z = 0$. A more careful analysis of the Fisher-Kolmogorov equation (Aronson and Weinberger 1978, Bramson 1983, 1987) shows that if the initial density $\varrho_0(x)$ decays as e^{-qx} for large x, then it travels with speed $c(q)$ for large t provided $q \leq 1$. For $q > 1$ the speed is always the minimal speed

$$c^* = \min_q c(q). \qquad (8.11)$$

In particular, an initial step travels with speed c^*. The mechanism behind is not difficult to understand. $\varrho_t = 0$ is an unstable solution of Equation (8.7), $\varrho_t = 1$ is stable. The slower the decay at infinity, the more effectively $\varrho_t(x)$ is broken away from zero and the faster the solution travels.

We have to issue one word of caution. Let us consider the Fisher-Kolmogorov equation with some different nonlinearity, say

$$\frac{\partial}{\partial t}\varrho_t = \frac{\partial^2}{\partial x^2}\varrho_t + (\lambda\varrho_t + \varrho_t{}^2)(1 - \varrho_t) \qquad (8.12)$$

with step initial conditions. From the large x decay, as above, the minimal speed is $c^*(\lambda) = 2\sqrt{\lambda}$. This result is valid however only for $\lambda \geq \lambda_c = \frac{1}{2}$. For $\lambda < \lambda_c$ the asymptotic velocity depends on the full steady solution. It no longer suffices to consider only the right tail. For our particular example $c^*(\lambda) = \sqrt{2}(\lambda + \frac{1}{2})$ for $0 \leq \lambda \leq \lambda_c$ (Ben-Jacob *et al.* 1985, van Saarloos 1989). To deal with this kind of situation the authors propose a principle of marginal stability (see also Dee and Langer 1983, Langer 1987, van Saarloos 1988, 1989).

Let us return to (8.6). Assuming a traveling front solution $w(x - c(q)t)$ with exponential decay e^{-qx} as $x \to \infty$ gives the direction dependent growth velocity

$$v(\mathbf{n}) = \min_q \left\{ \frac{1}{q} \sum_{\mathbf{e}, |\mathbf{e}| = 1} e^{-q \mathbf{n} \cdot \mathbf{e}} \right\}. \tag{8.13}$$

We arrived at the following recipe: Ignoring correlations one writes down the evolution equation for the average density. This could also be a discrete time iteration. The exponential ansatz yields then the direction dependent growth as the solution of a variational problem. As in Equation (8.12) it may happen however that the variational ansatz is valid only in a restricted range of parameters.

8.1 Shape anisotropy for the Eden model

As a first application of (8.13) we wish to compare the growth velocity along the lattice axis $(\mathbf{n} = (1, 0, \ldots, 0))$ and along the diagonal $(\mathbf{n} = 1/\sqrt{d}(1, 1, ..., 1))$ for the Eden model on a d-dimensional hypercubic lattice. The velocity is given by

$$v_0(d) = \min_{q > 0} \frac{2}{q} (\cosh q + d - 1) \tag{8.14}$$

for the lattice axis, and by

$$v_1 = \min_{q > 0} \frac{2\sqrt{d}}{q} \cosh q \tag{8.15}$$

for the diagonal. In arriving at (8.15) we have made the substitution $q \mapsto q\sqrt{d}$ in (8.13). It follows from (8.15) that the d-dependence of v_1 is trivial, $v_1 = \text{const} \cdot \sqrt{d}$. Taking the derivative of the right hand sides of (8.14) and (8.15) with respect to q we obtain the implicit equations

$$f(v_0) = 2d - 2 \tag{8.16}$$

and

$$f\left(\frac{v_1}{\sqrt{d}}\right) = 0, \tag{8.17}$$

where

$$f(x) := x \operatorname{arcsinh}(\tfrac{1}{2}x) - \sqrt{x^2 + 4}. \tag{8.18}$$

The solution of (8.17) is

$$v_1 \simeq 3.0177591\sqrt{d}. \tag{8.19}$$

Equation (8.16) can be solved analytically in the limit of large d. Using the asymptotics of (8.18), $f(x) \sim x\log x - x - x^{-1}$, we obtain

$$v_0(d) \approx \frac{2d}{\log d}\Big\{1 + \frac{\log(\log d)}{\log d} + O\Big(\frac{1}{\log d}\Big)\Big\}. \qquad (8.20)$$

The large d behavior coincides with the rigorous result (7.7) due to Dhar (1988). This is not completely surprising, as one would naïvely expect a mean field type approximation to become more accurate in high dimensions. As we go along we will encounter more situations in which our mean field theory appears to give a consistent description of the high dimensionality behavior. However we are not aware of any serious argument for why this should always be the case.

Together (8.19) and (8.20) imply that the shape anisotropy $a := v_0/v_1$ diverges as $a \approx 0.66\sqrt{d}/\log d$ for large d. However due to the large correction terms in (8.20) the asymptotics is approached very slowly. In Table 4 we show values of v_0 and a for small d. For $d = 2$ the anisotropy is 4.67% and it increases monotonously with d.

d	v_0	a
2	4.46685	1.0467
3	5.67295	1.0853
4	6.75370	1.1190
5	7.75405	1.1491
10	12.1058	1.2686
100	62.9996	2.0876

Table 4: Mean field estimates for the growth velocity v_0 and the shape anisotropy $a = v_0/v_1$ of Eden clusters on a d-dimensional hypercubic lattice.

For comparison with numerical simulations we first point out that our mean field approach makes no difference (with respect to the cluster shape) between the site and bond versions of the Eden model. For the site version the equation of motion corresponding to (8.5) reads

$$\frac{\partial}{\partial t}\tilde{\varrho}(\mathbf{x}) = \big(1 - \tilde{\varrho}_t(\mathbf{x})\big)\Big(1 - \prod_{\mathbf{e},|\mathbf{e}|=1}[1 - \tilde{\varrho}_t(\mathbf{x}+\mathbf{e})]\Big). \qquad (8.21)$$

Since our working assumption states that the growth velocity is determined only by the exponential tail of the density profile, we may linearize (8.21) and obtain the same expression (8.13) as for the bond version.

Hirsch and Wolf (1986) determine the shape anisotropy for the site version of the Eden model on the square lattice, finding $a \simeq 1.020$. Meakin *et al.* (1986a) use a bond version (version C in the notation of Jullien and Botet 1985) and obtain $a \simeq 1.025$ in $d = 2$. The mean field approach thus strongly overestimates the shape anisotropy, but it gives the correct order of magnitude. The numerical results published by Hirsch and Wolf (1986) for $d = 3$ indicate that the anisotropy is about twice as large than for $d = 2$ (we infer $a \simeq 1.043$ from their data) in accordance with the trend of Table 4. A systematic numerical study of the shape anisotropy in higher dimensions is so far lacking.

Let us finally note that our mean field estimates for v_0 and v_1 are in fact upper bounds to the true growth velocities. This can be seen by comparing Equations (8.16) and (8.17) to rigorous bounds derived by Dhar (1986, 1988). For the lattice axis Dhar (1988) proves that $v_0 \leq v_0^+$ where

$$f(v_0^+) = 2d - 3 \tag{8.22}$$

and $f(x)$ is given by (8.18), while for the diagonal it is shown that $v_1 \leq v_1^+$ with (Dhar 1986)

$$f\left(\frac{v_1^+}{\sqrt{d}}\right) = -\frac{1}{d}. \tag{8.23}$$

These bounds are derived by assuming that all infection paths from the origin to a given hyperplane are independent (Dhar 1986). Since $f(x)$ is a monotonously increasing function for $x > 0$, the solution to (8.22) is smaller than that to (8.16), and the solution to (8.23) is smaller than that to (8.17). Hence the mean field estimates are upper bounds also.

8.2 The faceting transition in the Richardson model

In this section we will use the mean field approach to explicitly check several of the scaling assumptions made in the treatment of the faceting transition in Section 7.2. Moreover we will gain some insight into the surprisingly rich structure hidden in simple equations like (8.13). Since faceting occurs in the diagonal direction, our starting point is the discrete time analogue of (8.6) with $\mathbf{n} = 1/\sqrt{d}(1,\ldots,1)$,

$$\varrho_{t+1}(z) - \varrho_t(z) = \tag{8.24}$$
$$p\big(1 - \varrho_t(z)\big)\Big[1 - \big(1 - \varrho_t(z-1)\big)^d\big(1 - \varrho_t(z+1)\big)^d\Big],$$

where p is the growth probability parameter (the Richardson model is described in Sections 4.1 and 7.2) and distances are measured in units of $1/\sqrt{d}$, $z = x\sqrt{d}$. We know already that faceting is related to an infection process which lives on the hyperplane $x_1+\ldots+x_d = t$, i.e. $z = t$ in (8.24). The occupation density $\sigma(t) := \varrho_t(t)$ of the facet plane evolves according to

$$\sigma(t+1) = p\Big(1 - \big[1 - \sigma(t)\big]^d\Big) \tag{8.25}$$

(note that $\varrho_t(z) = 0$ for $z > t$). This is simply the mean field version of directed percolation (Kinzel 1983): For $p < p_c = 1/d$ the only fixed point of (8.25) is at $\sigma = 0$, and it is approached exponentially fast in time, $\sigma(t) \sim e^{-t/\xi_t}$ with the temporal correlation length $\xi_t = -1/\log(pd) \sim (1 - p/p_c)^{-1}$ for $p \to p_c$. Hence the corresponding exponent is

$$\nu_t^{\text{MF}} = 1. \tag{8.26}$$

For $p > p_c$ a nontrivial fixed point $\sigma^* > 0$ appears. For p close to p_c, $\sigma^* \sim p - p_c$, leading to the order parameter exponent

$$\beta^{\text{MF}} = 1. \tag{8.27}$$

At $p = p_c$ $\sigma(t)$ relaxes according to a power law, $\sigma(t) \sim 1/t$ which is consistent with the general scaling law $\sigma(t) \sim t^{-\beta/\nu_t}$. We remark that the mean field value $p_c = 1/d$ is the leading term in the $1/d$-expansion for p_c as carried out by Blease 1977).

For $p > p_c$ the density profile $\varrho_t(z)$ has a jump of size σ^* at $z = t$. Hence the assumption of an exponential tail $\varrho_t(z) \sim e^{-qz}$ for $z \to \infty$ must break down as one passes p_c. To see what happens we derive the growth velocity v_1 along the diagonal from (8.24) in the manner described above. We obtain

$$v_1(p,d) = \min_q\{\tilde{v}_q(p,d), \tilde{v}_q(p,d)\}$$
$$= \frac{1}{q}\log(1 + 2pd\cosh q). \tag{8.28}$$

For large q, $\tilde{v}_q \approx 1+\log(pd)/q$. The limiting value $\tilde{v}_\infty = 1$ is approached from below for $p < p_c$ ($pd < 1$), but from above for $p > p_c$. Hence for

$p > p_c$ the minimum in (8.28) is located at $q^* = \infty$, which corresponds formally to a step profile. We shall see later that $1/q^*$ is in fact proportional to the growth shape curvature.

The divergence of the minimal value q^* in (8.28) for $p \to p_c$ is linked to the approach of the growth velocity $v_1 \to 1$. Indeed, for $p < p_c$ we have $\sigma(t) = \varrho_t(t) \sim e^{-q^*(t-v_1 t)} \sim e^{-t/\xi_t}$ and therefore

$$(1 - v_1)q^* \sim \xi_t^{-1}. \tag{8.29}$$

As q^* diverges for $p \to p_c$, we may determine the critical behavior of v_1 from (8.28) by expanding $\tilde{v}_q$ for large q,

$$\tilde{v}_q \approx 1 + \frac{\log pd}{q} + \frac{e^{-q}}{pdq}. \tag{8.30}$$

Taking the derivative with respect to q we obtain, with $\epsilon = 1 - p/p_c$,

$$(1 - v_1)\big(1 - \log(1 - v_1)\big) \sim \epsilon \tag{8.31}$$

which yields the critical behavior

$$1 - v_1 \sim \frac{\epsilon}{\log(1/\epsilon)} \tag{8.32}$$

and from (8.29) and (8.26)

$$q^* \sim \log(1/\epsilon). \tag{8.33}$$

The scaling theory of Kertész and Wolf (1989) predicts $1 - v_1$ and ξ_t^{-1} to vanish with the same exponent ν_t as $\epsilon \to 0$. Here we see that this is true up to logarithmic corrections. We also note that the derivative of v_1 with respect to p vanishes continuously at the transition.

To discuss the growth shape singularities we must determine the inclination dependence of the growth velocity close to the diagonal direction. As usual in mean field theory, the critical behavior is independent of dimension. Hence we may restrict ourselves to $d = 2$. We want to compute

$$v(u) = \min_q \tilde{v}_q(u),$$
$$\tilde{v}_q(u) = \frac{1}{q} \log\Big[1 + 2p\big(\cosh q(1 + u) + \cosh q(1 - u)\big)\Big]. \tag{8.34}$$

Note that here the growth velocity is measured in the direction of the diagonal rather than normal to the front as in (8.13). The two velocities differ by a factor of $\sqrt{1 + u^2}$ (cf. Section 2). As we are interested in situations where the minimum in (8.34) is located at a value $q^* \gg 1$ (close to p_c or close to $u = 0$ for $p > p_c$) the cosh's can be replaced by exponentials, whence

$$\tilde{v}_q(u) \approx 1 + \frac{1}{q} \log\left(2p \cosh qu + \mathrm{e}^{-q}\right). \tag{8.35}$$

In fact this amounts to considering a directed version of the Richardson model, in which infection propagates in the forward direction only.

We consider first the supercritical case, $p > p_c$. We know that $q^* = \infty$ for $u = 0$, thus $q^*(u)$ must diverge as $u \to 0$ (note that a front growing in a direction different from the diagonal, $u \neq 0$, is never faceted). Using (8.35) we find that $q^* \approx \lambda/|u|$ for $u \to 0$, where $\lambda = \lambda(p)$ is determined by

$$\lambda \tanh \lambda = \log(2p \cosh \lambda) \tag{8.36}$$

which has a solution only for $p > p_c = \frac{1}{2}$. For $p \to p_c$ from above λ vanishes as

$$\lambda \approx \left(2(p - p_c)\right)^{1/2}. \tag{8.37}$$

Setting $q = q^* = \lambda/|u|$ in (8.35) we obtain for $|u| \ll \lambda$

$$v(u) \approx 1 + |u| \tanh \lambda + \frac{|u|\,\mathrm{e}^{-\lambda/|u|}}{2p\lambda \cosh \lambda}, \quad (p > p_c). \tag{8.38}$$

The facet size is $c = \tanh \lambda$ and it vanishes for $p \to p_c^+$ as $(p - p_c)^{1/2}$. Comparing this with the generally expected behavior $c \sim (p - p_c)^{\nu_t - \nu_r}$, we conclude (using (8.26)) that

$$\nu_r^{\mathrm{MF}} = \frac{1}{2} \tag{8.39}$$

as is well known from other approaches (Harms and Straley 1982, Kinzel 1983). The surprising feature in (8.38) is the essential singularity in the next to leading term. It was shown in Section 2 that this term describes the growth shape close to the facet. We argued in Section 7.2 that it should be proportional to $|u|^{z_d - 1/\zeta_d - 1}$ in d dimensions. Thus the essential singularity reflects the fact that $\zeta/z = 0$ in our mean field theory. The corresponding growth shape is

$$g(c) - g(c + \Delta) \approx \frac{\lambda\Delta}{\log(1/\Delta)} \tag{8.40}$$

where $\Delta > 0$ and the facet boundary is located at $y = c$. Up to a logarithmic factor, the curved surface joins the facet linearly.

In the subcritical regime $(p < p_c)$ q^* is finite for $u = 0$, hence $q^*u \to 0$ for $u \to 0$. The $\cosh qu$ in (8.35) may then be expanded, and we obtain

$$\tilde{v}_q(u) \approx 1 - \frac{\epsilon - e^{-q}}{q} + \tfrac{1}{2}qu^2 \qquad (8.41)$$

for $\epsilon = 1 - p/p_c \ll 1$ and $|u| \ll 1$. Minimizing this relative to q we find the position $q^*(u)$ of the minimum to be

$$q^*(u) \approx q^*(0)\bigl(1 - \tfrac{1}{2}e^{q^*(0)}u^2\bigr). \qquad (8.42)$$

Inserting this into (8.41) it follows that the second derivative of $v(u)$ diverges on approaching the transition as

$$v''(0) \approx q^*(0) \sim \log\Bigl(\frac{1}{\epsilon}\Bigr), \quad (p < p_c) \qquad (8.43)$$

according to (8.33), and hence the curvature of the growth shape vanishes logarithmically, $g''(0) = -1/v''(0) \sim -1/\log(1/\epsilon)$. This is consistent with the scaling prediction (7.13), since $\nu_t^{\text{MF}} - 2\nu_r^{\text{MF}} = 0$.

At the transition point $p = p_c$ $q^*(u)$ diverges as $u \to 0$, however q^*u still vanishes and (8.41) with $\epsilon = 0$ can be used to determine the divergence. To leading order we find $q^*(u) \approx 2\log(1/u)$. Inserting into (8.41) we obtain the anomalous small u behavior of $v(u)$,

$$v(u) \approx 1 + u^2\log\Bigl(\frac{1}{|u|}\Bigr), \quad (p = p_c) \qquad (8.44)$$

and the corresponding growth shape $g(y) \approx 1 - y^2/\log(1/|y|)$. Again, this is consistent with (7.14) and (7.15), since $\nu_t^{\text{MF}}/\nu_r^{\text{MF}} = 2$.

Acknowledgements

Our understanding of growing surfaces has been formed through and benefited from many fruitful interactions. In particular, we are grateful to Bernard Derrida, Tim Halpin-Healy, Reinhard Lipowsky, Paul Meakin, Wolfgang Renz, Len Sander and Dietrich E. Wolf. We thank Paul Meakin for providing figures and Richard Stückl for his technical advice with TeX. H.S. thanks Claude Godrèche for the opportunity to present some of the material at the Summer School Beg-Rohu 1989. This work was supported by Deutsche Forschungsgemeinschaft.

References

Amar, J.G. and Family, F. (1990a), *Phys. Rev. Lett.* **64**, 543.

Amar, J.G. and Family, F. (1990b), *Phys. Rev.* **A41**, 3399.

Amar, J.G. and Family, F. (1990c), *Phys. Rev. Lett.* **64**, 2334.

Andreev, A.F. (1982), *Sov. Phys. JETP* **53**, 1063.

Aronson, D.G. and Weinberger, H.F. (1978), *Adv. Math.* **30**, 33.

Baiod, R., Kessler, D., Ramanlal, P., Sander, L. and Savit, R. (1988), *Phys. Rev.* **A38**, 3672.

Ball, R., Nauenberg, M. and Witten, T.A. (1984), *Phys. Rev.* **A29**, 2017.

Bausch, R., Dohm, V., Janssen, H.K. and Zia, R.K.P. (1981), *Phys. Rev. Lett.* **47**, 1837.

Baxter, R.J. (1982), *Exactly Solved Models in Statistical Mechanics* (Academic Press, London).

van Beijeren, H. (1977), *Phys. Rev. Lett.* **38**, 993.

van Beijeren, H., Kutner, R. and Spohn, H. (1985), *Phys. Rev. Lett.* **54**, 2026.

van Beijeren, H. and Nolden, I. (1987), in *Structure and Dynamics of Surfaces II*, ed. by W. Schommers and P. von Blanckenhagen, Topics in Current Physics **43** (Springer, Berlin).

Ben-Jacob, E., Brand, H., Dee, G., Kramer, L. and Langer, J.S. (1985), *Physica* **14D**, 348.

Bensimon, D., Shraiman, B. and Liang, S. (1984a), *Phys. Lett.* **102A**, 238.

Bensimon, D., Shraiman, B. and Kadanoff, L.P. (1984b), in *Kinetics of Aggregation and Gelation*, ed. by F. Family and D.P. Landau (Elsevier, Amsterdam).

Berker, A.N. and Ostlund, S. (1979), *J. Phys.* **C12**, 4961.

Blease, J. (1977), *J. Phys.* **C10**, 925.

Boldrighini, C., Cosimi, G., Frigio, S. and Grasso Nuñes, M. (1989), *J. Stat. Phys.* **55**, 611.

Bolthausen, E. (1989), *Commun. Math. Phys.* **123**, 529.

Botet, R. (1986), *J. Phys.* **A19**, 2233.

Bovier, A., Fröhlich, J. and Glaus, U. (1986), *Phys. Rev.* **B34**, 6409.

Bradley, R.M. and Strenski, P.N. (1985), *Phys. Rev.* **B31**, 4319.

Bramson, M. (1983), *Mem. Am. Math. Soc.* **285**.

Bramson, M. (1987), Lectures at Paris, unpublished.

Bramson, M. and Griffeath, D. (1980), *Math. Proc. Camb. Phil. Soc.* **88**, 339.

Bramson, M. and Griffeath, D. (1981), *Ann. Prob.* **9**, 173.

Brandstetter, H. (1991), *Diplomarbeit*, Universität München (unpublished).

Buff, F.P., Lovett, R.A. and Stillinger, F.H. (1965), *Phys. Rev. Lett.* **15**, 621.

Burgers, J.M. (1974), *The Nonlinear Diffusion Equation* (Reidel, Dordrecht).

Burkhardt, T.W. (1987), *Phys. Rev. Lett.* **59**, 1058.

Cardy, J.L. (1983), *J. Phys.* **A16**, L709.

Chakrabarti, A. and Toral, R. (1989), *Phys. Rev.* **A40**, 11419.

Chayes, J.T., Chayes, L. and Durrett (1986), *J. Stat. Phys.* **45**, 933.

Cheng, Z., Baiod, R. and Savit, R. (1987), *Phys. Rev.* **A35**, 313.

Chernoutsan, A. and Milošević, S. (1985), *J. Phys.* **A18**, L 449.

Chernov, A.A. (1963), *Sov. Phys. Cryst.* **7**, 728.

Chernov, A.A. (1984), *Modern Crystallography III: Crystal Growth*, Springer Series in Solid State Sciences **36** (Springer, Berlin).

Chorin, A.J. and Marsden, J.E. (1979), *A Mathematical Introduction to Fluid Mechanics* (Springer, Berlin).

Cole, J.D. (1951), *Quart. Appl. Math* **9**, 225.

Cook, J. and Derrida, B. (1989a), *J. Stat. Phys.* **57**, 89.

Cook, J. and Derrida, B. (1989b), *Europhys. Lett.* **10**, 195.

Cook, J. and Derrida, B. (1990), *J. Phys.* **A23**, 1523.

Cox, J.T. and Durrett, R. (1983), *Math. Proc. Camb. Phil. Soc.* **93**, 151.

Dee, G. and Langer, J.S. (1983), *Phys. Rev. Lett.* **50**, 383.

DeMasi, A., Presutti, E., Spohn, H. and Wick, D. (1986), *Ann. Prob.* **14**, 409.

Derrida, B. (1990), *Physica* **A163**, 71.

Derrida, B. and Griffiths, R.B. (1989), *Europhys. Lett.* **8**, 111.

Derrida, B. and Spohn, H. (1988), *J. Stat. Phys.* **51**, 817.

Devillard, D. and Stanley, H.E. (1989), *Physica* **A160**, 298.

Dhar, D. (1986), in *On Growth and Form*, ed. by H.E. Stanley and N. Ostrowsky (Martinus Nijhoff, Dordrecht), p. 288.

Dhar, D. (1987), *Phase Transitions* **9**, 51.

Dhar, D. (1988), *Phys. Lett.* **A130**, 308.

Diehl, H.W., Kroll, D.M. and Wagner, H. (1980), *Z. Phys.* **B36**, 329.

Domany, E. and Kinzel, W. (1981), *Phys. Rev. Lett.* **47**, 5.

Durrett, R. (1988a), *Mathematical Intelligencer* **10**, 37.

Durrett, R. (1988b), *Lecture Notes on Particle Systems and Percolation*, (Wadsworth and Brooks/Cole Advanced Books and Software, Pacific Grove CA).

Durrett, R. and Liggett, T.M. (1981), *Ann. Prob.* **9**, 186.

Eden, M. (1958), in *Symposium on Information Theory in Biology*, ed. by H.P. Yockey (Pergamon Press, New York).

Eden, M. (1961), in *Proceedings of the Fourth Berkeley Symposium on Mathematical Statistics and Probability*, ed. by F. Neyman, Vol. IV (University of California Press, Berkeley).

Edwards, S.F. and Wilkinson, D.R. (1982), *Proc. Roy. Soc. London* **A381**, 17.

Elskens, Y. and Frisch, H.L. (1985), *Phys. Rev.* **A31**, 3812.

Essam, J.W., Guttmann, A. J. and De Bell, K. (1988), *J. Phys.* **A21**, 3815.

Family, F. (1986), *J. Phys.* **A19**, L 441.

Family, F. (1990), *Physica* **A168**, 561.

Family, F. and Vicsek, T. (1985), *J. Phys.* **A18**, L75.

Feller, W. (1950), *An Introduction to Probability Theory and its Applications* (John Wiley, New York).

Ferrari, P.A., Kipnis C. and Saada, E. (1991), Annals of Prob. (in press).

Fisher, D.S. (1986), *Phys. Rev. Lett.* **56**, 1964.

Fisher, M.E. (1986), *J. Chem. Soc. Faraday Trans. 2* , **82**, 1569.

Fisher, R.A. (1937), *Ann. Eugenics* **7**, 355.

Forrest, B.M. and Tang, L.-H. (1990), *Phys. Rev. Lett.* **64**, 1405.

Forster, D., Nelson, D.R. and Stephen, M.J. (1977), *Phys. Rev.* **A16**, 732.

Frank, F.C. (1974), *J. Cryst. Growth* **22**, 233.

Freche, P., Stauffer, D. and Stanley, H.E. (1985), *J. Phys.* **A18**, L1163.

Friedberg, R. (1986), *Ann. Phys.* **171**, 321

Gärtner, J. and Presutti, E. (1990), *Annales de l'I.H.P.* (in press).

Garmer, M. (1989), Diplomarbeit, Universität München, unpublished.

Gaspard, P. and Wang, X.-J. (1988), *Proc. Nat. Acad. Sci. U. S. A.* **85**, 4591.

Gates, D.J. (1988), *J. Stat. Phys.* **52**, 245.

Gates, D.J. and Westcott, M. (1988), *Proc. Roy. Soc. London* **A416**, 443, 463.

Gilmer, G.H. (1980), *J. Cryst. Growth* **49**, 465.

Goldenfeld, N. (1984), *J. Phys.* **A17**, 2807.

Grassberger, P. and de la Torre, A. (1979), *Ann. Phys.* **122**, 373.

Griffiths, R.B. and Kaufman, M. (1982), *Phys. Rev.* **B26**, 5022.

Gross, R. (1918), Abhandl. math. phys. Klasse der Kgl. sächs. Ges. der Wiss. Leipzig, Bd. 35, Heft 4, p. 135.

Guo, H., Grossmann, B. and Grant, M. (1990), *Phys. Rev. Lett.* **64**, 1262.

Halpin-Healy, T. (1989a), *Phys. Rev. Lett.* **62**, 442.

Halpin-Healy, T. (1989b), *Phys. Rev. Lett.* **63**, 917.

Halpin-Healy, T. (1990a), *Phys. Rev. Lett.* **64**, 109 (E).

Halpin-Healy, T. (1990b), *Phys. Rev.* **A42**, 711.

Harms, B.C. and Straley, J.P. (1982), *J. Phys.* **A15**, 1865.

Henderson, D., Brodsky, M.H. and Chaudhari, P. (1974), *Appl. Phys. Lett.* **25**, 641.

Hirsch, R. and Wolf, D.E. (1986), *J. Phys.* **A19**, L251.

Hopf, E. (1950), *Comm. Pure Appl. Math.* **3**, 201.

Huse, D.A., Amar, J.G. and Family, F. (1990), *Phys. Rev.* **A41**, 7075.

Huse, D.A. and Henley C.L. (1985), *Phys. Rev. Lett.* **54**, 2708.

Huse, D.A., Henley, C.L. and Fisher, D.A. (1985a), *Phys. Rev. Lett.* **55**, 2924.

Huse, D.A., van Saarloos, W. and Weeks, J.D. (1985b), *Phys. Rev.* **B32**, 233.

Hyman, J.M., Nicolaenko, B. and Zaleski, S. (1986), *Physica* **D23**, 265.

Imbrie, J.Z. and Spencer, T. (1988), *J. Stat. Phys.* **52**, 609.

Janssen, H.K. and Schmittmann, B. (1986), *Z. Phys.* **B63**, 517.

Jullien, R. and Botet, R. (1985), *J. Phys.* **A18**, 2279.

Jullien, R., Kolb, M. and Botet, R. (1984), *J. Physique* **45**, 395.

Jullien, R. and Meakin, P. (1987), *Europhys. Lett.* **4**, 1385.

Jullien, R. and Meakin, P. (1989), *J. Phys.* **A22**, L219; ibid. L1115.

Kandel, D. and Domany, E. (1990), *J. Stat. Phys.* **58**, 685.

Kardar, M. (1985), *Phys. Rev. Lett.* **55**, 2923.

Kardar, M. (1987), *Nucl. Phys.* **B290**, 582.

Kardar, M., Parisi, G. and Zhang, Y.C. (1986), *Phys. Rev. Lett.* **56**, 889.

Kardar, M. and Zhang, Y. C. (1987), *Phys. Rev. Lett.* **58**, 2087.

Katz, S., Lebowitz, J.L. and Spohn, H. (1984), *J. Stat. Phys.* **34**, 497.

Kaufman, M. and Griffiths, R.B. (1984), *Phys. Rev.* **B30**, 244.

Kawasaki, K. and Ohta, T. (1982), *Progr. Theor. Phys.* **67**, 147; ibid. **68**, 129.

Kertész, J. and Wolf, D.E. (1988), *J. Phys.* **A21**, 747.

Kertész, J. and Wolf, D.E. (1989), *Phys. Rev. Lett.* **62**, 2571.

Kesten, H. (1986), *Lecture Notes in Mathematics* **1180** (Springer, Berlin).

Kida, S. (1979), *J. Fluid Mech.* **93**, 337.

Kim, J.M. and Kosterlitz, J.M. (1989), *Phys. Rev. Lett.* **62**, 2289.

Kim, J.M., Ala-Nissilä, T. and Kosterlitz, J.M. (1990), *Phys. Rev. Lett.* **64**, 2333.

Kinzel, W. (1983), in *Percolation Structures and Processes*, ed. by G. Deutscher, R. Zallen and J. Adler, *Ann. Isr. Phys. Soc.* **5**, 425.

Kolmogorov, A., Petrovsky, I., and Piskunov, N. (1937), *Bull. Univ. Moskou, Ser. Internat.*, Sec. **A1**, 1.

Krug, J. (1987), *Phys. Rev.* **A36**, 5465.

Krug, J. (1988), *J. Phys.* **A21**, 4637.

Krug, J. (1989a), *J. Phys.* **A22**, L769.

Krug, J. (1989b), *PhD dissertation*, Universität München; published as *Die Entstehung fraktaler Oberflächen* (Harri Deutsch, Frankfurt a.M. 1990).

Krug, J., Kertész, J. and Wolf, D.E. (1990), *Europhys. Lett.* **12**, 113.

Krug, J. and Meakin, P. (1989), *Phys. Rev.* **A40**, 2064.

Krug, J. and Meakin, P. (1991), **A43**, 900.

Krug, J. and Spohn, H. (1988), *Phys. Rev.* **A38**, 4271.

Krug, J. and Spohn, H. (1989), *Europhys. Lett.* **8**, 219.

Krug, J. and Spohn, H. (1990), *Phys. Rev. Lett.* **64**, 2332.

Kuramoto, Y. and Yamada, T. (1976), *Progr. Theor. Phys.* **56**, 724.

Langer, J.S. (1987), in *Chance and Matter*, ed. by J. Souletie, J. Vannimenus and R. Stora (North Holland, Amsterdam).

Leamy, H.J., Gilmer, G.H. and Dirks, A.G. (1980), in *Current Topics in Materials Science*, Vol. 6, ed. by E. Kaldis (North-Holland, Amsterdam).

Lebowitz, J.L. (1990), private communication.

Liang, S. and Kadanoff, L.P. (1985), *Phys. Rev.* **A31**, 2628.

Lieb, E.H. and Liniger, W. (1963), *Phys. Rev.* **130**, 1605.

Lieb, E.H. and Wu, F.Y. (1972), in *Phase Transitions and Critical Phenomena*, Vol. 1, ed. by C. Domb and M.S. Green (Academic Press, London).

Limaye, A.V. and Amritkar, R.E. (1986), *Phys. Rev.* **A34**, 5085.

Lipowsky, R. (1988), in *Random Fluctuations and Pattern Growth*, ed. by H.E. Stanley and N. Ostrowsky, NATO ASI Series **E157** (Kluwer Academic, Dordrecht).

Lipowsky, R. (1990), in *Fundamental Problems in Statistical Mechanics VII*, ed. by H. van Beijeren (Elsevier, Amsterdam).

Liu, D. and Plischke, M. (1988), *Phys. Rev.* **B38**, 4781.

Mandelbrot, B.B. (1982), *The Fractal Geometry of Nature* (W.H. Freeman, San Francisco).

Mandelbrot, B.B. (1986), in *Fractals in Physics*, ed. by L. Pietronero and E. Tosatti (North-Holland, Amsterdam).

McKane, A.J. and Moore, M.A. (1988), *Phys. Rev. Lett.* **60**, 527.

Marchand, J.-P. and Martin, Ph.A. (1986), *J. Stat. Phys.* **44**, 491.

Meakin, P. (1983), *Phys. Rev.* **B28**, 5221.

Meakin, P. (1986), in *On Growth and Form*, ed. by H.E. Stanley and N. Ostrowsky (Martinus Nijhoff, Dordrecht).

Meakin, P. (1987a), *CRC Crit. Rev. Solid State Mat. Sci.* **13**, 143.

Meakin, P. (1987b), *J. Phys.* **A20**, L1113.

Meakin, P. (1988a), in *Phase Transitions and Critical Phenomena*, Vol. 12, ed. by C. Domb and J.L. Lebowitz (Academic Press, London).

Meakin, P. (1988b), *Phys. Rev.* **A38**, 418.

Meakin, P. (1988c), in *Random Fluctuations and Pattern Growth*, ed. by H.E. Stanley and N. Ostrowsky, NATO ASI Series **E157** (Kluwer Academic, Dordrecht).

Meakin, P. (1988d), *Phys. Rev.* **A38**, 994.

Meakin, P. and Deutch, J.M. (1986), *J. Chem. Phys.* **85**, 2320.

Meakin, P. and Jullien, R. (1987), *J. Physique* **48**, 1651.

Meakin, P. and Jullien, R. (1989), *Europhys. Lett.* **9**, 71.

Meakin, P. and Jullien, R. (1990), *Phys. Rev.* **A41**, 983.

Meakin, P. and Krug, J. (1990), *Europhys. Lett.* **11**, 13.

Meakin, P., Jullien, R. and Botet, R. (1986a), *Europhys. Lett.* **1**, 609.

Meakin, P., Ramanlal, P., Sander, L.M. and Ball, R.C. (1986b), *Phys. Rev.* **A34**, 5091.

Medina, E., Hwa, T., Kardar, M. and Zhang, Y.C. (1989), *Phys. Rev.* **A39**, 3053.

Melrose, J.R. (1983), *J. Phys.* **A16**, 3077.

Métois, J.J., Spiller, G.D.T. and Venables, J.A. (1982), *Phil. Mag.* **A46**, 1015.

Mollison, D. (1972), *Nature* **240**, 467.

Mullins, W.W. (1959), *J. Appl. Phys.* **30**, 77.

Nadal, J.P., Derrida, B. and Vannimenus, J. (1984), *Phys. Rev.* **B30**, 376.

Nattermann, T. (1989), unpublished.

Nattermann, T. and Renz, W. (1988), *Phys. Rev.* **B38**, 5184.

Nauenberg, M. (1983), *Phys. Rev.* **B28**, 449.

Nauenberg, M., Richter, R. and Sander, L.M. (1983), *Phys. Rev.* **B28**, 1649.

Nieuwenhuizen, J.M. and Haanstra, H.B. (1966), *Philips Technische Rundschau* **27**, 177.

Nieuwenhuizen, T. (1989), private communication.

Parisi, G. and Zhang, Y.C. (1984), *Phys. Rev. Lett.* **53**, 1791.

Parisi, G. and Zhang, Y.C. (1985), *J. Stat. Phys.* **41**, 1.

Plischke, M. and Rácz, Z. (1985), *Phys. Rev.* **A32**, 3825.

Plischke, M., Rácz, Z. and Liu, D. (1987), *Phys. Rev.* **B35**, 3485.

Pomeau, Y. and Résibois, P. (1975), *Phys. Rep.* **19**, 63.

Renz, W. (1990), to be published.

Richardson, D. (1973), *Proc. Camb. Phil. Soc.* **74**, 515.

Rost, H. (1981), *Z. Wahrsch. verw. Geb.* **58**, 41.

Rottman, C. and Wortis, M. (1984), *Phys. Rep.* **103**, 59.

van Saarloos, W. (1988), *Phys. Rev.* **A37**, 211.

van Saarloos, W. (1989), *Phys. Rev.* **A39**, 6367.

van Saarloos, W. and Gilmer, G.H. (1986), *Phys. Rev.* **B33**, 4927.

Saito, Y. and Ueta, T. (1989), *Phys. Rev.* **A40**, 3408.

Savit, R. and Ziff, R. (1985), *Phys. Rev. Lett.* **55**, 2515.

Sawada, Y., Ohta, S., Yamazaki, M. and Honjo, H. (1982), *Phys. Rev.* **A26**, 3557.

Simon, B. (1976), *Ann. Phys.* **97**, 279.

Singer, H. and Peschel, I. (1980), *Z. Phys.* **B39**, 333.

Sivashinsky, G.I. (1977), *Acta Astronautica* **4**, 1177.

Smythe, R.T. and Wierman, J.C. (1978), *First Passage Percolation on the Square Lattice*, Lecture Notes in Mathematics **671** (Springer, Berlin).

Spohn, H. (1991), *Large Scale Dynamics of Interacting Particles*, Texts and Monographs in Physics (Springer, Berlin).

Stauffer, D. (1985), *Introduction to Percolation Theory* (Taylor and Francis, London).

Sutherland, B. (1968), *Phys. Lett.* **26A**, 532.

Szép, J., Cserti, J. and Kertész, J. (1985), *J. Phys.* **A18**, L 413.

Tang, C., Alexander, S. and Bruinsma, R. (1990), *Phys. Rev. Lett.* **64**, 772.

Vannimenus, J., Nickel, B. and Hakim, V. (1984), *Phys. Rev.* **B30**, 391.

Visscher, W.M. and Bolsterli, M. (1972), *Nature* **239**, 504.

Vold, M.J. (1959), *J. Colloid Sci.* **14**, 168.

Vold, M.J. (1963), *J. Colloid Sci.* **18**, 684.

Wang, X.-J. (1989), *Phys. Rev.* **A40**, 6647.

Williams, T. and Bjerknes, R. (1972), *Nature* **236**, 19.

Witten, T.A. and Sander, L.M. (1981), *Phys. Rev. Lett.* **47**, 1400.

Wolf, D.E. (1987), *J. Phys.* **A20**, 1251.

Wolf, D.E. and Kertész, J. (1987a), *J. Phys.* **A20**, L257.

Wolf, D.E. and Kertész, J. (1987b), *Europhys. Lett.* **4**, 651.

Wolf, D.E. and Kertész, J. (1989), *Phys. Rev. Lett.* **63**, 1191.

Wong, P.-Z. and Bray, A.J. (1987), *Phys. Rev. Lett.* **59**, 1057.

Wulff, G. (1901), *Z. Kristallogr. Mineral.* **34**, 449.

Yakhot, V. (1981), *Phys. Rev.* **A24**, 642.

Yakhot, V. and She, Z.-S. (1988), *Phys. Rev. Lett.* **60**, 1840.

Yamada, T. and Kuramoto, Y. (1976), *Progr. Theor. Phys.* **56**, 681.

Zabolitzky, J.G. and Stauffer, D. (1986), *Phys. Rev.* **A34**, 1523.

Zaleski, S. (1989), *Physica* **D34**, 427.

Zhang, Y.C. (1989), *J. Stat. Phys.* **57**, 1123.

Zwerger, W. (1981), *Z. Phys.* **B42**, 333.

INDEX